SOIL DYNAMICS AND EARTHQUAKE ENGINEERING

PROCEEDINGS OF THE CONFERENCE ON SOIL DYNAMICS AND EARTHQUAKE ENGINEERING / SOUTHAMPTON / 13-15 JULY 1982

Soil Dynamics & Earthquake Engineering

Edited by

A.S.CAKMAK / A.M.ABDEL-GHAFFAR

Princeton University, New Jersey, USA

C.A.BREBBIA

Computational Mechanics Centre, Southampton, UK

VOLUME TWO

A.A.BALKEMA / ROTTERDAM / 1982

The texts of the various papers in this volume were set individually by typists under the supervision of each of the authors concerned.

For the complete set of two volumes, ISBN 90 6191 253 9
For volume 1, ISBN 90 6191 254 7
For volume 2, ISBN 90 6191 255 5

Distributed in USA & Canada by: MBS, 99 Main Street, Salem, NH 03079
Printed in the Netherlands

Contents

VOLUME ONE

7 *Soil Dynamics and Ground Motion*

8 *Probabilistic Methods*

9 *Dams and Earth Slopes*

10 *Underground Structures*

6. Soil Structure Interaction

Soil Dynamics & Earthquake Engineering Conference / Southampton / 1982.07.13-15

Formulation of Torsional Soil-Structure Interaction of Structures with Base Masses and Wave Averaging Effects

P.C.LAM & R.J.SCAVUZZO
University of Akron , OH, USA

ABSTRACT

A closed-form solution of the torsional response of an N-mass structure with a large base mass to torsional seismic excitation of the foundation is developed by a semi-inverse method with the aid of Laplace and Hankel transforms. Large, commercial nuclear power plants are often constructed on thick reinforced concrete slabs with a mass that is often greater than the mass of the upper portions of the containment vessel and internal equipments. Thus, the base mass must be included in a realistic dynamic analysis of interaction effects.

Flexibility of the foundation is based on the response of the homogeneous elastic half-space to a time varying shear stress applied on the surface of the boundary of the half-space. By applying the normal mode theory and accounting for the base mass the torsional inertia torque at the base of the structure is expressed as a function of the foundation motion. Coupling these equations and superimposing the torsional free-field inputs with the transient half-space solution, the torsional interaction equation is derived. The interaction equation exhibiting the relationship between the structure and the soil is expressed in terms of an arbitrary base acceleration with the structure being excited by torsional free-field motions which are caused by wave averaging effects.

Numerical results are obtained by iteration for an idealized three-mass structural model subjected to the N-S component of the El-Centro California earthquake of 1940. Response for different soil stiffness and structural characteristics are studied. The significance of the interaction effects are evaluated by comparing the free-field response spectrum with the output acceleration response spectrum. Finally, the torsional response of a structure with a large base mass is compared to the response of a structure with no base mass and to lateral response caused by lateral excitations.

INTRODUCTION

During earthquake ground motion, soil-structure interaction is known to have significant effects upon building structures. The purpose of this paper is to present an analysis of torsional soil-structure interaction of structure with heavy base mass under the influence of torsional wave averaging free-field motions. Based on the previous investigation, (Scavuzzo, 1979) and Wolf's and Luco's works, the authors believe that this torsional soil-structure phenomenon will have pronounced effects on the response of the foundation and will increase the seismic loads at some locations of building structures. As a result, the problem of defining the true torsional foundation interaction effects is an important subject in the field of earthquake engineering.

The methods used herein are similar to those utilized previously (Lam, 1979). First, the torsional deflections of a half-space caused by a time-dependent shear force acting on the interval ($-r<x<r$, Figure 1) are obtained. Then by using normal mode theory (O'Hara et al., 1963) and accounting for the base mass, the foundation shearing force from the inertia loads of the structure can be expressed in terms of an arbitrary base acceleration. Coupling these equations in a manner to eliminate the base shear force yields an integro-differential equation which is then solved by numerical iteration techniques. Because of the base mass, the resulting torsional interaction equation is significantly different from the previous derived interaction equation.

THEORY

Free-field torsional inputs and foundation inertia moment

There are two basic sources of torsional excitation in building structures: (1) free-field motions caused by wave averaging effects and (2) torsional inertia moments caused by eccentric structural masses. In the first case, the wave lengths of typical seismic waves are often approximately equal to the length or width of building foundation. As a result the wave tends to twist the structure foundation. In this investigation torsional free-field motions developed previously (Lam, 1979) from wave averaging method are used as input excitations.

Often the center of gravity of a building structure does not coincide with the geometric center of the foundation. Thus, lateral motions of the structure foundation create torsional inertia moments which twist the foundation and also excite torsional vibration modes. In order to investigate the torsional structure interaction effects, an N-mass model (Figure 2) must be coupled to a homogeneous elastic half-space. The effects of torsion on the responses of the structures are studied by locating dynamic masses near the foundation. Inertia moments generated by the dynamic masses will resist the foundation rotation and excite the torsional mode of vibration.

For an N-mass structure subjected to an arbitrary torsional acceleration, the inertia moment about the vertical axis can be expressed by using the normal mode theory as

$$T(t) = \sum_{j}^{n\,=\,\text{modes}} Q_j \omega_j \int_0^t \ddot{\phi}(\xi) \sin\omega_j(t-\xi)d\xi + I_o\ddot{\phi}(t) \qquad (1)$$

In this formulation, the inertia moment is written in terms of the rotational base acceleration $\ddot{\phi}(t)$, the natural frequencies of torsional vibration ω_j, and the effective modal constants which are related to the concentrated mass m_j by

$$Q_j = \frac{\sum_{i=1}^{\#\text{Mass}} m_i(\bar{X}^y_{ij}\,\ell^x_i - \bar{X}^x_{ij}\,\ell^y_i)}{\sum_{i=1}^{\#\text{Mass}} m_i\,\bar{X}_{ij}^{\,2}} \qquad (2)$$

where $\bar{X}_{ij}$'s is the modal displacements of the N-mass structure, $\bar{X}_{ij}^k$ is the jth mode shape of the ith mass in the kth direction and ℓ^j_i is the location of the ith mass in the jth direction.

Solution to the Lamb problem

In order to derive the torsional soil-structure interaction expression, the response of an elastic half-space to an arbitrary time dependent torque applied over a rigid circular area on the elastic medium (Figure 1) is determined. The equation of motion for dynamic elasticity problem can be expressed as (Reissner et al., 1944).

$$\frac{\partial^2 v}{\partial r^2} + \frac{1}{r}\frac{\partial v}{\partial r} - \frac{v}{r^2} + \frac{\partial^2 v}{\partial z^2} = \frac{1}{c^2}\frac{\partial^2 v}{\partial t^2} \qquad (3)$$

where v is the circumferential displacement and c is the shear wave velocity which is related to the shear modulus of elasticity G and the density ρ of the half-space by $c^2 = G/\rho$. With the coordinate system introduced in Figure 2, the boundary conditions used in the solution are

$$\tau_{z\theta}(r,t) = \begin{cases} \dfrac{3rT(t)}{4\pi r_o^{\,4}} & r \le r_o \\ 0 & r > r_o \end{cases} \qquad (4)$$

By making use of transform methods, the solution for the circumferential displacement caused by a shear stress which varies arbitrarily with time is obtained. As derived previously (Lam, 1979) the angular displacement of the elastic half-space can be expressed as

$$\phi(t) = -\frac{3c}{2\pi^3 r_o^3 G}\int_0^t T(\tau)F(r,t-\tau)d\tau \tag{5}$$

The function F(r,t) is defined by

$$F(r,t) = \int_0^1\int_0^1 \frac{\beta(1-2\alpha^2)}{\sqrt{(1-\beta^2)(1-\alpha^2)}}\left[\frac{1}{\sqrt{r^2(\alpha+\beta)^2-(ct)^2}} - \frac{1}{\sqrt{r^2(\alpha-\beta)^2-(ct)^2}}\right]d\alpha d\beta \tag{6}$$

The angular displacement, $\phi(t)$, equals the sum of the displacement due to the inertia moments as given by Equation (5) and the free-field displacement ϕ_F (t), namely,

$$\phi(t) = -\frac{3c}{2\pi^3 r_o^3 G}\int_0^t T(\tau)\ F(r,t-\tau)d\tau + \phi_F(t) \tag{7}$$

Using Leibnitz's rule and assuming that the N-mass structure is initially at rest, the time derivative of Equation (7) can be written as

$$\dot{\phi}(t) = -\frac{3c}{2\pi^3 r_o^3 G}\left[\int_0^t T(\tau)\ \frac{\partial F}{\partial t}(r,t-\tau)d\tau + T(t)F(r,o)\right] + \dot{\phi}_F(t) \tag{8}$$

By substituting Equation (1) into (8),

$$\dot{\phi}(t) = -\frac{3c}{2\pi^3 r_o^3 G}\left[\sum_{j=1}^{N} Q_j\omega_j \int_0^t\int_0^\tau \ddot{\phi}(\xi)\sin\omega_j(\tau-\xi)\frac{\partial F}{\partial t}(r,t-\tau)d\xi d\tau + \sum_{j=1}^{N} Q_j\omega_j \int_0^t \ddot{\phi}(\xi)\sin\omega_j(t-\xi)F(r,o)d\xi + \int_0^t I_o\ddot{\phi}(\tau)\ \frac{\partial F}{\partial t}(r,t-\tau)d\tau + I_oF(r,o)\ddot{\phi}(t)\right] + \dot{\phi}_F \tag{9}$$

Rearranging Equation (9), the torsional interaction equation becomes

$$\ddot{\phi}(t) = -\sum_{j=1}^{N} \frac{Q_j\omega_j}{I_oF(r,o)} \int_o^t\int_o^\tau \ddot{\phi}(\xi)\sin\omega_j(\tau-\xi)\frac{\partial F(r,t-\tau)}{\partial t}\, d\xi d\tau$$

$$- \sum_{j=1}^{N} \frac{Q_j\omega_j}{I_o} \int_o^t \ddot{\phi}(\xi)\ \sin\omega_j(t-\xi)d\xi$$

$$- \frac{1}{F(r,o)} \int_o^t \ddot{\phi}(\tau)\frac{\partial F(r,t-\tau)}{\partial t} d\tau + \frac{2\pi^3 r_o^3 G}{3cI_oF(r,o)} \times$$

$$\int_o^t (\ddot{\phi}_F(\tau) - \ddot{\phi}(\tau))\ d\tau \qquad (10)$$

In Equation (10) the properties of the structure are defined in terms of the inertia of mass of the foundation, I_o, the effective modal masses M_j, the natural frequencies ω_j and the base half width r_o. The angular acceleration of Equation (10) is solved by numerical iteration technique.

DISCUSSION

Torsional structural response to the N-S component of the El-Centro earthquake

Numerical results of this investigation are based on the idealized 3-mass torsional structural model and free-field ground inputs developed previously (Lam, 1979). This relatively simple model consists of a base mass with inertia equal to 9.922×10^8 kg-m^2 and two dynamic masses of 178260 kg. The fixed-base frequency of the structural masses corresponds to the frequency of the peak input amplitude of the free-field ground motion. In addition, this frequency is varied in parametric studies in order to determine the significance of structure frequency on soil-structure interaction effects. Also, the ground density of 1600 kg/m^3 and shear wave velocities of 152.4, 304.8, 609.6, 1542.0, 3048.0 m/s are used in this analysis.

For the seismic design of structure, it must be emphasized that only the spectrum values at the fixed base frequency are used to calculate the peak shock loads. The maximum load is proportional to the acceleration spectrum response evaluated at the natural frequency of the vibratory mode under consideration (O'Hara, 1958). The significance of torsional interaction effects can be examined by comparing the response spectra of the free-field motion to the foundation motion at the fixed-base frequency. The following expressions define the input and output undamped angular response spectrum respectively:

$$S_i(\omega) = \frac{\omega}{g}\left[\int_0^t \ddot{\phi}_F(\tau)\sin\omega(t-\tau)\, d\tau\right]_{\text{Max over t}}$$

$$S_o(\omega) = \frac{\omega}{g}\left[\int_0^t \ddot{\phi}(\tau)\sin\omega(t-\tau)\, d\tau\right]_{\text{Max over t}} \tag{11}$$

In Figure 3, the spectrum response of the free-field angular acceleration curve (c=152.4 m/s) is shown as a solid line. In addition, the spectrum values determined from the base accelerations with and without base mass are plotted at various frequencies. These values are also listed in Table 1 and 2. The response ratio ($S_o(\omega)/S_i(\omega)$) versus r_o/c curves are plotted in Figure 4. As observed from Figure 4, the largest reduction in the structural response occurs for the softest soil (c = 152.4 m/s). Percentage reductions (1 - response ratio) of between 10.9 to 69.1% are calculated for torsional input without base mass. With base mass, percentage reductions of between 45.3 to 91% are calculated. Thus, the torsional interaction effects become more significant when base mass is included in the analysis and computing the seismic loads directly from the free-field response spectrum by neglecting the soil-structure interaction effects would create a considerable amount of error in the calculations.

Comparison of lateral and torsional structural responses

For the structural model driven by the N-S component of the El-Centro acceleration input, the maximum lateral response (Lam, 1980) for several soil stiffnesses are summarized in Table 3. Similar to the torsional case, the lateral soil-structure interaction effects are shown to be most pronounced at the highest fixed-base frequency, especially for the softest soil. The percentage reductions vary from 19% for the stiffest soil condition to 78% for the softest soil condition. Figure 5 shows the peak lateral responses for several soil stiffnesses at various fixed-based frequencies. By increasing the fixed-base frequency to 18 cps, the response ratio does not change very much even though the shear wave velocity increases from 152.4 to 609.6 m/s. For this soil-structure combination, the radiation damping is fairly constant. This similar trend is also observed for the torsional excitation problem.

Another important observation that can be made from the numerical studies for the analytical model without the heavy base mass is that the peak loads caused by torsional excitation are similar in magnitude to the seismic loads caused by the lateral mode of vibration. The addition of the heavy base mass reduces the peak torsional output acceleration and, thus, the shock loads of components rigidly attached to the base of the structure. This fact is determined by comparing the maximum acceleration of the two different modes of excitation. At the fixed-base frequencies of 6 and 6.75 cps, maximum spectrum values of 3.24 g and 3.03g (Table 3) are observed for lateral input motion. These inputs are the basis of the free-field torsional inputs. For the soil-structure mechanism investigated, the peak torsional response which occurs at 6 cps is calculated to be 3.65 g for the structural model without the base mass and .602g for the model with base mass. These values are obtained by multiplying the output spectrum value by the radius of the foundation ($S_0(\omega)xr_0$). The peak accelerations calculated at 6.75 cps are 2.89g and .604g for models without and with heavy base mass respectively.

The dynamic responses $\ddot{U}_{max}$ and $r_0\ddot{\phi}_{max}$ versus r_0/c are plotted in Figures 6-8. For the structural model without the heavy base mass, Figure 6 indicated that larger seismic load is generated from torsional mode than lateral excitation when $r_0/c > .034$. When heavy base mass is included in the model at a low fixed-base frequency (4 cps), larger seismic load from torsional excitation is observed as compared with lateral mode only if soil-condition is very stiff (c=3048 m/s). As fixed-base frequency increases, the range of r_0/c for which torsional excitation dominated increases (Figure 8). Hence, in the seismic analysis of structure, the effect of torsion on the dynamic response of the system may be very significant and must be evaluated.

CONCLUSIONS

The conclusions from this investigation are based on interaction of torsional base acceleration with torsional inertia moments and torsional ground motions excited from the particle motion under the foundation of lateral transverse waves. Based on numerical results, the principal conclusions are:

(1) By making use of normal mode theory and the solution of a half-space problem torsional interaction expression accounting for base mass effect between a structure and the half-space can be expressed by an integro-differential equation. For the torsional structure model driven by the N-S component of the El-Centro earthquake, results indicate the largest reduction of output acceleration for all soil conditions occurs when mode frequency is tuned to the frequency at which a peak occurs in the free-field acceleration spectrum curve.

(2) Results of this study show that the torsional soil-structure interaction effects are significant for all soil stiffnesses considered. In particular, when base mass is included in the analysis torsional responses remain fairly constant even though the soil stiffness increases.

(3) Comparisons of the torsional output accelerations with lateral responses that include heavy base mass should be conducted in order to accurately evaluate the significance of employing different analytical models in soil-structure interaction analyses.

(4) The addition of the heavy base mass reduces the peaks of the torsional acceleration substantially and, thus, decrease the shock loads of components rigidly attached to the foundation of the structure. Thus, computing seismic loads directly from free-field response spectrum by neglecting the torsional soil-structure interaction effects would create a considerable error.

(5) By using typical free-field inputs, the numerical results of this investigation indicate that peak torsional accelerations are similar in magnitude to maximum lateral accelerations. For certain soil-structure combinations the torsional response with base mass effect can still be larger than the lateral responses. This large amplification of torsional acceleration arises for the intermediate fixed-base frequencies, thus significant of torsional responses in structural design has to be accounted for in seismic analyses.

REFERENCES

Cunniff, P. and O'Hara, G. (1965) "Normal Mode Theory for Three-Dimensional Motion". NRL Report 6170, Naval Research Laboratory Washington, D.C.

Lam, P. and Scavuzzo, R. (1979) "Formulation of Torsional Soil-Foundation Interaction of Building Structures", Pressure Vessel and Piping Conference, ASME, San Francisco, CA.

Lam, P. and Scavuzzo, R. (1981) "Lateral-Torsional Structural Response from Free-field Ground Motion" Nuclear Engineering and Design 65, 269-281.

Luco, J. (1976) "Torsional Response of Structures for SH Waves: the Case of Hemispherical Foundation". Bulletin of the Seismological Society of America, Vol. 66, No. 1, 109-123.

Reissner, E. and Sagoci, H. (1944) "Forced Torsional Oscillations of an Elastic Half-Space, Journal of Applied Physics, Vol. 15, 652-662.

Scavuzzo, R., Raffopoulos, D. and Bailey, J. (1972) "Lateral Structure-foundation Interaction of Structures with Base Masses". Bulletin of the Seismological Society of America, Vol. 6, No. 2, 453-470.

Scavuzzo, R., Raftopoulos, D. and Bailey, J. (1973) "Application of Three-Dimensional Normal Mode Theory to Seismic Analysis of Nuclear Power Plants", Paper k 3/4, 2nd SMIRT, GERMANY.

Wolf, J. and Obernhuber, P. (1980) "Effects of Horizontal Travelling Waves in Soil-structure Interaction", Nuclear Engineering and Design, 57, 221-244.

TABLE 1: Torsional Response Spectra-No Base Mass [g/m]

Shear-wave velocity	Fixed-base frequencies											
	4 cps		6 cps		6.75cps		8 cps		10cps		18cps	
c(m/s)	$S_o(\omega)$	$S_1(\omega)$	$S_o(\omega)$	$S_1(\omega)$	$S_o(\omega)$	$S_1(\omega)$	$S_o(\omega)$	$S_1(\omega)$	$S_o(\omega)$	$S_1(\omega)$	$S_o(\omega)$	$S_1(\omega)$
152.4	0.098	0.110	0.097	0.249	0.087	0.229	0.097	0.162	0.063	0.165	0.051	0.165
304.8	0.133	0.140	0.163	0.259	0.130	0.219	0.110	0.151	0.072	0.155	0.048	0.160
609.6	0.097	0.096	0.240	0.256	0.189	0.234	0.208	0.190	0.158	0.195	0.059	0.155
1524.0	0.043	0.043	0.124	0.124	0.120	0.124	0.109	0.106	0.133	0.131	0.135	0.193
3040.0	0.022	0.022	0.064	0.064	0.064	0.064	0.056	0.056	0.073	0.071	0.121	0.124

TABLE 2: Torsional Acceleration Response Spectra-With Base Mass [g/m]

Shear-wave velocity	Fixed-base frequencies					
	4 cps	6 cps	6.75cps	8 cps	10cps	18cps
c(m/s)	$S_o(\omega)$	$S_o(\omega)$	$S_o(\omega)$	$S_o(\omega)$	$S_o(\omega)$	$S_o(\omega)$
152.4	.0301	.0214	.0190	.0203	.0203	.0278
304.8	.0430	.0337	.0263	.0243	.0261	.0350
609.6	.0521	.0395	.0362	.0353	.0293	.0396
1524.0	.0394	.0354	.0396	.0456	.0379	.0499
3040.0	.0245	.0350	.0311	.0363	.0330	.0283

TABLE 3: Lateral Acceleration Response Spectra [G's]

Shear-wave velocity c(m/s)	Fixed-base frequencies (cps)					
	4	6	6.75	8	10	18
Free-field	1.618	3.243	3.033	2.215	2.239	2.190
152.4	1.132	0.728	0.667	0.769	0.613	0.510
304.8	1.529	1.315	1.012	1.127	0.763	0.526
609.6	1.549	1.675	1.397	1.448	1.093	0.576
1524.0	1.573	2.124	1.661	1.745	1.411	0.808
3048.0	1.597	2.632	2.330	2.039	1.652	1.071

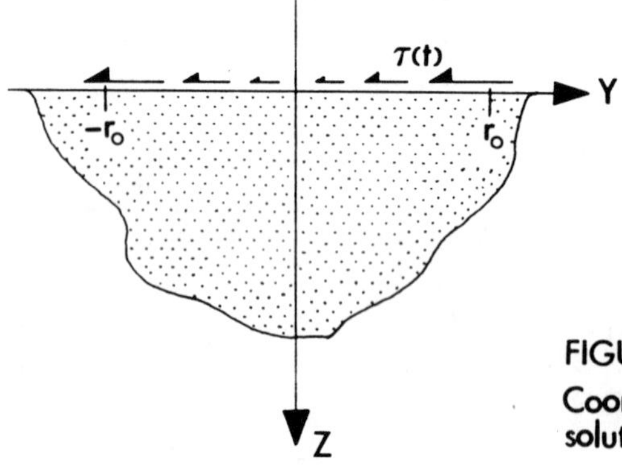

FIGURE 1:
Coordinate system used in the solution to the Lamb Problem

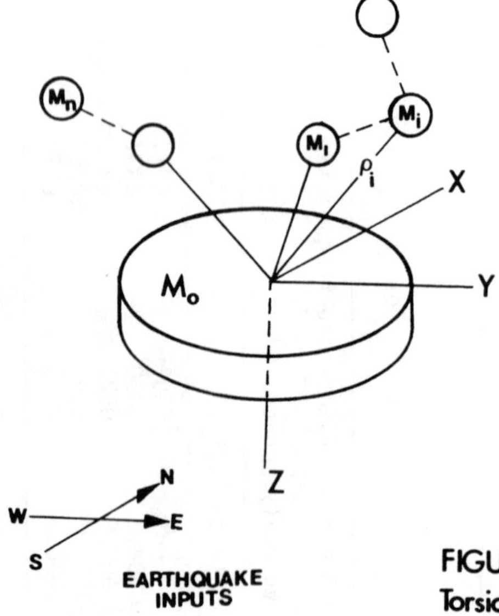

FIGURE 2:
Torsional N-mass structural model

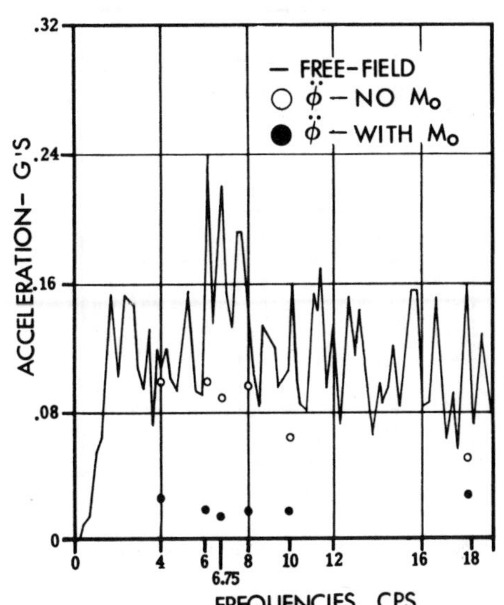

FIGURE 3: Maximum torsional spectrum response vs. frequency [c=152.4 m/s]

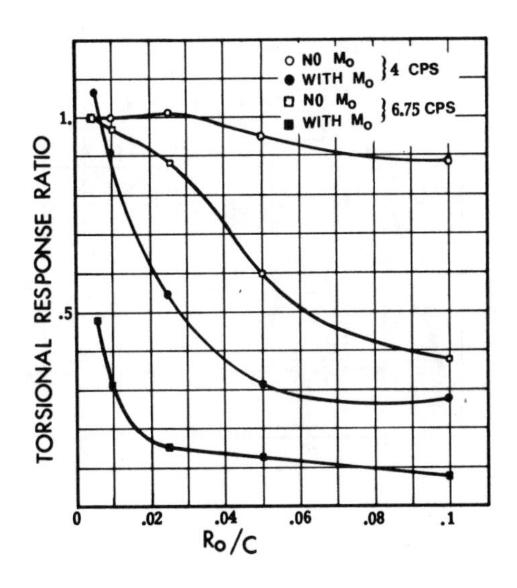

FIGURE 4: Torsional response vs R_o/C

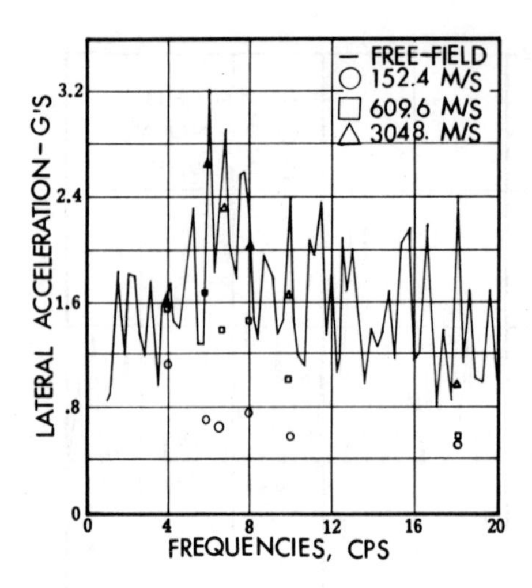

FIGURE 5: Lateral response vs. frequency

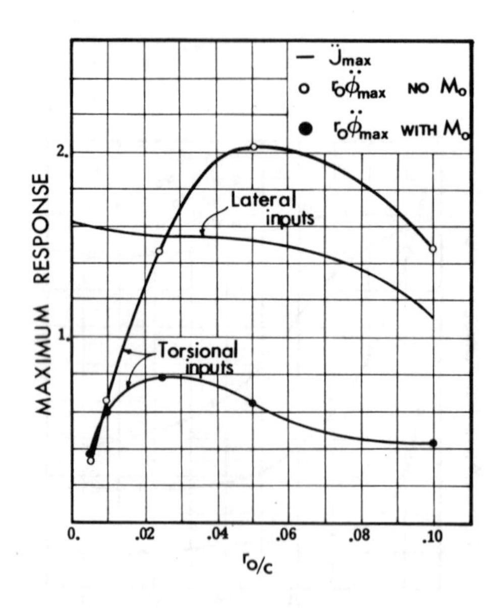

FIGURE 6: Comparion of maximum response vs $r_{o/c}$ [$\omega = 4$ cps]

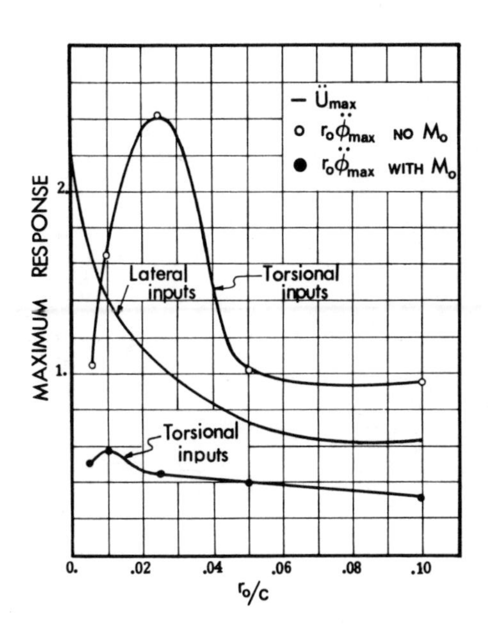

FIGURE 7: Comparison of maximum response vs r_o/c [ω = 10 cps]

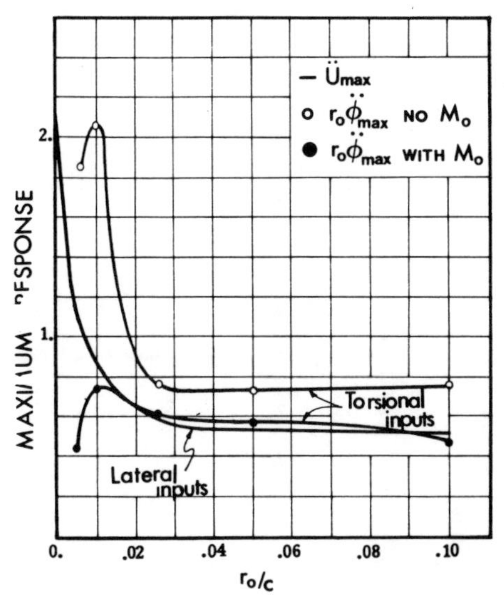

FIGURE 8: Comparison of maximum response vs r_o/c [ω = 18 cps]

Soil Dynamics & Earthquake Engineering Conference / Southampton / 1982.07.13-15

Dynamic Interaction of a Railroad-Bed with the Subsoil

WERNER RÜCKER
Federal Institute of Materials Testing, Berlin, Germany

SUMMARY

An analytical formulation is presented for the general three-dimensional solution of structure-soil-structure interaction problems. Any number of structures with rigid or flexible foundations of arbitrary shape and position in space resting on the surface of a linear elastic/viscoelastic homogeneous and in a well defined sense non-homogeneous halfspace can be considered. The method is applicated to the interaction problem of a railroad-bed of high speed rail vehicles. For a tie-soil-system with rigid ties resting on the surface of a homogeneous halfspace compliance functions are presented as a function of side length ratio and number of ties, which show the influence of the interaction effects on the vertical, horizontal and rocking displacements. Also amplitude curves are given for tie-soil-systems as a function of the number of ties, which indicate that an increasing number of ties produce an increasing number of maximas depending only on the distance between the ties. Finally, it is shown that the method proposed here furnishes results which are in good agreement with experiments made at a real railroad-bed.

INTRODUCTION

Running rail vehicles produce forces between the wheels and the rails which are transmitted to the ties. Due to the movement of the ties a three-dimensional surface wave field is initiated in the underlying soil which causes displacements of ties not directly loaded. A further coupling effect exists due to the connection of the ties by the rails. The dynamic behaviour of the tie-soil-tie system causes a feedback mechanism to the running rail vehicle which influences again the dynamic forces between rail and wheel. For a safety analysis of high speed rail vehicles it is therefore very important to analyze the dynamic of the tie-rail-tie system in detail. In this paper an analytical formulation is presented for the general three-dimensional solution of the interaction problem of tie-soil-tie systems with rigid or flexible ties at considerably less computational expense. The method is based on a Green's functions approach for a linear elastic/viscoelastic homogeneous halfspace. For problems concerning a non-homogeneous halfspace an approximate method is given. No limitations exist in frequency range and in geometry of the problems considered. The formulation can be used for the solution of general structure-soil-structure interaction problems.

ANALYTICAL FORMULATION OF THE INTERACTION PROBLEM

Rigid ties on homogeneous halfspace

A system of N rigid massless ties resting on the surface of a linear elastic/viscoelastic homogeneous halfspace with shear-modulus G, Poisson's ratio ν, mass density ρ and hysteretic damping β is considered (Fig.1). The system is excited by harmonic forces and moments $\underline{p}$, which produce reaction forces and moments $\underline{\bar{p}}$ in the contact area between the ties and the subsoil.

$$\underline{p}^T = \left[\underline{p}_x, \underline{p}_y, \underline{p}_z, \underline{m}_x, \underline{m}_y, \underline{m}_z\right], \tag{1}$$

$$\underline{\bar{p}}^T = \left[\underline{\bar{p}}_x, \underline{\bar{p}}_y, \underline{\bar{p}}_z, \underline{\bar{m}}_x, \underline{\bar{m}}_y, \underline{\bar{m}}_z\right]. \tag{2}$$

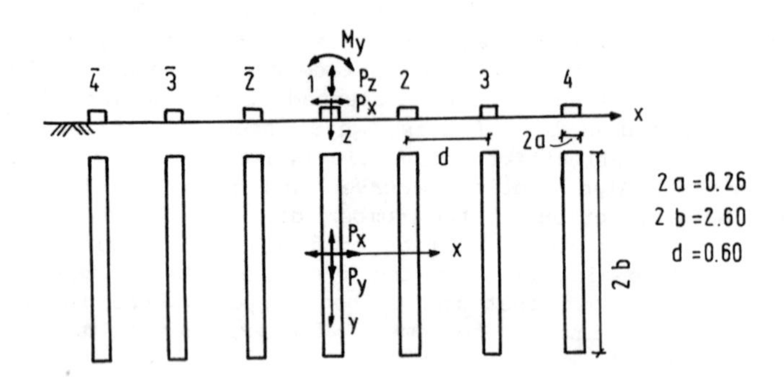

Fig. 1: Basic geometry-cross-section and plane view

The elements $\underline{p}_j$, $\underline{\bar{p}}_j$ of the column matrices $\underline{p}$, $\underline{\bar{p}}$ are also column matrices e.g.

$$\underline{p}_x^T = \left[p_x^{\,1}, \; \ldots p_x^{\,j} \ldots p_x^{\,N}\right]. \tag{3}$$

The forces and moments $\underline{\bar{p}}$ produce rigid body motions (displacements and rotations) of the N ties which are given by

$$\underline{u}^T = \left[\underline{u}_x, \underline{u}_y, \underline{u}_z, \underline{\varphi}_x, \underline{\varphi}_y, \underline{\varphi}_z\right]. \tag{4}$$

$\underline{u}$ is a column matrix of order (6N x 1) and the column matrices $\underline{u}_j$, $\underline{\varphi}_j$, j = (x,y,z) are of order (N x 1). The number 6 stands for the six degrees of freedom for each tie. The relation of the displacements $\underline{u}$ to the forces $\underline{\bar{p}}$ can be expressed by

$$\underline{u} = \underline{F}\ \underline{\bar{p}}, \tag{5}$$

where

$$\underline{F} = \begin{bmatrix} \underline{F}_{xx} & \underline{F}_{xy} & \underline{F}_{xz} & \underline{F}_{x\varphi x} & \underline{F}_{x\varphi y} & \underline{F}_{x\varphi z} \\ & \underline{F}_{yy} & \underline{F}_{yz} & \underline{F}_{y\varphi x} & \underline{F}_{y\varphi y} & \underline{F}_{y\varphi z} \\ & & \underline{F}_{zz} & \underline{F}_{z\varphi x} & \underline{F}_{z\varphi y} & \underline{F}_{z\varphi z} \\ & & & \underline{F}_{\varphi x\varphi x} & \underline{F}_{\varphi x\varphi y} & \underline{F}_{\varphi x\varphi z} \\ & & & & \underline{F}_{\varphi y\varphi y} & \underline{F}_{\varphi y\varphi z} \\ \text{Symm.} & & & & & \underline{F}_{\varphi z\varphi z} \end{bmatrix} \tag{6}$$

is the flexibility matrix of order 6N x 6N, with the NxN submatrices $\underline{F}_{jk}^{l,m}$, e.g.

$$\underline{F}_{xx} = \begin{bmatrix} f_{xx}^{1,1} & \cdots & f_{xx}^{1,N} \\ \vdots & & \vdots \\ f_{xx}^{N,1} & \cdots & f_{xx}^{N,N} \end{bmatrix} \tag{7}$$

For the evaluation of the elements $f_{jk}^{l,m}(a_o) = Re(f_{jk}^{l,m}(a_o) + iIm(f_{jk}^{l,m}(a_o))$ of the flexibility matrix $\underline{F}$, where the elements $f_{jk}^{l,m}$ depend on a non-dimensional frequency

$$a_o = a\omega / \sqrt{G/\rho} = a\ \omega/v_s, \tag{8}$$

where a is half the length of the ties, frequency ω and shear wave velocity v_s, the mixed boundary value problem of the N rigid ties on the surface of the halfspace must be solved. The displacement $v_{jk}^{l,m}$ of the l-th tie with area A_l in the direction j = (x,y,z) due to a harmonic excitation of the m-th tie with area A_m in the direction k = (x,y,z) can be written as

$$v_{jk}^{l,m}(x_l,y_l) = \overset{k}{\Sigma} \iint_{(A_m)} w_{jk}(\bar{r},\nu,\beta)\ \sigma_{zk}^{m}(\zeta_m,\eta_m)\ d\zeta_m\ d\eta_m. \tag{9}$$

σ_{zk}^{m} are the unknown contact stresses between tie m and the subsoil and

$$w_{jk}(\bar{r},\nu,\beta) = \frac{Q_k e^{i\omega t}}{G_1\ r}\ (Re(w_{jk}(\bar{r},\nu)+iIm(w_{jk}(\bar{r},\nu)) \tag{10}$$

are Green's functions expressing the horizontal and vertical displacements $w_j = (w_z, w_y, w_z)$ due to unit horizontal and vertical point forces $Q_k = (Q_x, Q_y, Q_z)$. Green's functions depend on Poisson's ratio ν and on a non-dimensional frequency

$$\bar{r} = r\ \omega/v_s = \sqrt{(x-\zeta_m)^2+(y-\eta_m)^2}\ \ \omega/\sqrt{G_1/\rho}. \tag{11}$$

r is the distance between the displacement at (x,y) due to a unit point force at (ζ_m, η_m) and $G_1 = G(1+2i\beta)$ is a complex modulus including hysteretic damping. For a Poisson's ratio of $\nu = 0.25$ absolute values of Green's functions are presented in Fig.2. Simple analytical expressions are given by Rücker (1982). The integral equation problem defined in equation (9) is solved approximately by subdividing each tie in a number equally spaced subregions and assuming a constant stress $\sigma_{zk} = q\ e^{i\omega t}$ within each subregion. Computing the displacements at the center of each subregion due to the constant stress within each other subregion a displacement influence matrix $\underline{S}$ is obtained. Imposing the displacement boundary conditions for a rigid tie at the center of each subregion a linear algebraic equation system for the stress coefficients q is obtained. After solving this equation system and substituting the stresses into the dynamic equilibrium conditions for each foundation one gets the force -displacement- relationship given in equation (5). It is important to notice that it is not necessary to repeat the computation for each element of matrix $\underline{S}$, since many of the coefficients can be derived from the others by a simple shift or by applying symmetry conditions.

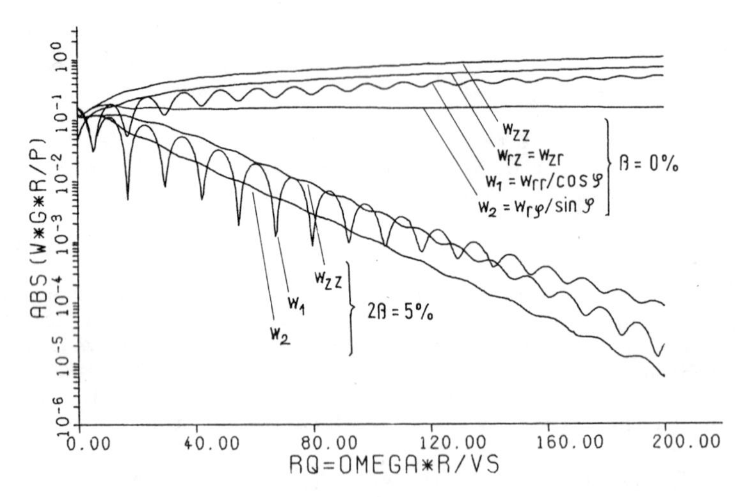

Fig. 2: Vertical w_{zz}, radial w_{rr}, tangential $w_{r\varphi}$, coupling w_{rz} displacement functions due to point forces.

Flexible ties on homogeneous halfspace

Using the procedure described above the halfspace reaction due to constant stresses $\underline{q}$ within each subregion of all ties is given by

$$\underline{S}\ \underline{q} = \underline{\tilde{u}}, \tag{12}$$

where $\underline{S}$ is the displacement influence matrix of order 3M x 3M, M = N x K, at which K is the number of subregions of each tie and $\underline{\tilde{u}}$ is the hyper column matrix of order 3M x 1 of the displacements at the center of each subregion of all ties. The number 3 stands for the displacement components in x-, y-, z-direction.

A ~ has been introduced in equation (12) in order to distinguish between the rigid body motions $\underline{u}$. Considering flexible ties resting on a halfspace the equilibrium conditions require that

$$\underline{G}(\underline{p}_z-\underline{q}_z) = \underline{\tilde{u}}_z, \tag{13}$$

where $\underline{G}$ is a displacement influence matrix of order M x M for the N flexible plates. The elements g_{zz} of matrix $\underline{G}$ can be evaluated easily from the well known analytical solutions of thin plate theory, Filonenko-Boroditsch (1967). For a flexible plate with dimension 2a x 2b which is forced by a constant stress $p_z = 1$ within a subregion "k" with dimension r_k x s_k and center coordinates (x_k, y_k) the displacement $u_z = g_{zz}^{j,k}$ at (x_j, y_j) can be evaluated by

$$g_{zz}^{j,k} = \frac{16}{\pi K} \sum_{l=1}^{\infty} \sum_{m=1}^{\infty} \frac{\sin\frac{l\pi r_k}{2a} \sin\frac{m\pi s_k}{2b} \sin\frac{l\pi x_k}{2a} \sin\frac{m\pi y_k}{2b}}{lm\left(\frac{l^2}{4a^2}+\frac{m^2}{4b^2}\right)^2} \sin\frac{l\pi x_j}{2a} \sin\frac{m\pi y_j}{2b}. \tag{14}$$

$K = E_p t_p^3/(12(1-\nu_p))$ represents the flexural rigidity of the plate, where E_p stands for Young's modulus, t_p for the thickness and ν_p for Poisson's ratio. As by the assumptions of the thin plate theory only the vertical displacements $\underline{\tilde{u}}_z$ are effected by elasticity of the foundation it is assumed in the following that matrix $\underline{G}$ and the column matrices $\underline{p}_z$, $\underline{q}_z$, $\underline{\tilde{u}}_z$ are of the same order as matrix $\underline{S}$ and column matrices $\underline{q}$, $\underline{\tilde{u}}$ by adding zero values at the positions which correspond to the other displacements. Finally, inverting equation (12) and substituting the resulting expression $\underline{q} = \underline{S}^{-1}\ \underline{\tilde{u}}$ into equation(13) a linear algebraic equation system

$$(\underline{G}^{-1}+\underline{S}^{-1})\ \underline{\tilde{u}} = \underline{p} \tag{15}$$

for the displacements $\underline{\tilde{u}}$ of the flexible plates is obtained.

Ties with mass

For a system with mass excited by external harmonic forces the equation of motion is given by

$$(\underline{F}^{-1} - \omega^2\underline{M})\ \underline{u} = \underline{p} \tag{16a}$$

or

$$(\underline{S}^{-1}+\underline{G}^{-1}-\omega^2\underline{M})\ \underline{\tilde{u}} = \underline{p} \tag{16b}$$

for rigid or flexible foundations, respectively. $\underline{M}$ is the mass matrix, which has different dimensions depending on the problem considered.

Non-homogeneous halfspace

For foundations resting on the surface of a halfspace the interaction effects are mainly a result of the propagation of the generated surface waves. Assuming that the halfspace has a shear modulus distribution G(z) and material density ρ(z), with $v_s(z) = \sqrt{G(z)/\rho(z)}$ due to the penetration depth of the surface wave for a particular frequency the velocity of the surface wave $v_R = \bar{\varkappa}\, v_s$ $-\bar{\varkappa}$ is the reciprocal value of the real root of Rayleigh's equation - is governed by the average of the elastic material constants between the surface and the penetration depth. Due to this behaviour an average shear wave velocity for each particular frequency can be evaluated

$$v_s(\omega) = \int_0^\infty v_s(\omega,\bar{z})\, W(\bar{z})d\bar{z} / \int_0^\infty W(\bar{z})d\bar{z} \tag{17}$$

where the amplitude distribution $W(\bar{z})$ of the surface wave is used as a weighting function. $\bar{z}$ is a non-dimensional depth described by

$$\bar{z} = z\ \omega/v_s(z)2\pi. \tag{18}$$

Assuming that the variation of the shear wave velocity with depth can be expressed as

$$v_s(z) = \sum_j k_j\ z^{nj}\ , \tag{19}$$

an analytical solution can be derived from equation (17), which reads as

$$v_s(\omega) = \sum_j \frac{k_j^{\,tj}\,\Gamma(t)\ (A\alpha^{-tj}-B\gamma^{-tj})}{A/\alpha\ -\ B/\gamma}\ \omega^{1-tj} \tag{20}$$

where

$$\alpha = \sqrt{\varkappa - h};\quad \gamma = \sqrt{\varkappa - 1};\quad h = \sqrt{\frac{1-2\nu}{2-2\nu}}$$
$$A = 2\varkappa^2 - 1;\quad B = 2\varkappa^2\alpha;\quad t_j = \frac{1}{1-n_j} \tag{21}$$

and the Gammafunction $\Gamma(t)$. The same procedure can be used to evaluate a frequency dependent material density $\rho(\omega)$ and material damping $\beta(\omega)$. The effect of non-homogeneity is then expressed in a frequency dependent shear modulus $G(\omega)$ and in the non-dimensional frequency $a_o = a\omega/v_s(\omega)$.

NUMERICAL RESULTS

In this section results are presented illustrating the interaction between rigid ties resting on the surface of a linear elastic homogeneous halfspace. In the first part compliance functions are shown in non-dimensional form defined by

$$f_{jj} = u_j\ Ga/P_j \tag{22}$$

$$j = (x,y,z)$$

$$f_{\varphi j\varphi j} = \varphi_j\ Ga^3/M_j \tag{23}$$

for the translation and rotation modes, respectively. In the second part amplitude curves are given for a tie-soil-system with different number of ties. In this representation physical units have been used for comparison with experiments shown in the last part. The units are (cps) for the frequency and (m) for the displacements. For the ties a mass of m = 290 kg and mass inertia moment of I_y = 2.375 kg m² has been assumed. The material constants of the soil are v_s = 200 m/s, ρ = 1.54 t/m³ and ν = 0.25. These constants are average values determined by cross-hole measurements made at the place of the experiments. As an example to demonstrate the accuracy of the proposed method vertical compliance functions for a rigid massless foundation with side length ratios $1 \leq b/a \leq 10$ were computed in the non-dimensional frequency range $0 \leq a_o \leq 10$. The comparison of the presented results shown in Fig.3 with the solution given by Savidis (1980) shows that the agreement is excellent. Compliance functions for all other rigid body motions are given by Rücker (1982) and therefore obmitted here.

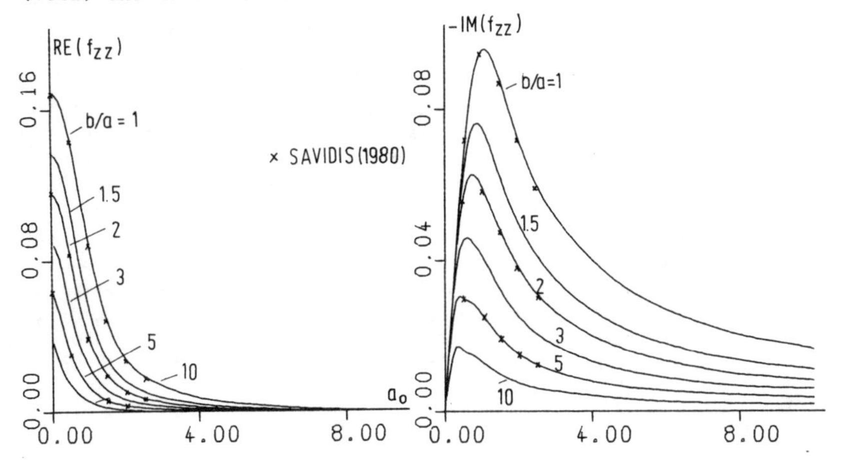

Fig.3 : Real and imaginary part of the vertical compliances for different side length ratio's.

In the following compliance functions for tie-soil-tie systems with different numbers of massless rigid ties are presented. The basic geometry of the problem is given in Fig. 1. Fig. 4 shows the influence of the presence of additional ties on the vertical and rocking compliances of tie "1". The results indicate that for the translation mode this feedback effect is relatively small and is influenced only by a limited number of ties, as the curves are nearly identical for systems with n≥3 ties. For the rotational mode in general a similar behaviour can be stated although the compliances seem to be more influenced by the number of ties in the higher range of the non-dimensional frequency a_o.

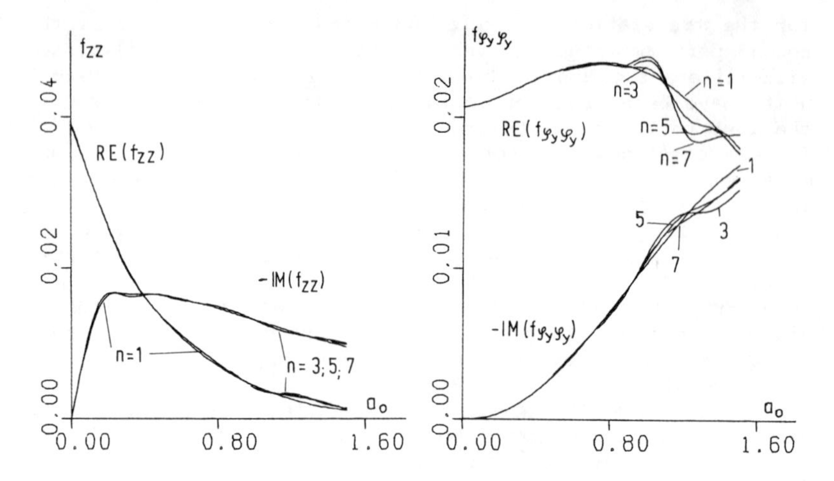

Fig.4: Vertical and rocking compliance functions of tie "1" in dependence on the number of additional ties.

For vertical, horizontal and rocking excitation of tie "1" the resulting compliances and coupling compliances are presented in the Fig. 5 - 8 for a seven-tie-system. For all translation modes a very high coupling occurs in the whole frequency range considered. For higher values of non-dimensional frequency a_o the real part of the coupling compliances tend to approach the compliance of tie "1". For the rocking mode (Fig.8) the coupling is relatively small in the lower frequency range but seems to increase with increasing frequency. An interesting effect is that for all excitation modes the compliances of the unloaded ties show an oscillatory behaviour. An investigation made on a

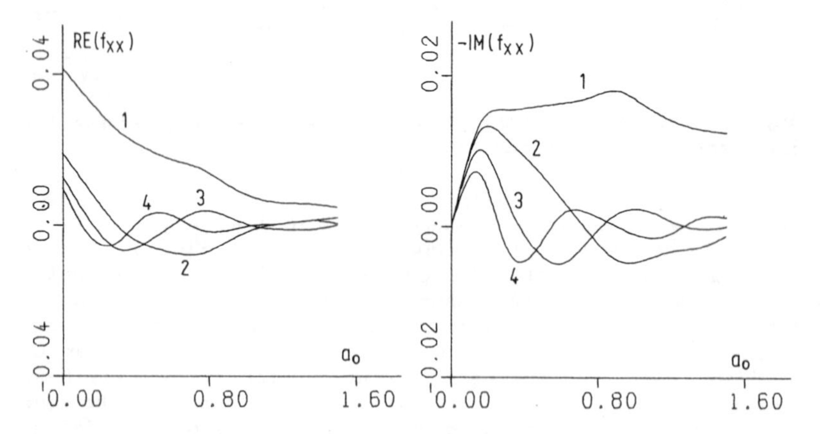

Fig. 5: Real and imaginary part of the horizontal compliance functions (x-direction) of a seven-tie-system

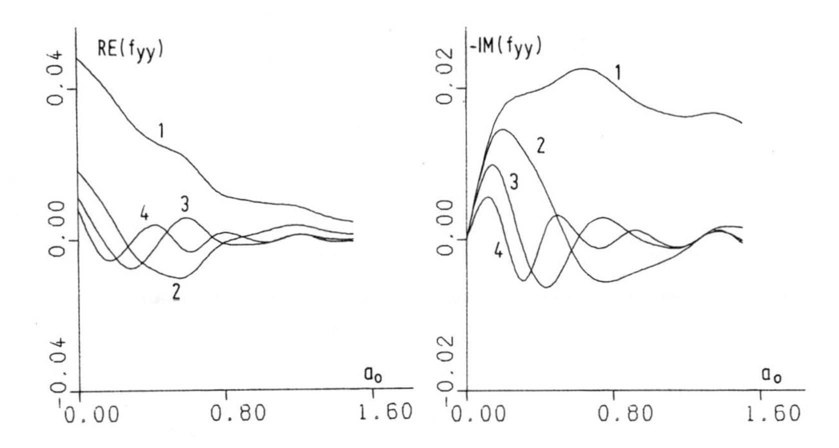

Fig. 6: Real and imaginary part of the horizontal compliance function (y-direction) of a seven-tie-system.

quite different side length ratio indicates that the "period" of the coupling compliances is effected only by the distance d between the ties and is proportional to $2/\pi d$.
It is important to note that a good approach can be obtained for the coupling compliances at a considerably less computational effort by regarding a one-tie-system only. Computing the resulting free field displacements $\bar{u}(x,y)$ outside the loaded tie and taking the average of the displacements over the area of the other ties a sufficient agreement with the exact solution can be reached as seen in Fig. 7 for the vertical compliance. The approximate solution can be obtained by a algebraic averaging in the quarter points of the area of tie "j"

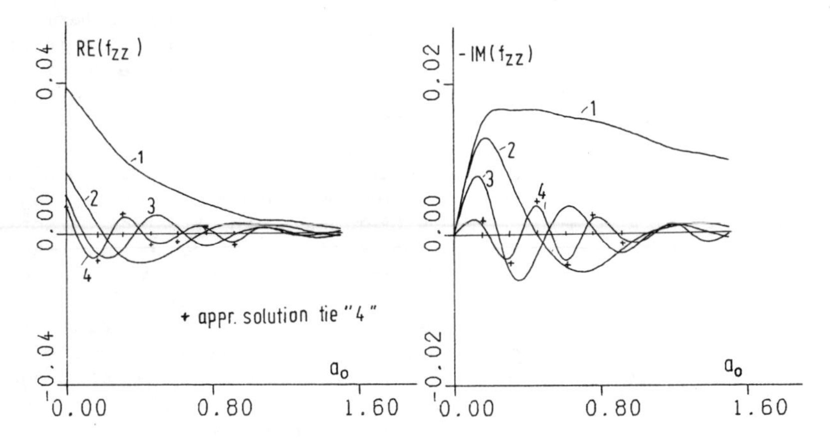

Fig. 7: Real and imaginary part of the vertical compliance function of a seven-tie-system

$$\bar{u}_j = \frac{1}{4}\sum^{4} u_j\left(\pm\frac{a}{4},\pm\frac{b}{4}\right); \quad \bar{\varphi}_y = \frac{2}{a}\sum^{2} u_j\left(-\frac{a}{4},\pm\frac{b}{4}\right)-u_j\left(+\frac{a}{4},\pm\frac{b}{4}\right) \qquad (24)$$

or more complex by an integral averaging

$$\bar{u}_j = \frac{1}{A_j}\iint \bar{u}(x_j,y_j)\,dx_j dy_j; \quad \bar{\varphi}_y = \iint \bar{u}(x_j,y_j)\,x_j dx_j dy_j \Big/ \iint x_j^2 dx_j dy_j . \qquad (25)$$

Both methods give nearly the same results. Of course, the feedback effect of the unloaded ties to the loaded tie is not included in such a simplified analysis.

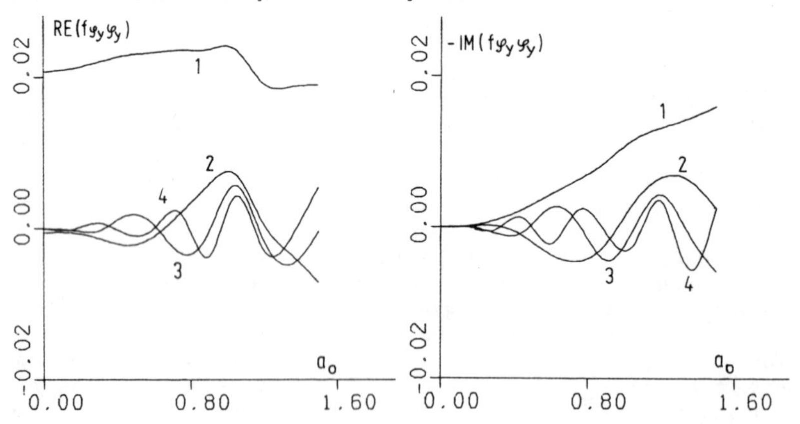

Fig. 8: Real and imaginary part of the rocking compliance function of a seven-tie-system

Illustrating the influence of the number of ties on the dynamic behaviour of the tie-soil-system amplitude curves are presented in the Fig. 9 - 11. Only tie "1" is excited by external harmonic forces and moments in order to demonstrate clearly the interaction effects. The influence of additional ties on the vertical amplitude of tie "1" can be recognized in Fig. 9. Comparing the amplitudes of the systems with N = 1, N = 3 and N = 7 ties it can be seen that an increasing number of ties produce an increasing number of local maximas with increasing values of the maximas. The shapes of the maxima tend to become sharper with increasing number of ties. Comparing the vibrations of the ties not directly loaded with the values of the loaded tie "1" it can be seen that the interaction induces vibrations of 40 %, 25 % and 20% for the second, third and forth tie in relation to tie "1". It is interesting to notice that the values of the frequencies at which maximas occur depend only on the distances between the ties, expressed by the relation $\lambda/d_j = \pi n$ at which n is an integer. For horizontal force excitation in the x- and y-direction the corresponding amplitudes for a seven-tie-system are presented in Fig. 10. In general a similar behaviour can be observed as for the vertical amplitudes. In relation to tie "1" the amplitudes of the unloaded ties "2-4" reach values of 45 %, 32 %, 23 % and 39 %, 23 %, 15 % for the x-direction and y-direction, respectively. For the x-direction resonance occurs at a higher

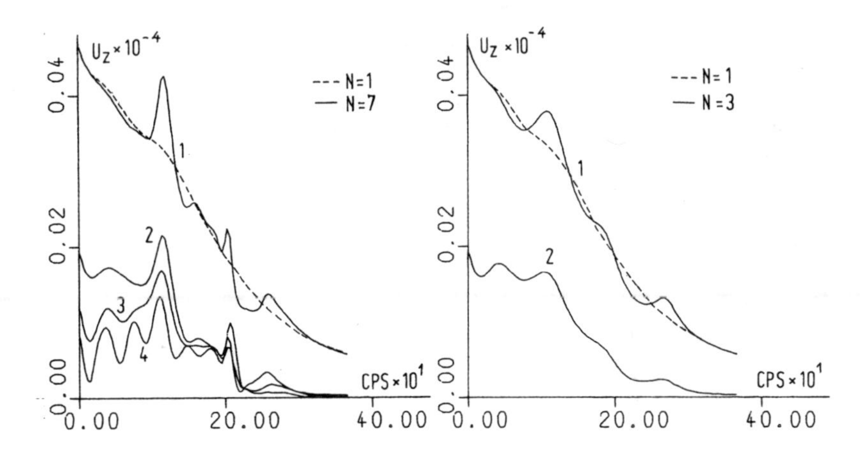

Fig. 9: Vertical amplitude curves of a system with N=1,N=3 and N=7 ties.

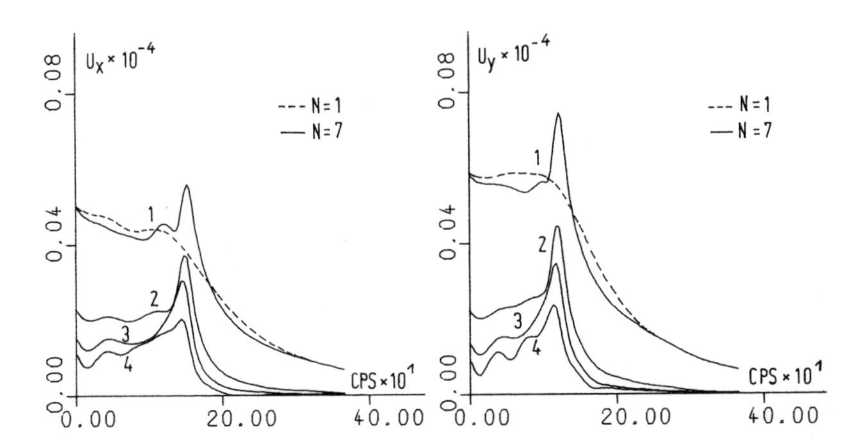

Fig. 10: Horizontal amplitude curves of a seven-tie-system.

frequency as for the y-direction, the amplitudes, however, are generally smaller.
For the amplitudes of the rocking mode shown in Fig. 11 in the low frequency range coupling seems to be not so important, but in the neighborhood of resonance a strong coupling exists. The presence of additional ties also induces a great amplification of the amplitude of tie "1". For a three-tie-system an amplification factor of 1.5 and for a seven-tie-system a factor of 3.3 can be observed.
A very important interaction effect is also the generation of modes which do not exist if a single tie is considered. In Fig. 12 this effect is demonstrated by the rotations of the unloaded

ties due to vertical excitation of tie "1". Comparing Fig. 12 with Fig. 9 it is seen that outside the resonance this rotational mode causes vertical displacements at the edges of the ties of about 65 % in the average in relation to the vertical displacements due to a vertical excitation. In resonance a value of 260 % can be reached.

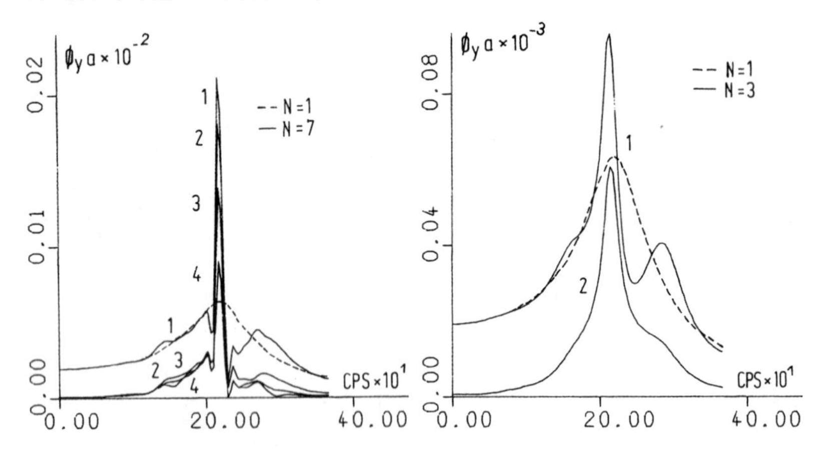

Fig. 11: Rocking amplitude curves of a seven- and three-tie-system.

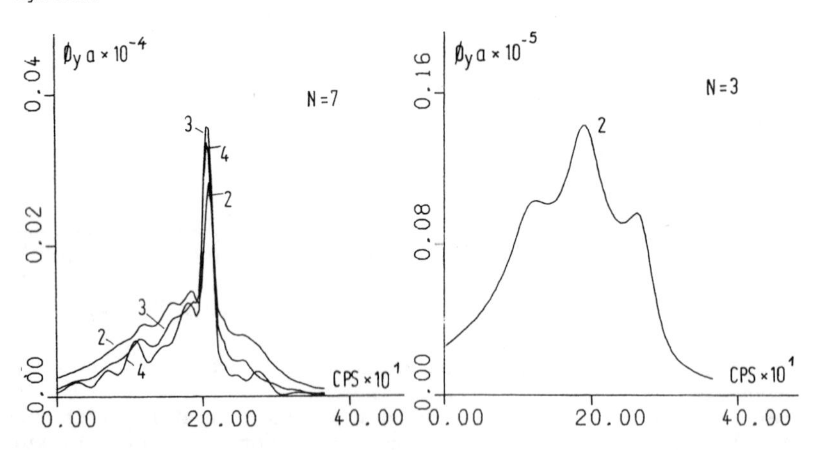

Fig. 12: Rotational displacements due to a vertical excitation of a seven- and three-tie-system

In the last part of this paper results are presented which show the comparison between theoretical results and experiments. In Fig. 13 the amplitude and phase for the vertical displacement of a rigid tie are given. One can see that a highly satisfactory agreement between theory and experiment can be achieved if a seven-tie-system is taken into account.

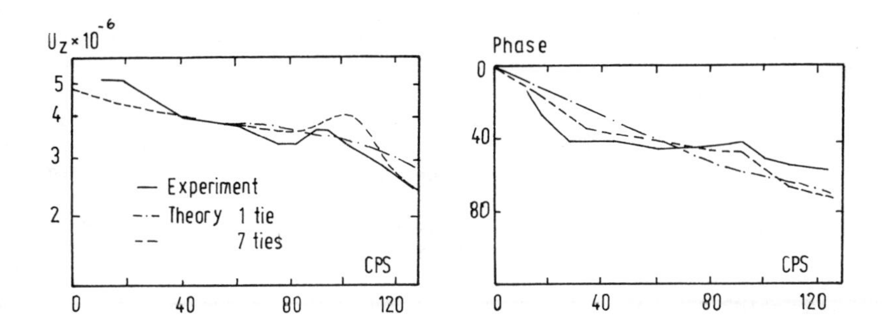

Fig. 13: Comparison between theory and experiment of the vertical displacement of a rigid tie.

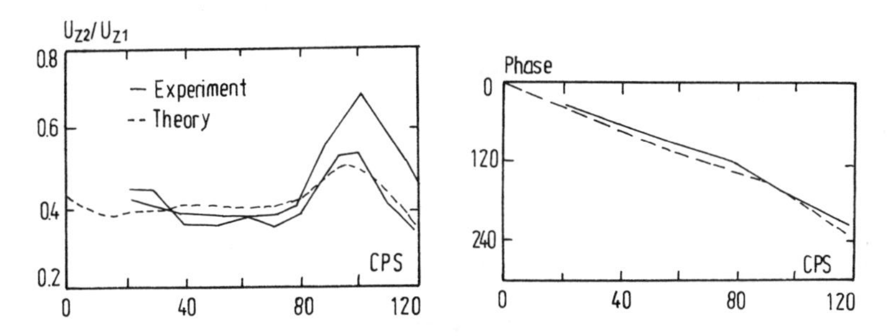

Fig. 14: Comparison of theoretical coupling with experiments of the vertical displacement of tie "2".

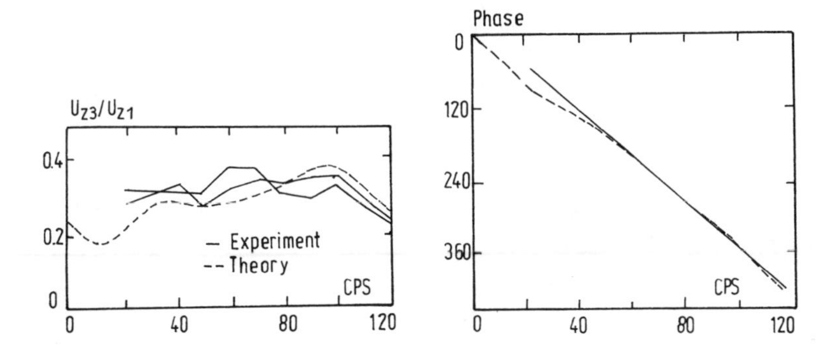

Fig. 15: Comparison of theoretical coupling with experiments of the vertical displacement of tie "3".

Fig. 14 and 15 show the comparison between the theoretical coupling of a seven-tie-system and experimental results obtained by measurements at a tie-system with "infinite" number of ties. It can be seen that the agreement is quite well. For investigations regarding the whole rail-tie-soil-system it seems therefore to be possible to reduce the infinite system by a system with a finite number of ties. The whole system can then be analyzed by using standard finite element techniques for the rails as has been shown by Rücker (1981).

CONCLUSIONS

An analytical formulation based on a Green's functions approach has been presented for a general three-dimensional solution of structure-soil-structure interaction problems. The structures can have rigid or flexible foundations resting on the surface of a linear elastic/viscoelastic homogeneous and non-homogeneous halfspace. There is no restriction as to the number of structures, their size, shape and frequency range. The method has been applicated to tie-soil-systems with different number of ties. For this particular system compliance functions have been presented which illustrate the feedback and coupling effects. From the studies described it seems that in the case of symmetry the feedback effects are less important compared with the coupling effects. This behaviour allows the use of a simplified interaction analysis by averaging the free field displacements due to excitation of a single-tie-system. The coupling effects are demonstrated by amplitude curves. From the results it appears that an increasing number of ties produce an increasing number of local maximas. The frequencies at which maximas occur depend only on a wavelength/tie-distance ratio. Another main effect of adjoining foundations is the generation of vibration modes which do not appear if interaction is not taken into account. Finally, it has been shown that the method proposed here furnishes results which are in good agreement with experiments made at a real railroad-bed.

REFERENCES

Filonenko-Boroditsch, M.M. (1967) Elastizitätstheorie, VEB Fachbuchverlag, Leipzig.

Rücker, W. (1981) Dynamische Wechselwirkung eines Schienen-Schwellen-Systems mit dem Untergrund. Research Report Nr. 78, BAM, Berlin.

Rücker, W. (1982) Dynamic behaviour of Rigid plates of Arbitrary shape on an Elastic Halfspace. accepted for publication in Earthq. Eng. & Struct. Dyn.

Savidis, S., W. Saarfeld (1980) Verfahren und Anwendung der dreidimensionalen dynamischen Wechselwirkung. Baugrundtagung, Mainz, 47-77.

Soil Dynamics & Earthquake Engineering Conference / Southampton / 1982.07.13-15

Effect of Water Table on Dynamic Response of Pile Group

R.C.VIJAYVARGIYA & D.S.DHARMADHIKARI
M.A.College of Technology, Bhopal, India

SUMMARY

Although a good amount of work has been done to study vibration characteristics of piles subjected to vibrations, meagre information is available on dynamic response of piles subjected to forced vibration under varying water table conditions (Richart, et al 1970). In this paper, an attempt has been made to investigate the effect of variation of water table on resonant frequency of pile group subjected to forced vertical variation in cohensionless soil. The field conditions are better simulated by considering the effect of water table which is generally encountered in case of pile foundations. Systematic model tests have been conducted in the laboratory on different pile groups. Water table was raised in stages. Effect of skin friction and pile spacing on resonant frequency of pile group are also investigated. The test results are analysed and useful conclusions are drawn.

INTRODUCTION

The pile foundations are often adopted in some of the following cases : (i) The total pressure on the soil, both static and dynamic is larger than the bearing capacity of the soil (ii) It is necessary to decrease the amplitude of free or forced vibrations and to increase the resonant frequency of the foundation-soil system (iii) It is necessary to decrease the residual dynamic settlement of foundation (Barkan 1962).

However, the assessment of resonant frequency of foundation-soil-system is one of the vital factor in the design of foundation subjected to vibratory forces. Resonance occurs when the operating frequency of a machine coinsides with the resonant frequency of the foundation-soil-system. At resonance the amplitudes of vibration of foundation are amplified to such a large extent that structural damage may occur. To avoid such damages an effort must be made to predict the resonant frequency of the foundation soil-system as precisely as possible, in order to eliminate the possibility of resonance. An attempt has been made to simulate the field conditions by considering the effect of water table on the resonant frequency of pile group subjected to forced vertical vibrations in cohensionless soil. The investigation involves laboratory tests on model pile groups. Different parameters affecting resonant frequency such as spacing of piles in a group, water table, skin friction have been considered in this investigation.

TEST PROGRAMME:

Laboratory tests were conducted on model piles of aluminium, 1.25 cm diameter and 55 cm length. Tests were carried in mild steel tank of internal dimensions 90 cm x 90 cm and 120 cm depth. A layer of vibration absorbing material was fixed on the interior surface of the tank. The tank was fitted with piezometer tube to indicate the level of water table. The dimensions of the piles were adopted so as to avoid the effects of boundary conditions.

Uniformly graded dry cohensionless Narmada sand was used for the test. Sand was deposited in layers by giving uniform vertical fall of 1.0 m. Water table was allowed to rise gradually from the bottom. This was achieved by admitting the flow of water from the bottom of the tank. The desired levels of water were marked from the piezometer tube. The water tables were raised in four equal stages. The first set of observations were recorded with the water table touching the tip of the pile. Subsequent tests were carried out by raising the water table in next two equal stages. Last set of tests was carried out by inundating the sand in the tank upto the top surface so that a very thin film of water was maintained at the surface. This stage, referred as the fully submerged case.

Aluminium plate, 1.25 cm thick with threaded holes to house pile shafts, was used as pile cap. In practice the spacing of piles in a group generally varies between two to six times diameter(d), of the piles. The spacing of piles in the group, adopted, was 3d, 4d, 5d and 6d respectively. Tests were carried out on single and square groups of piles. The skin friction of piles was altered by glueing water proof emery paper of three different grades i.e. 80, 100 and 220. Values of coefficient of friction for different grades of friction paper are given in table-1.

Table 1 : Coefficient of friction for different grades of paper

Grade of paper	Coefficient of friction
no paper	0.34
220	0.47
100	0.58
80	0.66

Tests for different grades of friction and pile spacings were carried out corresponding to each stage of water table.

An electromagnetic vibration generator of 1.5 Kg capacity was used for excitation of pile group. Frequency measurements were made directly on frequency meter. The resonant conditions were observed by recording the peak amplitudes on vibration meter, which in turn was connected to pile group through an electro-magnetic pick up and amplifier. Figure 1 shows photograph of the experimental set up and instrumentation.

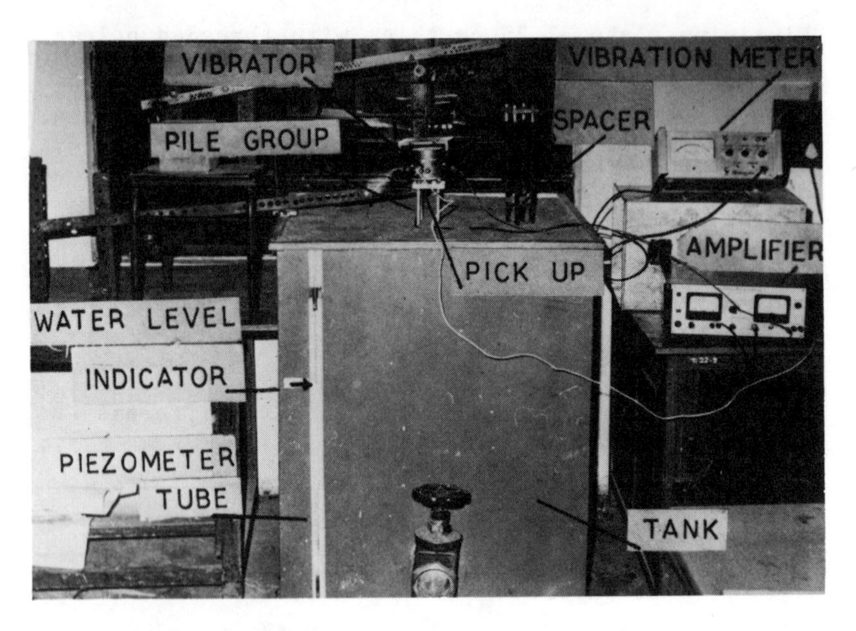

Fig.1 Photograph of the experimental set up.

RESULTS AND DISCUSSIONS

The effects of water table variation on resonant frequency of pile group were observed. Figure 2 shows the plot obtained between resonant frequency of different pile groups versus water table, for 80 grade friction paper. It is observed from the plot that resonant frequency of the pile group increases linearly with the rise in water table, for all pile groups. The nature of variation of increase in resonant frequency with the rise in water table is almost similar for different skin friction of all the pile groups under study. Considering a typical group of pile 6d x 6d (say), the observed resonant frequency when the water table just touching the tip of the pile is 147 c.p.s. When the water table is raised to 51 cm and the submerged condition was obtained, the observed resonant frequency increases to 167 cps. The increase in resonant frequency is about 13.6%. Increase in resonant frequency with increase in moisture have also been reported for block foundation, (Tailor C, et al 1967).

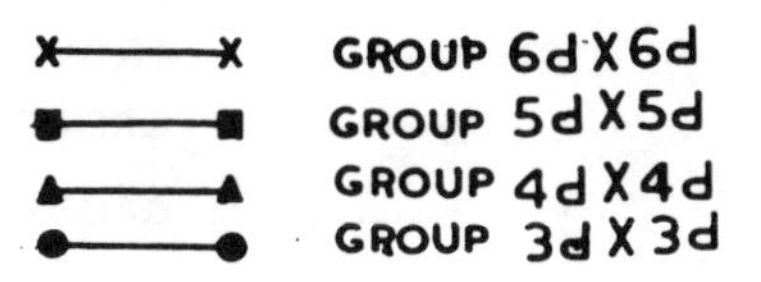

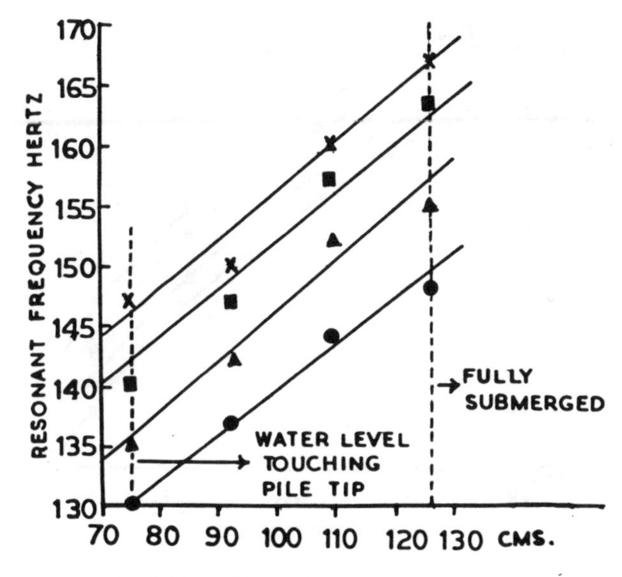

Fig.2 A typical plot of water level versus resonant frequency for coefficient of skin friction 0.66

Observations of resonant frequency in relation to the skin friction of the pile group, for each stage of water table has also been recorded. Figure 3, shows the plot of resonant frequency versus skin friction, for the first stage of water table i.e. water table just touching the pile tip. It can be concluded from the plot, that skin friction affects the dynamic response of the foundation-soil-system more predominently. Similar inferences hold good for the records of the subsequent three other stages of water tables. A typical group of pile, say 6d x 6d, with water table just at the tip of

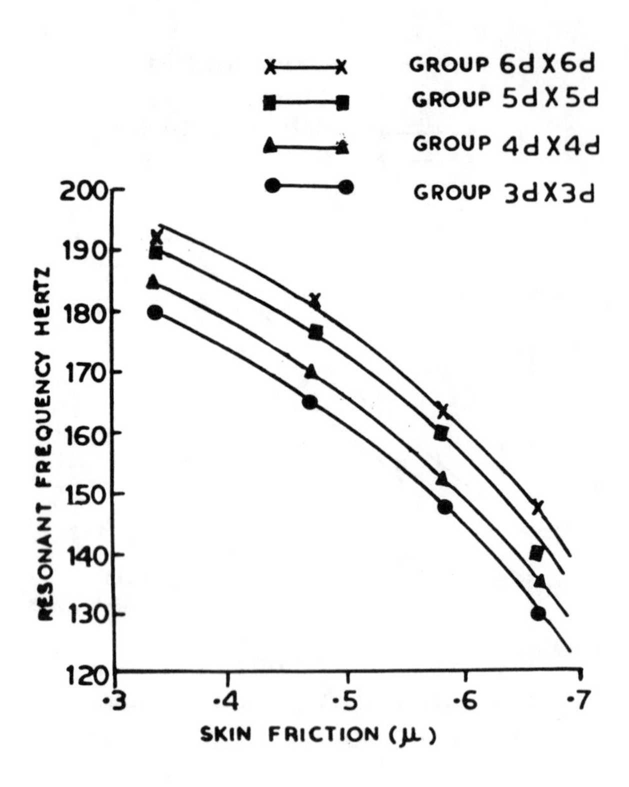

Fig. 3 A typical plot of skin friction μ versus resonant frequency for water table touching pile tip.

piles and coefficient of skin friction 0.34, records the resonant frequency of 192 c.p.s. For the same water table condition, when the coefficient of skin friction is changed to 0.66, the resonant frequency decreases to 147 c.p.s. Thus 50% increase in skin friction decreases the resonant frequency by about 23.5 percent.

The observations of resonant frequency in relation to the spacings of the different pile groups were studied. Plot of resonant frequency versus pile spacings has been obtained for each grade of skin friction. Figure 4, shows one of the such plot for no friction paper i.e. skin friction between the pile surface and sand.

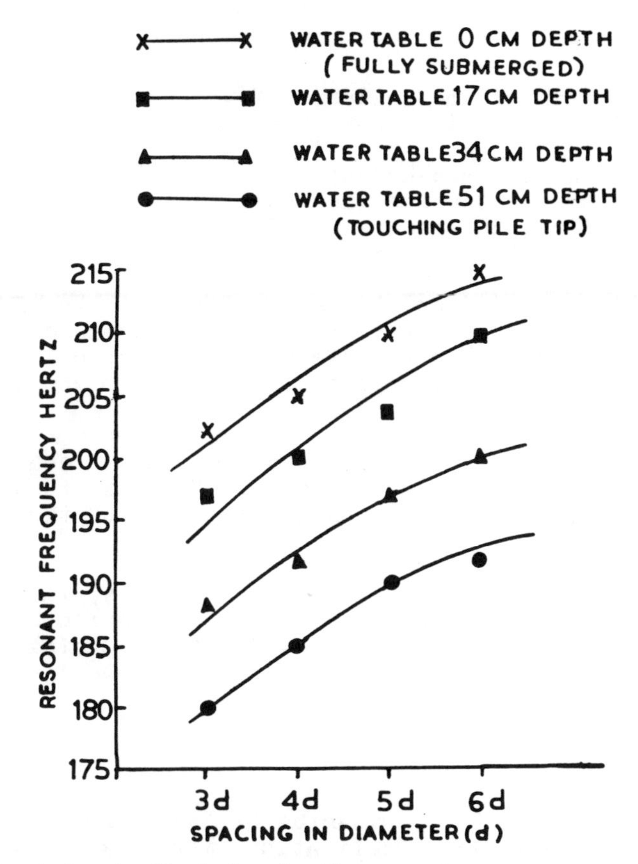

Fig.4 A typical plot of pile spacing versus resonant frequency for coefficient of skin friction 0.34.

It is infered from these plots that the resonant frequency of the pile group increases with the increase in the pile spacing of the foundation. However, the rate of increase appears slower as the spacing approaches the limiting value of 6d. Considering a particular value of coefficient of skin friction 0.34 (say) and, the fully submerged condition, it is observed that resonant frequency for spacing 3d x 3d is 202 c.p.s. If the spacing is increased to 6d x 6d the resonant frequency increases to 215 c.p.s. Thus the 50 percent increase in spacing registers about 6.5 % rise in resonant frequency of the foundation.

CONCLUSIONS

On the basis of the analysis of the test results, following conclusions have been drawn:-

(i) Resonant frequency of the soil-pile system increases with the rise in water table and need to be considered in the design of foundations subjected to forced vertical vibrations.

(ii) High tuned foundations are likely to be affected by decrease in water table while low tuned foundations are likely to be affected by increase in water table as the operating frequency tends to concide with the resonant frequency.

(iii) Skin friction affects the resonant frequency of the pile group significantly. Increase in skin friction reduces the resonant frequency of the pile group.

(iv) Increase in the spacing of the pile, increases the natural frequency of the pile foundation. The rate of increase appears slower as the spacing approaches the limiting value of 6d.

References

Barkan, D.D. (1962) Dynamics of Bases and Foundations, McGraw Hills Publication, New York.

Richart, F.E. Woods, R.D., Hall, J.R (1970) Vibrations of Soils & Foundations, Prentice-Hall, Inc., Englewood, New Jersy.

Tailor, C., Green, R., Kalita., U.C. (1967) Factors Affecting the Response of Machine Foundation, Proceedings of International Symposium on Wave Propogation and Dynamic Properties of Earth Material.

Soil Dynamics & Earthquake Engineering Conference / Southampton / 1982.07.13-15

Effect of Skin Friction on the Natural Frequency of Pile Groups Subjected to Vertical Vibrations

D.N.KRISHNAMURTHY, V.K.TOKHI & M.R.BEG
Maulana Azad College of Technology, Bhopal, India

SUMMARY
Model tests on aluminium pile groups are conducted to study their vibration characteristics under free vertical vibrations. Short piles at varying spacings are used. The skin friction of the pile surface is varied by gluing different grades of emery cloth round the piles. The pile groups were embedded in dry uniform sand contained in a large size masonry tank.

It can be concluded from the experiments that the natural frequency of vertical vibration of pile-soil system in dry sand can be substantially varied by varying 1) the skin friction on the pile surface, and 2) the spacing of the piles in the group. If under given conditions, resonance is likely to occur, the designer now has a method, as reported herein, of altering the natural frequency of the soil-pile system by altering the skin friction and spacing of the piles in the group.

INTRODUCTION

There are instances when machines or other foundations are supported on piles. These piles are subjected to different modes of vibrations depending upon the type of machine and the nature of the generated force. The behaviour of piles under static and dynamic loads is a complex phenomenon. Static loads act on piles when used as support for retaining walls, bridge abutments, piers, etc. Dynamic loads act in addition to static loads in the above cases when vehicular traffic passes over them. To be safe, resonance should be avoided and the amplitudes of vibrations kept within permissible limits in all machine foundations. The operating frequency of the machine is decided by mechanical design dictated by the function of the machine. In order to avoid resonance it should be possible to alter the natural frequency of the pile-soil system as desired.

Some work has been carried out to study the vibration characteristics of single and groups of piles subjected to vertical vibrations (Swiger, 1948; Barkan, 1962; Sridharan, 1964; Tokhi et al 1973 and 1974; Deshpande et al 1975; Novak and Grigg, 1976). Some work has also been reported on pile groups subjected to lateral vibrations (Krishnamurthy and Patki, 1980). In the present investigations tests have been carried out on model pile groups subjected to transient vertical vibrations.

The methods available to calculate the natural frequency of foundations under vibrations can be classified as under:

i) Empirical or Semi-empirical
ii) Analytical methods.
   a) Method based on semi-infinite elastic medium.
   b) Method based on mass-spring analogy.

In the present work, the method based on mass-spring analogy has been adopted. In this analogy the vibrating system is assumed to be one degree freedom system. The soil under vibration is assumed to act as a weightless spring. When this system is subjected to free vibrations, the natural frequency of the system is given by

$$f_n = \frac{1}{2\pi}\sqrt{\frac{K}{m}} \tag{1}$$

where K = Coefficient of stiffness, and
m = mass of vibrating system.

For pile foundations, the natural frequency, f_n, is given by

$$f_n = \frac{1}{2\pi}\sqrt{\frac{C_\delta}{m}} \tag{2}$$

where C_δ = coefficient of elastic resistance of pile,
and m = mass of pile foundation and superimposed load, if any; and of the soil vibrating with the pile-soil system.

C_δ represents the losd required to induce a unit elastic settlement of the pile. It can be determined by load versus elastic settlement curves of the pile, as the slope of the curve. In a pile group, the coefficient of elastic resistance, K_z, of the group is given by

$$K_z = n\, C_\delta\, \mu \tag{3}$$

and the natural frequency, f_n, by

$$f_n = \frac{1}{2\pi}\sqrt{\frac{n\,\mu\, C_\delta\, g}{W}} \tag{4}$$

where n = number of piles in the group

μ = factor depending upon the ratio of spacing to diameter.

K_z = Coefficient of elastic resistance of pile group, and

W = weight of pile and pile cap plus soil mass participating in vibration plus superimposed load.

$$f_n = \frac{1}{2\pi}\sqrt{\frac{K_z g}{W}} \qquad (5)$$

Reese and Matlock, 1956 proposed generalised solutions for laterally loaded piles, and introduced a linear dimension T, termed as relative stiffness factor. It is a relation between the stiffness of soil and flexural stiffness of piles, obtained as

$$T = \sqrt[5]{\frac{EI}{n_h}} \qquad (6)$$

where EI = Flexural modulus of pile, and
n_h = Coefficient of soil modulus variation.

The relative stiffness factor, T, provides an objective definition of long and short piles in terms of maximum depth coefficient Z_{max} given by

$$Z_{max} = \frac{L_e}{T} \qquad (7)$$

where L_e = depth of embedment of pile. A pile is long if $Z_{max} > 5$ and short if $Z_{max} < 2$.

TEST PROGRAMME AND SET UP

General

The behaviour of pile groups subjected to vertical vibrations is dependent on numerous factors, viz. soil and its properties, soil modulus variation, shear and moments acting on the piles, number and arrangement of piles, roughness of pile surface, superimposed load and relative stiffness factor.

In the present investigation, the variables considered are skin friction between the pile and soil, spacing of piles in the group, diameter of piles in the group and superimposed loads. Other parameters are kept constant.

Tests conducted

In order to study the effect of the above mentioned variables, tests were carried out for different combinations of variables and the natural frequency of pile group was measured under transient vertical vibrations.

Experimental Set Up

The general arrangement of the set up is shown in Fig 1. The set up consists of the following.

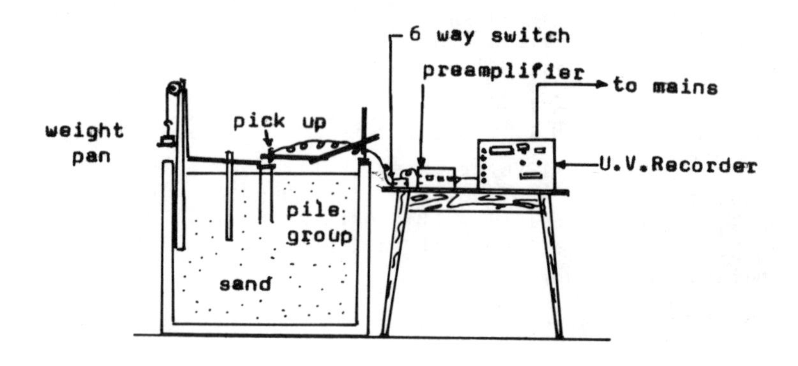

Fig 1 Experimental Set Up

1 Tank

The tests were carried out in a masonry tank of internal dimensions 1800 X 1800 X 1500 mm. Considering the size of piles used, it is assumed that there would be no edge effects.

2 Soil

Air dried uniform sand having an effective size (D_{10}) of 0.34 mm, maximum grain size (D_{max}) of 2.41 mm and uniformity coefficient of 2.34 and density of 16.08 kN/m^3 was used.

3 Vibration recording arrangement

Free vibrations were induced by striking the pile cap lightly with a pointed wooden mallet made to strike the geometrical centres of the pile cap. The vibrations were recorded on a photographic paper with a ultra-violet recorder.

4 Pile models

a) Pile shaft Solid circular aluminium rods of 19 mm dia x 252 mm long comprising T - 1 series and 25 mm dia x 312 mm long comprising T - 2 series were used. The aluminium rods used as piles have flexural stiffness of 47.18 $kN\text{-}m^2$ and 149.12 $kN\text{-}m^2$ for T - 1 and T - 2 series respectively.

b) Pile caps 12 mm thick aluminium plates were used as pile caps. In practice the spacing of piles in a group generally varies between two to six times the diameter. The same range of spacing was used in the present investigation. Fig 2 shows the 15 groups formed with different spacings. To ensure rigidity of connection the piles were screwed tightly to the caps.

5 Emery cloth
For changing skin friction, emery cloth of grades 60, 80 and 100 was glued on the pile surface with araldite. One set of observation was taken with plain piles.

Recording
The vibrations were picked up with the help of an electro-dynamic pick-up and recorded with the help of an ultra violet recorder on a photographic paper via a junction box and a pre-amplifier. A small tin strip was fixed on the top surface of the pile cap so as to provide a magnetic surface. The paper speed was 250 mm per second in the U.V recorder.

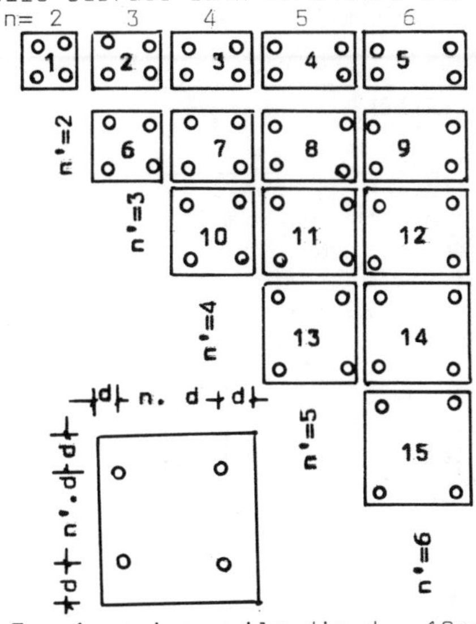

T - 1 series, pile dia d = 19mm
T - 2 series, pile dia d = 25mm

Fig 2 Diagram showing pile spacings.

Compaction of sand
The sand inside the tank was compacted in five equal layers at pre-determined resonant frequency of 1000 r.p.m for 3 minutes duration with a LA oscillator by placing it at 9 uniformly spaced positions.

Coefficient of friction
Coefficient of friction between emery cloth and the sand was obtained by shear box test. The coefficients of friction between emery cloth and sand are 0.27, 0.38 and 0.45 for 100, 80 and 60 grades. The coefficient of friction between the plain pile and sand was found to be 0.1.

Determination of Natural Frequency
The pile group was driven in sand by using a wooden mallet. This method was preferred to jacking as in practice the piles are driven rather than jacked down. The pile cap was kept free of sand by 50 mm. The gap between the

pick-up and the magnetic strip of the pile cap was adjusted to about 1 mm. The pile group was subjected to vertical vibrations by impact of a light wooden mallet, as described earlier. Simultaneously the vibrations were recorded on photographic paper with the U.V recorder. The record was read by a travelling microscope and the natural frequency of free vibrations of the pile group was calculated.

ANALYSIS AND INTERPRETATION OF RESULTS

1 Variation in coefficient of elastic resistance of pile groups

1.1 Coefficient of elastic resistance versus spacing The variation in coefficient of elastic resistance of the pile group with the spacing is shown in Figs 3 and 4. It is seen that the coefficient of elastic resistance of the pile group increases with spacing of the piles in the groups for both the series.

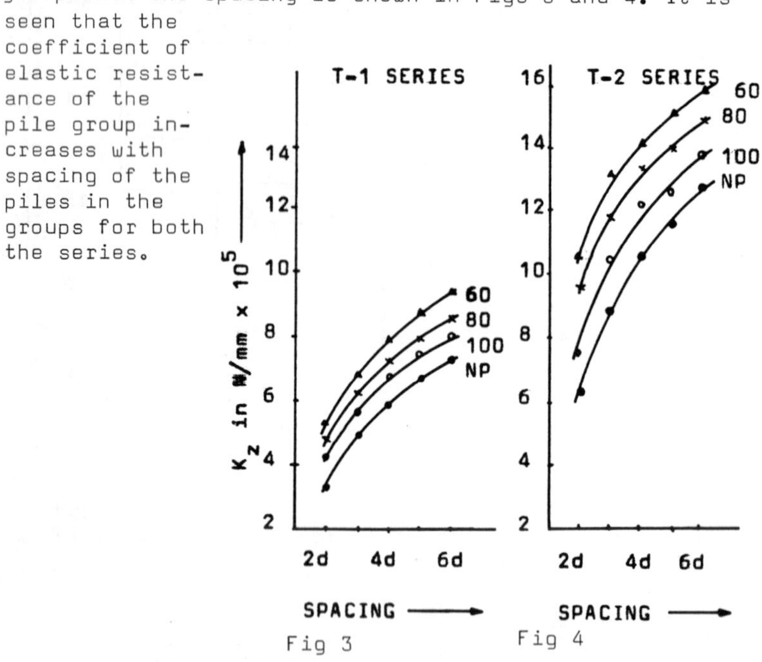

Fig 3 Fig 4

Figs 3 and 4 Coefficient of elastic resistance versus spacing of piles (Squre groups)

1.2 Coefficient of elastic resistance of the pile group versus skin friction The variation in coefficient of elastic resistance of pile group with the skin friction is shown in Figs 5 and 6. The coefficient of elastic resistance increases with the increase in skin friction on the pile surface. The value of the coefficient of elastic resistance is more in the case of T - 2 series than for T - 1 series. Higher values of coefficient of elastic resistance in case of T - 2 series is attributed to

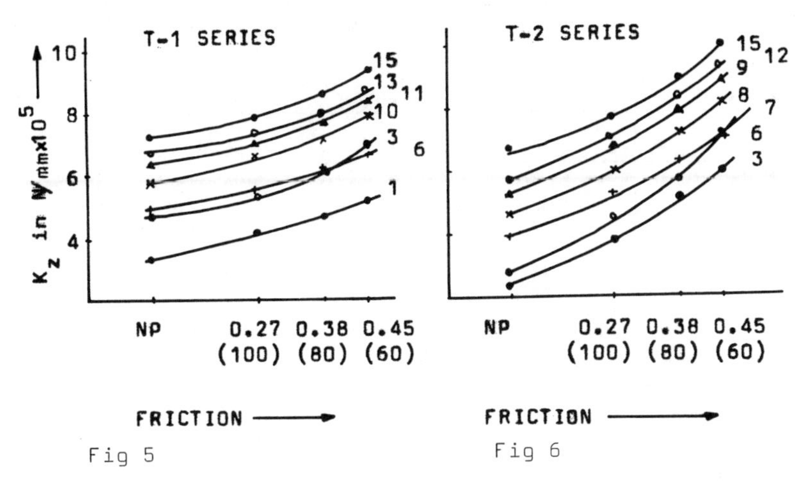

Fig 5 Fig 6

Figs 5 and 6 Coefficient of elastic resistance versus skin friction

greater surface area. The surface area of embedded piles of T - 2 series is 1.78 times the surface area of T - 1 series and the ratio of coefficient of elastic resistance of pile groups of T - 2 series to that of T - 1 series for the corresponding group is also found to be around 1.78.

1.3 Coefficient of elastic resistance of pile group versus area and perimeter of the group The variation in the coefficient of elastic resistance of the pile group with area and perimeter of the group in T - 1 and T - 2 series are shown in Figs 7 and 8. It is seen from

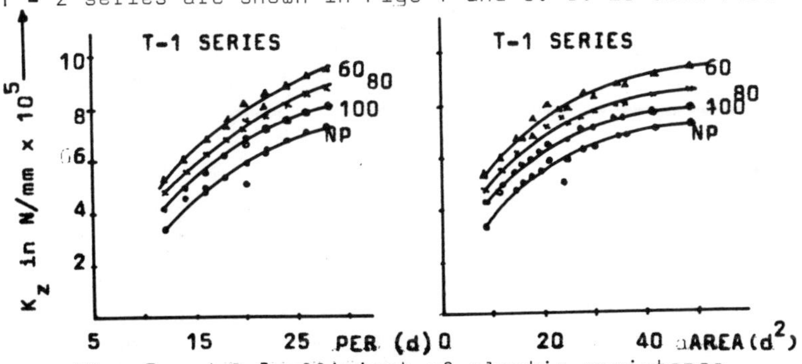

Figs 7 and 8 Coefficient of elastic resistance versus perimeter and area of pile groups

these figures that the coefficient of elastic resistance of the group increases with the increase in area and perimeter of the pile group. It is noted that the pile groups having the same perimeter, irrespective of the spacing between the piles, have the same coefficient of elastic resistance.

2 Variation of Natural Frequency of pile groups

2.1 Natural frequency versus diameter of pile group The variation in the natural frequencies of the groups of piles in T - 1 and T - 2 series are given in Table 1. It is seen that the natural frequency of T - 1 series is

Table 1

Natural frequency of vertical vibration of pile groups with different skin friction (no load) in H_z.

	T - 1 series				T - 2 series			
Group No	NP	100	80	60	NP	100	80	60
1	355	326	300	282	305	285	268	260
2	357	332	302	284	310	285	270	260
3	360	330	302	281	314	293	272	262
4	362	330	310	280	319	297	277	264
5	368	340	309	284	321	297	275	264
6	362	333	306	280	317	297	277	264
7	362	329	308	280	324	298	280	265
8	367	340	310	292	328	305	284	267
9	372	340	314	292	335	308	287	268
10	370	340	315	290	328	305	284	265
11	368	342	315	295	337	310	288	270
12	370	347	329	300	345	317	295	273
13	377	345	315	301	344	316	295	268
14	381	342	327	302	350	320	298	269
15	385	354	325	306	355	327	302	269

greater than that of T - 2 series indicating that the natural frequency is a function of diameter of the piles in the group. This could be expected as the increase in the diameter of piles in the group would result in an increase in the surface area of the pile group, which in turn would increase the weight of the pile group and the amount of soil mass participating in vibrations.

2.2 <u>Natural frequency versus skin friction</u> The variation in the natural frequency versus skin friction is shown in Figs 9 and 10. It is seen that the natural frequency

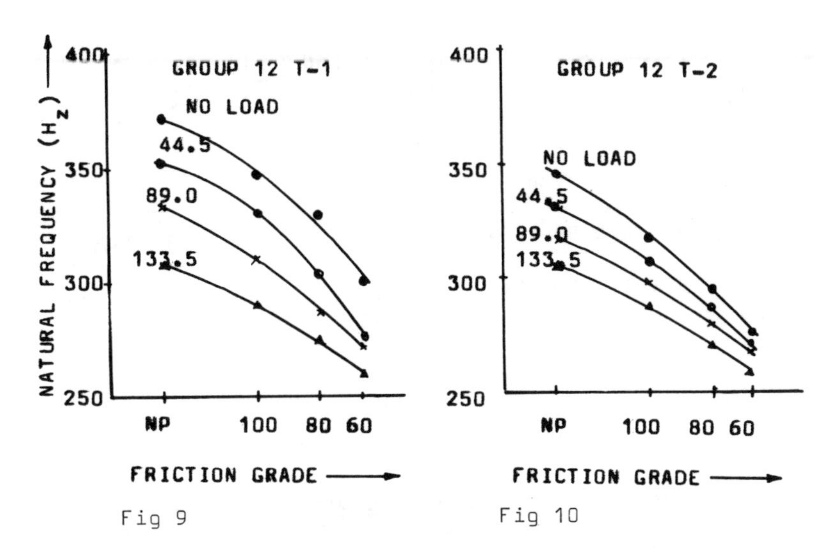

Fig 9 Fig 10

Figs 9 and 10 Natural frequency versus skin friction for different static loads.

decreases as the skin friction on the pile group increases. This can be attributed to the increase in the mass ratio at a faster rate than the coefficient of elastic resistance of the group.

2.3 <u>Natural frequency versus static load</u> The variation in the natural frequency versus static load is shown in Figs 11 and 12. It can be seen that the natural frequency decreases as the static load increases. It is also seen that the variation in the natural frequency with static load is nearly linear.

2.4 <u>Natural frequency versus spacing of piles in the group</u> The variation in the natural frequency versus spacing of piles in the group (for T-1 and T-2 series) is shown in Figs 13 and 14. It can be seen that the natural frequency increases as the spacing of the piles in the group increases both for T - 1 and T - 2 series.

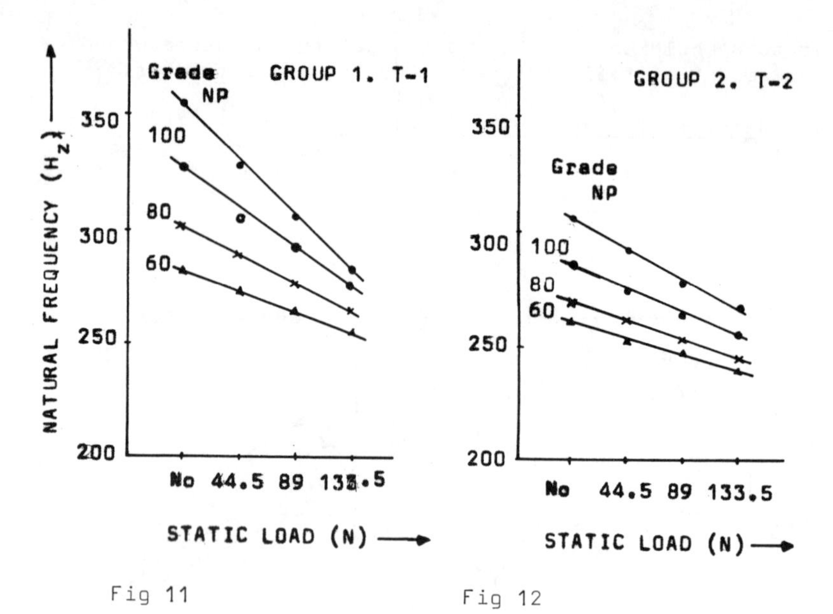

Fig 11

Fig 12

Figs 11 and 12 Natural frequency versus static load.

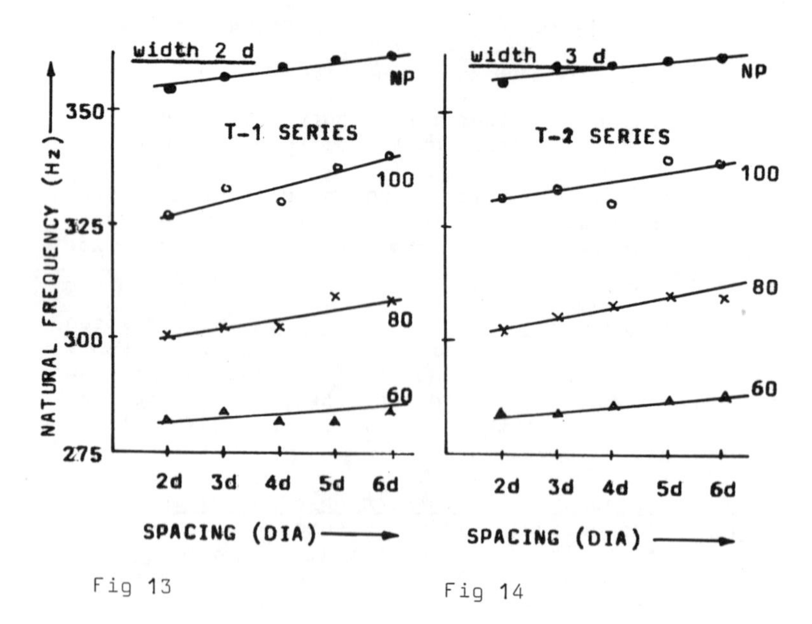

Fig 13

Fig 14

Figs 13 and 14 Natural frequency versus spacing.

2.5 Natural frequency versus plan area and perimeter of pile group The variation in the natural frequency versus plan area and perimeter of the pile group (no load) is shown in Figs 15 and 16. It can be seen that the natural frequency increases with increase in plan area and perimeter of the group.

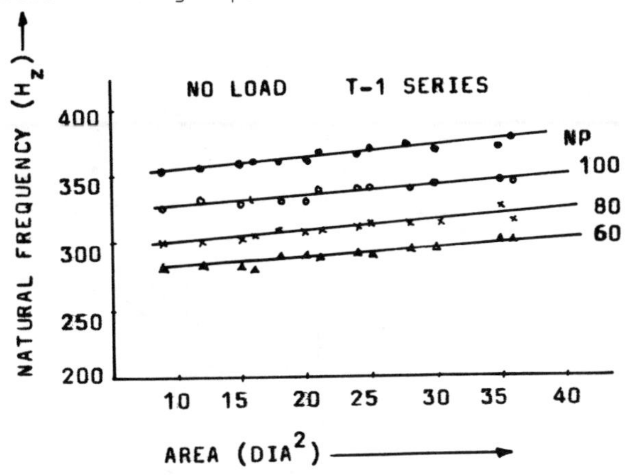

Fig 15 Natural frequency versus plan area of group.

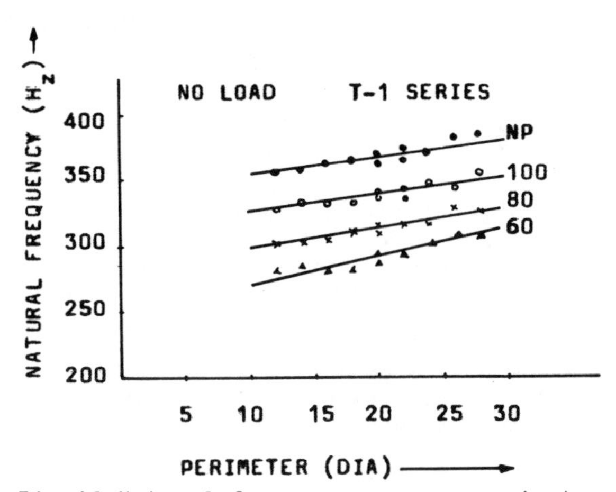

Fig 16 Natural frequency versus perimeter of group.

2.6 Mass ratio versus skin friction The variation in the mass ratio (defined as the mass of soil participating in vibration with pile group, M_s, to the mass of piles and the cap, M_p) versus friction of pile group is shown in Fig 17. It can be seen that the soil mass participating with the pile group increases with the increase in the friction on the surface of the piles. It also increases with the increase in spacing of piles in the group but decreases with the increase in static load on the pile group. The increase in the spacing of piles in the group would reduce the overlap and will result in an increase in the mass of soil participating in the vibrations.

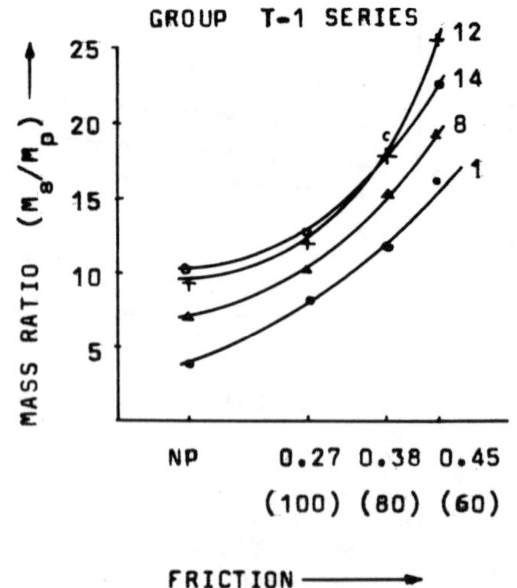

Fig 17 Mass ratio versus skin friction

EFFECT OF SUBMERGENCE ON NATURAL FREQUENCY

The effect of submergence on the natural frequency of pile groups have been studied by carrying out tests on pile groups under submerged conditions. Both short and long piles have been used during the tests. The results(reported in detail elsewhere) show that the natural frequency of pile groups subjected to free vertical vibrations, increases with submergence for short as well as long piles.

CONCLUSIONS

The effect of skin friction on the natural frequency of pile groups subjected to vertical vibrations was studied. The results obtained lead to the following conclusions.

1. Coefficient of elastic resistance of a pile group is a funtion of diameter of the piles in the group. It increases with the spacing between the piles and the friction on the surface of piles.

It is directly proportional to the embedded surface

area irrespective of the spacing of the piles in the group.
Groups having equal perimeter and skin friction have nearly the same value of coefficient of elastic resistance.

2. Natural frequency of a pile group is a function of the diameter of the pile.
It increases with the increase in spacing of piles, area and perimeter of the group.
Pile groups having the same perimeter and friction have nearly equal natural frequency.

ACKNOWLEDGEMENTS

The authors wish to express their gratitude to the authorities of Maulana Azad College of Technology, Bhopal for the facilities provided during the present invetigation.

REFERENCES

Bansal,R.K. (1982) Effect of skin friction on the natural frequency of pile groups subjected to vertical vibrations. Thesis (M.Tech), Bhopal University, Bhopal.

Barkan,D.D. (1962) Dynamics of Bases and Foundations. Mc Graw Hill Book Co.

Beg,M.R. (1979) Effect of skin friction on the natural frequency of pile groups subjected to vertical vibrations. Thesis (M.Tech), Bhopal University, Bhopal.

Deshpande, P.A., Vijay vargiya, R.C., and Tokhi, V.K.(1975) Effect of skin friction on the natural frequency of pile groups. Proc. Speciality session (IGS) Asian Regional Conference on Soil Mechanics and Foundation Engineering, Bangalore, India, pp 133-135.

Krishnamurthy, D.N., and Patki, S.D.(1980) Effect of skin friction on the natural frequency of pile groups subjected to lateral vibrations, Proc. 6th South East Asian Conf. on Soil Engineering, May 1980, Taipei, Taiwan, (R.O.C) pp 223-235.

Novak,M. and Grigg,R.F. (1976) Dynamic experiments with small pile foundations. Journal, Canadian Geotechnical Journal,Vol.13, No.4 pp 372-385.

Reese,L.C. and Matlock,H. (1956) Non-dimensional solution for laterally loaded piles with soil modulus, assumed proportional to depth. Proc. Eigth Texas Conf. on Soil Mechanics and Foundation Engineering, Austin.

Sridharan,A.(1964) Natural frequency of pile foundations.

Symposium on Bearing Capacity of piles, Central Building Research Institute, Roorkee, India.

Swiger,M.F. (1948) Effect of vibration on piles in loose sand. Proc. Second International Conference on Soil Mech. and Foundation Engineering, Rotterdam, pp 141.

Tokhi,V.K., Gupta,R and Sagar,A. (1973) Effect of skin friction on the natural frequency of friction pile. Journal, J.M.A.College of Technology, Bhopal.

Tokhi,V.K., Gupta,R and Dua,A.K. (1974) Studies on Natural frequency of friction pile subjected to free vertical vibrations. 16th Annual Technical session, Indian Geotechnical Society, Warangal.

Soil Dynamics & Earthquake Engineering Conference / Southampton / 1982.07.13-15

An Evaluation of Soil Damping Techniques Used in Soil Structure Interaction Analysis of a Nuclear Power Plant

THOMAS A.NELSON
Lawrence Livermore National Laboratory, USA

INTRODUCTION

A prediction of dynamic soil properties at the site of a nuclear power plant plays a very important role in the seismic analysis of the facility. Conventional modal analysis procedures can accommodate virtually any range of equivalent elastic soil stiffness which is used to characterize the site. However, high radiation damping associated with energy dissipation in the soil half-space is difficult to accomodate in an elastic modal solution to the dynamic problem. Several methods are available to combine the soil damping with the structural damping in a composite modal damping coefficient. However, even with this convenient representation, the resulting large fractions of critical damping can make modal solutions to the problems suspect. This paper is based on experience gained in this area during studies performed for the Nuclear Regulatory Commission involving seismic analyses of power plants.

SIMPLIFIED METHODS OF REPRESENTING SOIL

Many methods are available for including the effects of soil-structure interaction when conducting seismic analyses of power plant structures. These range from relatively simple lumped parameter methods to more sophisticated, and sometimes quite expensive, finite element procedures. For the analyses described in this paper, there was significant motivation for finding resonably accurate, simplified procedures for accounting for soil-structure interaction. In all cases the structures already existed and not a great deal of information was available about the soil properties.

---

*This work was supported by the United States Nuclear Regulatory Commission under a Memorandum of Understanding with the United States Department of Energy.

Because of this uncertainty, it was desirable to conduct several analyses with different soil properties. These variations in soil modulus also were designed to bound the possible variations in modulus as a result of soil strain as shown, for example, in Seed & Idriss, 1970. Because of familiarity with elastic modal superposition analysis, it is natural for the analyst to choose a procedure employing this analysis method to predict the seismic response of structures. It is possible to use this method if frequency-independent, lumped parameters are used to represent the soil impedences. The problem which arises when this technique is used stems from the large values of apparent damping ratios associated with soil radiation damping.

The simplest forms for representing soil parameters are those derived for a rigid footing resting on the surface of an elastic half space such as those given by Richart, Hall, & Woods, 1970. These parameters are frequency independent and have been used in many comparative studies which have demonstrated their suitability, see for example, Tsai, et.al., 1974 and Johnson, et.al., 1977. The parameters calculated are, equivalent spring stiffnesses, K_h, K_r, and K_v for horizontal, rotational, and vertical directions, respectively, and C_h, C_r, and C_v representing the corresponding damping coefficients. The radiation damping, sometimes called geometric damping, is a measure of the energy loss due to wave propagation into the soil medium and is a function of the geometry of the foundation-to-soil interface and the characteristics of the underlying soil medium. These damping coefficients could be used directly in a time-history analysis using direct-integration of the equations of motion. However, the usual procedure is to convert the damping coefficients to damping ratios in order to predict the effect of damping on the structural response. This practice may have begun because engineers are familiar with damping ratios as applied to material damping in structures. As discussed by Tsai, 1980, this practice may introduce further uncertainties in the seismic analysis. The common formulation used is that for a single-degree-of-freedom system, such that the damping ratio, $D_h = C_h/2\sqrt{km}$. It is usually assumed that $k=K_h$ and m= the total structural mass. Note that a given soil with a given foundation geometry will have a unique damping coefficient but not a unique damping ratio. The ratio is dependent on the mass of the structure. Because this ratio doesn't have a well defined physical meaning, its magnitude is sometimes misinterpreted. A structure with a large foundation area to weight ratio will exhibit large damping ratios, sometimes in excess of 100% of critical damping. This large value can cause concern for the analyst who wishes to conduct a modal superposition analysis. The first temptation is to reduce this value to a magnitude that is more familiar to a structural engineer. This obviously would introduce error.

However, the radiation damping ratio associated with the soil is not directly applicable to the modal superposition analysis which accepts modal damping specification only. Therefore, a method must be used which will combine damping from two different materials.

APPLICATION OF SIMPLIFIED METHODS TO MODAL ANALYSIS

The structural model employed with a modal analysis using frequency-independent soil parameters usually consists of a series of vertical beam elements with masses lumped at floor levels and concentrated spring and dashpot elements attached to the base to represent the soil. For a given seismic input, the dynamic equations of motion of such a system can be solved rigorously by three methods: 1.) in the frequency domain, 2.) by direct integration in the time domain or 3.) using a complex eigenvalue analysis method. Because engineering judgment plays such a large part in verifying the validity of structural models, the engineer usually likes to conduct a modal analysis of the soil-structure system so that the mode shapes and frequencies can be inspected. This type of solution is approximate since the resulting non-proportional damping matrix does not allow for modal decomposition. The accuracy of the resulting solution depends on the method used to select equivalent composite modal damping values. Three methods have commonly been used: 1.) Bigg's method (commonly referred to as the strain-energy approach or the stiffness-proportional method), Hanson, 1969; 2.) a modified Bigg's method as proposed in Roessett, et.al., 1973; or 3.) a method proposed by Tsai, 1974. In Bigg's method, the portion of the soil radiation damping contributing to a given mode is determined by the ratio of strain energy in the soil springs in that mode relative to that in the remainder of the structural model. Roesset's method includes an additional scale factor for the soil damping contribution based on a ratio of structure frequency to soil frequency. Tsai's method uses transfer functions calculated in the frequency domain and modal properties from an eigenvalue analysis. The modal damping values are back-calculated by matching responses predicted by the two methods at a point within the soil-structure system. Johnson, et.al., 1977, showed that the latter two methods give results similar to a direct integration technique. For the stiffness proportional method in that study, the damping was arbitrarily cut-off to 10% and, thus, was not directly comparable. However, in certain cases, Biggs' method gave similar results to the other methods. Because the stiffness proportional method has found such common use, it was used in two of the studies reported in this paper.

CASE STUDIES

Three power plants were analyzed for the U.S. Nuclear

Regulatory Commission (NRC) for which soil-structure interaction was an important consideration. The first was the Palisades nuclear power plant for which the containment building housing a pressurized water reactor (PWR) was analyzed (Nelson, et.al., 1981). The second was the Oyster Creek nuclear power plant (Murray, et.al, 1980). Both the reactor building and turbine building were analyzed for this plant which uses a boiling water reactor (BWR). Since the turbine building had the larger calculated damping ratio, it was chosen for comparison herein. These two nuclear plants were analyzed for an earthquake with a peak ground acceleration of about 0.2g. The third power plant structure analyzed for seismic loads was the El Centro Steam Plant, Unit 4 (Murray, Nelson, et.al., 1980). This plant is an oil-fired conventional plant that experienced the October 15, 1979, Imperial Valley Earthquake.

The two nuclear plants were analyzed as part of the NRC's Systematic Evaluation Program; a part of which is to investigate the seismic safety of older plants. The philosphy employed in this program was to perform a screening analysis in order to identify significant weaknesses in the structures or equipment housed in them. Therefore, it was not desirable to perform sophisticated analyses which might prove to be long and costly. Rather, the intent was to perform reasonably conservative simple analyses. In order to provide guidance for this effort, the NRC asked a Senior Seismic Review Team chaired by N.M. Newmark to formulate guidelines for soil-structure interaction analyses of these plants. The resulting guidelines are published as part of the Oyster Creek report. N.C. Tsai provided further guidance in Tsai, 1980. He showed that as the soil damping increased in a two-degree-of-freedom system, the structure response decreased while the system vibrated at the primary frequency of the combined system. As the soil damping increased further, to a damping ratio of one or greater, the structure response started to increase and the frequency shifted to that of the fixed base structure. One lesson from his example is that at sufficiently high damping, the structural response does not continue to decrease and it may be unconservative to arbitrarily limit the radiation damping values.

The final guidelines specify that analyses incorporating a range of soil shear moduli should be used in order to account for the uncertainty in the soil properties. The final in-structure response spectra are taken as smooth envelopes of the resulting spectra from these analyses. By using this procedure, one should conservatively capture all the important frequencies which could be transmitted through the soil at various soil strain levels. For radiation damping, composite modal damping calculated by techniques as described herein are suggested. However, if a composite modal damping ratio exceeds 20%, the guidelines state that a confirmatory analysis must be performed.

For both of the nuclear plants studied, confirmatory analyses were performed and form the basis for the current discussion. The approach used was to construct a fairly detailed three-dimensional structural model which accounted for asymmetry and provided for many locations to generate in-structure spectra for subsequent equipment analysis. For confirmatory analyses, simplified models could then be used.

Palisades containment building

The containment building at the Palisades plant has a radius of 60 feet and rests on a foundation 150 feet above bedrock. Since the site is relatively shallow, spring constants and damping were calculated using methods that consider a layered site such as those in Luco, 1974, Kausel, 1973 and Kausel, 1977. The major effect in considering layering was a reduction in soil radiation damping in the vertical direction from 60%, as predicted by a half space method, to less than 10% for the layered methods. Composite modal damping values were calculated using the method in Roesset, 1973. The resulting modal damping ratios which include soil material damping are shown in Table 1. For this plant, the maximum modal damping ratio was limited to 20% so that the damping for mode 3 was reduced. The effect of this limit on damping is shown in Figure 1. The case 1 and case 2 curves result from a direct integration time history analysis in which the soil damping is represented by discrete dashpots attached to the base of the structure. In case 1, the full theoretical soil damping is employed and in case 2, 75% of the theoretical damping is used, as suggested in the guidelines mentioned earlier. The in-structure spectra generated by a direct modal method, labeled "Response-spectrum analysis," use modal analysis results with composite modal damping limited to 20%. The response-spectrum analysis results gave conservative estimates for moments and shears in the structure, but, as shown in Figure 1, the in-structure spectra were not conservative at all frequencies. In the period range where the modal analysis was unconservative, between 0.06 and 0.12 seconds, the unconservatism may be due to limiting the modal damping to 20% or to the fact that the mode at 0.08 seconds comprises the deformation of the concrete internal structure, with little soil displacement.

TABLE 1. Modal damping ratios for the Palisades containment building (median soil case).

Mode	Frequency, Hz	Damping, % of critical
1	2.06	8
2	4.79	10
3	5.78	20
4	12.85	10
5	17.74	4
6	20.02	6

Since composite modal damping tends to spread the damping throughout the soil-structure system, the apparent damping at this natural period may be greater than that resulting in the direct integration analysis. Note that using only 75% of the theoretical damping did not appreciably affect the results. The final spectra, when made to envelop the results from the range of soil properties considered and having the peaks broadened, envelop the spectra resulting from the confirmatory analyses.

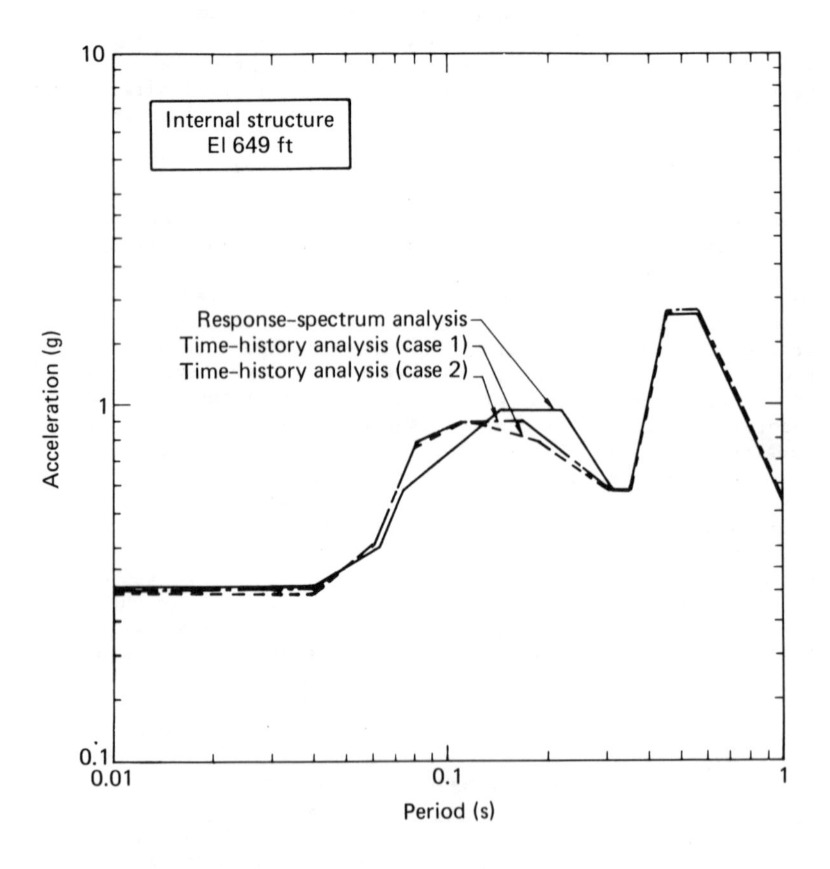

Figure 1. Comparison of in-structure response spectra (3% damping) for the Palisades containment building generated by modal analysis and direct time-history integration.

Oyster Creek turbine building
The turbine building at the Oyster Creek plant has base dimensions of 270' x 170'. It lies on a deep soil site with depth to bedrock of about 1700 feet. In this case, the half space formulations for stiffness and damping represent the site quite well. The shear wave velocity at the site is estimated to be 1250 feet per second. The modal properties of the building, the soil spring-mass system (assuming a rigid building), and the combined system are shown in Table 2. The composite modal damping values as calculated by the stiffness proportional method are shown in the last column.

Table 2. Modal properties and damping for the Oyster Creek turinbe building.

	(1) Fixed-Base Bldg.		(2) Soil-Rigid Bldg.		(3) Combined Soil-Bldg.	
	T_i SEC	D_i DAMP	T_i SEC	D_i DAMP	T_i SEC	D_i DAMP
MODE 1 undamped	.0714	10%	.217	80%	.225	76.8%
damped *	.0718		.362		.352	
MODE 2 undamped	.027	10%	.083	100%	.079	84.0%
damped	.027		N/A		.146	
MODE 3 undamped					.040	28.9%
damped					.042	

* damped T_i = undamped $T_i/\sqrt{1-D_i^2}$

The resulting in-structure response spectra are shown in Figures 2 and 3. Since the soil radiation damping values are large for this structure, an essentially fixed base model (referred to as the high stiffness model in Figure 2) was analyzed to compare frequency content with the flexible base cases. As can be seen in Figure 2, the damping is not large enough to exhibit the effect described by Tsai, 1980, wherein the frequency of the soil-structure system shifts to the fixed-base frequency of the structure. Also illustrated in Figure 2 is the range of soil moduli considered. The confirmatory analyses (Dahspot analysis) conducted by direct integration methods show that the modal analysis gave results in the correct range of magnitude but were slightly deficient in the higher frequency region. In addition, the modal analysis results at different elevations of the building were essentially the same since the primary contributing mode was a rigid-body lateral translation mode. The dashpot analysis results showed increased response at higher elevations.

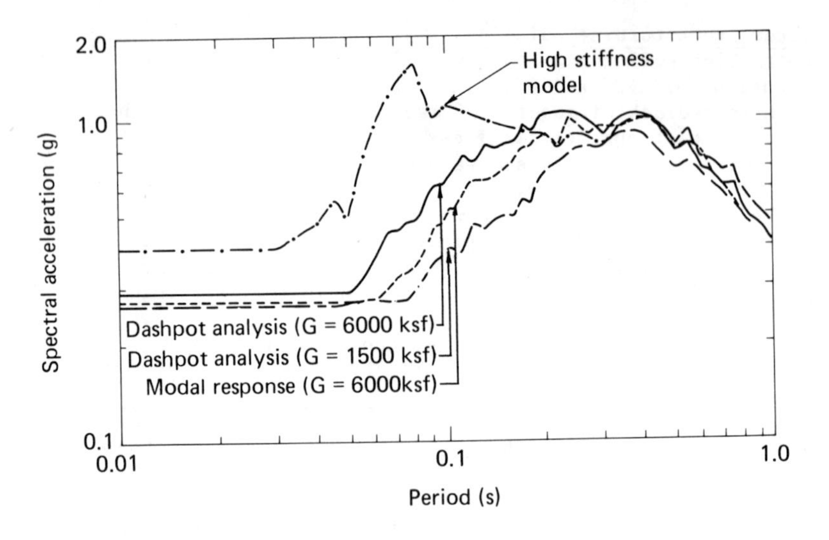

Figure 2. Comparison of spectral curves (3% damping) for the Oyster Creek turbine building at the operating floor, El. 46'

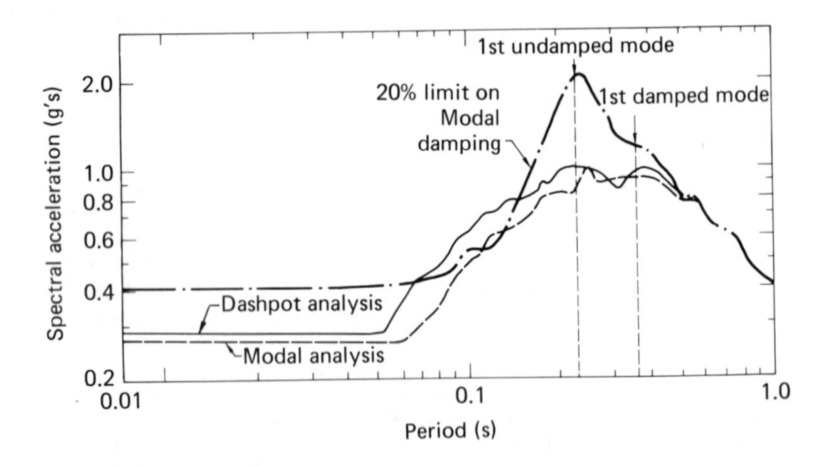

Figure 3. Spectra comparison (3% damping) showing the effect of limiting modal damping to 20% for the Oyster Creek turbine building, El.46'.

The spectrum obtained from a modal analysis with the modal damping limited to 20% is shown in Figure 3. This limitation had a dramatic effect in this case, producing a much higher peak at the lateral-rigid-body mode and broadening the response in the higher frequency region. Thus, reducing the soil damping contribution tends to increase the rigid-body lateral response and reduce portions of the medium frequency region which characterizes the properties of certain critical components in a nuclear power plant. For this plant, the final spectra used for equipment analysis were broadened to envelop the dashpot analysis results.

El Centro Steam Plant, Unit 4

The El Centro Steam Plant provided a unique opportunity for seismic study. It experienced a significant earthquake (0.5g peak horizontal acceleration) with very little damage and had several recording stations nearby which recorded the free field earthquake time history. The purpose in analyzing this plant was to estimate the response of the equipment contained therein in order to show the capacity of the equipment in a real earthquake environment. Unfortunately there were no strong motion recorders in the plant at the time of the earthquake so an analysis of the building was required to provide an estimate of the in-structure response spectra.

The foundation of unit 4 measures approximately 204' x 96' and is composed of a honey comb-like foundation. The soil at the site consists of alluvial deposits composed primarily of stiff to hard clay layers.

Again, the stiffness proportional method was used to compute composite modal damping. The calculated soil damping is shown in Table 3 and the first ten frequencies and modal damping are shown in Table 4. Because the composite modal damping values were quite large, the concern regarding the applicability of modal analysis again was raised. By group concensus, it was decided that modal analysis should be applicable at least up to 40% modal damping, so a test case with 40% damping cutoff was analyzed. The resulting in-structure spectra comparison is shown in Figure 4.

The flexible base case, without any damping cutoff, looked quite reasonable. This case was very similar to the cutoff case with only slightly reduced response. The use of high damping was considered appropriate because the purpose of the analysis was to estimate equipment capacity and, thus, overestimation of response was undesirable. It was concluded that the dynamic model with the specified composite damping was a resonable representation. This conclusion is evidenced by:

- The low level of damage observed at the plant.
- The close relationship of the design to predicted base shears.
- The small displacement of the turbine pedestal predicted by the analysis and evidenced in the earthquake.

TABLE 3. Summary of soil damping for the El Centro Steam Plant.

Direction	Damping ratio, % of critical
N-S translation	94.5
Rotation about E-W axis	77.8
E-W translation	89.2
Rotation about N-S axis	68.6
Vertical translation	152[a]
Rotation about vertical axis	48.8

[a]100% was used in the dynamic analysis.

TABLE 4. Natural frequencies and composite modal damping for the El Centro Steam Plant

Mode No.	Natural frequencies, Hz	Composite modal damping %
1	1.92	43.4
2	2.02	54.4
3	2.69	32.0
4	3.69	32.7
5	4.43	79.6
6	4.75	50.3
7	4.90	78.7
8	5.01	42.7
9	5.47	61.4
10	6.16	12.3

CONCLUSION

Studies were performed for a class of soil-structure interaction problems involving stiff buildings on soils with moderate to soft stiffnesses where the radiation damping ratio ranged from 40 to greater than 100% of critical damping. The results indicate that elastic modal superposition analysis can provide reasonable and efficient prediction of response of structures founded in soil to seismic loading. Thus, readily available software may be conveniently applied to this class of problems. Soil damping in these analyses can be accounted for by using available composite modal damping calculation techniques. The results compared were in-structure response spectra which are most sensitive to variations in damping. It has been shown that other response quantities such as story moments and shears were not appreciably affected by the approximations inherent

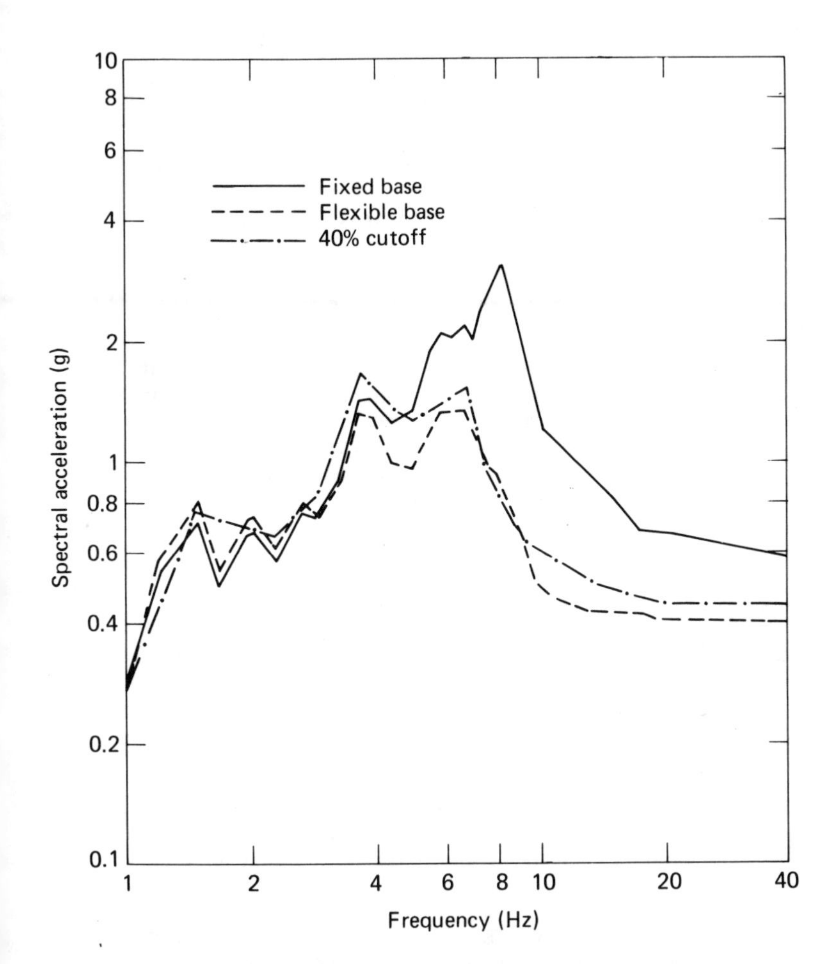

Figure 4. Response spectra (5% damping) at El. 969' in the El Centro Steam Plant comparing three modal analyses results.

in the modal analyses. A possible defect in the simplified techniques is that the composite modal damping techniques tend to overestimate damping in certain modes. However, it has been shown that arbitrarily limiting damping can result in artificially high response at certain frequencies and low response at others. Thus, if the analyst finds that modal analysis is not applicable for a problem with high soil damping, the answer is not to put artificial limits on the composite modal damping but rather to use a more rigorous method.

DISCLAIMER

REFERENCES

Hansen, R.J. ed., (1969) "Seismic Design for Nuclear Power Plants," The MIT Press, Cambridge, Mass.

Johnson, J.J., Wesley, D.A., Almajan, I.T., (1977) "The Effects of Soil-Structure Interaction Modeling Techniques on In-Structure Response Spectra," Transactions, 4th conference on Structural Mechanics in Reactor Technology, Vol. K(a), K2/12, San Francisco.

Kausel, E., and Ushijima, R., (February 1979) Vertical and Torsional Stiffness of Cylindrical Footings, MIT Research Report R79-6, Dept. of Civil Engineering, Massachusetts Institute of Technology.

Kausel, E., Whitman, R.V., Elsabee, F., and Morray, J.P., (1977) "Dynamic Analysis of Embedded Structures," in Transactions, Fourth International Conference on Structural Mechanics in Reactor Technology, San Francisco, Paper K2/6, Vol. K(a).

Luco, J.E., (1974) "Impedance Functions for a Rigid Foundation on a Layered Medium," Nucl. Engin. Design, Vol. 31.

Murray, R.C., et.al, (1981) "Seismic Review of the Oyster Creek Nuclear Power Plant as Part of the Systematic Evaluation Program," Lawrence Livermore National Laboratory, Livermore, CA, UCRL-53018, NUREG/CR-1981.

Murray, R.C., Nelson, T.A., et.al., (1980) "Equipment Response at the El Centro Steam Plant During the October 15, 1979 Imperial Valley Earthquake," Lawrence Livermore National Laboratory, Livermore, CA, UCRL-53005, NUREG/CR-1665.

Nelson, T.A., et.al., (1981) "Seismic Review of the Palisades Nuclear Power Plant Unit 1 as Part of the Systematic Evaluation Program," Lawrence Livermore National Laboratory, Livermore, CA, UCRL-53015, NUREG/CR-1833.

Richart, F.E., Jr., Hall, J.R., and Woods, R.D., (1970) "Vibrations of Soil and Foundations," Prentice Hall, Inc.

Roesset, J.M., Whitman, R.V. and Dobry, R., (March 1973) "Modal Analysis for Structures with Foundation Interaction," J. of Structural Division, ASCE.

Seed, H.B., Idriss, I.M., (1970) "Soil Moduli and Damping Factors for Dynamic Response Analyses," University of California, EERC 70-10.

Tsai, N.C., (April 1974) "Modal Damping for Soil-Structure Interaction," J. of Eng. Mech. Division, ASCE, Vol. 100, No. EM2, 323-341.

Tsai, N.C., (1980) "The Role of Radiation Damping in the Impedance Function Approach to Soil-Structure Interaction Analysis," Lawrence Livermore Laboratory, Livermore, California, UCRL-15233.

Tsai, N.C., Nichott, Swatta, and Hadjian, (July 1974) "The Use of Frequency-Independent Soil-Structure Interaction Parameters," Nuclear Engineering and Design, Vol. 31, No. 2, pp. 168-183.

Soil Dynamics & Earthquake Engineering Conference / Southampton / 1982.07.13-15

Model Tests on Pile Foundations Under Liquefied Soils

GOPAL RANJAN, SWAMI SARAN & P.S.SANDHU
University of Roorkee, India

SYNOPSIS

A model study using 7.8 mm diameter piles arranged in square pattern in groups of 4 to 16 piles has been carried out. The spacing between the piles has been varied. Tests have been carried out on a horizontal steady state vibration table with a tank of size 105 cm x 60 cm x 61 cm (high) mounted on it. Locally available sand deposited at initial relative densities of 20 percent and 30 percent has been used. Various parameters namely, magnitude of acceleration; number of cycles, size of group, spacing of piles and initial density of deposit have been varied and the settlement of single pile/pile group has been observed. Analysis of the test data indicates that the total magnitude of settlement of a pile group increases initially with increasing number of cycles and tends to become constant. The magnitude and rate of settlement of pile group also increases with increase in acceleration and decrease in initial relative density.

INTRODUCTION

The failure of the ground due to loss of strength during an earthquake could be a major cause of destruction. In cohesionless soils, the loss of strength may be due to an increase in pore pressure. Thie phenomenon is termed as liquefaction which can occur in case of loose saturated sandy deposits. In case of pile foundations supporting the super-structure significant amount of displacements are likely to occur once the surrounding/supporting soil looses the strength. Studying the failure of several bridges (during Nigata earthquake of 1964) founded on piles, Fukuoka (1966) observed that due to liquefaction the horizontal resistance of pile foundation reduces. Thus a structure supported on pile foundations moves horizontally and severe deformations and destruction has taken place. Also Kishida (1966) after study of failure of structures during Nigata earthquake concluded that if piles of appropriate length have their tip resting in strata with SPT-N value greater than 25, the degree of damage suffered by such structures was less. Ross et.al. (1969) collected and analysed the data from several bridges damaged during 1964 Alaska earthquake. It was concluded that liquefaction of soils adjacent to bridge foundations played a major role in the damage. Bridges founded in saturated sand and silts sustained severe displacement of pile supported foundations even at the places of high penetration resistance.

These observations reveal that:
(i) piles used as foundations of structures have also failed if soil supporting pile fails by liquefaction. If soil does not liquefy, the piles are not likely to fail.
(ii) piles driven through loose deposits (which could liquefy) into denser deposits behave no better than those which do not penetrate the denser deposit.

These conclusions are based on limited studies related to behaviour of piles driven in loose to moderately dense sands and silts that liquefied during earthquakes and their comparison with piles driven in strata which did not liquefy. Piles in gravelly strata have behaved relatively better. The available studies thus suggest that liquefaction of surrounding soil has to be prevented to ensure pile stability.

Liquefaction characteristics of a soil depend upon a large number of factors (Prakash, 1981). Most of the studies reported in the available literature have been restricted to the factors that influence liquefaction potential of a sand. Attempts have also been made to study the failure of structures after an earthquake. However, no attempt seen to have been made to study the behaviour of piles in liquefied sand deposits.

In the present investigation small scale model pile groups of vertical piles have been tested in sand deposited in a container mounted on a horizontal steady-state vibration table. The influence of several variables e.g. number of piles, spacing, initial relatives density, acceleration, number of cycles on the settlement pattern of pile groups have been studied. Based on the study conclusions have been drawn.

DEVELOPMENT OF TEST PROGRAMME

Model pipe piles (7.8 mm outer diameter) have been used with three different group of piles having 4, 9 and 16 piles arranged in a square pattern. The spacing between piles in a group was varied from two to five times the pile diameter. Tests were carried out on locally available sand(C_u = 1.9;

D_{10} = 0.082, e_{max} = 0.86, e_{min} = 0.48) placed at initial relative densities of 20 percent and 30 percent. Tests were carried out in a horizontal steady-state vibration table with a tank size 105 cm x 60 cm x 61 cm high, mounted on it. The sand was deposited at a required density by 'Rainfall Method' (Sandhu, 1980). Pile-groups were pushed gradually into the prepared sand bed. In all the tests the embedded length of pile was maintained constant (33 cm). Steady-state vibrations were imparted to the table with the help of a motor. The table was subjected to accelerations of 15 percent, 20 percent, 25 percent and 30 percent. The frequency of vibration was kept constant at 4 cps and amplitudes were carried to give the desired acceleration. It was assumed that the maximum pore pressure develops in about 10 cycles. This rise was obtained with the help of piezometers. The maximum pore pressures were later corrected with the help of automatic pressure pick ups (Gupta, 1977). The settlement of group after a given number of cycles were recorded through specially designed arrangements (Sandhu, 1980). Complete set-up is shown in Fig. 1.

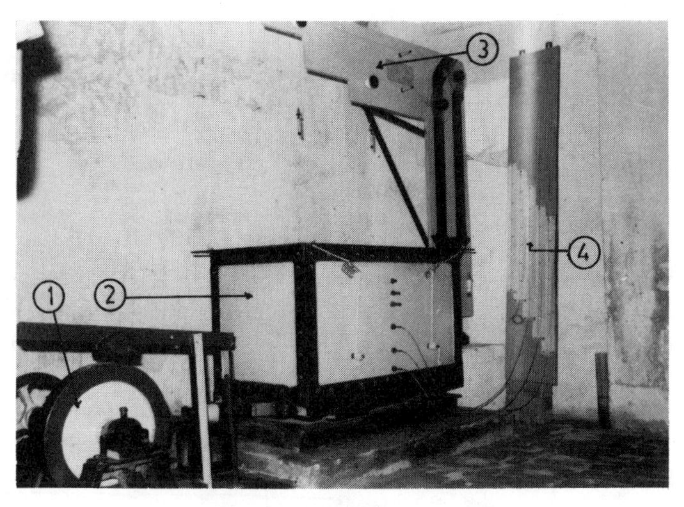

Fig.1 Test Set-up

TEST DATA

Figure 2 is the plot between number of cycles and vertical settlement of four pile group with piles at spacing of twice the pile diameter with sand deposit at 20 percent initial relative density. Data obtained from tests at accelerations of 15 percent g, 20 percent g, 25 percent g and 30 percent g has been plotted. Similarly plot for sand deposit at 30 percent initial relative density has been shown in Fig.3. Tests were also carried out on single piles and pile groups

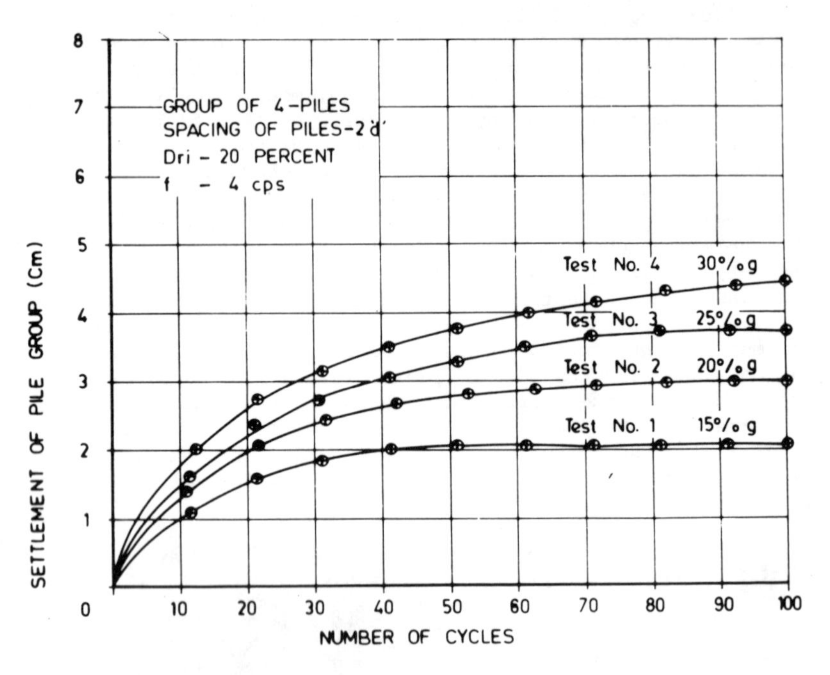

Fig.2 Number of cycles vs settlement of pile group
(initial relative density 20 percent)

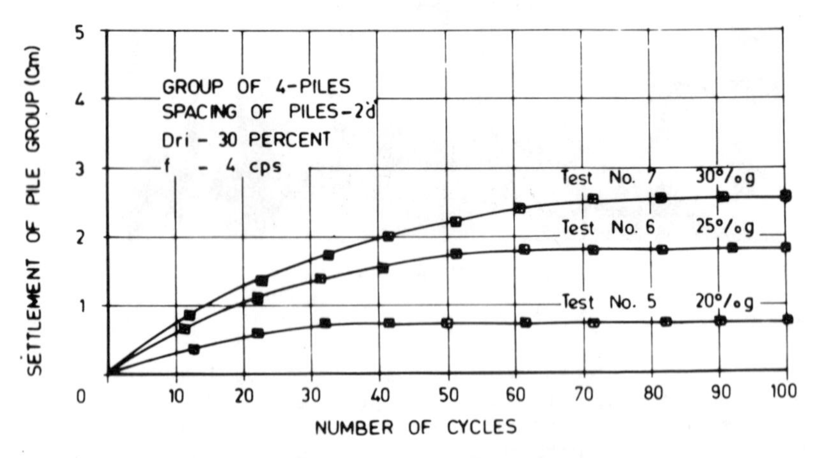

Fig.3 Number of cycles vs settlement of pile group
(initial relative density 30 percent)

of 9 and 16 piles arranged in a square pattern. The spacing between the piles was varied from twice the pile diameter to five times the pile diameter. Piles were placed by pushing. Two initial relative densities, namely 20 percent and 30 percent were used. Groups were subjected to acceleration levels varying from 15 percent to 30 percent g. The piles were subjected to a maximum of 100 cycles.

INFLUENCE OF VARIOUS PARAMETERS

Effect of Number of Cycles on Settlement of Pile Groups

Figs. 2 and 3 indicate that with all other parameters being constant, the vertical settlement of the group increase with the increase in number of cycles. Initially the settlement increase rapidly but tends to become constant at larger number of cycles. Similar trend is noted at higher values of accelerations and other variables like initial relative densities accelerations, number of piles in group and spacing.

The possible reason for such a behaviour is that, once the table is set to motion at a particular acceleration, the process of liquefaction starts. After a certain number of cycles there is initial liquefaction followed by partial liquefaction and subsequently complete liquefaction. Once complete liquefaction has occurred the process of consolidation starts due to dissipation of pore-pressure, resulting in relatively much less increase in the amount of settlement and sand settles in a dense state, when the table motion is stopped. With increase in initial relative density, the increase in pore-pressure decreases and chances of liquefaction are reduced. At a particular acceleration, after every interval of number of cycles, the deposit changes to a relatively denser state, the chances of liquefaction are reduced, which reduces the rate of settlement of pile group.

Further, it is noted (Figs. 2 and 3) that as the magnitude of acceleration increases, the rate and magnitude of settlement of pile-group also increases. This is due to the fact that the time required for the soil mass to liquefy at lower magnitudes of acceleration (say 15 percent g) is more than that the time required for the soil to liquefy at higher acceleration (say 30 percent g). Hence, at a particular time larger settlements would result for larger accelerations.

The magnitude and rate of settlement of pile-group increases as the initial relative density of deposit decreases. As discussed earlier, this is because with the increase in initial relative density, the increase in pore-pressure decreases under vibrations and chances of liquefaction are reduced. Hence, more number of cycles would be required to liquefy the soil with higher initial relative density and hence at a particular value of number of cycles, the settlement of pile-group would be less for higher initial relative density.

Effect of Spacing of Piles in a Group on Number of Cycles

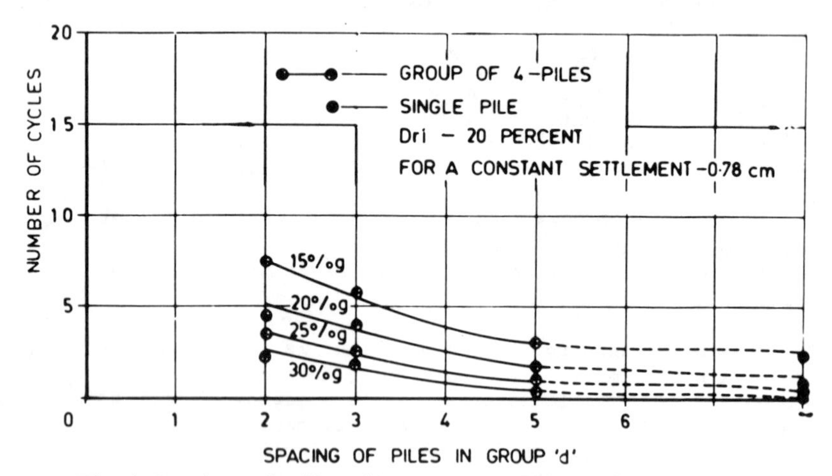

Fig.4 Spacing of piles in group vs number of cycles(initial relative density 20 percent)

At a constant settlement of 7.8 mm(i.e. equal to pile diameter) for different spacing of the piles in the group the number of cycles at a constant acceleration have been noted. Figure 4 shows the plot between spacing of piles in group and number of cycles for group of 4-piles, with sand at an initial relative density of 20 percent for different magnitudes of accelerations. The points corresponding to same magnitude of settlement for single pile (i.e. a group having infinite spacing) are also plotted. The figure indicates that,as spacing of piles in group decreases, the number of cycles to attain a constant settlement increases. Similar behaviour is noted for all values of acceleration tested. Such a behaviour is possible due to the group action which is more pronounced at small spacings. When the spacing between piles in a group is small, the soil is confined and behaves as a group. The mass increases and hence the number of cycles required for a constant settlement increase. At larger spacings the behaviour tends to individual action. Similar trend was noted in other groups and other initial relative density.

Effect of Spacing of Piles on Settlement

To study the influence of spacing of piles on settlement at a given number of cycles the plot between spacing of piles in group and settlement for group of 4-piles at 10 cycles for deposit at an initial relative density of 20 percent has been made (Fig.5). Values for different magnitudes of accelerations have been picked and plotted. It can be seen from these curves that as spacing of piles in a group increases, the settlement also increases for all accelerations. Such a behaviour is due to the fact that with increasing spacing, larger soil zone is getting excited.

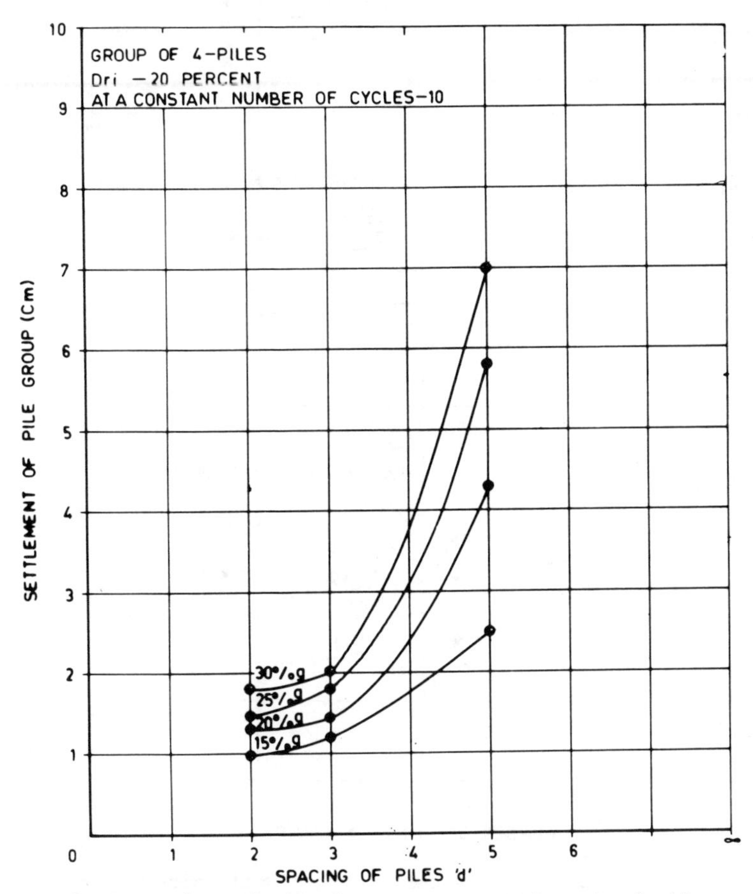

Fig.5 Spacing of pile in group vs settlement of pile group(initial relative density 20 percent)

Similar trend for other groups of 9-piles and 16-piles and with sand of 30 percent initial relative density has been noted.

Effect of Number of Piles in a Group on Settlement

Figures 6a, b and c show the plots between the number of piles in a group and the corresponding settlement of the

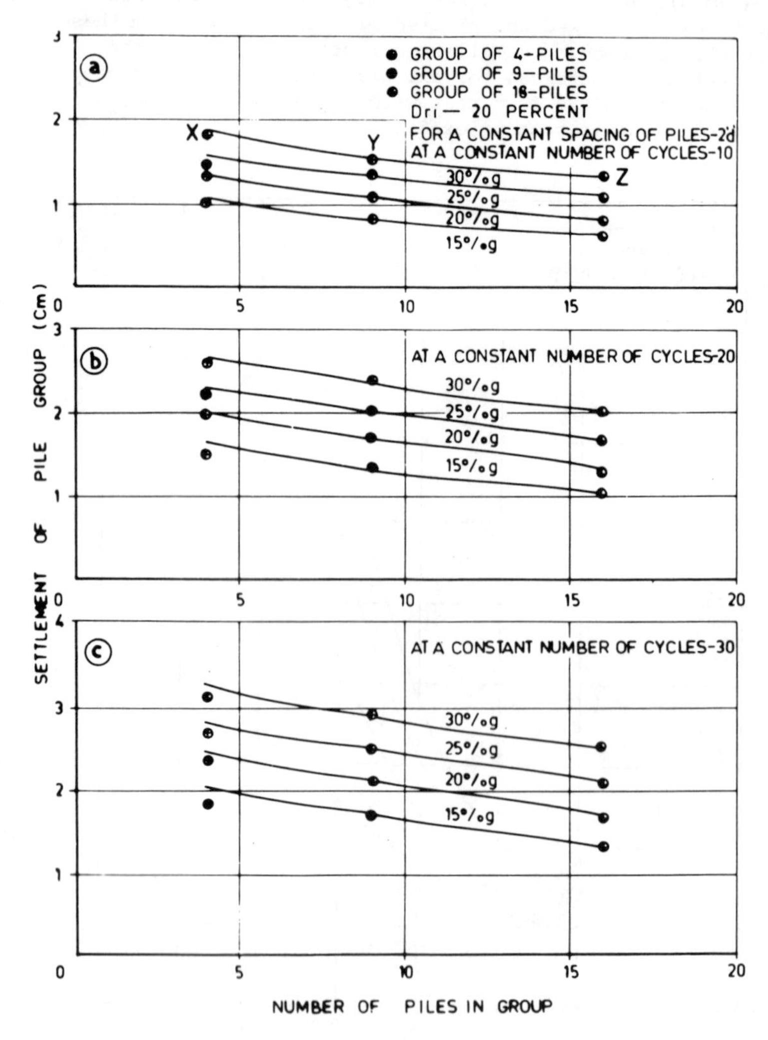

Fig.6 Number of piles vs settlement of pile group (initial relative density 20 percent spacing 2d)

group (at a constant spacing of twice pile diameter) subjected to cycles of motion of 10, 20 and 30 respectively. Points X, Y and Z (Fig. 6a) correspond to the observed experimental values for the settlement of pile groups of 4, 9 and 16-piles, at a spacing of twice pile diameter. The cycles of motion being 10 and initial relative density of deposit being 20 percent. Figure 7 shows plot for groups of 4, 9 and 16-piles at 5d spacing of piles. Similar trend was noted for groups in sand deposit at the relative density of 30 percent.

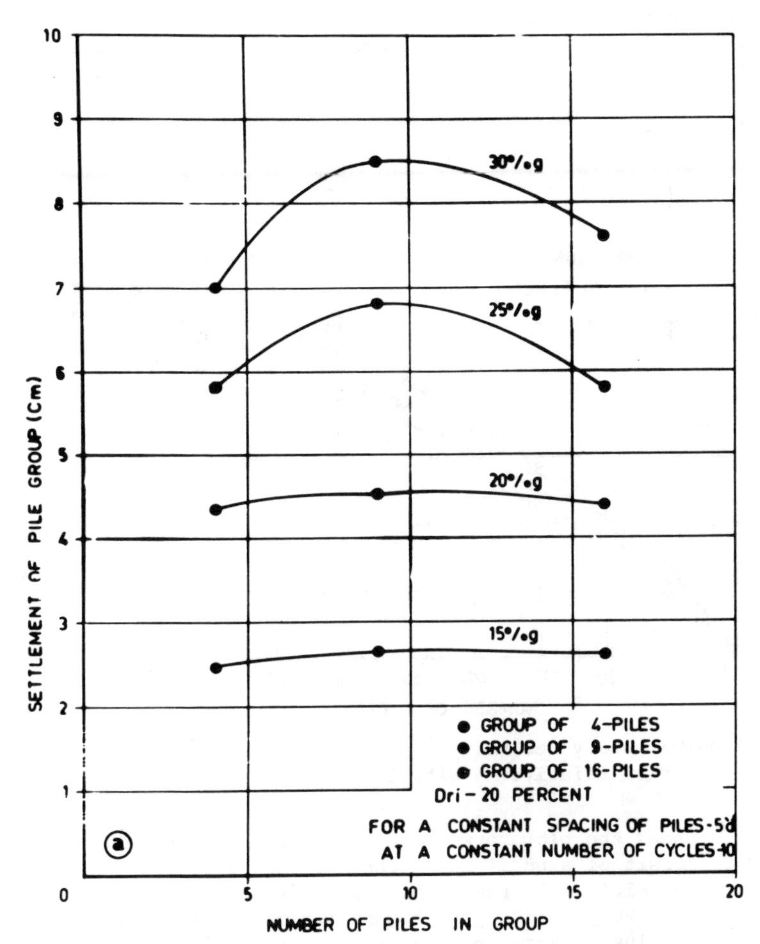

Fig.7 Number of piles vs settlement of pile group (initial relative density 20 percent, spacing 5d)

It can be seen from Fig.6, that at a spacing of 2d for a particular acceleration and for a certain number of cycles of motion, the settlement of pile groups decreases as the number of piles in the group increase. However, the trend at spacing of 5d for the groups tested is noted to change (Fig.7). This change in behaviour in settlement with respect to spacing of piles may be attributed to the fact that, at a lower spacing of piles in the group i.e. 2d, the effect of interference of piles in the group is significant

The pile group tends to behave as a block. Also, the pushing of piles in the deposit tends to alter the initial density to a relatively denser state. However, a larger spacings of 5d the behaviour of the pile group tends to change to that of individual action. Also, at smaller spacings as the number of piles in the group increases, the size of the group increases resulting in an increase of frictional forces which are likely to cause reduction in settlement.

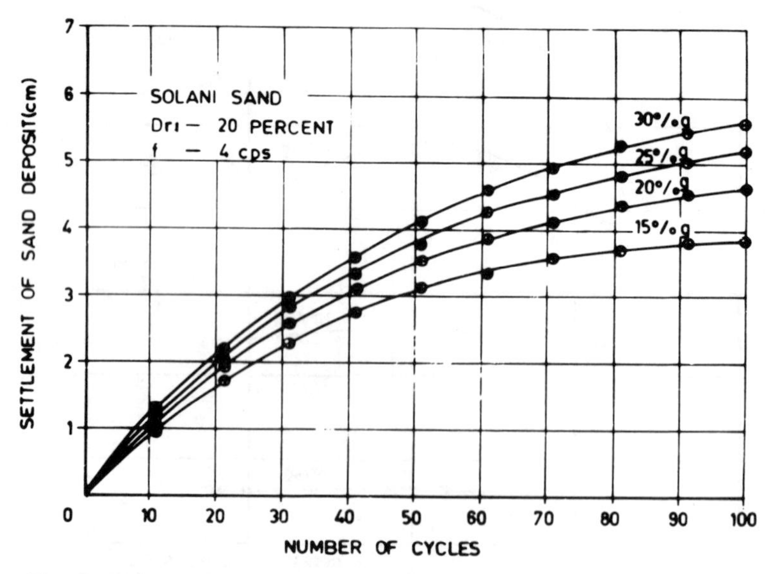

Fig.8 Number of cycles vs settlement of sand deposit (initial relative density 20 percent)

Figure 8 shows the plot between number of cycles and settlement of deposit of sand alone (without piles) at various values of accelerations the initial relative density of deposit being 20 percent. The figure indicates that for a particular value of acceleration, with increase in number of cycles, the magnitude of settlement of deposit increases and tends to attain a constant value at higher values of number of cycles. Similar plots with initial relative density of 30 percent are shown in Fig. 9.

It can be seen from these figures, that the magnitude and rate of settlement of sand deposit increases with increase in acceleration and number of cycles but decreases with the increase in relative density. Similar observations have been reported by Gupta(1977).

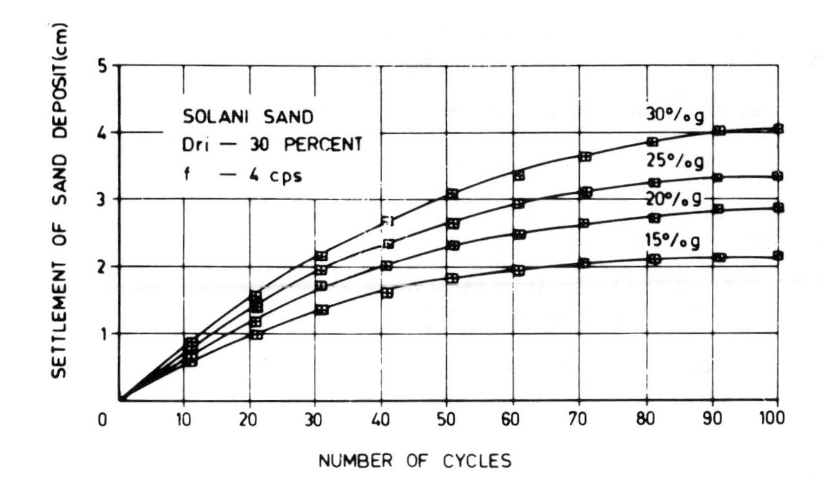

Fig.9 Number of cycles vs settlement of sand deposit (initial relative density 30 percent)

Considering net settlement of pile group as the difference between the settlement of pile group and the settlement of sand surface at a particular value of initial relative density, acceleration and number of cycles it is interesting to note that at smaller number of cycles the group shows positive net settlement indicating that group settles more. However, at large number of cycles the net settlement of group is negative indicating that sand deposit settles more. This behaviour is possibly due to the fact that at lesser number of cycles the friction and point bearing for group is reduced due to partial liquefaction but at larger number of cycles the whole deposit liquefies, the density increases at pile tips and group settlement reduces.

CONCLUSIONS

The analysis of data indicates that the number of cycles, acceleration, initial relative density, spacing and number of piles influence the magnitude and rate of settlement of pile group. Total settlement increases initially with number of cycles and tends to become constant whereas the rate of settlement decreases as number of cycles increase. Increase in acceleration results in increase of settlement. At all accelerations to achieve a constant settlements the number of cycles of motion increase as spacing of piles in a group decreases.

REFERENCES

Fukuoka, M. (1966) Damage to Civil Engineering Structures. Soils and Foundations, Japan, VI, 2:45-52.

Gupta, M.K.(1977) Liquefaction of Sands During Earthquakes, Ph.D. thesis, University of Roorkee, Roorkee, India.

Kishida, H. (1966) Damage to Reinforced Buildings in Nigata City with Special Reference to Foundation Engineering. Soils and Foundations, Japan, VI, 1:71-88.

Prakash, S. (1981) Soil Dynamics. McGraw Hill Book Co., New York.

Ross, G.A. et. al. (1969) Bridge Foundation Behaviour in Alaska Earthquake. Proc. ASCE, 95, SM:4.

Sandhu, P.S.(1980) Behaviour of Pile Foundations Under Liquefied Soils. M.E. thesis, University of Roorkee,Roorkee, India.

Soil Dynamics & Earthquake Engineering Conference / Southampton / 1982.07.13-15

Soil-Pipeline Interaction Through a Frictional Interface During Earthquakes

T.AKIYOSHI
Kumamoto University, Japan

K.FUCHIDA
Yatsushiro College of Technology, Japan

INTRODUCTION

There is still considerable attention to the interaction between soil and pipes during earthquakes. So far most of the interaction analyses have been conducted for steady-harmonic earthquakes and made clear the important features of the buried pipe response to earthquakes, under the assumption that the shear force acting at the soil-pipe interface does not yield and therefore of very small vibrational amplitudes[Toki and Takada(1974), Ugai(1979), and Parnes(1981)]. However it seems to be difficult to hold such assumptions during earthquakes because the spectral displacement amplitudes dominate in the low frequency range in general.

This paper deals with the interaction of imperfectly bonded soil-pipe system during earthquakes, based on the existing solutions. The friction at the interface is assumed to be Coulomb and linearized in terms of the slip displacement and the velocity amplitudes[Miller(1977) and Akiyoshi(1982)]. Analysis is first made for steady-harmonic earthquakes(=plane P- and S-waves) and the slip displacement is represented in closed form which involves every parameter of earthquakes, soil and pipes in which the break-loose condition for slip is compared with Ugai(1978)'s solution. The frequency response function is investigated for the earthquakes with the flat acceleration spectra, and then used for the formulation of pipe and soil strains to randomly vibrating earthquakes.

In this study the following assumptions are adopted:

(1) Soil is linear, homogeneous and isotropic infinite medium, and pipes are long elastic rods without joints.
(2) Frictional stress distributes uniformly around the pipe, and slip occurs when the boundary shear stress equals the frictional one.
(3) Earthquakes are correlated plane P- and S-waves with the same spectral distributions.

GENERAL FORMULATION AND SOLUTION

When an earthquake propagates toward a pipeline as shown in Figure 1, part of the incident wave will be reflected at the surface of the pipeline. Thus the wave in soil is generally represented by superposing the incident and outgoing waves.

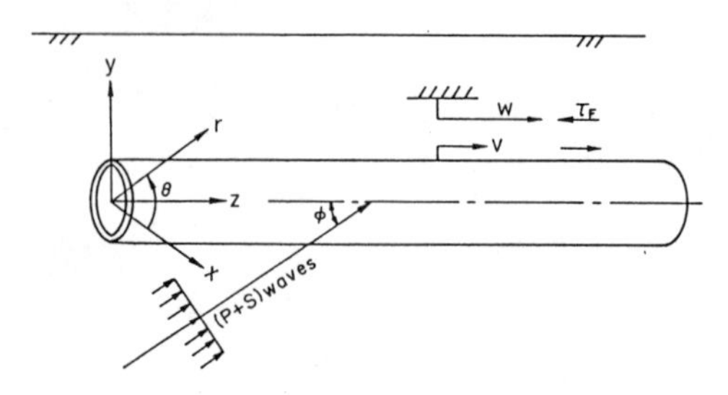

Figure 1 Geometry of problem

Consider first a steady-harmonic P-wave Ψ_p;

$$\Psi_p = w_1 e^{i(\omega t - k_l x \sin\phi - k_l z \cos\phi)} \tag{1}$$

where w_1 = displacement amplitude of incident P-wave, $k_l = \omega/v_l$ = wave number of P-wave, ω = circular frequency, $v_l = \sqrt{[(\lambda+2\mu)/\rho]}$ = velocity of P-wave, ϕ = incident angle to pipe axis, ρ = mass density of soil, λ, μ = Lamé's constants, $i = \sqrt{(-1)}$.

Axial vibration of soil-pipeline system

For the axial vibration of the soil-pipe system the radial and axial displacements, w_r and w_z, may be sufficient to analyze the interaction which yields[Ugai(1979)]

$$w_r = A\frac{1}{k_l^2} q_l H_1^{(2)}(q_l r) + Bik_l\cos\phi\frac{1}{k_s^2} H_1^{(2)}(q_s r) \tag{2}$$

$$w_z = w_1\cos\phi \sum_{m=0}^{\infty} (2)_m (-i)^m J_m(k_l r\sin\phi)\cos(m\theta)$$

$$+ Aik_l\cos\phi \frac{1}{k_l^2} H_0^{(2)}(q_l r) + B\frac{1}{k_s^2}\frac{1}{r}\frac{\partial}{\partial r}[rH_1^{(2)}(q_s r)] \tag{3}$$

where $(2)_m = 1(m=0)$ or $2(m=1,2,\ldots)$, A, B = unknown coefficients, r, θ = cylindrical coordinates, $J_m(\ )$ = Bessel function of order m of the first kind, $H_0^{(2)}(\ )$, $H_1^{(2)}(\ )$ = Hnakel functions of order zero and one of the second kind, $k_s = \omega/v_s$ = wave number of S-wave, $v_s = \sqrt{(\mu/\rho)}$ = velocity of S-wave, $\mu = G$ =shear modulus of soil and $q_l = k_l\sin\phi$, $q_s = k_s h_s$, $h_s = \sqrt{[1-(v_s/v_l)^2\cos^2\phi]}$. Further in equations(2) and (3), exponential term $exp[i(\omega t - k_l z\cos\phi)]$ is omitted.

Assuming that the radial displacement w_r in equation (2) vanishes at the interface as it is known to be small[Toki and

Takada(1974), Parnes(1981)], then one of the unknown coefficients in equation (3) can be eliminated. Moreover the displacement spectra of earthquakes are generally dominated in the low frequency range. Thus for the case of r_0(=pipe radius) < 1 meter and $v_l \simeq 400$ m/sec, then nondimensional frequency $q_l r_0 = k_l r_0 sin\phi = r_0 \omega sin\phi / v_l < 1$ and therefore

$$J_0(q_l r_0) > J_1(q_l r_0) > J_2(q_l r_0) > \ldots..$$

In view of those relations, equation (3) may be represented in the following approximate form

$$w_z = w_{z1} + w_{z2} = (w'_{z1} + w'_{z2}) e^{i(\omega t - k_l z cos\phi)} \tag{4}$$

where

$$w'_{z1} = w_1 cos\phi \; J_0(k_l r sin\phi) \qquad \text{(=incident wave)} \tag{5}$$

$$w'_{z2} = B\left(\frac{v_s}{\omega}\right)^2 H_1^{(2)}(q_s r_0) F(r) \qquad \text{(=outgoing wave)} \tag{6}$$

$$F(r) = \frac{k_l^2 cos^2\phi}{q_l} \frac{H_0^{(2)}(q_l r)}{H_1^{(2)}(q_l r_0)} + \frac{q_s H_0^{(2)}(q_s r)}{H_1^{(2)}(q_s r_0)} \tag{7}$$

Using equation (4) the shear stress τ_{rz} at the interface is obtained as

$$\tau_{rz}|_{r=r_0} = \mu \frac{\partial w_z}{\partial r}\Big|_{r=r_0} = -[\alpha_1 w_{z1} + \alpha_2 w_{z2}] \tag{8}$$

where

$$\alpha_1 = \mu k_l sin\phi \frac{J_1(q_l r_0)}{J_0(q_l r_0)}, \qquad q_l r_0 = r_0 \omega sin\phi / v_l \tag{9}$$

$$\alpha_2 = \frac{\rho \omega^2}{F(r_0)} \tag{10}$$

In the above expressions α_2 in equation (10) has been called "resistance factor" of soil by Nogami and Novak(1976).

Formulation of Slippage When the slip at soil-pipe interface occurs, the frictional stress τ_F equals the boundary stress τ_{rz} in equation (8):

$$\tau_F = \tau_{rz}|_{r=r_0} \tag{11}$$

For a broad class of frictional models, the frictional stress τ_F depends only on the slip displacement u and the velocity amplitude $\dot{u} = du/dt$ which can be replaced with an equivalent system[Caughey(1963), Miller (1977)]. Thus if Coulomb friction as shown in Figure 2 is assumed at the soil-pipe interface, τ_F is approximately represented by [Akiyoshi(1982)]

$$\tau_F = c_e \dot{u} = \frac{4\tau_s}{\pi \omega U} \dot{u} \tag{12}$$

where $c_e = 4\tau_s / \pi\omega U$ = equivalent viscous coefficient,

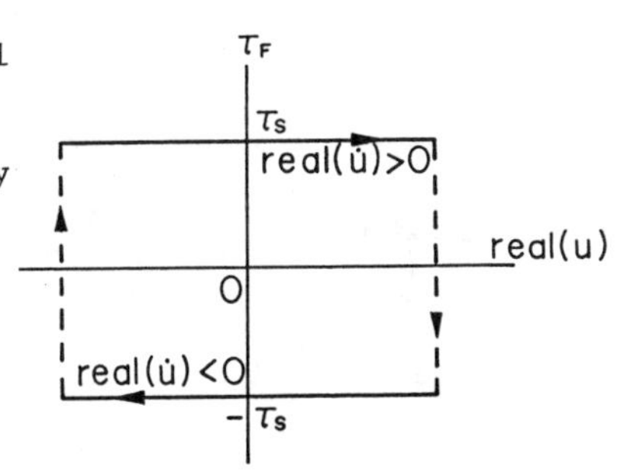

Figure 2 Frictional stress-slip displacement relation in Coulomb friction

τ_s = slip stress, U = real amplitude of the slip displacement u.

Under the steady-harmonic excitation the slip displacement u will take the following expression:

$$u = Ue^{i(\omega t - k_l z\cos\phi - \phi_U)} \quad (13)$$

where ϕ_U = phase difference of the slip displacement u. Thus substituting equations (8), (12) and (13) into equation (11) leads to

$$\alpha_1 w_{z1} + \alpha_2 w_{z2} = -i\frac{4\tau_s}{\pi}e^{i(\omega t - k_l z\cos\phi - \phi_U)} \quad (14)$$

<u>Axial interaction</u> The governing equation of axially vibrating pipe is written by

$$m\frac{\partial^2 v_z}{\partial t^2} = ES\frac{\partial^2 v_z}{\partial z^2} + 2\pi r_0 \tau_{rz}|_{r=r_0} \quad (15)$$

where v_z = axial displacement of pipe, m = unit length mass of pipe, E = Young's modulus of pipe, S = cross sectional area of pipe, r_0 = radius of pipe and τ_{rz} = shear stress acting on the interface which is given by equation (8).

Now define the slip displacement u with the difference between the soil displacement w_z and the pipe one v_z;

$$u = w_z - v_z \quad (16)$$

Substituting equations (4), (8), (13), (16) into equation (15) leads to

$$M_0 Ue^{-i\phi_U} + 2\pi r_0 i\frac{4\tau_s}{\pi}e^{-i\phi_U} = M_0[(1-\frac{\alpha_1}{\alpha_2})w'_{z1} - i\frac{4\tau_s}{\pi\alpha_2}e^{-i\phi_U}] \quad (17)$$

which will be separated into the real and imaginary parts;

$$(U+\frac{4\tau_s}{\pi}s_2)\cos\phi_U + \frac{4\tau_s}{\pi}(\frac{2\pi r_0}{M_0}+s_1)\sin\phi_U = (1-\alpha_1 s_1)w'_{z1} - \alpha_1 s_2 w'_{1I} \quad (18)$$

$$\frac{4\tau_s}{\pi}(\frac{2\pi r_0}{M_0}+s_1)\cos\phi_U - (U+\frac{4\tau_s}{\pi}s_2)\sin\phi_U = \alpha_1 s_2 w'_{1R} + (1-\alpha_1 s_1)w'_{1I} \quad (19)$$

where $M_0 = -m\omega^2 + ESk_l^2\cos^2\phi$, w'_{1R}, w'_{1I} = respectively real and imaginary part of w'_{z1}, a, b = respectively real and imaginary part of α_2, and $s_1 = a/(a^2+b^2)$, $s_2 = b/(a^2+b^2)$.

Sum of squares of equations (18) and (19) leads to

$$U = -\frac{4\tau_s}{\pi}s_2 + \sqrt{[\{(1-\alpha_1 s_1)^2 + b^2\alpha_2^2\}|w'_{z1}|^2 - (\frac{4\tau_s}{\pi})^2(\frac{2\pi r_0}{M_0}+s_1)^2]} \quad (20)$$

only if $|w'_{z1}| > w_{cr}$, and otherwise $U = 0$ where

$$w_{cr} = \frac{4\tau_s}{\pi}\sqrt{[(\frac{2\pi r_0}{M_0}+s_1)^2 + s_2^2]}/\sqrt{[(1-\alpha_1 s_1)^2 + \alpha_1^2 s_2^2]} \quad (21)$$

Ugai(1978) also presented a similar approximate solution of the critical slip displacement w_{cr} which neglects the pipe mass and the imaginary part s_2 in equation (21). However it is not valid because the imaginary part is not small enough to be neglected compared with the real part s_1[Akiyoshi(1982)].

Phase difference ϕ_U is also obtained from equation (17). So using U and ϕ_U to equations (4), (13) and (16) yields

$$w_z = H_{z1}(\omega)\, w_1 \cos\phi\, e^{i(\omega t - k_l z \cos\phi_U)} \tag{22}$$

where $H_{z1}(\omega)$ = frequency response function of axially vibrating pipe to P-wave which yields

$$H_{z1}(\omega) = (k'_{z2} - k_{z1}) J_0(k_l r_0 \sin\phi)/(M_0 + k'_{z2}) \tag{23}$$

where $k_{z1} = 2\pi r_0 \alpha_1$, $k'_{z2} = 2\pi r_0 \alpha'_2$ = apparent stiffness of soil per unit length, and

$$\alpha'_2 = \alpha_2 [1 - (1 - \frac{\alpha_1}{\alpha_2}) M_0 U/(M_0 U + i \frac{4\tau_0}{\pi} \frac{M_1}{\alpha_2})], \quad M_1 = M_0 + 2\pi r_0 \alpha_2.$$

Thus the axial strain ε_{a1} of the pipe is obtained using equation (22) as

$$\varepsilon_{a1} = \frac{\partial v_z}{\partial z} = (-ik_l \cos\phi)\, H_{z1}(\omega) w_1 \cos\phi\, e^{i(\omega t - k_l z \cos\phi)} \tag{24}$$

Lateral vibration of soil-pipeline system

Consider the lateral component (x-axis direction) of P-wave in equation (1). In this case, refering to Figure 1, the radial and tangential components of the soil displacements may be sufficient to analyze the transverse interaction of the soil-pipe system. Thus to the lateral component of P-wave in equation (1), superposed waves w_r, w_θ respectively in the radial and tangential direction are given by[Ugai(1979)]

$$w_r = w_1 \sin\phi \sum_{m=0}^{\infty} \frac{(2)}{2} m(-i)^{m-1} [J_{m-1}(q_l r) - J_{m+1}(q_l r)] \cos(m\theta)$$
$$+ \sum_{m=0}^{\infty} [C \frac{\partial H_m^{(2)}(q_l r)}{\partial r} + D \frac{m}{r} H_m^{(2)}(q_s r)] \cos(m\theta) \tag{25}$$

$$w_\theta = -w_1 \sin\phi \sum_{m=0}^{\infty} (-i)^{m-1} \frac{2m}{q_l r} J_m(q_l r) \sin(m\theta)$$
$$- \sum_{m=0}^{\infty} [C \frac{m}{r} H_m^{(2)}(q_l r) + D \frac{\partial H_m^{(2)}(q_s r)}{\partial r}] \sin(m\theta) \tag{26}$$

where C, D = unknown coefficients and the exponential term $exp[i(\omega t - k_l z \cos\phi)]$ is also neglected in the expressions above.

Denoting the lateral displacement of the pipe as

$$v_x = v'_x\, e^{i(\omega t - k_l z \cos\phi)} \tag{27}$$

then the boundary conditions of displacement continuity at the soil-pipe interface may be written by

$$w_r|_{r=r_0} = v'_x \cos\phi\, e^{i(\omega t - k_l z \cos\phi)} \tag{28}$$

$$w_\theta|_{r=r_0} = -v'_x \sin\phi\, e^{i(\omega t - k_l z \cos\phi)} \tag{29}$$

Using the boundary conditions (28) and (29) to equations (25) and (26), unknown coefficients C and D are represented in terms of the displacement amplitude v'_x of the pipe. Then exerting force p'_x of the soil to the unit length pipe can be represented by

$$p'_x = \rho\pi r_0^2 w_1 \omega^2 h_s^2 \sin\phi \; K_{x1} - \rho\pi r_0^2 v'_x \omega^2 h_s^2 \; K_{x2} \tag{30}$$

where

$$K_{x1} = \frac{2[J_0(q_l r_0)(R_s-2)+J_1(q_l r_0)(R_l-R_s)/(q_l r_0)]}{R_l R_s - R_l - R_s}$$

$$K_{x2} = \frac{R_l+R_s-4}{R_l R_s - R_l - R_s}, \quad R_l = q_l r_0 \frac{H_0^{(2)}(q_l r_0)}{H_1^{(2)}(q_l r_0)}, \quad R_s = q_s r_0 \frac{H_0^{(2)}(q_s r_0)}{H_1^{(2)}(q_s r_0)}$$

Lateral interaction The governing equation of the lateral vibration of the pipe can be written by

$$EI \frac{\partial^2 v_x}{\partial z^4} + m \frac{\partial^2 v_x}{\partial t^2} = p'_x e^{i(\omega t - k_l z\cos\phi)} \tag{31}$$

where I = geometrical moment of inertia of the pipe.

Substituting equations (27) and (30) into equation (31) yields

$$v_x = H_{x1}(\omega) w_1 \sin\phi \; e^{i(\omega t - k_l z\cos\phi)} \tag{32}$$

where $H_{x1}(\omega)$ = frequency response function of laterally vibrating pipe to P-wave which yields

$$H_{x1}(\omega) = \frac{\rho\pi r_0^2 \omega^2 h_s^2 K_{x1}}{EIk_l^4 \cos^4\phi - m\omega^2 + \rho\pi r_0^2 \omega^2 h_s^2 K_{x2}} \tag{33}$$

Thus the bending strain ε_{b1} of the pipe is obtained by

$$\varepsilon_{b1} = -r_0 \frac{\partial^2 v_x}{\partial z^2} = r_0 k_l^2 \cos^2\phi \; v_x \tag{34}$$

Therefore the total pipe strain ε_1 to P-wave is represented by adding the axial strain ε_{a1} in equation (24) to the bending one ε_{b1} in equation (34);

$$\varepsilon_1 = \varepsilon_{a1} + \varepsilon_{b1} = G_1(\omega) \; w_1 e^{i\omega t} \tag{35}$$

where $G_1(\omega)$ = frequency response function of the total pipe strain to P-wave which yields

$$G_1(\omega) = -ik_l \cos\phi \; H_{z1}(\omega)\cos\phi + r_0 k_l^2 \cos^2\phi \; H_{x1}(\omega)\sin\phi \tag{36}$$

It is clear that the similar procedures to P-wave will apply to S-wave in which the total strain ε_2 will take the form

$$\varepsilon_2 = G_2(\omega) \; w_2 e^{i\omega t} \tag{37}$$

where w_2 = displacement amplitude of incident S-wave, and $G_2(\omega)$ = frequency response function of the total pipe strain to S-wave which yields

$$G_2(\omega) = -ik_s \cos\phi \; H_{z2}(\omega)\sin\phi - r_0 k_s^2 \cos^2\phi \; H_{x2}(\omega)\cos\phi \tag{38}$$

where $H_{z2}(\omega)$, $H_{x2}(\omega)$ = frequency response functions respectively of the axial and lateral pipe displacements to S-wave, and $k_s = \omega/v_s$ = wave number of S-wave, v_s = velocity of S-wave.

Therefore when the pipe is subjected simultaneously to P- and S-waves, the total strain ε of the pipe is written by adding equation (35) to (37);

$$\varepsilon = \varepsilon_1 + \varepsilon_2 = G_1(\omega)\, w_1 e^{i\omega t} + G_2(\omega)\, w_2 e^{i\omega t} \tag{39}$$

where $z = 0$ is used for both P- and S-waves.

Random vibration analysis

Consider that a pipeline is subjected to randomly vibrating P- and S-waves. In this case the steady-harmonic waves $w_1 exp(i\omega t)$ and $w_2 exp(i\omega t)$ in equation (39) are replaced respectively with the random variables $w_1(t)$ and $w_2(t)$. So if the random processes $w_1(t)$ and $w_2(t)$ are stationary, the mean power σ_p^2 of the pipe strain can be written by[Robson(1963)]

$$\sigma_p^2 = \sigma_{p1}^2 + \sigma_{p2}^2 + 2\sigma_{p12}^2 \tag{40}$$

where σ_{p1}^2, σ_{p2}^2 = mean powers of the pipe strain respectively due to P- and S-waves, σ_{p12}^2 = cross mean power due to correlated both waves, which yields

$$\sigma_{p1}^2 = \int_0^\infty |G_1(\omega)|^2 S_1(\omega) df \tag{41}$$

$$\sigma_{p2}^2 = \int_0^\infty |G_2(\omega)|^2 S_2(\omega) df \tag{42}$$

$$\sigma_{p12}^2 = \int_0^\infty Re[G_1^*(\omega) G_2(\omega)] S_{12}(\omega) df \tag{43}$$

where $S_1(\omega)$, $S_2(\omega)$ = displacement power spectral density functions respectively of P- and S-waves, $S_{12}(\omega)$ = cross power spectral density function, and $G_1^*(\omega)$ = complex conjugate of $G_1(\omega)$.

Now define the following transform:

$$W_1(\omega) = \frac{1}{T}\int_0^T w_1(t) e^{-i\omega t} dt, \quad w_1(t) = T\int_0^F W_1(\omega) e^{i\omega t} df \tag{44}$$

$$W_2(\omega) = \frac{1}{T}\int_0^T w_2(t) e^{-i\omega t} dt, \quad w_2(t) = T\int_0^F W_2(\omega) e^{i\omega t} df \tag{45}$$

where T = duration of earthquakes, $F = 1/\Delta t$ = upper limit of frequency integration, Δt = time step, and $\Delta f = 1/T$ = frequency step in which the spectral displacements $W_1(\omega)$ and $W_2(\omega)$ in equations (44) and (45) are related to the displacement power spectral density functions $S_1(\omega)$, $S_2(\omega)$ and $S_{12}(\omega)$ as

$$S_1(\omega) = T|W_1(\omega)|^2 \tag{46}$$

$$S_2(\omega) = T|W_2(\omega)|^2 \tag{47}$$

$$S_{12}(\omega) = TW_1^*(\omega) W_2(\omega) \tag{48}$$

NUMERICAL RESULTS

Numerical results are presented for the strains of typical pipes [ductile cast iron pipes and PVC pipes] embedded in an elastic medium. All computations are performed for the standard values of parameters of earthquakes, pipes and soil which are listed in Table 1, except for specified parameters in the diagrams.
In the table, mass ratio $\overline{m} = \rho_p S/\rho\pi r_0^2$ = ratio of pipe mass to soil mass of the unit length, ρ_p = mass density of pipe, S = cross sectional area of pipe, ρ = mass density of soil, r_0 =

pipe radius, and I = geometrical moment of inertia of pipe.

Table 1 Dimensions of earthquakes, soils and pipes

	Ductile cast iron pipes			PVC pipes		
	Low	Stand.	High	Low	Stand.	High
ϕ(deg.)	0	45	90	0	45	90
$\sigma_{\ddot{w}1}$(m/s^2)	0.5	1.0	3.0	0.5	1.0	3.0
$\sigma_{\ddot{w}12}$(m/s^2)	0	0.5	1.5	0	0.5	1.5
v_s(m/s)	50	200	500	50	200	500
v_l/v_s	2	2	3	2	2	3
τ_s/G	10^{-6}	10^{-4}	10^{-2}	10^{-6}	10^{-4}	10^{-2}
r_0(m)	0.01	0.05	1.00	0.01	0.05	0.20
v_p(m/s)		4000			1400	
$\overline{m}$		0.50			0.25	
I/Sr_0^2		0.50			0.50	

(m = meter, s = second)

Frequency response

Consider two typical acceleration power spectral densities of earthquakes(P- and S-waves) as shown in the left diagram of Figure 3. One is concentrated in the narrow frequency range of 1.5 to 2.5 c/s[=case (A)] and another is uniformly spreaded over 0 to 10 c/s[= case (B)] in which both root mean squares (RMS) of acceleration are assumed to be equal to 1 meter/sec^2. Cross power spectral density $S_{12}(\omega)$ of earthquakes are in general complex values, but the imaginary part is not considered in the random analysis. So if $S_1(\omega)$, $S_2(\omega)$ and $S_{12}(\omega)$ are given as acceleration spectra, equations (46) and (47) provide the following discretized spectral displacements;

$$|W_1(\omega)| = \sqrt{[S_1(\omega)\Delta f]}/\omega^2, \quad |W_2(\omega)| = \sqrt{[S_2(\omega)\Delta f]}/\omega^2$$

which depends on frequency step Δf or duration T of earthquakes. Discretized displacements are plotted in the right diagram of Figure 3 for Δf = 0.2 c/s.

Figure 4 is the diagram of the break-loose conditions for slip for nondimensional slip stress $\overline{\tau}_s = \tau_s/G = 10^{-4}$ in which the axial displacement component w'_{z1}[=equation (5)] of P-wave is compared with the critical slip displacement w_{cr} in psuedo-acceleration. $\omega^2 w_{cr}$ is independent of frequency for small-diameter pipes. However as r_0 increases $\omega^2 w_{cr}$ decreases in the low frequency range, which will cause slip for the case like (A).

Figure 5 is the illustrations of the frequency response functions(FRF) $H_{x1}(f)$ and $H_{z1}(f)$ of the pipes respectively in the lateral and axial directions to P-wave for $\overline{\tau}_s = 10^{-4}$. The axial FRF $H_{z1}(f)$ which is the solid lines in the figure decreases with increasing frequency which means that the input loss is marked for large r_0. Trenches appearing in the axial

FRF are due to the slip for the case (A) input. However the lateral FRF $H_{x1}(f)$ which is the dotted lines in the figure is independent of f and r_0 which means no interaction in the lateral vibration.

Figure 6 is the diagram of the absolute values of the axial strain ε_a(=solid lines) and bending strain ε_b (=dotted lines) for $\overline{\tau}_s = 10^{-4}$. ε_a decreases with increasing frequency, while ε_b is independent of frequency. As radius r_0 increases ε_a decreases and ε_b increases, but it is noted that ε_b is negligible for $r_0 < 1$ meter compared with ε_a.

Random vibration

Stochastic investigations of the pipe and the soil strains are conducted to specific earthquakes where the acceleration power spectral density function is shown in Figure 7. For the sake of simplicity it is assumed through computations that the acceleration spectral densities $S_{\ddot{w}1}(\omega)$, $S_{\ddot{w}2}(\omega)$ respectively of P- and S-waves, and the cross spectral density $S_{\ddot{w}12}(\omega)$ have the similar distributions; $S_{\ddot{w}1}(\omega) = S_{\ddot{w}2}(\omega)$ and $S_{\ddot{w}12}(\omega) = \beta S_{\ddot{w}1}(\omega)$, β = a constant.

Figure 8 is the diagram of RMS of the pipe strain versus incident angles for $\sigma_{\ddot{w}1} = \sigma_{\ddot{w}2} = 1$ m/s^2, and $\sigma_{\ddot{w}12} = 0.5$ m/s^2 where $\sigma_{\ddot{w}1}$, $\sigma_{\ddot{w}2}$ are the acceleration RMSs respectively of P- and S-waves, and $\sigma_{\ddot{w}12}$ is the root of acceleration cross mean power. In the diagram σ_{p1}, σ_{p2} are RMSs of the pipe strain respectively due to P- and S-waves, and σ_{p12} is the root of the cross mean power of the pipe strain in which it is assumed that the real part of σ_{p12} is positive and the imaginary one is zero. Thus the total RMS σ_p of the pipe strain is about 10 % greater than σ_p' without correlated term σ_{p12}, and σ_p is dependent of σ_{p1} for small incident angle ϕ and of σ_{p2} for large ϕ.

RMS of the pipe strain is also plotted in Figure 9 for $\sigma_{\ddot{w}1} =$ 1.0 m/s^2(=solid lines) and $\sigma_{\ddot{w}1} = 1.5$ m/s^2(=dotted lines). As the slip stress $\overline{\tau}_s$ decreases, slip arises in the region of small incident angle. It may be noted that the slip is caused mainly by P-wave in view of the last figure.

Figure 10 is the plots of the effect of the correlation coefficient $c_{12} = \sigma^2_{\ddot{w}12}/(\sigma_{\ddot{w}1}\sigma_{\ddot{w}2})$ of P- and S-waves on RMS of the pipe strain. It is shown that c_{12} has linear effect on increasing RMS strain.

Figure 11 is the diagram showing the effect of the pipe radius r_0 on RMS strain of ductile cast iron(DCI) pipe in which a result of PVC pipe is plotted for comparison with the broken lines. When the mean powers $\sigma^2_{\ddot{w}1}, \sigma^2_{\ddot{w}2}$ of the incident acceleration waves are large, the slip is possible even for the small-diameter DCI pipe, while not for PVC pipe.

Figure 12 is the diagram denoting the effect of S-wave velocity

v_s of the soil on RMS strains of DCI and PVC pipes respectively shown with the solid and dotted lines for r_0 = 0.05 meters. Decreasing v_s which is equivalent to decreasing soil stiffness increases the possibility of the slip of DCI pipes. Therefore it follows that the slip depends on the wave velocity ratio v_p/v_s (v_p = volumetric wave velocity of pipe and v_s = shear wave velocity of soil) as well as the slip stress τ_s.

Figure 13 is the diagram of the effect of the slip stress $\overline{\tau}_s = \tau_s/G$ on RMS strains of DCI and PVC pipes of r_0 = 0.05 meters which are plotted respectively with the solid and dotted lines in the diagram. For such relatively small-diameter pipes as r_0 = 0.05 meters, DCI pipes have the possibility of the slip between $\overline{\tau}_s = 10^{-5}$ and 10^{-6}, while PVC pipes have no chance to slip.

In Figure 14 RMS strain σ_p of the pipe is plotted versus RMS strain σ_s of the soil for $\overline{\tau}_s = 10^{-5}$ and various r_0 in which σ_{p0} is the RMS strain of the pipe due to the existing solutions which do not account for the slip. In the figure σ_p (or σ_{p0}) and σ_s are plotted respectively with the solid and dotted lines. It is noted that the small-diameter pipes follow the circumferential soil unless the boundary shear stress yields. However if the slip occurs, σ_p becomes independent of the soil strain σ_s and is kept small for small r_0.

In Figure 15 RMS of the pipe strain is plotted against RMS of the soil strain due to the accelerograms of the duration of T = 20.48 seconds which have been recorded on the ground surface in Japan and U.S.A. It is also shown that the slip works effectively for lowering the strain of the large-diameter and stiff pipe.

CONCLUSIONS

Dynamic interaction between soil and pipelines during earthquakes which involves slippage at the frictional interface has been theoretically investigated, using the existing solutions of elastic waves in the far field. Results obtained are summarized as follows:

(1) Slip consideration at the soil-pipe interface enables to eliminate the restriction of the small amplitude assumed in the previous seismic response analyses.
(2) Most of slips are caused by P-waves. Therefore the maximum slip is expected when P-wave propagates along the pipe axis. Further the slip in general tends to occur for large-diameter pipe, large stiffness ratio of pipe to soil, small slip stress, and large power of earthquakes.
(3) For small-diameter pipes, interaction effect on pipe strain increases with the diameter, but is very small. Hence the pipe strain is almost equal to the soil one unless the slip occurs. If the slip occurs the pipe strain is restricted below the soil

strain during earthquakes.
(4) Pipe strain increases linearly with the correlation coefficient of P- and S-waves.

ACKNOWLEDGMENT

The authors are grateful for the financial support by Science Research Grant No. 1089830 in Japan. They also acknowledge the critical discussions and usuful advices of Professor S. Takada, Kobe University, Japan.

REFERENCES

Akiyoshi, T. (1982) Soil-Pile Interaction in Vertical Vibration Induced through a Frictional Interface. Int. J. Earthqu. Eng. Struct. Dyn., 10(in print).

Caughey, T. K. (1963) Equivalent Linearization Techniques. J. Acout. Soc. America, 35, 11: 1706-1711.

Miller, R. K. (1977) An Approximate Method of Analysis of the Transmission of Elastic waves through a Frictional Boundary. J. App. Mech., ASME, 44: 652-656.

Nogami, T. and M. Novak (1976) Soil-Pile Interaction in Vertical Vibration. Int. J. Earthqu. Eng. Struct. Dyn., 4: 277-293.

Parnes, R. and P. Weidlinger (1981) Dynamic Interaction of an Embedded Cylindrical Rod under Axial Harmonic Forces. Int. J. Soilds Struct., 17: 903-913.

Robson, J. D. (1963) An Introduction to Random Vibration. Edinburgh Univ. Press, Edinburgh.

Toki, K. and S. Takada (1974) Earthquake Response Analysis of Underground Tubular Structure. Part 2. Bull. Disas. Prev. Res. Inst., Kyoto Univ., 221: 107-125.

Ugai, K. (1978) Dynamic Analysis of Underground Pipelines under the Condition of Axial Sliding. Proc. JSCE, 272: 27-37(in Japanese).

Ugai, K. (1979) A Theoretical Study on the Dynamic Modulus of Earth Reaction for Buried Pipe. J. JSSMFE, 21, 4: 93-102(in Japanese).

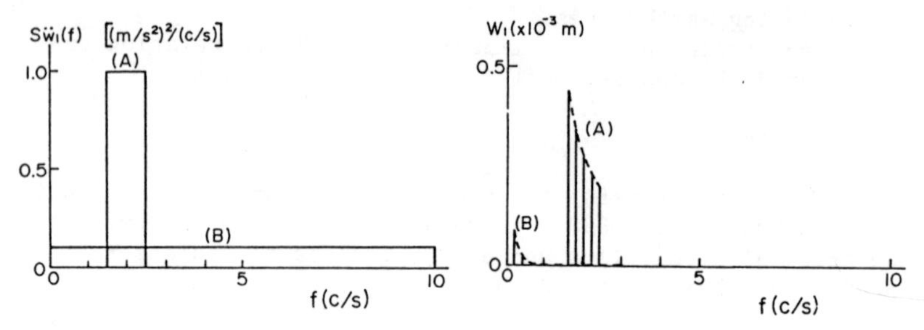

Figure 3 Power spectral density functions and discretized spectral displacements of P-wave

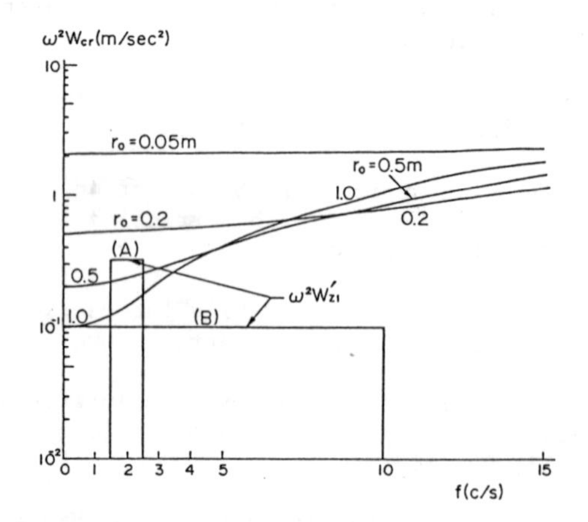

Figure 4 Comparison of critical slip displacement w_{cr} with spectral displacement w'_{z1}

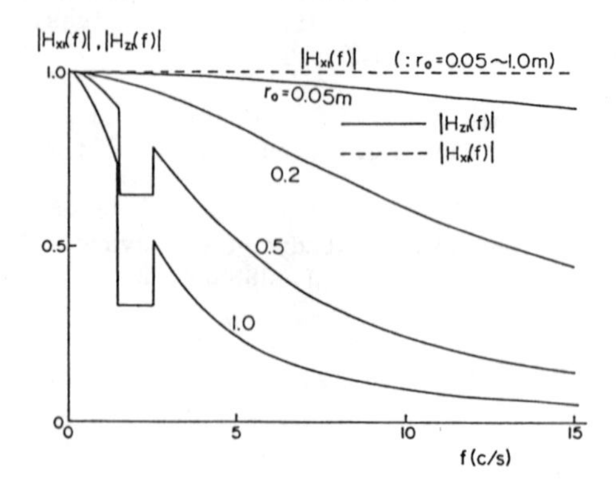

Figure 5 Frequency response functions in lateral and axial directions

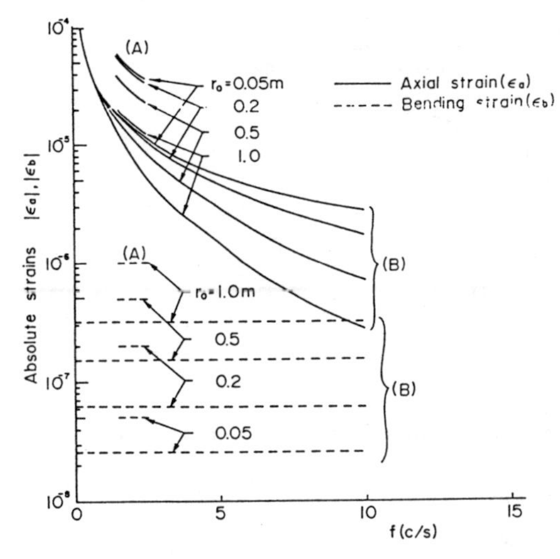

Figure 6 Axial and bending strains of pipes in frequency

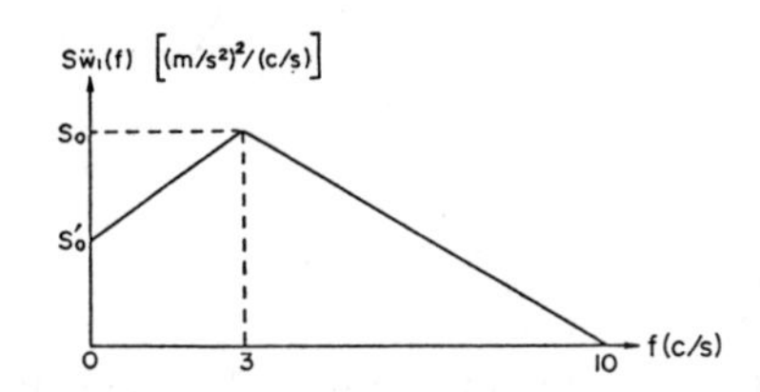

Figure 7 Power spectral density function for computation

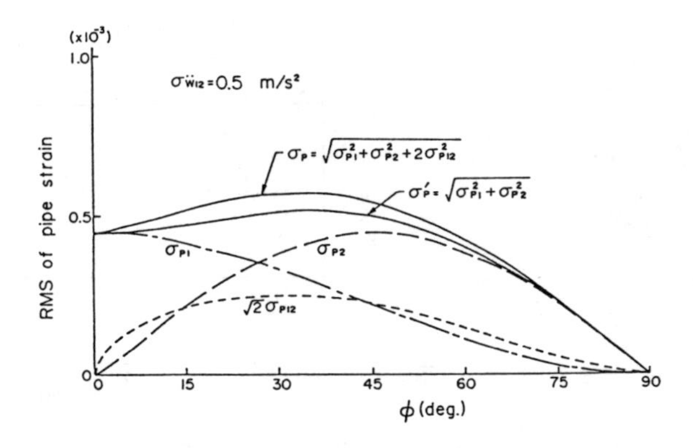

Figure 8 Root mean square of pipe strain versus incident angle

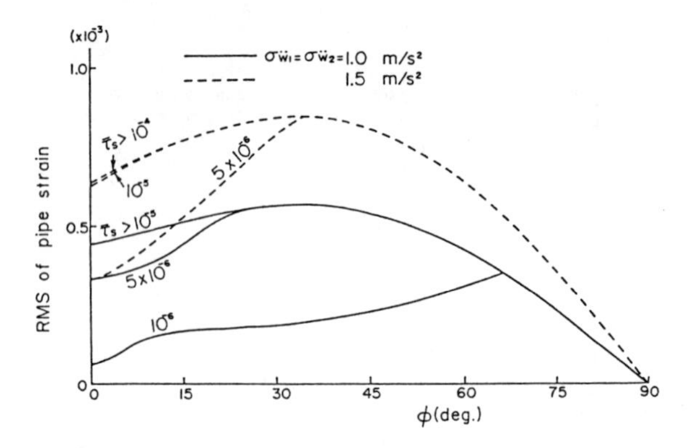

Figure 9 Root mean square of pipe strain versus incident angle

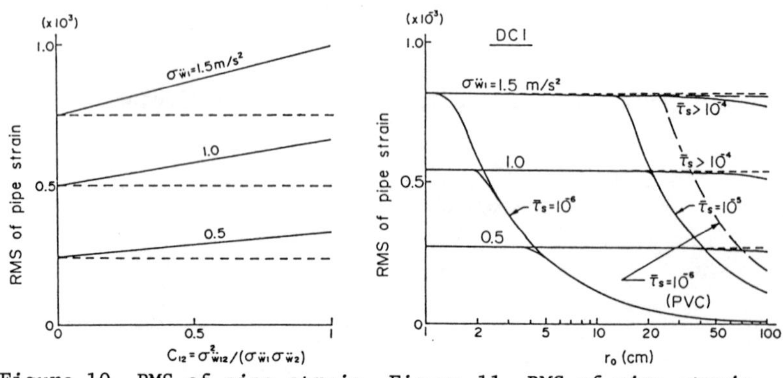

Figure 10 RMS of pipe strain versus P- and S-wave correlation

Figure 11 RMS of pipe strain versus pipe radius

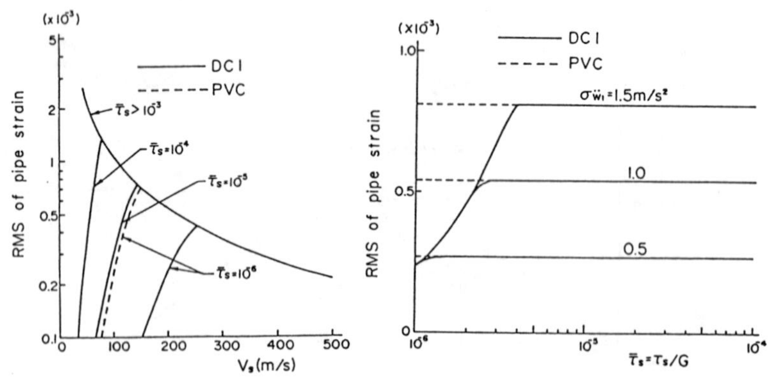

Figure 12 RMS of pipe strain versus S-wave velocity in soil

Figure 13 RMS of pipe strain versus slip stress

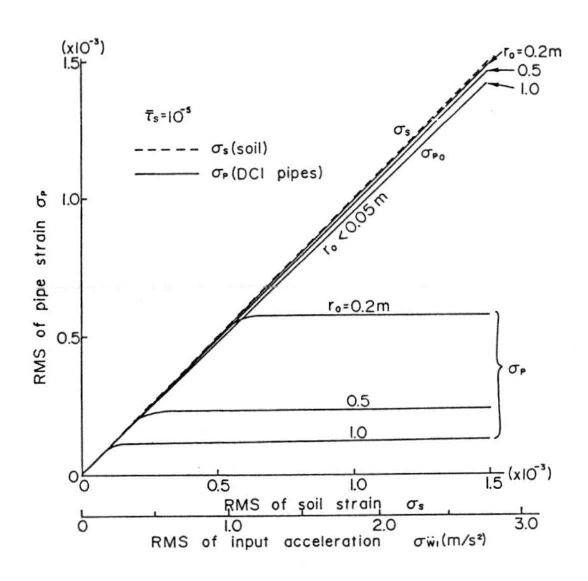

Figure 14 RMS of pipe strain versus RMS of soil strain

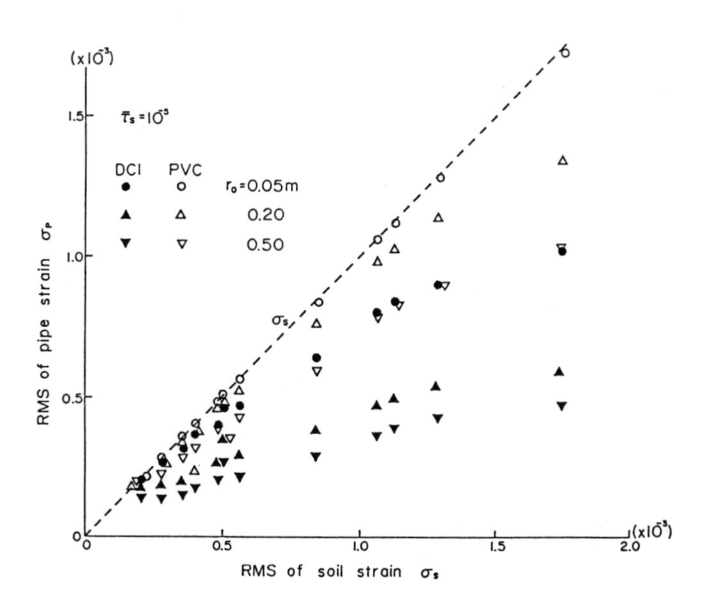

Figure 15 RMS of pipe strain versus RMS of soil strain

Soil Dynamics & Earthquake Engineering Conference / Southampton / 1982.07.13-15

Three-Dimensional Pile-Soil-Pile Interaction Analysis for Pile Groups

V.M.TRBOJEVIC & J.MARTI
Principia Mechanica Ltd., London, UK

A procedure is presented for conducting dynamic analyses of dense pile groups based on the complex response method in the frequency domain. The soil is treated as a linear viscoelastic material with hysteretic damping. A three-dimensional soil dynamic stiffness matrix relating points corresponding to pile nodes is derived from a geometrically axisymmetric finite element model. Solutions are obtained for loads in two Fourier harmonics, simulating horizontal and vertical forces applied at points along the axis of the model. The responses to these forces represent the complex flexibility coefficients of the soil. The total soil complex flexibility matrix is evaluated by superimposing the displacement field for all pile nodes. The inverse of the flexibility matrix gives the dynamic stiffness matrix of the soil. Pile dynamic stiffnesses are then superimposed to those of the soil system and all degrees of freedom except those associated with pile heads are eliminated by a simple reduction procedure. The resulting impedance functions fully incorporate pile-soil-pile interaction. The three-dimensional structural analysis then follows in the traditional fashion. To demonstrate the effects of the pile-soil-pile interaction, the dynamic stiffness and damping coefficients of a foundation consisting of 2 x 2 piles are calculated and compared for different pile radii, distance between piles and soil profiles.

1. INTRODUCTION

Piles have been used for many years to improve some of the characteristics of structural foundations in weak soils, particularly in respect of their stiffness properties and bearing capacity. In exchange for these advantages, pile foundations are usually more complex and costly. An important consideration when assessing the desirability of a pile foundation for nuclear structures is the increased difficulty of the necessary design and safety calculations. The analysis of the soil-structure and pile-soil-pile interaction should be approached in principle as a three-dimensional non-linear problem. Determination of the response of such a system to an earthquake would require great computational effort, firstly because the idealisations used in such analysis would result in a large number of degrees of freedom, and secondly, the solution would have to be evaluated successively at many different times. This may be possible with a powerful computer and no constraints on computing time. However, in practice, some simplifying approximations must be made. The state-of-the-art opens two major possibilities:

- The non-linearity of the pile-soil interaction is accounted for usually by means of the so-called p-y curves or non-linear subgrade reaction moduli, Penzien et al [1]. The pile-soil-pile interaction must then be neglected. This method is very popular in analysis of offshore pile structures.
- Pile-soil-pile interaction is duly accounted for by using superposition methods. However, the non-linearity of soil behaviour must be neglected.

The second method is based on single pile analyses. It can rely on continuous models and theory of elasticity (Novak [2]), as well as in discrete models, such as those used in the finite element method (Blaney et at [3]). These can be generalisd then by superposition for the analysis of pile groups.

None of the above routes is entirely satisfactory. But as pile spacings under nuclear facilities and particularly under the reactor building are usually very small (3 or 4 pile diameters), the pile-soil-pile interaction should be taken into account. In this light, the second possibility appears more realistic. Based on this reasoning, a computer program was developed in which pile-soil-pile interaction and radiation damping are properly modelled at the price of a linearly viscoelastic soil with hysteretic damping. Such procedure allows carrying out the analysis by the complex response method in the frequency domain. The approach selected is a simplified and possibly improved version of that by Wolf and von Arx [4]. The complete analysis includes the following steps:

- Fourier transform analysis of the seismic input motion
- Calculation of the complex impedance functions for the basemat for the set of frequencies of interest
- Calculation of the transfer functions of the seismic input motion
- Dynamic analysis of the structure and the soil-pile foundation
- Calculation of pile displacements and stresses due to the obtained basemat motion
- Inverse Fourier transform of the results

Mathematical formulation of the complex response method in the frequency domain is well known and may be found in many textbooks, e.g. Clough and Penzien [5]. The calculation of the impedance and transfer functions for a rigid basemat by the finite element method is presented in the following section.

2. CALCULATION OF THE IMPEDANCE AND TRANSFER FUNCTIONS

The calculation incorporated in the program for Pile-soil-pile INTERaction (PINTER) is described here due to its special interest. Since the analysis is linear and utilises the principle of superposition, the soil is considered first. Displacements $\mathbf{r}_s$ at points in the soil medium are related to the forces acting at those points $\mathbf{R}_s$ by a simple relationship

$$\mathbf{R}_s = \mathbf{S}_s \mathbf{r}_s \tag{1}$$

where $\mathbf{S}_s$ is the complex dynamic stiffness matrix of the soil. Note that in the complex response method all displacements and forces are complex

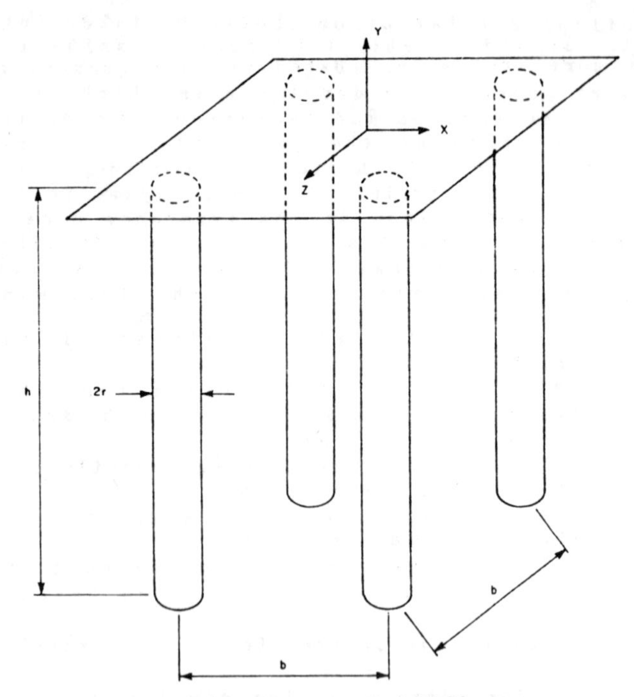

FIG.1 FOUNDATION WITH 2 x 2 PILES

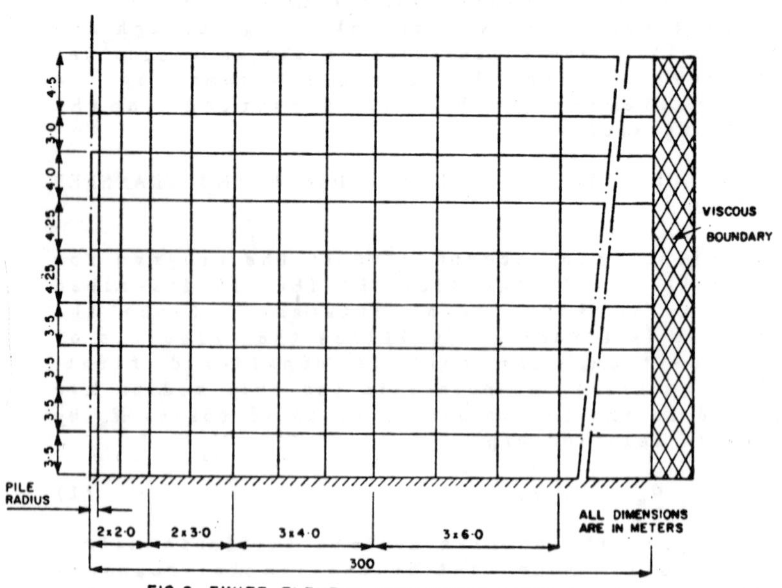

FIG.2. FINITE ELEMENT MESH OF SOIL LAYER

amplitudes and should be multiplied by $e^{i\omega t}$, where ω is the frequency of excitation. This has been omitted here for simplicity. The form of the complex dynamic stiffness matrix is as follows

$$\mathbf{S} = - \omega^2 \mathbf{M} + (1+2iD) \mathbf{K} \tag{2}$$

where $\mathbf{M}$ = mass matrix (real)
$\mathbf{K}$ = stiffness matrix (real)
ω = circular frequency
D = ratio of hysteretic damping

Since this is a general expression all indices are omitted. To obtain the complex soil stiffness $\mathbf{S}_s$, different methods may be applied. With no constraints on computer time and capacity, a straightforward finite element discretisation could be applied. However, to reduce the size of the problem, it is possible to utilise a two-dimensional solution to produce a fully three-dimensional complex flexibility matrix of the soil which inverted gives the desired stiffness. Therefore the soil is idealised as an axisymmetric continuum of isotropic viscoelastic material which consists of horizontal layers. The soil properties may vary with depth, but remain constant within each layer. The finite element discretisation of the soil makes use of a four-noded toroidal element. The variation of the nodal parameters around the circumference of the element is described by the Fourier series. Thus utilising a series of two-dimensional solutions which are decoupled, full three-dimensional effects are obtained. The boundary of this axisymmetric model may be fixed or non-reflecting (viscous) but a more complicated consistent boundary can be included as well. The presence of a single pile at the centre of the axisymmetric mesh is simulated by forcing the cross-section at imaginary pile nodes to be rigid. Unit translational force (in the first harmonic) and unit vertical force (in the zero harmonic) are applied in turn to all nodes of the imaginary pile. Hence the calculated displacement amplitudes represent the complex flexibility coefficients. For each pile in a group, these flexibility coefficients are calculated for all pile nodes by simple interpolation. This yields the complex soil flexibility matrix which relates forces and displacements as follows

$$\mathbf{r}_s = \mathbf{F}_s \mathbf{R}_s \tag{3}$$

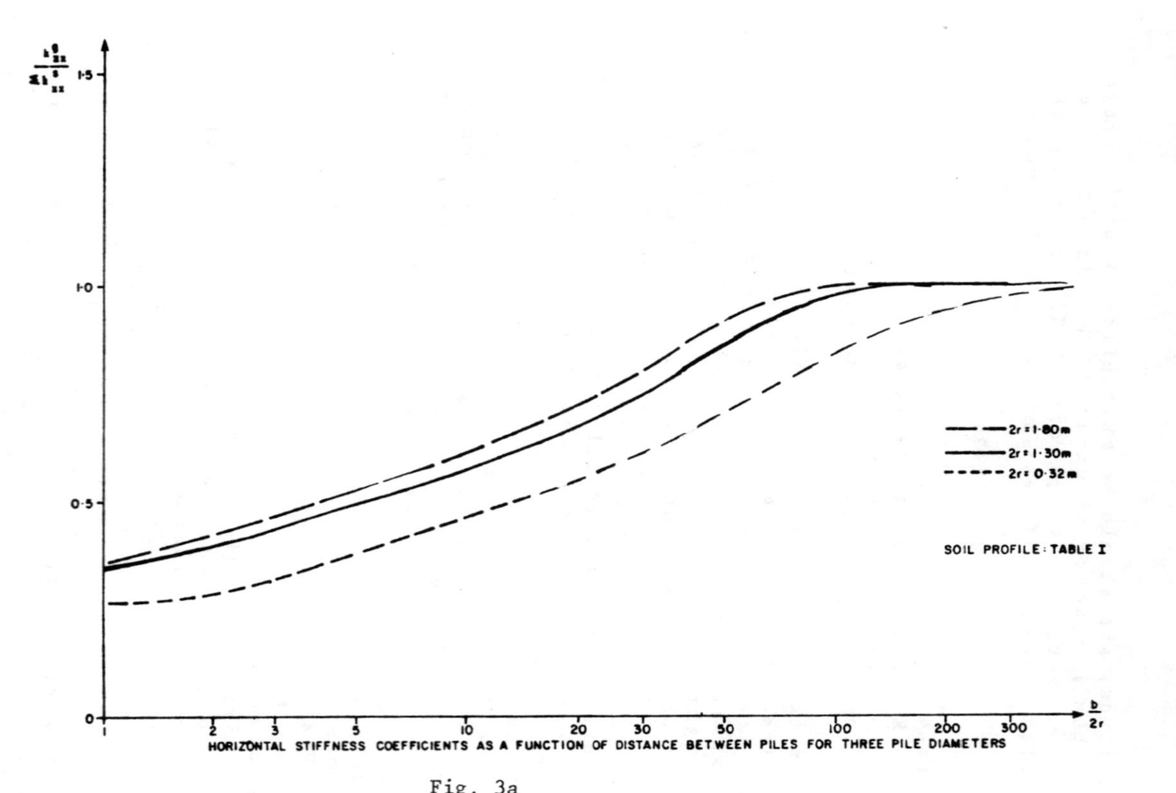

2r = 1·80m
2r = 1·30m
2r = 0·32m
SOIL PROFILE: TABLE I
1·5
1·0
0·5
0
1
2
3
5
10
20
30
50
100
200
300
b/2r
HORIZONTAL STIFFNESS COEFFICIENTS AS A FUNCTION OF DISTANCE BETWEEN PILES FOR THREE PILE DIAMETERS

Fig. 3a

The complex stiffness matrix of the soil $\mathbf{S}_s$ is obtained by inversion of this flexibility matrix. Note that there are no restrictions on the position of pile nodes in the program; piles may be vertical or battered, floating or end-bearing. The more generalised version of the program calculates the soil flexibility matrix for any given set of points and this may be used to calculate impedance functions of embedded foundations. In other words, eqs. (1) and (3) relate soil displacements and forces for a set of points which, for a pile foundation, happen to coincide with pile axes.

The soil displacements $\mathbf{r}_s$ are relative displacements, i.e. the total displacements are

$$\mathbf{r} = \mathbf{r}_s + \mathbf{r}_f \tag{4}$$

where $\mathbf{r}_f$ are free-field displacements. Denoting forces acting on pile heads by $\mathbf{R}_h$, the equilibrium equation for pile foundation follows

$$\mathbf{S}_p \; \mathbf{r} + \mathbf{R}_s = \mathbf{R}_h \tag{5}$$

where $\mathbf{S}_p$ is the complex dynamic stiffness matrix of all piles. Matrix $\mathbf{S}_p$ is of the form given in eq.(2). The piles are modelled by a two-noded three-dimensional beam element with three displacements and three rotations at each node. Since the soil stiffness coefficients at each node correspond to three displacements, to avoid incompatibility, it has been assumed here that there were no moments applied at pile nodes other than pile heads. This enables all rotations except those of pile heads to be eliminated by simple static condensation without any loss of accuracy in pile performance. Substituting eqs.(1) and (4) into eq.(5) yields

$$(\mathbf{S}_p + \mathbf{S}_s) \; \mathbf{r} = \mathbf{R}_h + \mathbf{S}_s \; \mathbf{r}_f \tag{6}$$

The second term on the right-hand side corresponds to the forces produced by a free-field motion. It should be noted at this stage that cavities may also be included in the calculation by substituting $\mathbf{S}_s$ with

$$\mathbf{S}_{sc} = \mathbf{S}_s - \mathbf{S}_c \tag{7}$$

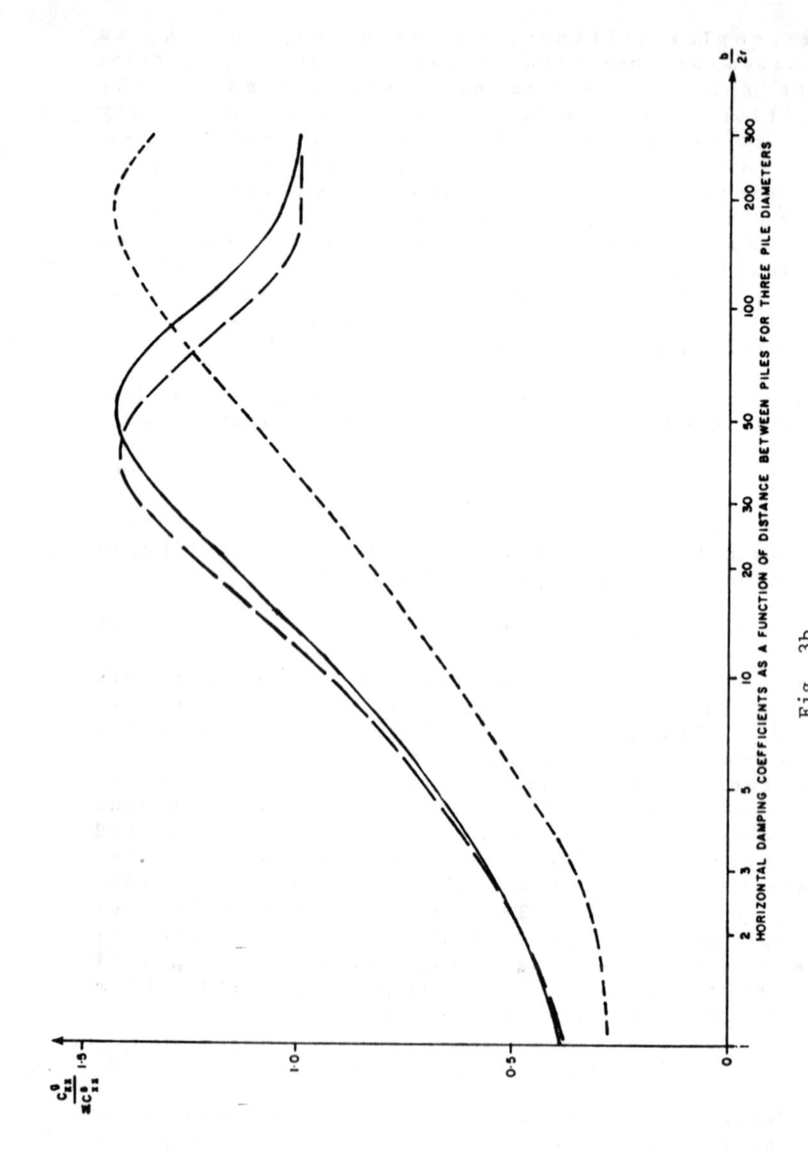

HORIZONTAL DAMPING COEFFICIENTS AS A FUNCTION OF DISTANCE BETWEEN PILES FOR THREE PILE DIAMETERS
b/2r
2
3
5
10
20
30
50
100
200
300
0
0·5
1·0
1·5

Fig. 3b

where $\mathbf{S}_c$ is the complex dynamic stiffness matrix of cavities. The static condensation is now performed to eliminate all variables which do not correspond to pile heads. This results in the following

$$\mathbf{G}_h \mathbf{r}_h = \mathbf{R}_h + \mathbf{R}_f \tag{8}$$

where $\mathbf{G}_h$ is the complex dynamic stiffness matrix for pile heads and $\mathbf{R}_f$ is the vector of the amplitudes of the reactions at pile heads induced by a free-field motion and by fixing pile heads (vectors $\mathbf{S}_s \mathbf{r}_f$ reduced to pile heads).

The dynamic stiffness matrix of pile heads may now be coupled to the elastic basemat or superstructure. If the basemat is assumed to be rigid then pile head displacements $\mathbf{r}_h$ may be expressed in terms of the basemat displacements $\mathbf{r}_o$ as follows

$$\mathbf{r}_h = \mathbf{T} \mathbf{r}_o \tag{9}$$

where $\mathbf{T}$ is a transformation matrix. The displacement transformation eq.(9) is now introduced into eq.(8) to give

$$\mathbf{Q}_o \mathbf{r}_o = \mathbf{Z}_h + \mathbf{Z}_f \tag{10}$$

where $\mathbf{Q}_o$ is 6 x 6 impedance matrix of the basemat defined as

$$\mathbf{Q}_o = \mathbf{T}^t \mathbf{G}_h \mathbf{T} \tag{11}$$

and

$$\mathbf{Z}_h = \mathbf{T}^t \mathbf{R}_h \tag{12a}$$

$$\mathbf{Z}_f = \mathbf{T}^t \mathbf{R}_f \tag{12b}$$

The transfer functions to the basemat are calculated as follows

$$\mathbf{r}_o = \mathbf{Q}_o^{-1} (\mathbf{Z}_h + \mathbf{Z}_f) \tag{13}$$

The impedance functions eq.(11) and the forces eq.(12) are coupled to the superstructure and the solution is completed. Once the motion of the basemat and the the superstructure is determined it is necessary to calculate forces and moments on the piles. The program for this task is called

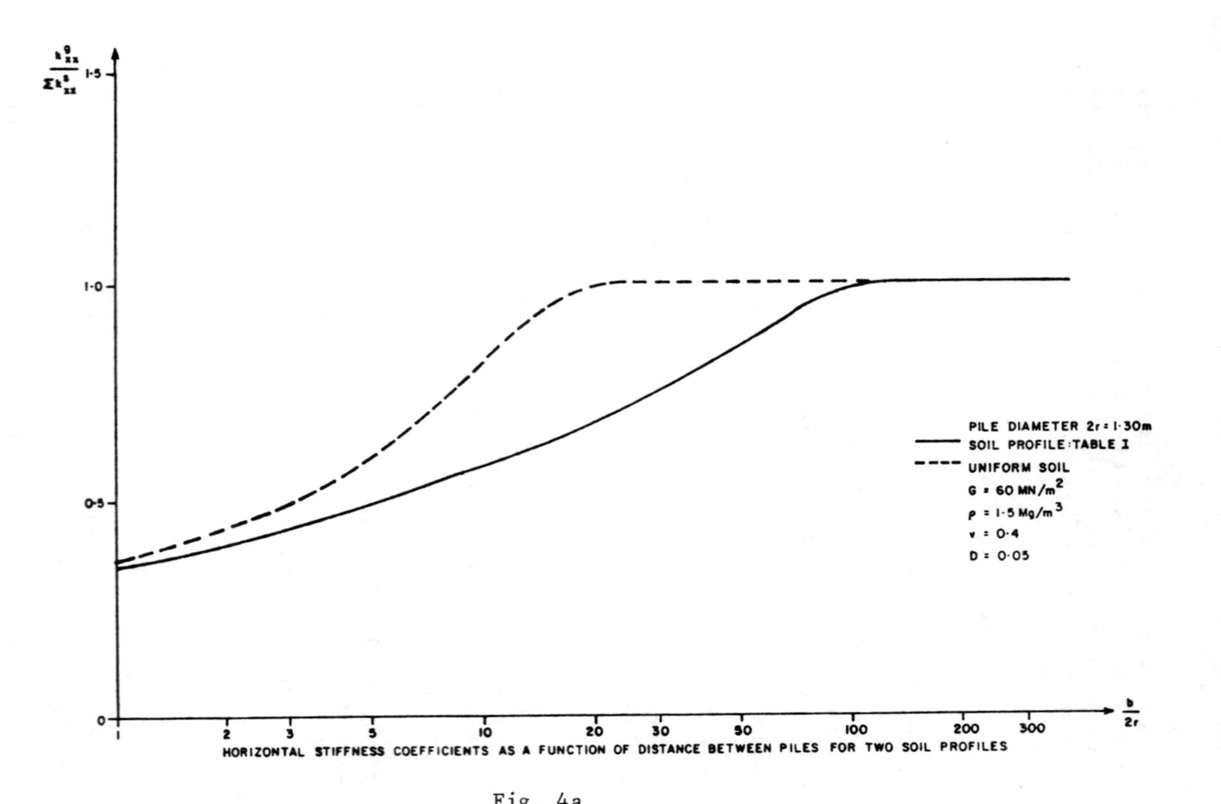

PILE DIAMETER 2r = 1·30m
SOIL PROFILE: TABLE I
UNIFORM SOIL
G = 60 MN/m²
ρ = 1·5 Mg/m³
ν = 0·4
D = 0·05
1·5
1·0
0·5
0
1
2
3
5
10
20
30
50
100
200
300
b/2r
HORIZONTAL STIFFNESS COEFFICIENTS AS A FUNCTION OF DISTANCE BETWEEN PILES FOR TWO SOIL PROFILES

Fig. 4a

PILFOR and basically follows the described procedure in the reversed order.

3. EXAMPLE

The procedure of the previous section is applied to the calculation of dynamic stiffness coefficients for a 2 x 2 pile foundation for different pile radii, distance between piles and soil properties.

A stratified layer of soil of a depth of 34m resting on rigid bedrock is analysed first. The material properties are listed in Table I (following Wolf [6]). The ratio of hysteretic

Table I Free-Field Soil Properties

Depth	Layer	SHEAR MODULUS (GN/m^2)	MASS DENSITY (Mg/m^3)	POISSON'S RATIO (-)
0.0m to -4.5m	SAND	0.435	1.95	0.42
-4.5m to -7.5m	CLAY	0.027	1.68	0.49
-7.5m to -11.5m	TRANSITION	0.944	2.10	0.35
-11.5m to -20.0m	RESIDUAL 1	1.022	2.05	0.36
-20.0m to -27.0m	RESIDUAL 2	1.556	2.05	0.34
-27.0m to -34.0m	WEATHERED ROCK	3.018	2.40	0.43

damping for all layers is 0.05. Piles are made of concrete with a modulus of elasticity $E=30GN/m^2$, a mass density $\rho=2.5Mg/m^3$, a ratio of hysteretic damping $D=0$ and a length $h=34m$. Three pile diameters are considered: $2r=0.32m$, 1.30m and 1.80m. The distance b (Fig.1) is varied. The finite element idealisation of soil layer is presented in Fig.2. The overall radius of the axisymmetric soil mesh is 300m, soil is bonded to the bedrock and the far-field is represented by the viscous boundary of Lysmer and Kuhlemeyer [7]. The frequency of excitation is 4Hz. The dynamic

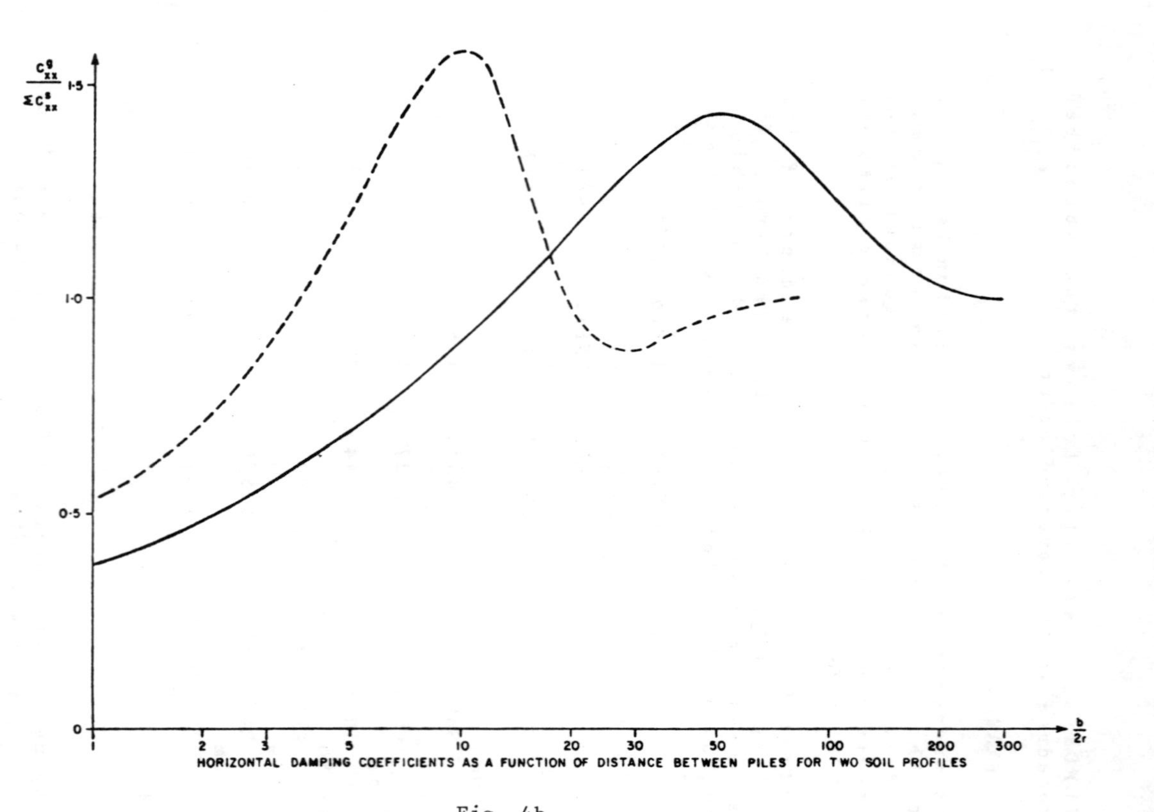

$\frac{C^g_{xx}}{\Sigma C^s_{xx}}$
1·5
1·0
0·5
0
1
2
3
5
10
20
30
50
100
200
300
$\frac{b}{2r}$
HORIZONTAL DAMPING COEFFICIENTS AS A FUNCTION OF DISTANCE BETWEEN PILES FOR TWO SOIL PROFILES

Fig. 4b

stiffness coefficients are presented in Fig.3. The notation is the same as in ref.[6], and [8], therefore k_{xx}^{g} is the stiffness coefficient (real part) for a pile group when pile-soil-pile interaction is taken into considration, k_{xx}^{s} stands for a single pile, hence Σk_{xx}^{s} denotes the stiffness coefficient for a pile group neglecting pile-soil-pile interaction. The damping coefficients (imaginary part) are denoted by c_{xx}^{g}, c_{xx}^{s} and Σc_{xx}^{s} respectivcly. It is interesting to note the effect of the pile stiffness on the pile-soil-pile interaction. There is a remarkable difference between piles of 2r=0.32m and 2r=1.30m; this was not reported by Wolf [6], whose corresponding curves nearly coincide.

For a particular pile of a diameter 2r=1.3m the interaction effects may be neglected for a distance-to-diameter ratio exceeding 100 for a vibration in the horizontal direction. Intuitively this ratio seems to be very high and some kind of 'non-linearity' proposed by Novak [9], would be desirable. To show the influence of the soil properties for the pile group of a diameter 2r=1.30m, the calculation is repeated for a weak soil.

Pile properties are the same as before, while the soil layer is of uniform properties: shear modulus $G=60MN/m^2$, mass density $\rho=1.5Mg/m^3$, Poisson's ratio $\nu=0.4$, ratio of the hysteretic damping D=0.05. The finite element mesh and all the other data is the same as before. The results are given in Fig.4.

From the simple example presented here the following conclusion may be suggested:

- Interaction effects are stronger for softer soil media
- Interaction effects are stronger for more flexible piles
- Interaction effects are neglible for large pile distance to diameter ratios
- Radiation damping, in general, increases with pile distance to diameter ratios

4. REFERENCES

[1] Penzien, J., Scheffley, C.F. and Parmelee, F.A. (1964) 'Seismic Analysis of Bridges on Long Piles', ASCE J. Eng. Mech. Div., Vol.90, No.EM3, pp.223-254, June.

[2] Novak, M. (1974) 'Dynamic Stiffness and Damping of Piles', Soil Mechanics Research Report SM1-74, University of Western Ontario, Canada.

[3] Blaney, G.W., Kausel, E. and Roessett, J.M. (1976) 'Dynamic Stiffness of Piles', proc. 2nd Int. Conf. on Numerical Methods in Geomechanics, Blacksburg, Virginia, USA, Vol.II, p.1001.

[4] Wolf, J.P. and von Arx, G.A. (1978) 'Impedance Function of a Group of Vertical Piles', proc. ASCE Specialty Conf. on Earthquake Engineering and Soil Dynamics, Pasadena, California, USA, Vol.II, p.1024.

[5] Clough, R.W. and Penzien, J. (1975) 'Dynamics of Structures', McGraw-Hill Book Co., New York, NY.

[6] Wolf, J.P. (1979) 'Dynamic Stiffness and Seismic Input Motion of a Group of Battered Piles', Nuclear Eng. and Design, 54, pp.325-335.

[7] Lysmer, J. and Kuhlemeyer, R.L. (1969) 'Finite Dynamic Model for Infinite Media', ASCE J. Geotech. Eng. Div., Vol.95, No.EM4, pp.859-877, August.

[8] Trbojevic, V.M. et al (1981) 'Pile-Soil-Pile Interaction Analysis for Pile Groups', 6th Int. Conf. on Structural Mechanics in Reactor Technology, Paris, August.

[9] Novak, M. and Sheta, M. (1980) 'Approximate Approach to Contact Effects of Piles', Dynamic Response of Pile Foundations: Analytical Aspects, ASCE, Hollywood, FL, October 30.

7. Soil Dynamics and Ground Motion

Soil Dynamics & Earthquake Engineering Conference / Southampton / 1982.07.13-15

A State-of-the-Art Report on Soil Dynamics and Earthquake Engineering at PWRI

EIICHI KURIBAYASHI, TOSHIO IWASAKI & TADASHI ARAKAWA
Japanese Ministry of Construction, Tokyo

ABSTRUCT

The Public Works Research Institute has been in charge of civil engineering technology except building engineering technology in Japan. The most purposes are as follows,
1. Land Use Planning
2. Water Resources
3. Highways and Transportations
4. Preventive Measures to Floods and Tidal Waves
5. Erosion Control
6. Earthquake Disaster Prevention
In the Field of Earthquake Disaster Prevention these past two decades, the Institute has concerned with earthquake safety of highway construction, rivers and coasts. In 1976, the Institute established Earthquake Disaster Prevention Department in order to refer public opinions. After that research works on earthquake safety measures have been conducted.
This paper describes past research efforts and future research programs briefly.

1. TSUKUBA SCIENCE CITY

Tsukuba Science City was planned to relocate the national research and educational institutions from Metropolitan Tokyo and its surrounding area in a comprehensive science city, an integrated and well-organized academic city with an ideal enviroment for advanced research activities and education. Its construction is aimed specifically at meeting the demands of today for enhancement of scientific technology and academic research and advancement of education as well as ensuring the balanced development of the entire Capital Region.
The city is located about 60 km to the northwest of Tokyo. It covers an area of 2,700 ha extending 18 km north to south and

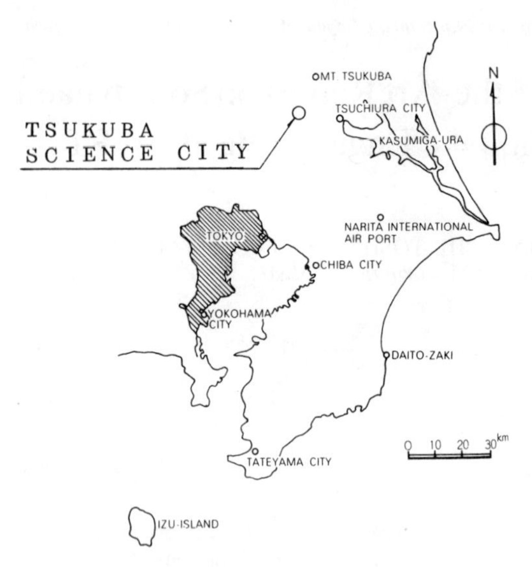

Fig.1 Location Map of Tsukuba Science City

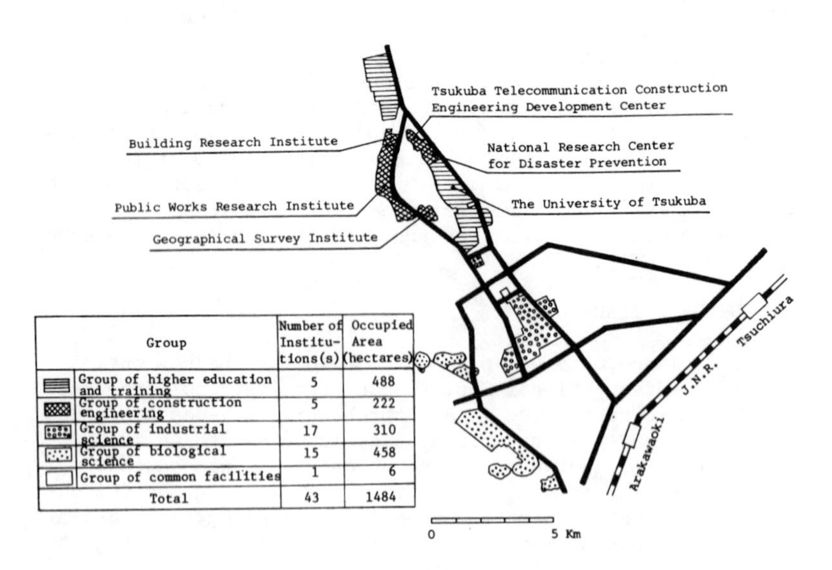

Group	Number of Institu-tions(s)	Occupied Area (hectares)
Group of higher education and training	5	488
Group of construction engineering	5	222
Group of industrial science	17	310
Group of biological science	15	458
Group of common facilities	1	6
Total	43	1484

Fig.2 Plan of Tsukuba Science City

Fig.3 Aerial View of Public Works Research Institute

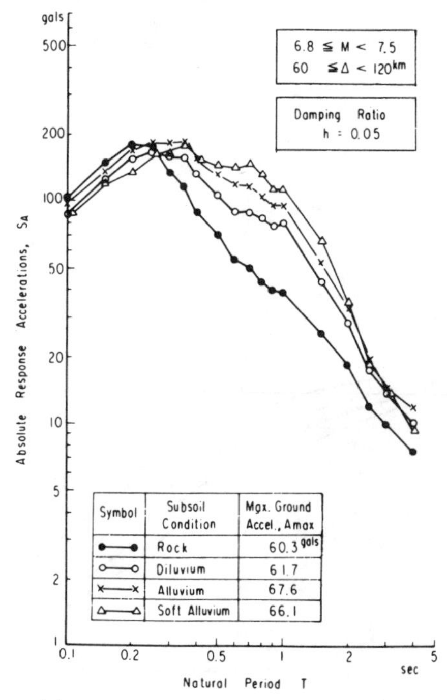

Fig.4 Examples of Design Seismic Motion

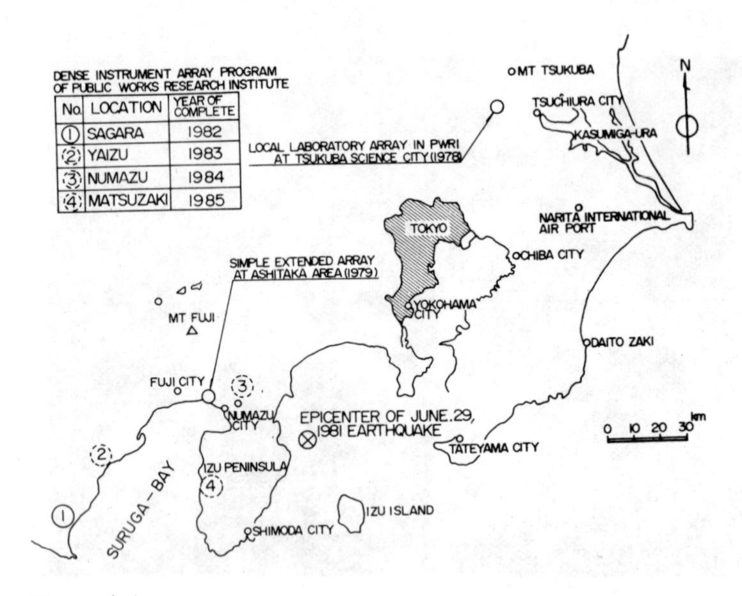

Fig. 5 (a) Locations of Dense Strong-Motion Array by PWRI

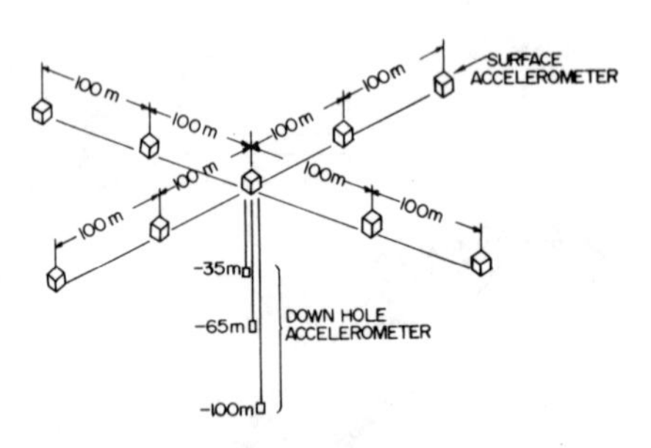

Fig.5 (b) Proposed Configuration of Local Laboratory Array

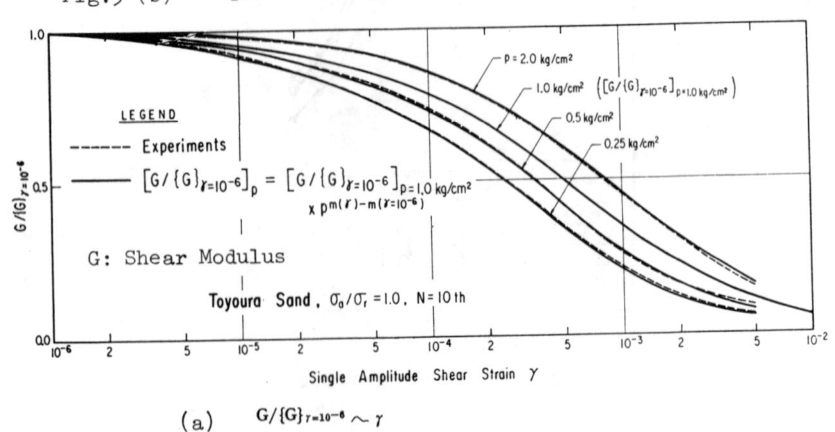

(a) $G/\{G\}_{\gamma=10^{-6}} \sim \gamma$

5 km east to west, and embraces six towns in Ibaraki Prefecture as shown in Fig.1.
Until 1980 a total of 43 national education and research institurions were relocated in the City, and they are classified into five groups based on their functions. That is:

1. Group of higher education and training
2. Group of construction engineering
3. Group of industrial science
4. Group of biological science
5. Group of Comon facilities

Fig. 2 shows the overall layout of these institutions. The population of the City, now standing at about 120000, is expected to increase to about 220000 in the future, inclusive of the population increment resulting from the development of surrounding areas.

2. PUBLIC WORKS RESEARCH INSTITUTE, MINISTRY OF CONSTRUCTION

The Public Works Research Institute of the Ministry of Construction was established in 1922. After then, for about 60 years, the Institute has worked ixtensively in many areas, including river engineering, dam construction, erosion control problems, waste water treatment, highway engineering, structural engineering and construction mechineries. As a national research center of civil engineering, it has significantly contributed to the development of civil engineering and to the achievement of consturction works.
In March 1979, the Institute was relocated and consolidated the research and development facilities at Tsukuba Sceince City, which previously situated around Tokyo area, and it had research equipment amplified and updated to one of the highest levels currently available.
The institute is situated in the northen part of the City, stretching over three towns, and it covers an area of about 126 ha. Fig. 3 shows an aerial view of the Institute. The Institute consists of 48 facilities, which are buildings, test plots and structures. These facilities can be classified into five categories by their functions. That is:

1. Main Building and the facilities for common use
2. Research facilities relating to geotechnical, soil and foundation engineering
3. Research facilities relating to structural engineering
4. Research facilities relating to hydraulic and water quality engineering
5. Research facilities relating to road and trafic engineering

In these facilities, for example, the prototype loop Test Track (6150 m long), Large Structural Universal Testing Machine (Maximum load:3000 ton) and River Model Test Yard are included.

3. FIELD OF EARTHQUAKE DISASTER PREVENTION

In the field of Earthquake Disaster Prevention these past two decades, the Institute has concerned with earthquake safety of

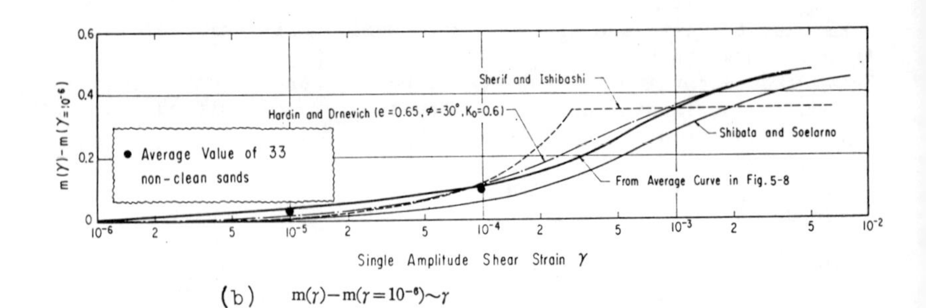

(b) $m(\gamma)-m(\gamma=10^{-6})\sim\gamma$

Fig.6 An Example of Dynamic Deformation Characteristics of Sand

Table.1 Engineering Standards Heretofore Formulated

Names of Engineering Standards	Publication Year	Publishing Organization
Specifications for Earthquake Resistant Design of Highway Bridges	1971	Japan Road A ssociation
Ministerial Ordinances and Notifications Specifying the Technicai Standards of Facilities of Petroleum Pipeline Projects	1972, 1973	Ministry of International Trade and Industry,Ministry of Trans-port,Ministry of Construction and Ministry of Home Affairs
Honshu-Shikoku Bridge Aseismic Design Guideline	1967	Honshu-Shikoku Bridge Authority
Submerged-prefabricated Tunnel Aseismic Design Guideline (Draft)	1975	Japan Society of Civil Engineers
Trans-Tokyo Bay Road Design Specifications (Draft)	1975	Japan Road Association
A Proposal for Earthquake Resistant Design Method	1977	Ministry of Construction
Specification for Earthquake Resistance Method on Water Supply Institutions	1979	Japan Water Supply Association
PartV Earthquake Resistance Design, Specifications for Highway Bridges	1980	Japan Road Association

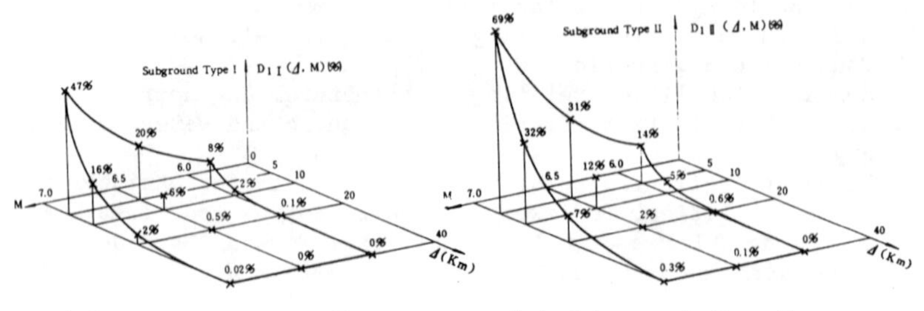

(a) Subground Type I (b) Subground Type II

Fig.7 Relationship between Magnitude,Epicentral Distance and Equivalent Ratio of Razed House

highway construction, rivers and coasts. In 1976 the Institute established Earthquake Disaster Prevention Department. This Department is composed of two divisions, i.e., Ground Vibration Division and Earthquake Engineering Division, and it is conducting research and technical guidances concerning earthquake ground motions, stability of grounds during earthquakes, sesimic design and preventive measures to earthquake disasters. Typical Subjects concerned with earthquake disaster prevention are follows.

3.1 Observation of Strong-Motion Earthquakes

Observations and analyses of strong motions on a countrywide basis are carried out in order to know seismic behavior of subsurface grounds, free-field motions and response characteristics of structures. As of March, 1979, the number of strong motion accelerographs installed at public civil engineering structures related to the works of the Ministry of Construction counts 262, and the number of the stations is 132.
Analyses of the records obtained provide the basic information for structural designing, such as maximum accelerations, response spectra, etc. Fig.4 shows examples of response spectral curves analyzed for different conditions. Seismic forces evaluated by these strong-motion records have been employed in establishing criteria for aseismic design of civil engineering structures constructed and maintained by the Ministry. The Institute is in charge of the strong-motion earthquake observation network and publication of strong-motion records.

3.2 Dense Instrument Arrey Program of The Institute

Basing upon the knowledges accumulated through past strong motion observations, it is widely recognized that ground motions are significantly dependent on source characteristics, path condition between the source and the observation station, and local geological and topological conditions. In order to investigate at each individual site is not enough, and installation of dense instrument arrey is considered inevitably necessary.
Recognizing these requests in solving the important problems, and preparing for the practical observations, the Institute had deployed a local laboratory array consisting of 20 accelerometers in the Institute and a simple extended array consisting Of 5 accelerographs at Ashitaka area in Shizuoka Prefecture in 1979. The Institute is going to deploy four local laboratory arreys around the Suruga Bay in Shizuoka Prefecture within four years starting in 1981. Fig.5 shows locations and proposed configuration of Dence Strong-Motion Instrument Array by the Institute.

3.3 Study on Stability of Grounds and Earth Structures

Evaluation of dynamic deformation characteristics of soils, liquefaction potential of sandy layers and experiment and analysis on dynamic soil-structure interactions are carried out in order to obtain reasonable procedures for assessing the

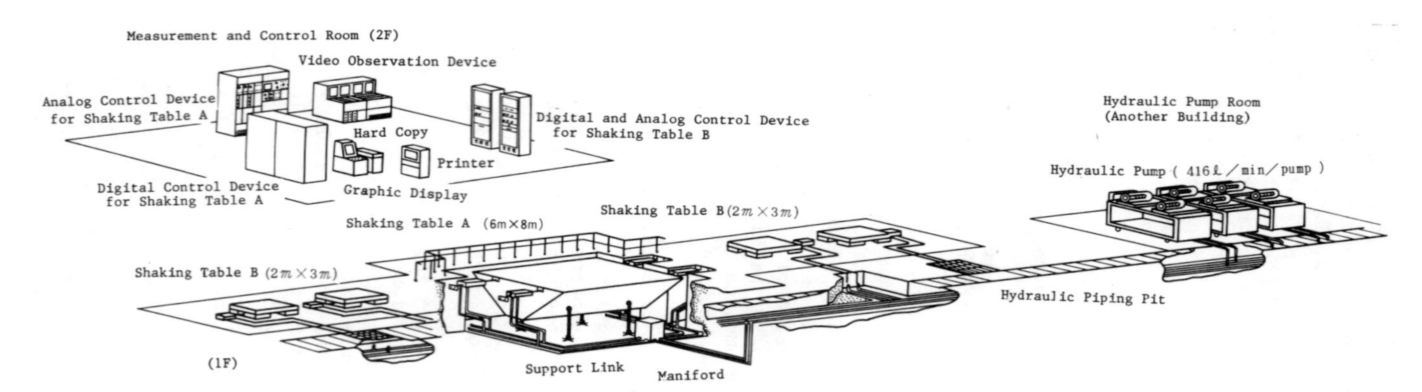

(a) An Bird's-eye View of Shaking Table Systems

	Table Size	Maximum Londing Capacity	Maximum Stroke	Maximum Velocity	Maximum Accelertion	Direction
Shaking Table-A	6m 8m(1table)	100ton	±75mm	±60cm/sec	±700gal:Under Maximum Loading	Horizontally
Shaking Table-B	2m 3m(4tables)	25ton			±2,500gal:Under No Loading	One Direction

	Control Of Input Motions	Excitation Motions	Control Item	Frequency Range	Control System
Shaking Table-A	Analog Control and Digital Control	Sinusoidal Waves, Trianguler waves, Rectanguler Waves, Earthquake Waves	Acceleration Velocity Displacement	DC~30 HZ	Electro-Hydraulic Servo System
Shaking Table-B				DC~50 HZ	

(b) Main Specification of Shaking Tables

Fig.8 Shaking Table Systems

earthquake resistance of grounds, earth structures and foundation structures. Evaluation of shear deformation characteristics hysteretic damping characteristics and dynamic strengths of soils over a wide range of shear strain (10^{-6}~10^{-2}) is made with use of resonant colum apparatus, cyclic torsional shear apparatus, and cyclic triaxial compression apparatus. Fig.6 shows an example of dynamic deformation characteristics of sand. Dynamic soil-structure interactions are also investigated with use of shaking tables. The results of the investigation are used in developing procedures for estimating the stability of grounds and earth structures, aiming to rationalize seismic design procedures of these structures.

3.4 Study on Improvement of The Aseismic Design of Civil Engineering Structures

For the establishment of rational and effective aseismic design methods for civil engineering structures, the Institute is conducting seismic response analyses of bridges, submerged tunnels etc. To verify the results of such analysis, vibration tests with prototype or scaled-down model by using Earthquake Engineering and Vibration Laboratories.
Based upon the study described above, the Institute is formulating engineering standards concerning to the aseismic design of various structures. Table 1 shows engineering standards heretofore formulated.

3.5 Study on Restore and Rehabilitation Method for Earthquake Disaster, and Retrofitting Method for Existing Structures

Japan is one of the famous earthquake-hazardous countries in the world and suffered from major seismic disasters many times in its history. Those disaster experiences have urged the development of aseismic design methods for civil engineering structures. As a result, major damages to structures such as total collapse have decreased and correspondingly human life has become to be kept in safe.
However, those minor damages or partial failures as observed in recent earthquakes in Japan have yet been expected to occur in future earthquakes. The measure to be taken for those structures is a matter of significant concern to government jurisdictions.
In El-Asnam, Algeria earthquake and South Italy earthquake occurred consecutively in 1980, there were observed the aftershock damage to buildings the resistance of which had been weakened by the initial shock and also observed the magnification of disaster caused by the delay of urgent helps of post-earthquake inspection and repair.
These facts suggest the importance of the programs mentioned on the face.

3.6 Study on Evaluation of Earthquake Disasters

Formulation of an earthquake disaster prevention plan presupposes the evaluation of the earthquake disaster risk potentials by a rational methods. It is the first step to clarificate the

Fig.9 An Example of Shaking Table Tests on a High-rized Bridge

Fig.10 An Example of Shaking Table Tests on Soil Liquefaction

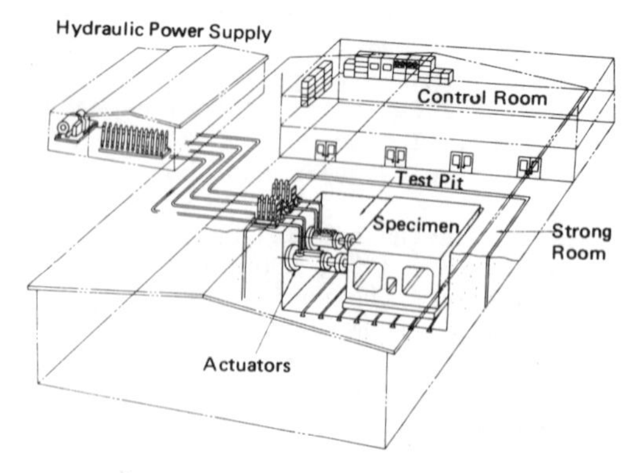

Fig.11 (a) An Bird's-eye View of the Laboratory—The diagram illustrates schematically a vibration test of submerged tunnel conducted in the test pit.

correlation of macroscopic characteristics of earthquake disasters with the magnitude, epicentral distance and ground conditions.
An example of the relationship is shown in Fig.7, where the earthquake damage ratios of wooden dwelling houses are used as an index of the earthquake disaster potential.

3.7 Study on Evaluation Models for Long Term Effects of Earthquake Losses

An earthquake brings not only physical losses such as structural damages but also economic losses and social damages in and adjacent the quake-hit area. Basing upon this point, in addition to the evaluating the direct losses, the Institute has developed methods of evaluating the indirect losses which the economic entities i.e. household sector, bussiness sector and governmental sector are affected by the earthquakes.

4. EXPERIMENTAL FACILITIES FOR EARTHQUAKE ENGINEERING

In the Institute there are many large scale facilities concerning to soil mechanics, structural engineering and earthquake engineering. Due to the limitation of space, only the features of earthquake engineering test facilities are described.

4.1 Earthquake Simulator Facility

The earthquake simulator facility is composed of vibration laboratory, control and data processing room and oil pump room as shown in Fig.8(a). In the vibration loboratory, new two type shaking tables are installed, i.e., electrohydraulic type shaking tables with maximum loading capacities of 100 tons (a shaking table A) and 4x25 tons (4 shaking tables B). The shaking table A is designed to be excited by four electrohydraulic type actuators in one derection horizontally. The four shaking tables B are also designed to be excited by actuators independently with arbitrary input motions. The main specifications of performance for shaking tables are summarized in Fig.8(b). By employing the earthquake simulator facility the following research subjects have been performed since 1979.
1) Seismic tests on highway bridges
2) Seismic tests on submerged tunnels
3) Seismic tests on soil-structure interaction
4) Liquefaction tests on saturated sandy layers
Some examples of Shaking Table Tests are shown in Figs.9 and 10.

4.2 Earthquake Engineering Laboratory

The earthquake engineering laboratory consists of test pit, control and data processing room and hydraulic power supply room as shown in Fig.11(a). In the test pit (25m x 10m x 5m), two actuators with maximum loading capacity of 125 tons are installed. The main specifications of performance are summarized in Fig.11(b). By employing this facility, the following

Items	Perfomances
(1) Maximum Excitation	±125 t/1 Actuater
(2) Maximum Stroke	±125 mm
(3) Maximum Speed	±1 m/sec.
(4) Frequency Range	DC~30Hz
(5) Excitation Motions	Sinusoidal Waves Trianguler Waves Rectanguler Waves Ramp Waves Earthquake Motions
(6) Maximum Excitation Time	10 sec. (At Maximum Velocity)
(7) Control of Input Motions	Analog Control and Digital Control
(8) Control Item	Acceleration,Velocity or Displacement
(9) Exciter	Electro-hydraulic Servo System
(10) Hydraulic Power Supply	Main Pump 300 ℓ/min x 2 with 200 ℓ/10sec×6 Accumulators (3 Accumulators /1 Actuator)×2

Fig.11 (b) Main Specification of Test Apparatus

Fig.11 Earthquake Engineering Laboratory

Fig.12 An Example of Tests on the Earthquake Response Characteristics of Concrete Pier

researches have been performed since 1980.
1) Dynamic strengths and deformabilities of steel members and reinforced concrete members
2) Dynamic properties of rock and soil
3) Fatigue properties under limited cyclic loading with large amplitudes
Fig.12 shows an example test by this facility.

REFERENCES

1. Sakagami, Y. (1980), "An Outline of The Public Works Research Institute, Ministry of Construction", Civil Engineering in Japan, Japanese Society of Civil Engineers, Vol.19, PP.147-160
2. PWRI (1979), "A Brief Introduction to Research Activities on Earthquake Disaster Prevention"
3. Okubo, T., Iwasaki, T. and Kawashima, K. (1981), "Dense Instrument Array Program of The Public Works Research Institute and Preliminary Analysis of Some Records" The 13th Joing Meeting U.S.-Japan Panel on Wind and Seismic Effects, U.J.N.R.
4. Kuribayashi, E., Hirosawa, M. and Murota, T. (1981), "Development of Post-Earthquake Measures for Building and Structures Damaged by Earthquakes" 2nd Joint Meeting, U.S.-Japan Cooperative Research Program on Repair and Retrofit
5. Kuribayashi, E., Ueda, O. and Tazaki, T. (1981), "An Economic Model of Long Term Effects of Earthquake Losses" The 13th Joint Meeting, U.S.-Japan Panel on Wind and Seismic Effects, U.J.N.R.
6. PWRI (1980), "Manual of the Public Works Research Institute, Ministry of Construction"
7. Iwasaki, T., Tatsuoka, F. and Takagi, Y. (1980), "Experimental Study on Dynamic Deformation Characteristics of Soils (II)", Report of PWRI, Ministry of Construction Vol.153 (in Japanese)

Soil Dynamics & Earthquake Engineering Conference / Southampton / 1982.07.13-15

The Effects of Earthquake Wave Dispersion on the Response of Simple Dynamic Structural Models

BRUCE D.WESTERMO
San Diego State University, CA, USA

INTRODUCTION

The dynamic response of a structure to strong ground motion generated by faulting at a distance clearly depends on the characteristics of the geologic medium through which the seismic waves travel. The scattering, diffraction, and attenuation of the waves caused by the inhomogeneity of the geologic structure can significantly alter the time history of the ground motion at a site compared to that produced at the source. Inhomogeneities of a given dimension will influence waves with wavelengths of equal dimension, thus the medium will be dispersive, i.e. the wave velocity will be dependent upon its temporal frequency. In terms of the ground motion experienced at a site, the dispersive nature of the travel path geology will dictate the arrival times of the seismic energy with a specific frequency component. This time-frequency dependence could be of significance to the structural response, especially in systems where the resonant frequency is amplitude dependent. Deriving the dispersive nature of a particular site is not a simple task and therefore the value of its role in aseismic design must also be assessed.

This paper examines the significance of the dispersion to linear, single (SDOF) and multiple degree of freedom (MDOF) and bilinear systems with the purpose of defining the variation in the response due to lack of information about the dispersive characteristics of a site and to indicate what dynamic system parameters will produce the most favorable response when the dispersion is understood.

The strong ground motion at a site will be assumed to be due to wave groups propagating through the medium with a set of group velocities, $U_m(\omega)$, where ω is the frequency of the group and M is the mode number. The amplitude of excitation at the source is partitioned among the total number of modes of propagation that are defined for the medium. The modes correspond to the different characteristics of propagating waves such as body, Love, Rayleigh, and other surface waves. With a large enough number of such modes any realistic medium can be accurately modelled with respect to linear, elastic wave propagation.

SINGLE DEGREE OF FREEDOM SYSTEMS

With this model the first objective will be to define the dispersion, $U(\omega)$, which will create the largest possible response of a linear, single degree of freedom system with all other characteristics of the problem (i.e. fourier amplitude of the acceleration, oscillator natural frequency, damping, etc.) held constant. The excitation at the site produced by this dispersion is defined as the critical excitation. For an oscillator with natural frequency, ω_n, critical damping ratio, ξ , and base acceleration, g(t), the system's dynamic equation is

$$\ddot{x} + 2\xi\omega_n\dot{x} + \omega_n^2 x = g(t), \qquad (1)$$

where x(t) is the displacement response of the system and a dot over the variable denotes differentiation with respect to time. Following Drenick (1970) the critical excitation acceleration, g(t), is the member of an unbounded class of excitations corresponding to the fixed parameters chosen which satisfy

$$\int_0^T g^2 dt = C_o, \qquad (2)$$

where T is the duration of the excitation and C_o is an arbitrary constant. Variational calculus shows that the critical excitation will be

$$g = a_o\dot{x}, \qquad (3)$$

where a_o is an arbitrary constant related to C_o in equation (2). Thus equation (1) becomes

$$\ddot{x} + (2\xi\omega_n - a_o)\dot{x} + \omega_n^2 x = 0, \qquad (4)$$

and the critical excitations can be derived from the solutions of this equation. The advantage of looking at the response in terms of its worst excitation is that the original inhomogeneous problem is transformed into a homogeneous one. This also applies to nonlinear systems providing the damping term is linear. Physically, bounded excitations, g(t), are sought and thus a_o is restricted to

$$a_o = 2\xi\omega_n, \qquad (5)$$

revealing from equation (4) that the critical excitations of equation (1) have a frequency equal to ω_n. It can be shown that the critical excitations with frequencies other than ω_n are given by

$$g = a_o\dot{x} - a_1 x, \tag{6}$$

where a_1 is an arbitrary constant. With this equation (1) becomes

$$\ddot{x} + (2\xi\omega_n - a_o)\dot{x} + (\omega_n^2 + a_1)x = 0, \tag{7}$$

and it is seen that a_1 controls the amplitude and frequency of the response because equation (5) must still be satisfied. The frequency of the critical excitation, ω_{ce}, is

$$\omega_{ce} = \sqrt{\omega_n^2 + a_1} \quad . \tag{8}$$

The ratio of the response, x(t), to the critical excitation, (i.e. the transfer function), $H(\omega)$, can be calculated from equations (6) and (7) and is given by

$$H(\omega) = \frac{|x|_{max}}{|g|_{max}} = \left\{(\omega_n{}^2 - \omega^2)^2 + (2\xi\omega\omega_n)^2\right\}^{-\frac{1}{2}}, \tag{9}$$

where ω is the frequency of the excitation. The phase of the response, ϕ , is

$$\phi(\omega) = -\tan^{-1}\left\{\frac{2\xi\,\frac{\omega}{\omega_n}}{1 - \left(\frac{\omega}{\omega_n}\right)^2}\right\} \quad . \tag{10}$$

Inspection of equations (9) and (10) show that these results are identical to the harmonic response transfer function for the system in equation (1), Thompson (1965), for the same definition of the phase.

The response of the system in equation (1) may be written in terms of the frequency domain variables as

$$x(t) = \frac{1}{2\pi}\int_{-\infty}^{\infty} G(\omega)e^{i\alpha(\omega)}\ R(\omega)e^{i\phi}e^{-i\omega t}d\omega\ , \tag{11}$$

where $G(\omega)$ is the fourier transform amplitude of g(t) with phase α and $R(\omega)$ is the fourier transform amplitude of the transfer function for equation (1) with ϕ being its phase. If T_m is defined as the time at which the maximum value of x(t) occurs, the maximum response possible for a given $G(\omega)$ exists when

$$\alpha = b_o + \omega T_m - \phi, \tag{12}$$

where b_o is an arbitrary constant. This is consistent with the results of equations (9) and (10) since the definition of T_m is independent of the response so T_m may be taken as zero. It should be noted that the system is linear and therefore the phase producing the maximum response is independent of the fourier amplitude of the excitation.

To define the influence of the dispersion on the critical response as evaluated from equation (11), the acceleration, g(t), is modelled as a collection of K discrete frequency bands of waves, travelling in M modes, each band containing the frequencies in the range $\omega_n - \Delta\omega \leq \omega \leq \omega_n + \Delta\omega$. Following Wong (1978) the fourier amplitude of the acceleration is given by

$$G_k(\omega) = \sum_{m=1}^{M} B_{km} e^{i((\omega-\omega_k)t_{km} - \beta_k)}\ , \tag{13}$$

where ω_k is the center frequency of the kth band, t_{km} is the arrival time of the kth band for the mth mode, and β is the phase due to the source mechanisms. B_{km} is the amplitude corresponding to the kth group and m^{th} mode. Attention is now focused on the case of a single propagating mode, M=1, and the source phase, β , will be assumed to be constant. Applying the conditions of equations (12) and (13) to equation (11) indicates the maximum response occurs when

$$(\omega-\omega_k)t_k = a_o + \omega T_m - \phi \ , \tag{14}$$

and differentiation with respect to the frequency over each band gives

$$t_k = T_m - \left.\frac{d\phi}{d\omega}\right|_{\omega_k} \ . \tag{15}$$

Being that t_k is defined as the arrival time for the kth frequency group, the group velocity, U, is given by

$$U(\omega_k) \ = \ R/t_k \ , \tag{16}$$

where R is the source to site distance. Thus equation (15) indicates the dispersion, $U(\omega)$, yielding the maximum possible response, denoted as the critical dispersion, is given by

$$U(\omega) \ = \ R(T_m - \frac{d\phi}{d\omega})^{-1} \ . \tag{17}$$

A set of these dispersion curves calculated for R=10 and 100 km and for various values of T_m are shown in Figure 1. These curves show that the dispersion causes the excitation components with frequencies close to the natural frequency of the system to arrive before the remaining frequency components because it is these frequencies contributing to the maximum response. Physically, the critical dispersion filters out the noncontributing frequencies from arriving at T_m.

The phase, α, yielding the maximum response is a function of the dynamic systems parameters, ω_n and ξ, and therefore, so is the critical dispersion function. To define how a dispersive mode that is critical to the response at a particular natural frequency influences the response at other frequencies, examples of the psuedovelocity response, PSV, for several different types of dispersive modes where calculated. Fourier amplitudes, $G(\omega)$, empirically estimated for the earthquake magnitude, M, and epicentral distance, R, as supplied in Wong (1978) were applied to equation (11) along with the particular dispersion dependent phase, α to produce the PSV response curves shown in Figure 2. Here the uppermost curve, labeled the critical response, is the maximum response at all frequencies, ω_n, such that equation (12) is satisfied. Because each

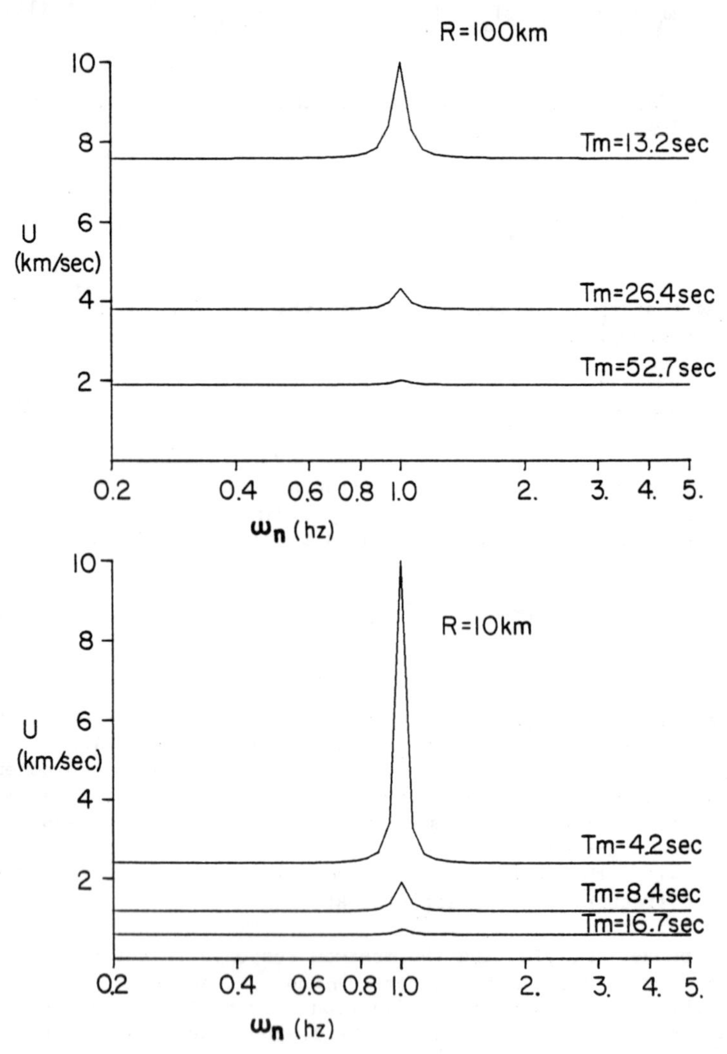

Figure 1. Dispersion curves for the maximum response of a 5% critically damped linear oscillator with $\omega_n = 1$ hz. For various times of maximum response, T_m.

dispersion curve can be critical for only one specific frequency, this curve represents the upperbound to the response which would be produced by an infinite number of modes each having a critical dispersion relation for a distinct natural frequency. The critical single mode dispersion is that defined by equation (12) with R=50 km and a 5% critical damping. The nondispersive single mode curve is the response to a

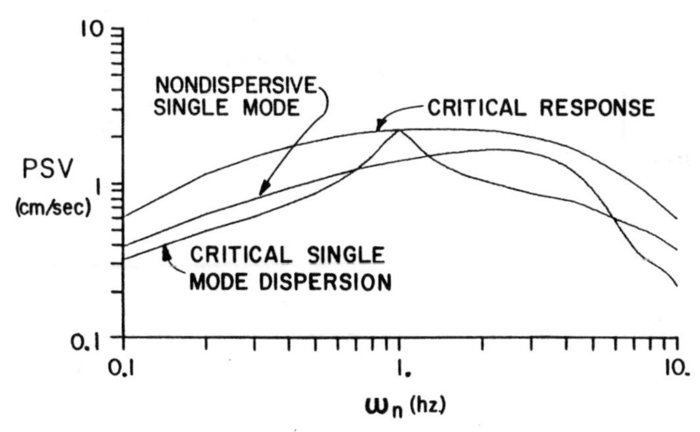

Figure 2. PSV response of a 5% critically damped SDOF system to an empirically defined magnitude 7, R = 50 km earthquake for the three phase functions listed.

propagation system where the group velocity is not frequency dependent. Figure 2 shows, as expected, that at the natural frequency of the critical dispersive mode the response is the critical response. Away from this frequency, however, the response is as much as 60% less than the critical response. The nondispersive propagation mode response is greater than the critical single mode for a range of frequencies away from 1hz. This may be due to the fact that the large R critical modes are relatively nondispersive at these frequencies and thus are more closely approximated by a totally nondispersive mode. The nondispersive response is the dispersion producing the maximum response averaged over all frequencies. This may prove useful in simplifying the characteristics needed for developing response models incorporating the geophysical characteristics of the solution.

To define the critical responses dependence on the source to site distance, R, the response to critical modes for three distances calculated for magnitude 7, R = 20 km are shown plotted in Figure 3. The R = 10 km and R = 50 km curves represent the response to dispersion functions producing maximum response at their R values. The R = 50 km curve shows a marked response reduction at the critical system's phase with that produced by the dispersion. The response curves plotted in Figure 2 show the extremes of the dispersion influence for the number of wave modes, M, used. The critical single mode dispersion being the maximum response for a single wave mode while the critical response represents the maximum response possible for a virtually infinite number of wave mode dispersion.

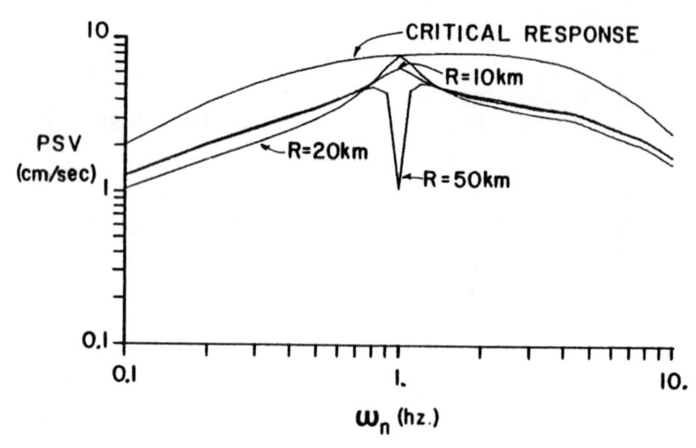

Figure 3. PSV response of a 5% critically damped SDOF system at R = 20 km to a magnitude 7 earthquake shown plotted for the three dispersive single mode systems described in the text.

TWO DEGREE OF FREEDOM SYSTEMS

To examine the MDOF systems response interplay with the dispersion, consider a two degree of freedom linear oscillator with response, $\underset{\sim}{x}(t)$, given by

$$x(t) = \begin{Bmatrix} x_1 \\ x_2 \end{Bmatrix} = \frac{1}{2\pi} \int_{-\infty}^{\infty} G(\omega) e^{i\alpha} \underset{\sim}{A}(\theta) \begin{Bmatrix} z_1 \\ z_2 \end{Bmatrix} d\omega \quad , \tag{18}$$

where

$$\underset{\sim}{A}(\theta) = \begin{bmatrix} \sin\theta & \cos\theta \\ -\cos\theta & \sin\theta \end{bmatrix} \quad . \tag{19}$$

The system has natural vibrational modes z_1 and z_2 and a modal participation matrix, $\underset{\sim}{A}(\theta)$, and the amount of the relative coupling of the modes is dictated by the value of θ . The modal participation influence is, naturally, the unique feature of MDOF systems. The critical phase, $\alpha(\omega)$, can be calculated from the representation in equation (18) using the same concepts that derived equation (12). Plots of these critical phases are shown in Figure 4. The dispersion relations producing the maximum response at R are again given by equation (17) and are shown in Figure 5 for R = 50 km and R = 100 km. The group velocities are highest at the two resonant frequencies of the system, and they are lowest over the frequency range where the phase, α , decreases with frequency (Figure 4). The frequencies for which this occurs are dependent on the relative participation of each mode, θ .

A comparison of the response this dispersion creates relative

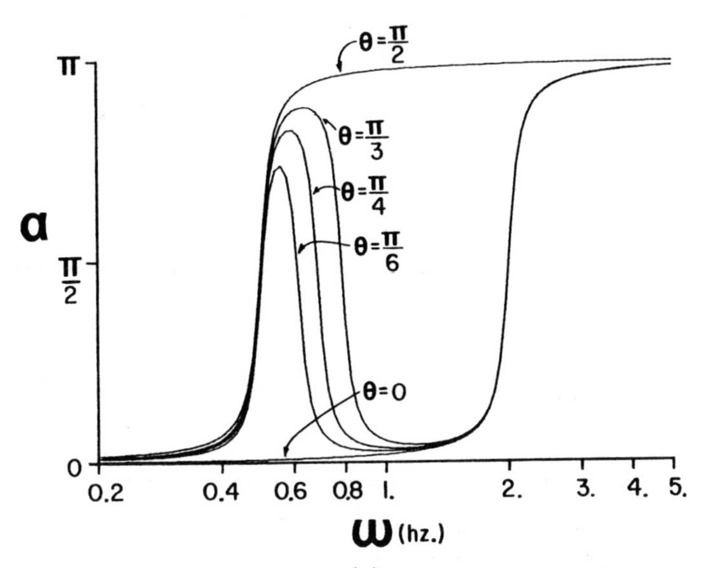

Figure 4. The critical phase, $\alpha(\omega)$, which produces the maximum response of x_1 for various degrees of modal participation, θ. $\omega_1 = 0.5$, $\omega_2 = 2.0$ hz, $\xi = 0.05$.

to the overall maximum response and a nondispersive mode is given in Figure 6. Again, the critical single mode response equals the critical response when the systems characteristics match those defining the dispersion mode, however, at the point where ω_1 equals the second natural frequency of the dispersion, $\omega_2 = 2.4$ hz, the response is quite close to the critical value. The separation in the natural frequencies ω_1 and ω_2 is large enough, in this example, that the phase at these frequencies is not strongly dependent on the modal participation (Figure 4). For large frequency separation in the natural modes, the phase variation around the natural frequencies is the most dominant in producing the response. Therefore, the particular features of the critical SDOF dispersion are the important features in this problem. The general features of Figure 6 are not unlike those of Figure 2 except that the response in Figure 6 show a smaller variation among the three dispersion models used. It appears that this is due to the MDOF systems response dependence on a wider range of excitation frequencies than a SDOF system. The nondispersive mode response is greater than the critical single mode for the frequencies away from the natural frequencies. The nondispersive mode is itself a critical dispersion response assuming the maximum response over a range of frequencies is the variational criteria. The nondispersive mode response for the MDOF system is closer to the critical response that it is for the SDOF system (Figure 2). Basically, there is an overall reduction of the dispersion dependence for the MDOF system compared to the

SDOF results.

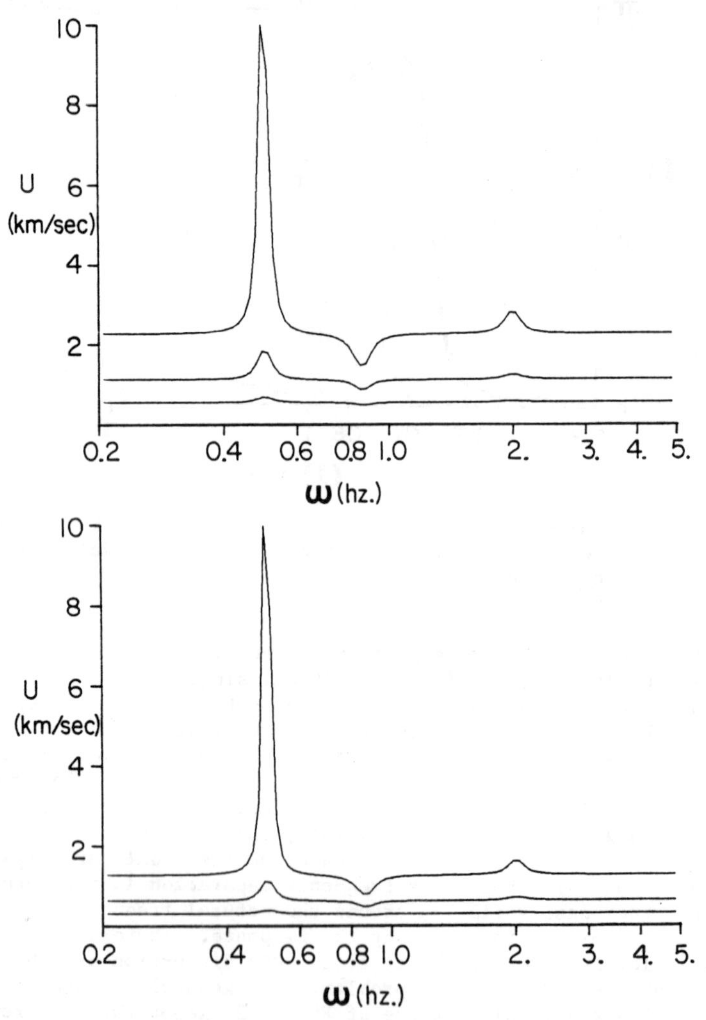

Figure 5. The MDOF critical dispersions for R = 100 km (top) and R = 50 km (bottom) at $\theta = 3\pi/8$.

THE BILINEAR SDOF DISPERSION

Consider the scaled, bilinear dynamic system

$$\ddot{x} + 2\xi\omega_n\dot{x} + \omega_n^2 f(x) = g(t), \tag{20}$$

where

$$f(x) = \begin{cases} x & ; |x| \leq 1 \\ \mathrm{sgn}(x); & |x| > 1 \end{cases} \quad . \tag{21}$$

The damping term in equation (20) is linear and so the critical excitations are of the form in equation (6) except the response equation is now

$$\ddot{x} + (2\xi\omega_n - a_o)\dot{x} + (\omega_n^2 + a_1)f(x) = 0. \tag{22}$$

The phase for the solutions have been calculated numerically from the solutions of the initial value problem, equation (22), and are shown for four values of the peak excitation amplitude in Figure 7. The peak acceleration values correspond to maximum response amplitudes of 2, 4, 8 and, 16 times the elastic limit, $x = 1$. These curves represent the phase functions for a peak acceleration, g_m, independent of the frequency, ω_n. The solutions of equation (22) show that the critical excitations for $a_1 < -\omega_n^2$ are unbounded, representing the transient critical excitations, and are noncontributing to the steady state solutions. The dispersion curves for these phase functions are shown in Figure 8 for the same four acceleration amplitudes. For case a and b the system spends comparable time in and out of the elastic range hence the system is influenced more by frequencies at and below its linear natural frequency. The larger amplitude solutions (c and d) display a significant portion of the response inelastically and the dispersion shows a more marked dependence on the lower frequency velocities only. Here, the system is a softening one so the critical excitation frequency decreased with increasing amplitude. At frequencies higher than the linear natural frequency the solutions are linear, and the critical group velocities are close to nondispersive. Work is now being carried on by the author to calculate the particular amplitude dependent dispersive modes for realistic forier amplitudes and to examine how the response varies with these dispersions.

CONCLUSIONS

The use of the concept of defining the maximum response of a system to a specific class of excitations has been used here to establish the upperbound of the response dependence upon various dispersion models used in producing the input accelerations. The particular effects of the dispersion are dependent on the fourier acceleration amplitudes chose and thus only general comments on the dependence are possible, but the results do indicate that the influence of the dispersion characteristics, here $U(\omega)$, is significant.

ACKNOWLEDGEMENTS

This research was supported by NSF Grant No. CEE-8105781.

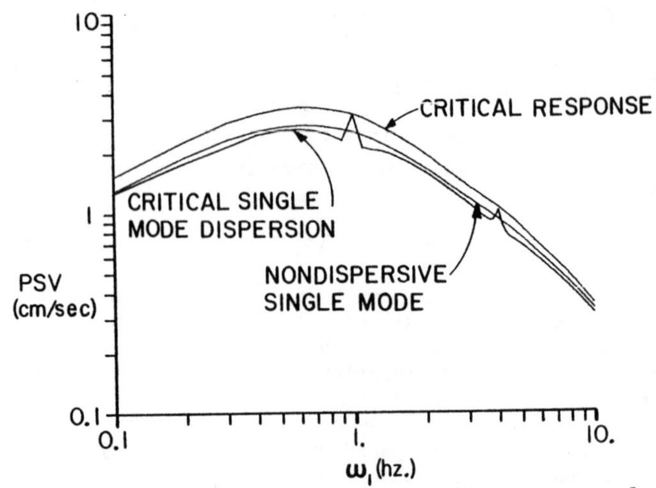

Figure 6. The magnitude 5, R = 10 km, PSV response of x_1 for a MDOF system with $\omega_2/\omega_1 = 4$, $\xi = 0.05$, and $\omega_2 = 4$ hz, for the 1) ideal critical dispersion, 2) a single mode critical dispersion for R = 10 km, and 3) a nondispersive mode.

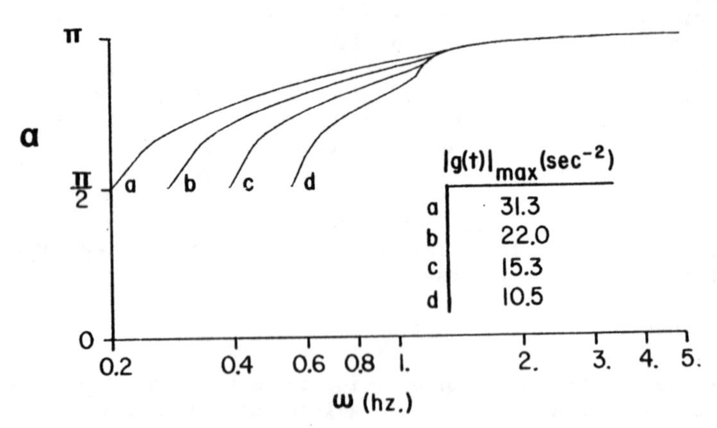

Figure 7. The single mode critical phases of the bilinear system for various levels of the peak acceleration, g_m.

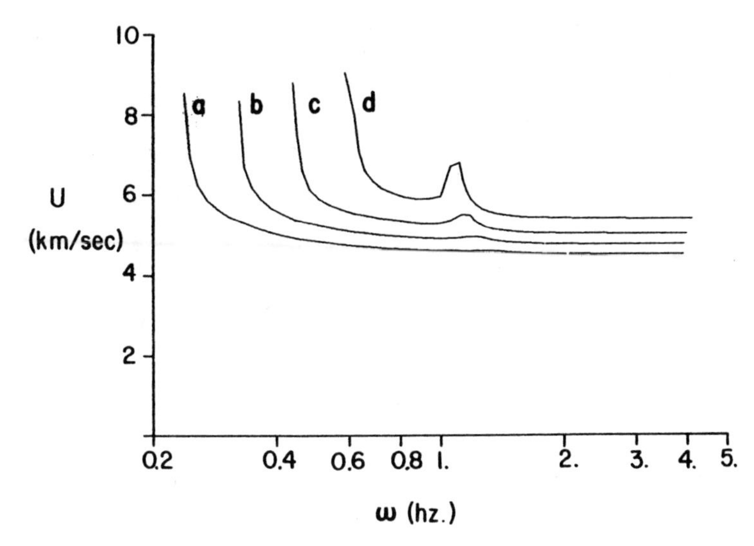

Figure 8. The critical dispersion functions based on the critical phase, α , of Figure 7.

REFERENCES

Drenick, R.F. (1970) Model-free design of aseismic structures. Jl. of EMD, ASCE, EM4:483-493

Thomson (1972) Theory of Vibrations. Prentice-Hall, New Jersey.

Wong, H.L. and M.D. Trifunac (1978) Synthesizing realistic ground motion accelerations. Dept. of Civil Eng., Univ. of Southern Calif., USA, Report No. CE 78-07.

Soil Dynamics & Earthquake Engineering Conference / Southampton / 1982.07.13-15

Boundary Effects on Pore Pressure Buildup in Loose Saturated Sand

PHILIPPE P.MARTIN & MOHAMAD H.ABEDI-HA
University of Illinois, Urbana-Champaign, USA

INTRODUCTION

Extensive laboratory programs have been under way for several years to determine the liquefaction characteristics of sand specimens. With the assumption of undrained field behavior, it has been possible to use directly laboratory test results to assess the liquefaction potential of soil deposits.

Seed et al. (1976) developed a simple procedure to analyze the variations of pore pressure in a saturated soil profile under seismic excitation, taking into account the volume change and diffusion characteristics of the material in the soil profile. The first writer et al. (1979) showed that the assessment of the liquefaction potential in a soil profile by the simple procedure was in good agreement with the results obtained by more fundamental approaches. The procedure was also verified experimentally by Yoshimi (1977). The most important conclusion to be drawn from such analyses was that the drainage conditions at the boundaries of a potentially liquefiable layer might substantially reduce the rate of pore water pressure buildup in that layer under seismic excitation. Drainage conditions are related to both the permeability and the rebound characteristics of the soil. The range of variation of permeabilities in sands is much wider however than the range of stiffnesses, which might lead to the belief that drainage characteristics are related to permeability alone.

After presenting some background information on the permeability and rebound characteristics of saturated sands, this paper analyzes their influence on the boundary drainage conditions of a potentially liquefiable layer of loose sand.

PERMEABILITY OF SANDS

Notwithstanding the extreme variability of permeability as a soil characteristic, some correlation has been found to exist between grain-size characteristics and permeability coefficients. A general relationship between the 20% size (D_{20}) for a soil and the coefficient of permeability was prepared by Creager et al. (1947) which may be expressed as:

$$k, \text{ in cm/sec} = 77\ (D_{20})^{2.32} \tag{1}$$

where D_{20} = the 20% size, in cm.

Another empirical relationship is Hazen's formula which may be expressed in terms of D_{20} (in cm) as:

$$k, \text{ in cm/sec} = 85 - 130\ (D_{20})^{2} \tag{2}$$

The relationships expressed by Equation 1 for average field values and those computed by Equation 2, obtained from data on uniform loose sands, are shown in Figure 1. It can be seen that coefficients of permeability of fine and medium sands range from 10^{-3} cm/sec for the finer to 2 cm/sec for the coarser, i.e. a ratio of 1 ÷ 2,000. Note that considerable care should be exercised in using this figure for the assessment of the permeability of natural soil deposits where stratification and thin layers of finer-grained soil may dramatically reduce the vertical permeability.

REBOUND CHARACTERISTICS OF SANDS

There is a scarcity of literature on the rebound and compressibility characteristics of saturated sands under changes of effective stress. Lee and Albeisa (1974) measured the volumetric strains due to reconsolidation of triaxial test samples that had been subjected to excess hydrostatic pore pressure increases caused by cyclic loading or static loading. From the study of their data, Seed et al (1976) determined the variations of the coefficient of volume compressibility defined by the expression

$$m_v = \frac{d\varepsilon_v}{d\sigma'} \tag{3}$$

in which $d\varepsilon_v$ = the change in volumetric strain corresponding to a small change in effective stress $d\sigma'$. They concluded that the two factors that influenced the value of the coefficient m_v the most were the relative density of the soil and the peak pore pressure ratio, $r_u = \Delta u/\sigma'_o$, i.e. the ratio of maximum excess pore pressure, Δu, and the confining pressure, σ'_o. It was found that for pore pressure ratios greater than 60%, the values of m_v were influenced to a great extent by the relative density. Figure 2 shows the measured variations of the normalized coefficient of volume compressibility, i.e., the ratio m_v to m_{v_o} in which m_{v_o} = the coefficient of volume compressibility at low pore pressure for different relative densities. The analytical expression relating m_v/m_{v_o}, r_u and relative density in the simple procedure takes the graphical form shown in Figure 3.

Finn et al (1977) developed rebound data in one-dimensional unloading on crystal silica sand at a relative density of 45%. The first writer et al (1979) showed that their data fitted the expression

$$\frac{E_{r_o}}{E_r} = \frac{1}{(1 - r_u)^m} \quad (4)$$

where E_{r_o} is the rebound modulus for low pore pressure ratios r_u and m a constant function of relative density.

For purposes of comparison between the expression of m_v/m_{v_o} given by Seed et al (1976) and the ratio E_{r_o}/E_r given by Equation 4, the ratio E_{r_o}/E_r has been plotted in Figure 3 for values of $m = 0.3$ and 0.7. It can be seen that for pore pressure ratios ranging between 0 and 0.9 both ratios m_v/m_{v_o} and E_{r_o}/E_r increase in a similar fashion with increasing pore water pressures.

The analysis of the data obtained by Lee and Albeisa (1974) indicates that at low pore pressure ratios, values of m_{v_o} typically range between 1×10^{-6} sq ft/lb and 2×10^{-6} sq ft/lb (21×10^{-6} 1/kPa and 42×10^{-6} 1/kPa) for average densities and grain sizes of sands.

More recently, Ismael and Vesic (1981) published the results of triaxial consolidated drained compression tests on sands at 70% relative density. The one-dimensional coefficient of volume compressibility m_{v_o} is related to the initial tangent modulus E_i and Poisson's ratio ν by the expression:

$$m_{v_o} = \frac{1}{E_i} \frac{(1 - 2\nu)(1 + \nu)}{1 - \nu} \quad (5)$$

For values of Poisson's ratio ranging between 0.3 and 0.33 as measured in the tests, Equation 5 yields the relationship

$$m_{v_o} \simeq (0.7) \frac{1}{E_i} \quad (6)$$

Measured E_i values ranged between 1,210 psi and 5,000 psi (8600 kPa and 35,700 kPa) but for one low value at 585 psi. According to Equation 6, the values of m_{v_o} computed from the test results obtained by Ismael and Vesic ranged from 1×10^{-6} sq ft/lb to 4×10^{-6} sq ft/lb, which is in agreement with Lee and Albeisa's results.

G. Martin et al (1975), experimenting on dry crystal silica sand at 45% relative density, obtained values of the coefficient of volume rebound, $m_{r_o} = 1/E_{r_o}$ ranging between

4×10^{-7} sq ft/lb and 8×10^{-7} sq ft/lb for confining pressures ranging between 7 and 30 psi. These values are roughly one half the values of the coefficient of volume compressibility stated previously. Hendron (1963) also ran experiments on the one-dimensional consolidation and rebound of dry sands. Values of m_r backfigured from his experiments range between 4 and 5×10^{-7} sq ft/lb at a confining pressure of 70 psi.

These few data indicate that there is no clear correlation between either m_{v_o} or m_{r_o} and confining pressure in the range of pressures extending from 5 to 70 psi. These coefficients of volume change under loading or unloading range from a high of 4×10^{-6} sq ft/lb to a low of 4×10^{-7} sq ft/lb, i.e. a ratio of 1 ÷ 10.

ANALYSIS OF BOUNDARY EFFECTS ON POTENTIALLY LIQUEFIABLE LAYER

The one-dimensional soil model used to study the boundary effects on the liquefaction potential of a layer of loose sand is shown in Figure 4. It consists of a layer of sand, 10.2 m thick confined between two impermeable strata. The sand is mostly dense (D_r = 80%) except for a seam, 20 cm thick of loose sand (D_r = 45%) parting the dense sand into two 5 m thick layers. The computer program APOLLO (First writer et al, 1978) was used to predict the excess pore water pressure redistribution within the model during the cyclic loading. The parameters of the analysis were the duration and number of equivalent cycles of excitation, the coefficients of volume change of loose sand and dense sand and the permeability characteristics of the sand.

A short excitation of 5 seconds duration (Earthquake I) was assumed to generate 2 1/2 uniform cycles of loading that would bring the loose sand layer to the point of initial liquefaction under undrained conditions. The analysis was also performed with a longer excitation of 10 seconds duration (Earthquake II) producing 5 uniform cycles of loading that would also bring the loose sand layer to initial liquefaction under undrained conditions.

Table 1 summarizes the loading characteristics and the undrained resistance characteristics of the sand used in the analysis. The ratios of volume change characteristics at low pore pressure between loose and dense sands were varied from 1 to 1000, the practical range extending from 1 to 10 as mentioned above. The coefficient of volumetric change at low pore pressure ratios for the loose sand was taken as 2×10^{-5} m^2/kN in all analyses and varied from 2×10^{-5} m^2/kN to 2×10^{-8} m^2/kN for the dense sand. The permeability characteristics of the dense sand covered the range of 1×10^{-4} cm/sec (fine silty sand) to 1×10^{-1} cm/sec (medium sand). The coefficient of permeability of the loose sand was taken as twice the permeability of the dense sand in all the analyses.

Table 1 Earthquake Characteristics and Undrained Liquefaction Characteristics

	Number of Uniform Cycles	Duration	Number of Cycles to Initial Liquefaction Under Undrained Conditions	
			Loose Sand (D_r = 45%)	Dense Sand (D_r = 80%)
Earthquake I	2 1/2	5 secs	2 1/2	50
Earthquake II	5	10 secs	5	50

The soil properties used in the analyses are summarized in Table 2.

Table 2 Synopsis of Soil Properties Used in the Analyses

	Loose Sand	Dense Sand
Permeability		
Silty Sand	2×10^{-4} cm/sec	10^{-4} cm/sec
Very fine Sand	2×10^{-3} cm/scc	10^{-3} cm/sec
Fine Sand	2×10^{-2} cm/sec	10^{-2} cm/sec
Medium Sand	2×10^{-1} cm/sec	10^{-1} cm/sec
Volume change coefficient	2×10^{-5} m^2/kN	2×10^{-5} m^2/kN to 2×10^{-8} m^2/kN

Peak excess pore pressure redistribution

The peak pore pressure ratios that have developed in the sand layer by the end of the cyclic loading are shown in Figure 5. Vertically, Figure 5 shows the effect of permeability on the pore pressure redistribution, while horizontally it shows the effect of the relative stiffnesses of dense and loose sand. The maximum pore pressure ratio in the loose sand layer does not exceed 0.65 for the cases shown in Figure 5. The study of Figure 5 calls for the following remarks:

1. The higher the stiffness ratio between the dense sand and the loose sand, the higher the pore pressures in the whole layer.

2. The higher the permeability, the greater is the amount of pore pressure redistribution within the entire layer.

For a stiffness ratio of 1 between the loose and dense sand, the pore pressure redistribution causes a decrease of the peak pressure in the loose sand without a pore pressure increase in the dense sand. For a stiffness ratio of 10, the decrease of peak pressure in the loose sand caused by the redistribution is accompanied by an increase of the pore pressures in the dense sand, which are then higher than that developed in the same number of cycles under undrained conditions.

Peak pore pressure buildup in the loose sand

The empirical pore pressure buildup curve that is used by the analytical model is shown in Figure 6. It relates the excess pore pressure ratio, r_u, to the cycle ratio, i.e. the ratio of the number of uniform applied stress cycles N to the accumulative number of cycles required to cause liquefaction, N_ℓ. The two loading sequences (Earthquakes I and II) were chosen so that the total number of applied stress cycles, N_{eq}, would cause liquefaction under undrained conditions (N_{eq} = N_ℓ, undrained). The number of cycles, N_ℓ, required to cause liquefaction of the loose sand layer in the analytical model may be obtained in Figure 6 from the value of the cycle ratio N_{eq}/N_ℓ which corresponds to the peak pore pressure ratio, r_u that has developed in the layer by the end of the excitation. The variation of the ratio $N_\ell/N_{\ell,\ undrained}$ in terms of the stiffness of the dense sand layer is shown in Figure 7 for various permeabilities. The boundary effect on the increase in the number of cycles required to cause liquefaction of the loose sand layer is the largest for values of the stiffness ratio likely to be found in the field, i.e. ranging between 1 and 10. In these analyses, the rebound properties of the dense sand are significant in avoiding liquefaction of the loose sand for permeabilities of the order of 10^{-3} cm/sec and higher.

Factor of safety against liquefaction of the loose sand

A typical liquefaction curve for a uniform loose sand at a relative density, D_r = 45%, is shown in Figure 8. By using the curve shown in Figure 8 it is possible to determine the cyclic stress ratio which will cause liquefaction after N_ℓ cycles. If the loose sand layer were in an undrained condition, the values of the stress ratios induced by Earthquake I and Earthquake II would be read off Figure 8 as 0.27 and 0.19, corresponding to N_ℓ = 2 1/2 and N_ℓ = 5, respectively.

The boundary effects on the pore pressure buildup in the loose sand increase the number of cycles to liquefaction, N_ℓ in all cases, which is tantamount to a reduction of the cyclic stress ratio induced by the excitation. The factor of safety against liquefaction of the loose sand may be defined as the ratio of the cyclic stress levels that would produce liquefaction under an undrained condition and with a boundary effect. The variations of the factor of safety thus defined are shown in Figure 9 in terms of permeability for the two stiffness ratios of 1 and 10 between dense and loose sand. It may be seen that for a silty sand (K = 10^{-4} cm/sec), the rebound in the dense sand yields factors of safety of 1.12 and 1.30 for stiffness ratios of 10 and 1, respectively. As it was shown in Figure 5, more pore pressure redistribution takes place in sands of higher permeabilities, therefore yielding higher values of the factor of safety.

CONCLUSION

The effects of the boundary conditions on the liquefaction potential of a layer of loose sand that would liquefy under undrained conditions have been analyzed. The results of the analysis have shown that both the permeability and the rebound characteristics of the medium bordering the loose sand layer are instrumental in reducing the amount of pore pressure build-up and therefore in increasing the resistance of the loose sand to liquefaction. Experimental evidence shows that stiffnesses of sands range within one order of magnitude and the analytical results suggest that even the densest sands are compliant enough to induce some amount of excess pore pressure redistribution in the field under an earthquake excitation.

REFERENCES

Creager, W. P., Justin, J. D. and Hinds, J. (1947) Engineering for Dams. Vol. III Earth, Rock-Fill, Steel and Timber Dams, John Wiley and Sons, Inc., New York.

Finn, W. D. L., Lee, K. W., and Martin, G. R. (1977) An Effective Stress Model for Liquefaction. Journal of the Geotechnical Engineering Division, ASCE, 103, GT6, June.

Hendron, A. J. H., Jr. (1963) The Behavior of Sand in One-Dimensional Compression, Ph.D. Thesis, University of Illinois, Urbana.

Ismael, N. F., and Vesic, A. S. (1981) Compressibility and Bearing Capacity. Journal of the Geotechnical Engineering Division, ASCE, 107, GT12, December.

Lee, K. L., and Albeisa, A. (1974) Earthquake Induced Settlements in Saturated Sands, Journal of the Geotechnical Engineering Division, ASCE, 100, GT4, April.

Martin, G. R., Finn, W. D. L., and Seed, H. B. (1975) Fundamentals of Liquefaction Under Cyclic Loading. Journal of the Geotechnical Engineering Division, ASCE, 101, GT5, May.

Martin, P. P., and Seed, H. B. (1978) APOLLO - A Computer Program for the Analysis of Pore Pressure Generation and Dissipation in Horizontal Soil Layers During Cyclic or Earthquake Loading, EERC Report No. UCB/EERC78-21, University of California, Berkeley.

Martin, P. P., and Seed, H. B. (1979) Simplified Procedure for Effective Stress Analysis of Ground Response. Journal of the Geotechnical Engineering Division, ASCE, 105, GT6, June.

Seed, H. B., Martin, P. P., and Lysmer, J. (1976) Pore Water Pressure Changes During Soil Liquefaction. Journal of the Geotechnical Engineering Division, ASCE, 102 GT4, April.

Yoshimi, Y., et al. (1977) Soil Dynamics and Its Application to Foundation Engineering, State-of-the-Art Report, Proceedings IX International Conference on Soil Mechanics and Foundation Engineering, Tokyo.

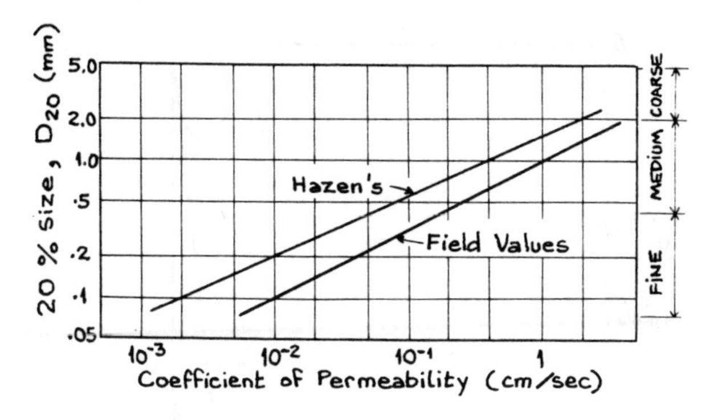

Figure 1 Empirical Relationships between Permeability and Grain Size

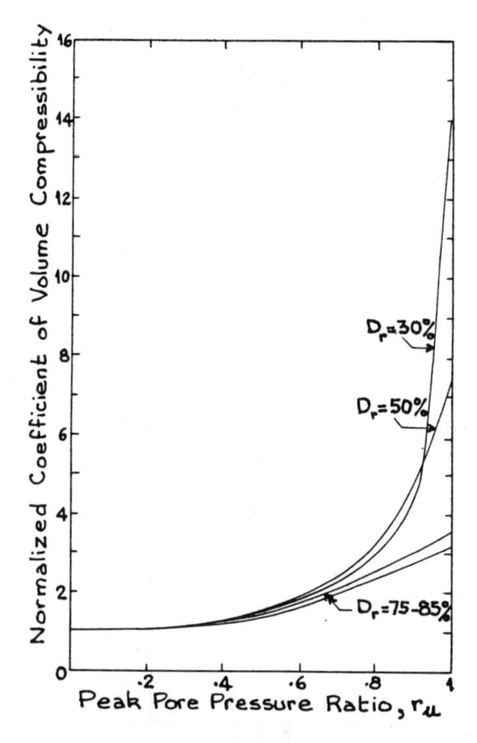

Figure 2 Effect of Pore Pressure Buildup on Compressibility of Sands [after Lee and Albeisa (1974)]

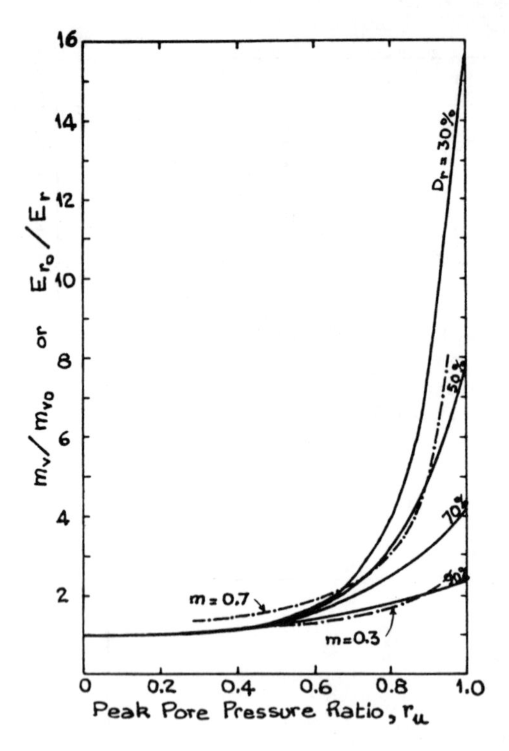

Figure 3 Theoretical Relationships between Compressibility or Rebound of Sands and Pore-Pressure Buildup

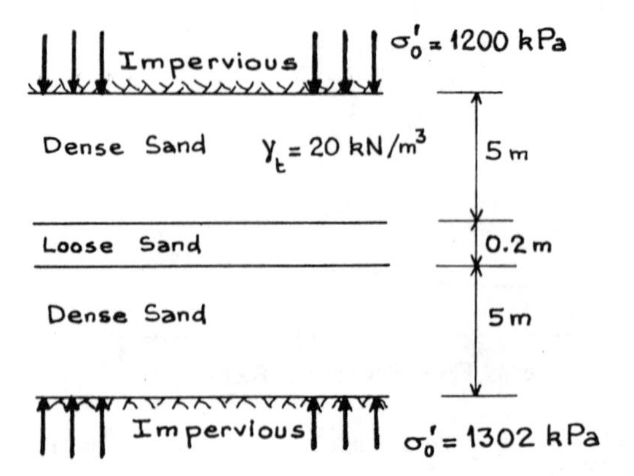

Figure 4 One Dimensional Soil Model

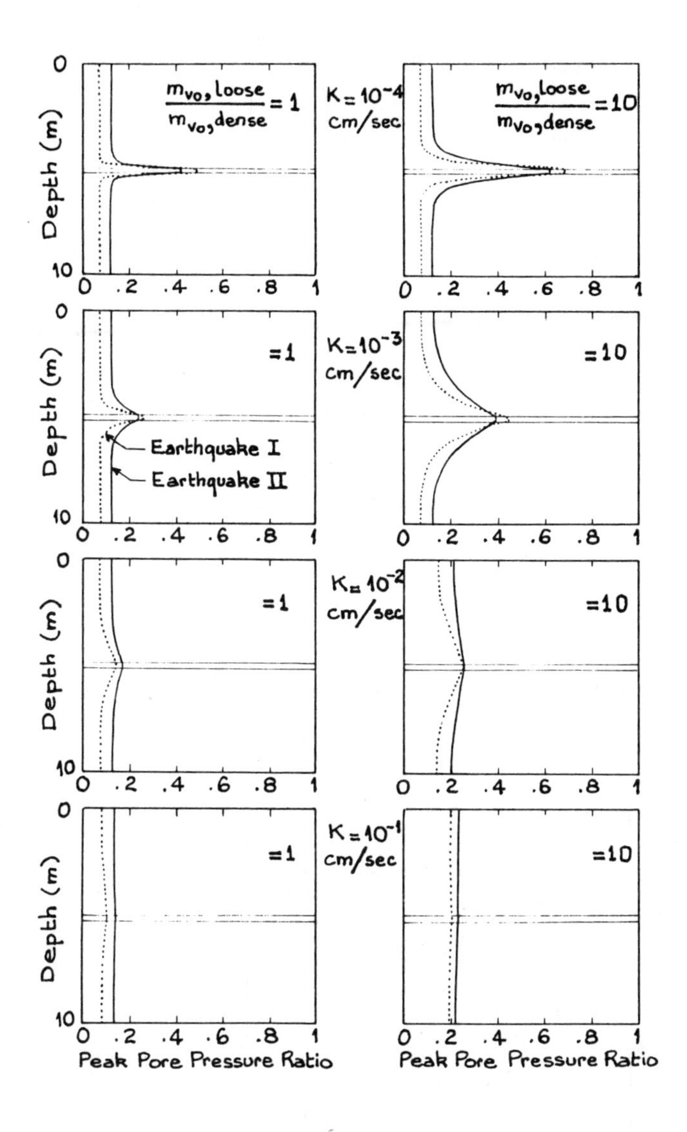

Figure 5 Peak Excess Pore Pressure Redistribution

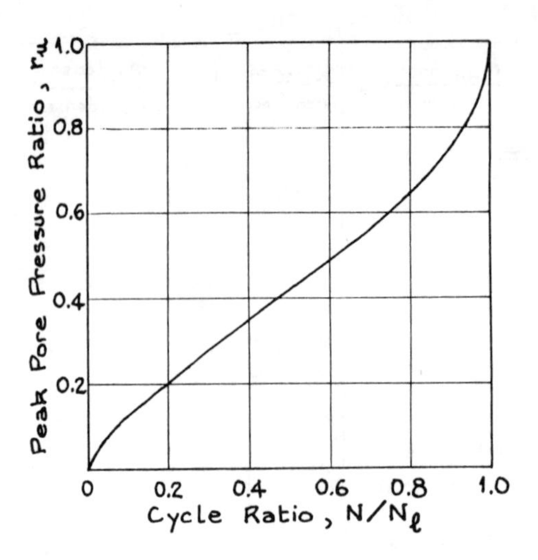

Figure 6 Theoretical Relationship between Undrained Pore Pressure Buildup and Number of Loading Cycles

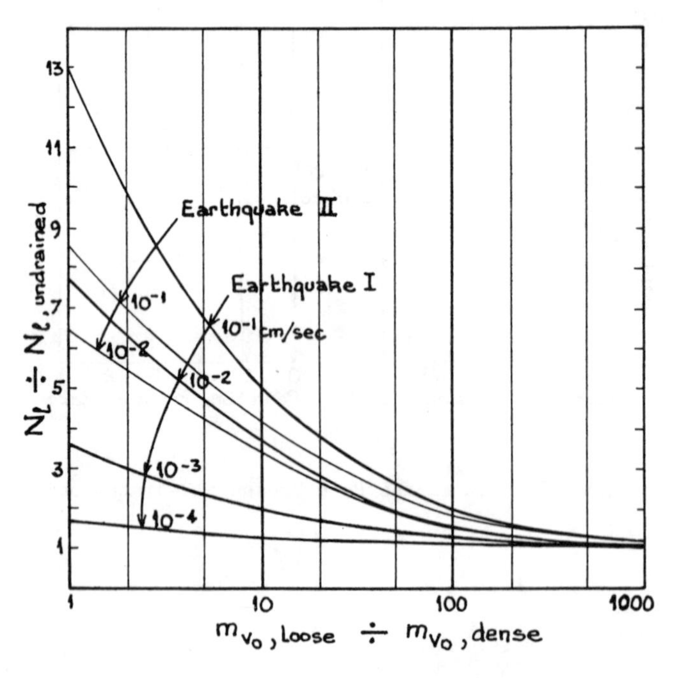

Figure 7 Effect of Stiffness Ratio on Liquefaction Potential of Loose Sand

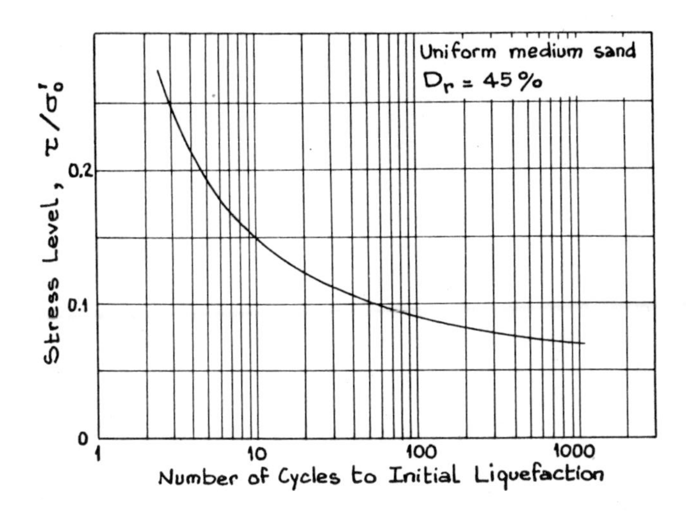

Figure 8 Typical Liquefaction Curve for Uniform Loose Sand

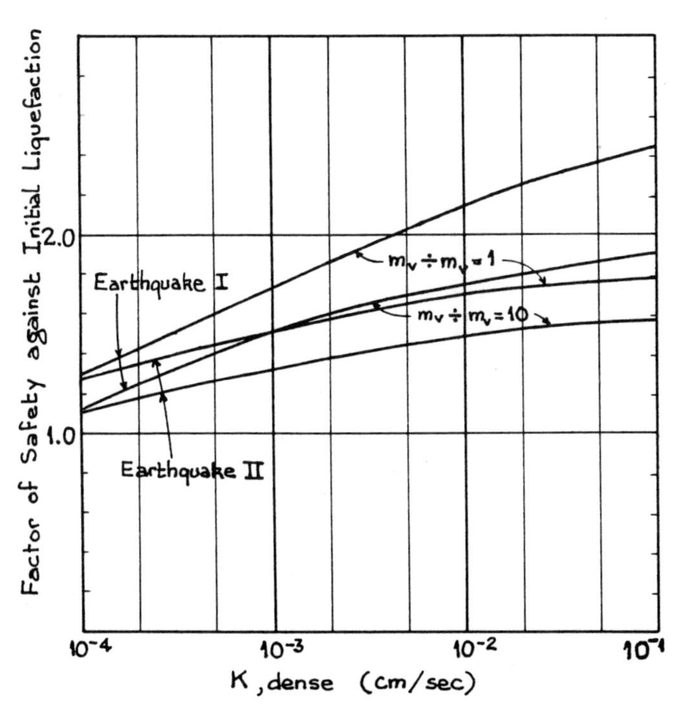

Figure 9 Boundary Effects on the Factor of Safety against Liquefaction of a Layer of Loose Sand

Soil Dynamics & Earthquake Engineering Conference / Southampton / 1982.07.13-15

Geological Materials Model Fits by Parameter Identification Techniques

S.F.MASRI & G.A.BEKEY
University of Southern California, Los Angelos, USA

F.B.SAFFORD
Agbabian Associates, El Segundo, CA, USA

R.E.WALKER
Army Corps of Engineers, Vicksburg, MS, USA

INTRODUCTION

Computation of wave propagation and groundshock caused by explosive sources and earthquakes has become increasingly important in recent years. The need for high accuracy from such computations has led to the development of mathematical models which represent the behavior of soil materials more realistically than the approximate models used in the past. One of the major problems of complex and realistic models, however, is the fact that they contain many parameters which are difficult or impossible to measure directly. Hence, the parameter values must be estimated from field measurements of the significant variables in conjunction with computer implementation of the model equations. A prescribed series of laboratory and field measurements are made and used as inputs to a computer model. Then, identification methods are used to adjust the model parameters such that the computed groundshock effects result in response values which approximate the measured values to within appropriate tolerances.

The general mathematical models of material properties specify the mathematical equations of state for earth materials. An equation of state for a certain material consists of a set of rules that interrelates the constitutive and/or thermodynamic properties of the material such as stress, strain, density, pressure, specific internal energy. The equations also express the stress state in terms of those material properties and previous history of loading.

Realistic mathematical models of material properties are invariably complicated and depend upon numerous parameters. This complication is due to the behavior of earth materials which is highly nonlinear and to permit the equations of state to apply to a wide range of materials. However, such representations are convenient for computer code applications since calculations for different media can be made by simply changing parameters through input data.

SOIL MODELS

The following are among the present soil models that are being widely used in research studies involving ground shock computations and earthquake engineering:

- The cap model, a continuum model based on the classical incremental theory of plasticity (DiMaggio and Sandler, 1971).
- The multisurface plasticity model that simulates hysteretic effects (Ghaboussi and Karshenas, 1977; Prevost, 1978; Baladi and Rohani, 1979).
- The endochronic model, an alternative to plasticity models, that uses intrinsic time as a basis for measuring the memory of past deformation history (Valanis and Read, 1978).

Other models are suggested by Valera et al. (1978) and Isenberg, Vaughn, and Sandler (1978).

All the above listed models contain material constants. The constants are determined by trying to match as closely as possible the results predicted by theory based on these models with experimental results from uniaxial strain and triaxial compression tests.

Since real earth materials sometimes exhibit instability and often do not appear to respond in a unique way, no practical model can represent such a material in full detail. The best that can be accomplished is to employ a consistent theory which (1) is simple enough to be applicable with available computers, (2) does not contain built-in mathematical uncertainties, and (3) fits available experimental data (normally with considerable scatter) within acceptable margins of error.

In specific applications, data is given in the form shown in Figures 1 and 2, where ε and σ are the vertical strain and stress, respectively, obtained from a uniaxial test, and where

J_1 = First stress invariant ($\Sigma\sigma_i$)

J_2' = Second reduced stress (stress deviation) invariant $= \frac{1}{6}\,\Sigma\,(\sigma_i - \sigma_j)^2$

σ_i, σ_j = Principal stress components

The stress invariants depend on ε, σ, the loading history, as well as numerous other soil parameters which are to be identified.

Once a candidate mathematical model is selected, the problem is then to determine the "best" set of parameters which will result in a loading/unloading curve (Fig. 1) and stress path (Fig. 2) that will simultaneously match both sets of experimental data such as those in Figures 1 and 2, as well as possible. Thus, the identification problem is reduced to a search in parameter space which will yield the extremum (minimum) value of a suitable criterion function which measures the goodness of fit.

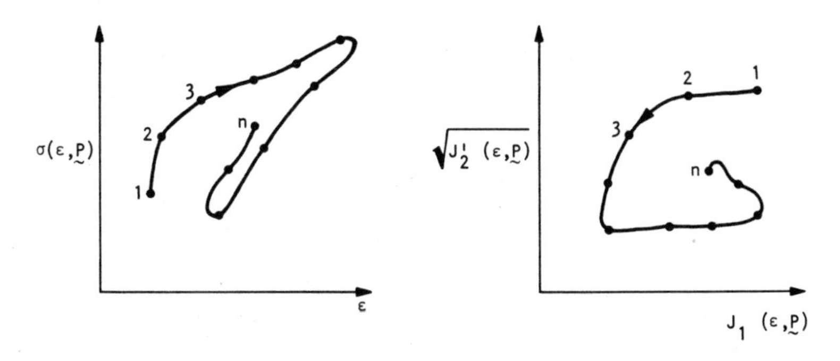

FIGURE 1. REPRESENTATIVE STRESS-STRAIN CURVE

FIGURE 2. REPRESENTATIVE STRESS PATH

<u>Soil cap model</u>

The cap model used to represent soils in this study is described in Weidlinger Associates (1974). This model employs a yield surface illustrated in Figure 3, which combines both strain hardening and ideal plasticity. The ideally plastic modified Drucker-Prager criterion represents the ultimate shear strength of the material, and it is associated with fracture or sustained plastic flow in laboratory experiments. The form of the yield criterion is

$$f_1 = (J_1, J_2') = 0 \tag{1}$$

As the plastic volumetric strain increases or decreases, the strain-hardening cap expands or contracts, respectively. The cap is denoted by

$$f_2 = (J_1, J_2', \varepsilon^P) = 0 \tag{2}$$

where ε^P is plastic strain.

It is assumed that elastic behavior is governed by

$$d\sigma_{ij} = Kd\varepsilon_{kk}\delta_{ij} + 2G\,(d\varepsilon_{ij} - \frac{1}{3}d\,\varepsilon_{kk}\delta_{ij}) \qquad (3)$$

where

$d\sigma_{ij}$ = Stress increment tensor

$d\varepsilon_{ij}$ = Strain increment tensor

$d\varepsilon_{kk}$ = Linear dilatation ($= d\varepsilon_{11} + d\varepsilon_{22} + d\varepsilon_{33}$)

δ_{ij} = Kronecker delta (= 1 if i = j; = 0, if i ≠ j)

K = Bulk modulus which is a function of pressure

G = Constant shear modulus

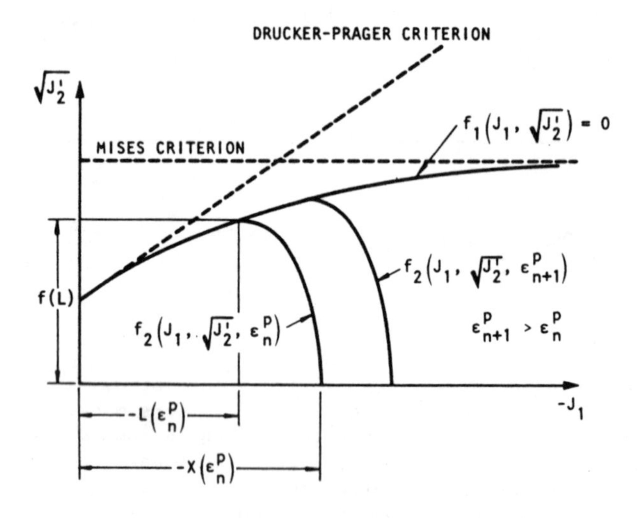

FIGURE 3. MATHEMATICAL MODEL HAVING A FIXED FRACTURE OR FAILURE CRITERIA (f_1) AND A MOVEABLE OR STRAIN-HARDENING CAP (f_2) (Isenberg, 1972)

The basic cap model used in this study consists of the following parts: (1) a variable bulk modulus, (2) a constant shear modulus, (3) a fracture surface, and (4) a cap. The cap model is defined by the following equations and an associated flow rule.

Bulk and shear modulus

$$K = B_1\{1 - B_2\exp(B_3J_1) - B_4\exp(B_5J_1)\} \tag{4}$$

$$G = \text{Constant}$$

where B_1 through B_5 = empirical coefficients

Fracture surface (f_1)

$$f_1 = 0 = \sqrt{J_2'} - [Y_1 - Y_2\exp(Y_3J_1)] \tag{5}$$

Cap (f_2)

$$f_2 = 0 = (L - J_1)^2 + Y_4^2\,J_2' - (L - x)^2 \tag{6}$$

$$X = \frac{1}{Y_5}\ln[(I_1^{\,P} + Y_7)/Y_6 + 1] \tag{7}$$

$$L + Y_4[Y_2\exp(Y_3L) - Y_1] - x = 0 \tag{8}$$

$$I_1^{\,P} = \varepsilon_1^P + \varepsilon_2^P + \varepsilon_3^P \tag{9}$$

where

$I_1^{\,P}$ = First invariant of plastic strain = $\sum_{i=1}^{3} \varepsilon_1^{\,P}$

L = Distance to the center of the elliptical cap

X = Intersection of elliptical cap with J_1 axis

Y_1 through Y_7 = Empirical coefficients

In the application of this model to soils (Isenberg, 1972) the cap movement is controlled by the change of plastic volumetric strain; hence, strain hardening can be reversed.

The computer implementation of this model results in a code which, when supplied with a strain time history, will generate the corresponding state of stress, and subsequently the stress invariants, at every point in the stress path.

In view of the complexity of the aforementioned model, its computer implementation has required a substantial effort and resulted in a large computer code that performs a significant number of computations for every incremental change in the strain level in order to determine the corresponding state of stress (Agbabian Associates, 1972).

PARAMETER IDENTIFICATION TECHNIQUES

Background

Mathematical models of nonlinear systems containing large numbers of unknown parameters are only of limited usefulness since the estimation of numerical values for these parameters is so difficult. In practice, it is customary to linearize the models, so that well understood linear methods can be applied to the identification problem. Alternatively, the number of parameters can be drastically limited.

However, there are physical problems (such as soil models) with intrinsic nonlinearities which cannot be linearized without destroying their usefulness. In such cases, the use of random search techniques has proven useful because such methods are capable (at least in principle) of locating all the local minima within a constrained region of a parameter space.

Models of soils for use in soil dynamics and earthquake engineering computations are highly nonlinear and contain many unknown parameters. As many as fifty significant unknown parameter values are needed to obtain a model which yields good agreement with experimental data. Such model fitting has been done in the past by laborious manual trial and error methods. This paper describes the development and use of an adaptive random search algorithm to achieve the parameter identification.

The basic random search algorithm

Strictly speaking, random search refers to techniques in which successive trial values of the parameter vector $\underset{\sim}{\alpha}$ are selected at random in the m-dimensional parameter space, the criterion function $J(\underset{\sim}{\alpha})$ is evaluated, and any parameter set yielding an improvement in the goodness of fit measure is retained. In effect, this technique is one of searching for a needle in an m-dimensional haystack. Each trial parameter vector is independent from the previous one and hence no assumptions are made concerning the topology of the criterion surface. It is evident that such a "brute force" method has a major virtue, namely that it is unaffected by the presence of multiple minima. It converges with probability 1 to the global minimum--however, the convergence time may be astronomical with large numbers of parameters, since each parameter value trial requires a new evaluation of $J(\underset{\sim}{\alpha})$.

The basic algorithm for minimization of a criterion function $J(\underset{\sim}{\alpha})$, where $\underset{\sim}{\alpha} = (\alpha_1, \alpha_2, \ldots, \alpha_m)^T$ is a vector of unknown parameters, proceeds as follows:

a. An initial parameter value $\underset{\sim}{\alpha}^o$ is estimated and $J(\underset{\sim}{\alpha}^o)$ is evaluated.

b. Trial points $\underset{\sim}{\alpha}^i \in \Omega_\alpha$, where Ω_α is the given permissible region in the m-dimensional parameter space, are selected from an appropriate probability density function defined over Ω_α.

c. A successful point $\underset{\sim}{\alpha}^{i+1}$ is one for which $J(\underset{\sim}{\alpha}^{i+1}) < J(\underset{\sim}{\alpha}^i)$

The sequence $J(\underset{\sim}{\alpha}^i)$ thus converges to the global minimum in Ω_α. Rather than using "pure random search," most algorithms are based on a "random creep" procedure in which exploratory steps are confined to a hypersphere centered about the latest successful point $\underset{\sim}{\alpha}^i$. However, convergence by such procedures may be extremely slow, since no allowance is made for variations in the nature of the criterion function surface as the search progresses towards a minimum.

Several procedures have been tried in the past to circumvent slow convergence. However, such schemes are highly directional and lose some of the flexibility of the random search procedure, as in the case of narrow "canyons" or "ridges" in the criterion function surface. Rastrigin (1963) has compared a random search where each step is random in direction but fixed in length (fixed-step-size random search), with a fixed-step-size gradient search, and showed that under certain conditions the random search is superior. It is intuitively evident, however, that one would obtain even better performance if the step size of the random search procedure were optimized at each step of the iteration. If the steps are too small, the average improvement per step will also be small and convergence time will be lengthened. If the steps are too large, they may overshoot the minimum; and the probability of improvement will again be too small. Hence, some method of adapting the step size to the local behavior of $J(\underset{\sim}{\alpha}^i)$ is indicated.

Schumer and Steiglitz (1968) have tested one approach to adaptive step-size random search. They used a fixed-step-size algorithm, where the trial vector $\Delta\alpha^i$ is of length ℓ and distributed uniformly over the hypersphere of radius ℓ and an incremental step of size $\ell(1 + a)$, where $0 < a < 1$ are taken in the same direction. The step size that produces the larger improvement is used as the nominal length ℓ for the next iteration. If no improvement occurs, ℓ is incrementally reduced. Thus, the algorithm is capable of adapting its step size as a function of the search.

<u>Adaptive random search algorithm</u>

The algorithm described here is another approach to the determination of the optimal step size. Rather than a fixed-length step, steps used are random in both length and direction. Hence the adaptation described below is based on the selection of the optimal variance of the step-size distribution as the search progresses. Large variances are desirable in the early, exploratory portions of a search. However, in the vicinity of a local optimum, a smaller value of σ will decrease the probability of overshoot.

The algorithm for the adaptive random search consists of alternating sequences of a global random search with a fixed value for the step size variance followed by searches for the locally optimal σ. At selected iteration intervals, the wide-range search is reintroduced to prevent convergence to local minima. The steps in the adaptive search are:

a. Select a starting estimate of the normalized parameter vector $\underset{\sim}{\alpha}^0 = (\alpha_1^0, \alpha_2^0, \ldots, \alpha_m^0)$.

b. Let $\underset{\sim}{s}$ be a sequence of standard deviations, $\underset{\sim}{s} = \{s_1, s_2, \ldots, s_K\}$, selected to cover as wide a range as desired.

c. Set σ, the standard deviation of the random number generator $N(\sigma)$, to s_1.

d. Starting with $\underset{\sim}{\alpha} = \underset{\sim}{\alpha}^0$ and $\sigma = s_1$, perform N iterations of the global random search; store the values of $\underset{\sim}{\alpha} = \underset{\sim}{\alpha}^N$ and $J(\underset{\sim}{\alpha}^N)$.

e. Repeat steps c and d with $s_k = s_2, s_3, \ldots, s_K$ and store the results.

f. From the above, determine $k = k*$, i.e., the value of σ_k for which

$$J(\underset{\sim}{\alpha}^N, \sigma_{k*}) = \min_k J(\underset{\sim}{\alpha}^N, \sigma_k)$$

g. Starting with $\underset{\sim}{\alpha}^i$ equal to the parameter vector obtained with $\sigma = s_{k*}$, set the iteration counter to $i = N$ and proceed to perform Q iterations of global random search with $N(\sigma_{k*})$.

h. Repeat steps c to g at predetermined increments of the iteration counter.

i. The search will be terminated when an absolute limit on the iteration counter i is reached, or when the tolerance on $J(\underset{\sim}{\alpha})$ is satisfied.

Further details concerning this method are given in Masri, Bekey, and Safford (1980).

APPLICATION TO SOIL MODEL FITS

Given n discrete data points with coordinates

$$[\varepsilon_i, \sigma_i(\varepsilon_i, \underset{\sim}{P})] \quad \text{and} \quad \left[J_{1_i}(\varepsilon_i, \underset{\sim}{P}), \sqrt{J'_{2_i}(\varepsilon_i, \underset{\sim}{P})}\right]$$

of the type shown in Figures 1 and 2, where

ε_i = Value of independent variable (strain), $i = 1, 2, \ldots, n$

$\underset{\sim}{P}$ = Parameter vector whose m components correspond to the parameters to be identified

$\sigma_i(\varepsilon_i, \underset{\sim}{P})$ = Digitized values of stress, $i = 1, 2, \ldots, n$

$J_{1_i}(\varepsilon_i, \underset{\sim}{P})$ = Digitized values of stress invariant J_1

$J'_{2_i}(\varepsilon_i, \underset{\sim}{P})$ = Digitized values of stress invariant J'_2

The criterion function J used in the present example is a measure of the goodness of fit of the three state variables σ, J_1, and J'_2 to the corresponding model quantities:

$$J(\underset{\sim}{\alpha}) = \sqrt{W_1 S_1 + W_2 S_2 + W_3 S_3} \tag{10}$$

where

$$S_1 = \sum_{i=1}^{n} [\sigma(\varepsilon_i) - \hat{\sigma}(\varepsilon_i)]^2 \tag{11}$$

$$S_2 = \sum_{i=1}^{n} [J_1(\varepsilon_i) - \hat{J}_1(\varepsilon_1)]^2 \tag{12}$$

$$S_3 = \sum_{i=1}^{n} \left[\sqrt{J'_2(\varepsilon_i)} - \sqrt{\hat{J}'_2(\varepsilon_i)}\right]^2 \tag{13}$$

W_1, W_2, W_3 = Error weighting functions

n = Number of data points

$\varepsilon, \sigma, J_1, J'_2$ represent measurement data and $\hat{\sigma}, \hat{J}_1, \hat{J}'_2$ are the corresponding quantities obtained by specifying the sequence of strain increments $\Delta \varepsilon_i = \varepsilon_{i+1} - \varepsilon_i$, $i = 1, 2, \ldots, n - 1$.

The values of all the m parameters which characterize the model are used in determining the state of stress in accordance with the predictions of our mathematical model.

The random search algorithm was used to fit soil data by performing the following steps:

1. The discretized values of the stress/strain curve of Figure 1 and the stress path curve of Figure 2 (furnished by Walker, 1976) were used as data.

2. A judicious choice was made for the soil cap model parameters which were subsequently used as initial parameter values to start the random search algorithm. These initial values are listed in Column 4 of Table 1. The computer code parameters appearing in Column 2 of Table 1 correspond to the parameters appearing in Equations 4 to 9.

3. Using the initial values given in Table 1 in conjunction with the computer code for the cap model, the "initial solution" that is listed in Table 2 was obtained. Figure 4 shows a comparison of this "initial solution" and the experimental data. Note that this "initial solution" was obtained by merely substituting the parameter values given in Table 1 into the material package subprograms. The optimization section of the parameter identification code was not exercised in this step.

4. Using equal error weighting functions $W_1 = W_2 = W_3 = 1$ in the criterion function defined by Equation 10, the deviation error between the "initial solution" and the given data was 0.272.

5. Again using the "initial values" listed in Table 1 as a starting estimate to initiate the parameter identification code and then operating the code for a few hundred iterations, an optimum set of parameters was found. These parameters are listed in Column 5 of Table 1. The optimized solution is shown in Figure 5 and its corresponding deviation error is 0.070 ($\cong$ 1/4 of the starting error).

DISCUSSION

As with all parameter identification problems, no attempt was made to identify all 36 of the model parameters appearing in Table 1. Many of these values could be obtained from the literature to the desired accuracy. The use of externally determined parameters, in part, is a critical consideration for identification methods in arriving at values that reflect

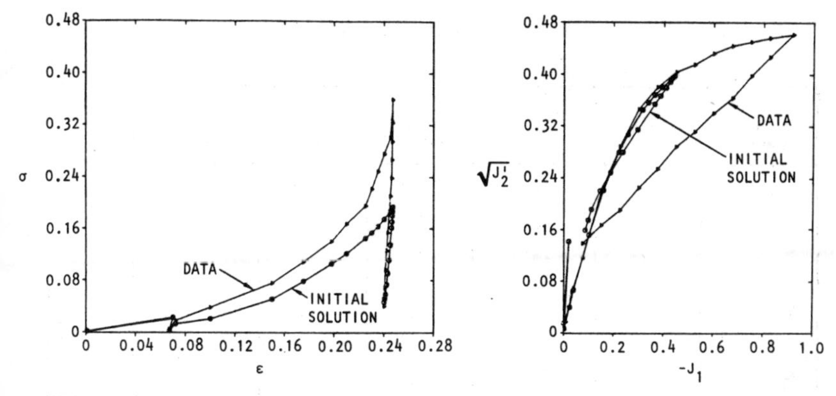

FIGURE 4. COMPARISON OF EXPERIMENTAL DATA AND INITIAL SOLUTION

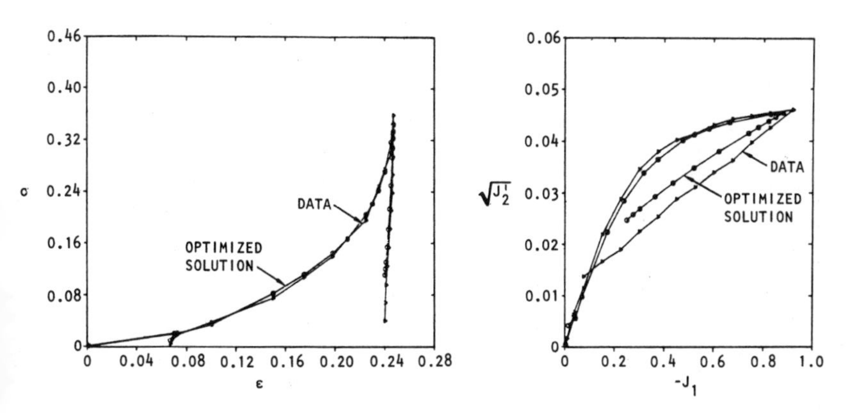

FIGURE 5. COMPARISON OF EXPERIMENTAL DATA AND OPTIMIZED SOLUTION

closely the appropriate physical quantities. This is to say of 36 parameters, 19 parameters chosen externally force the remaining 17 parameters to values consistent with the cap model and reasonable physical reality. Only 17 parameters (as indicated in Column 6 of Table 1) were optimized by the random search program. It is important to note that the selection of parameters to be identified is an important part of many engineering problems. It is important to verify (by sensitivity analysis) that errors in the parameters which were not identified would not cause unacceptable variations of the dependent varibles. Such a study was performed in this case.

The guidelines suggested in (Masri, Bekey, and Safford, 1980), were used to select values for the random search, such as the standard deviations of the steps which ranged from 10^{-3} to 10), the bounds on the system parameters, the number of searches to be performed with each standard deviation, and so forth. It

TABLE 1. PARAMETER VALUES

Cap Model Parameters	Computer Code Parameter	Coefficient Index C (1)	Initial Values	Optimized Values	Parameter C_i Optimized, i	% Change
A	Y_1	C_{61}	0.047	0.047		
B	Y_3	C_{63}	4.5	4.5		
C	Y_2	C_{62}	0.0456	0.0456		
R_0	Y_{13}	C_{73}	2.0	2.24	73	12
R_1	Y_{14}	C_{74}	2.5	2.25	74	-10
R_2	Y_{15}	C_{75}	5.0	4.54	75	- 9
R_3	Y_{16}	C_{76}	0	0		
R_4	Y_{17}	C_{77}	0	0		
R_5	Y_{18}	C_{78}	0	0		
W	Y_4	C_{64}	0.237	0.237		
D	Y_5	C_{65}	4.96	4.54	65	- 8.5
α	Y_6	C_{66}	1.0	1.0		
D_1	Y_7	C_{67}	100.0	100.0		
D_2	Y_8	C_{68}	1000.0	1000.0		
W_1	Y_{10}	C_{70}	50.0	45.46	70	- 9.1
D_F	Y_{11}	C_{71}	31.0	31.0		
W_2	Y_{12}	C_{72}	-0.009	-0.009		
D_3	Y_9	C_{69}	0.5	0.5		
K_{EI}	B_1	C_{21}	123.2	139.2	21	13
$K_S \Leftrightarrow K_i$	B_9	C_{29}	7.050	7.050		
δ	B_2	C_{22}	-0.900	-0.958	22	6.4
β	B_3	C_{23}	1.25	1.36	23	8.8
K_1	B_4	C_{24}	1.0	0.92	24	- 8.0
K_2	B_5	C_{25}	1000.0	1078.0	25	7.8
K_3	B_6	C_{26}	450.0	418.	26	- 7.1
K_4	B_7	C_{27}	0.0015	0.0015		
K_5	B_8	C_{28}	0	0		
G_{EI}	G_1	C_1	24.0	21.8	1	- 9.2
$G_S \Leftrightarrow G_i$	G_9	C_9	0.968	1.057	9	9.2
γ	G_2	C_2	-0.9	-0.996	2	11.0
η	G_3	C_3	4.0	3.58	3	-10.5
G_1	G_4	C_4	0	0		
G_2	G_5	C_5	16.0	13.95	5	-12.8
G_3	G_6	C_6	5000.0	5510.0	6	10.2
G_4	G_7	C_7	0	0		
G_5	G_8	C_8	0	0		

TABLE 2. INITIAL SOLUTION

Data Point No.	Strain	Stress Data	Computed Stress Value	J_1 Data	Computed J_1 Value	$\sqrt{J_2'}$ Data	Computed $\sqrt{J_2'}$ Value
1	-1.000-03	-1.669-03	-1.609-03	-3.000-03	-2.444-03	5.774-04	6.876-04
2	-7.000-02	-2.097-02	-1.446-02	-3.890-02	-1.977-02	6.928-03	6.818-03
3	-6.700-02	-5.000-03	4.855-04	-9.090-03	3.673-03	1.732-03	6.400-04
4	-7.200-02	-1.828-02	-9.913-03	-3.290-02	-1.901-02	6.371-03	3.097-03
5	-1.000-01	-3.841-02	-1.690-02	-7.490-02	-3.133-02	1.155-02	5.592-03
6	-1.500-01	-7.552-02	-4.247-02	-1.500-01	-8.321-02	2.194-02	1.276-02
7	-1.750-01	-1.080-01	-6.817-02	-2.249-01	-1.367-01	2.887-02	1.958-02
8	-1.980-01	-1.400-01	-9.443-02	-3.000-01	-1.943-01	3.464-02	2.569-02
9	-2.100-01	-1.680-01	-1.090-01	-3.749-01	-2.281-01	3.811-02	2.853-02
10	-2.250-01	-1.960-01	-1.301-01	-4.500-01	-2.784-01	4.041-02	3.232-02
11	-2.300-01	-2.220-01	-1.378-01	-5.249-01	-2.977-01	4.157-02	3.344-02
12	-2.350-01	-2.490-01	-1.463-01	-6.000-01	-3.189-01	4.328-02	3.463-02
13	-2.400-01	-2.760-01	-1.555-01	-6.749-01	-3.424-01	4.444-02	3.583-02
14	-2.450-01	-3.010-01	-1.657-01	-7.500-01	-3.688-01	4.503-02	3.703-02
15	-2.460-01	-3.270-01	-1.680-01	-8.249-01	-3.746-01	4.563-02	3.732-02
16	-2.470-01	-3.596-01	-1.702-01	-9.180-01	-3.805-01	4.619-02	3.756-02
17	-2.468-01	-3.240-01	-1.642-01	-8.249-01	-3.653-01	4.272-02	3.669-02
18	-2.466-01	-2.950-01	-1.583-01	-7.500-01	-3.507-01	3.982-02	3.585-02
19	-2.463-01	-2.670-01	-1.498-01	-6.749-01	-3.296-01	3.635-02	3.461-02
20	-2.460-01	-2.389-01	-1.417-01	-6.000-01	-3.094-01	3.404-02	3.341-02
21	-2.450-01	-2.110-01	-1.173-01	-5.249-01	-2.489-01	3.118-02	2.971-02
22	-2.440-01	-1.830-01	-9.622-02	-4.500-01	-1.972-01	2.887-02	2.639-02
23	-2.430-01	-1.540-01	-7.811-02	-3.749-01	-1.532-01	2.540-02	2.342-02
24	-2.420-01	-1.260-01	-6.286-02	-3.000-01	-1.185-01	2.250-02	2.024-02
25	-2.410-01	-9.683-02	-5.022-02	-2.249-01	-9.218-02	1.903-02	1.688-02
26	-2.405-01	-6.910-02	-4.466-02	-1.500-01	-8.092-02	1.672-02	1.532-02
27	-2.400-01	-4.083-02	-3.955-02	-7.490-02	-7.074-02	1.386-02	1.383-02

should be noted that the total number of iterations was selected arbitrarily (for computer time consideration) as 250. A larger number of total iterations would have further improved the fit of the computed solution to the data, shown in Figure 5b.

The approximately 75% reduction in the criterion function was achieved by changes in the values of the 17 parameters which ranged from -13% to +13%.

CONCLUSION

We have demonstrated the usefulness of adaptive random search to the identification of large numbers of unknown parameters in highly nonlinear contemporary models of geological materials. Once these parameters have been identified, the models are completely quantitied and can be used to predict soil response to a variety of forcing functions.

ACKNOWLEDGEMENT

This study was supported in part by the U.S. National Science Foundation.

REFERENCES

Agbabian Assoc. (AA). (1972) A Study of Different Methods of Coding a Capped Model, R-7134-2174. El Segundo, CA: AA, Jan.

Baladi, G.Y. and Rohani, B. (1979) "Elastic-Plastic Model for Saturated Sands," Jnl Geotech. Div., ASCE, 105:GT4, Apr.

DiMaggio, F.L. and Sandler, I. (1971) "Material Model for Granular Soils," Jnl Eng. Mech. Div., ASCE, 97:EM3, Jun, pp 935-950.

Ghaboussi, J.G. and Karshenas, M. (1977) "On the Finite Element Analysis of Certain Material Nonlinearities in Geomechanics," presented at Int. Conf. on Finite Elements in Nonlinear Solids and Structural Mechanics, Geilo, Norway, Aug.

Isenberg, J. (1972) Nuclear Geoplosics, A Sourcebook of Underground Phenomena and Effects of Nuclear Explosions, Part II: Mechanical Properties of Earth Materials, DNA 1285H2. Washington, DC: Defense Nuclear Agency, Nov.

Isenberg, J.; Vaughan, D.K.; and Sandler, I. (1978) Nonlinear Soil-Structure Interaction, EPRI Report NP-945. New York: Weidlinger Assoc., Dec.

Masri, S.F.; Bekey, G.A.; and Safford, F.B. (1980) "A Global Optimization Algorithm Using Adaptive Random Search," Applied Math. and Computation, 7, pp 353-375.

Prevost, J.H. (1978) "Anisotropic Undrained Stress-Strain Behavior of Clays," Jnl Geotech. Div. ASCE, 104:GT8, Aug.

Rastrigin, L.A. (1963) "The Convergence of the Random Search Method in the Extremal Control of a Many Parameter System," Automation and Remote Control, Vol. 24, pp 1337-1342.

Schumer, M.A. and Steiglitz, K. (1968) "Adaptive Step Size Random Search," IEEE Trans. on Automatic Control, Vol. AC-13, No. 3.

Valanis, K.C. and Read, H.E. (1978) "A Theory of Plasticity for Hysteretic Materials; I: Shear Response," Jnl Comp. and Struct., 8, pp 503-510.

Valera, J.E. et al. (1978) Study of Nonlinear Effects on One-Dimensional Earthquake Response, EPRI Report NP-865. Dames and Moore and SAI, Inc., Aug.

Walker, R. (1976) Private communication from U.S. Army Engineers, Waterways Experiment Station, Vicksburg, Mississippi.

Weidlinger Associates. (1974) A Generalized Cap Model for Geological Materials, DNA 3443T. Nov.

Soil Dynamics & Earthquake Engineering Conference / Southampton / 1982.07.13-15

Specifying Earthquake Motions in Engineering Design

E.L.KRINITZSKY & W.F.MARCUSON III
Waterways Experiment Station, Vicksburg, MS, USA

ABSTRACT

The selection of earthquake motions is dependent on the engineering analysis to be performed. A pseudostatic analysis uses a coefficient obtained from an appropriate map, while a dynamic analysis is made from accelerograms assigned to a site or from specified response spectra. Each type of analysis requires significantly different input motions. All selections of motions must allow for a paucity of representative strong motion records, especially near field motions from earthquakes of Magnitude 7 and greater, and an enormous spread in the available data. One must project data and bracket its spread in order to fill in the gaps and assure that there will be no surprises. Meanwhile, each locality may have differing special characteristics in its geology, seismic history, attenuation, recurrence, interpreted maximum events, etc.; thus, each part of the investigation must entail differences in decision levels. In some cases, a "least work" approach may be suitable, simply assuming the worst of several possibilities and testing for it. Since there are no standard procedures to follow, multiple approaches are useful. For example, peak motions at a site may be obtained from several methods that involve magnitude of earthquake, distance from source, and corresponding motions. Concurrently, motions may be assigned from other charts on the basis of earthquake intensity. There are various interpretations for duration, recurrence, effects of site conditions, etc. The various interpretations can be compared. Probabilities can be assigned, however, they present very serious problems when they are extrapolated beyond their data base.

Ultimately, what is reasonable is a subjective judgment and should depend, to an important extent, on the consequences

---

* Geologist, ** Chief, Geotechnical Laboratory, Corps of Engineers, Waterways Experiment Station, Vicksburg, MS 39180 USA.

of failure. Usually, a mean plus one standard deviation of possible variants in the data puts one in a good position. Where failure presents no hazard to life, lower numbers may be justified providing there are cost benefits and the risk is acceptable to the owner. Where there is a large hazard to life, i.e., a dam above an urbanized area, one may wish to use values that approximate the very worst. Each selection then has to be appropriate for its particular set of circumstances.

INTRODUCTION

What is done in selecting earthquake motions is first dependent on the engineering analysis to be performed. Essentially, there are two categories of analyses: pseudostatic and dynamic.

Pseudostatic Analysis

A pseudostatic analysis treats the earthquake loading as an inertial force that is applied statically to a structure, or a structural component, at the center of mass. The analysis determines the ability of the structure to sustain that additional load. The magnitude of this inertial force is determined as the product of the structural mass and a seismic coefficient. Ideally, the seismic coefficient is a ratio between the acceleration for an appropriate spectral content and response in a structure with that of the ground. Each coefficient has to be determined for a particular type of structure. Historically, seismic coefficients have been chosen by structural engineers on the basis of experience and judgment. Sometimes the coefficients are modified by factors that represent changes in local foundation conditions or differences in a structure.

To obtain a seismic coefficient, one uses a map that was created for the purpose. It is a geographic area, anything from a continent to a city, and is contoured or zoned to provide appropriate coefficients for any part of the area. Sometimes a coefficient is spoken of as if it were a value for acceleration. Properly it is a dimensionless unit. In no case can it be related readily to acceleration from a strong motion instrument.

Where a pseudostatic analysis is to be done, no geologic or seismologic investigation is needed. Only in exceptional cases, where there is a question of differential ground displacement along a fault at a site, is a geologic examination warranted.

Dynamic Analysis

A dynamic analysis tests a structure by applying a cyclical load approximating that of an earthquake. The shaking may be applied as a wave traveling vertically from bedrock through soil and into a structure. The objective is to test for possible structural damage. Examinations are made of such factors as failure in concrete from excessive peak stresses, the buildup of strain in soils beyond acceptable limits, and, in the case of saturated granular soils, the possibility of failure by liquefaction.

There are two general approaches to doing dynamic analyses. They determine the way earthquake motions are specified and used.

One approach begins with acceleration values which may be modified by factors for given structural components and are then entered directly into standard curves for smoothed response spectra. The accelerations are commonly taken from maps of geographic areas containing acceleration contours. The applicability and usefulness of such maps should be judged on an individual basis.

A notable set of maps was made by the Applied Technology Council (1978) for the United States. These maps present, (1) contours of horizontal acceleration in terms of 90 percent probability of not being exceeded in 50 years; (2) effective peak accelerations suitable for entering smoothed response spectra; and (3) an effective peak velocity-related acceleration coefficient.

Another approach begins by generating appropriate accelerograms for a site. Values are specified for peak horizontal acceleration, velocity, and displacement, and a duration of shaking is assigned. The motion must be identified as representing the ground surface, rock outcrop or bedrock surface.

The first category is nonsite specific and is best for analysis of concrete and steel structures. The second is site specific and is best for soils. The most widely used method in the nonsite specific category is that of the Nuclear Regulatory Guide 1.60. It was produced by combining the spectral components from a selected group of 37 earthquake records. Accelerograms, if needed, can be produced by fitting them to the smooth response spectra. Beginning with accelerograms in the alternative approach, smooth response spectra can be produced from them when needed.

SPECIFYING MOTIONS

Motions that are specified must be based on the following relationships:

a. The presence or absence of identifiable active faults capable of producing earthquakes.

b. Estimated maximum magnitudes for these earthquakes.

c. The boundaries for zones of seismic activity in which maximum earthquakes are assigned and floated throughout the zones.

d. The types of faulting that produce these earthquakes (including focal depths) and the character of surface displacement.

e. The peak motions (particle acceleration, velocity, displacement), as well as duration and predominant period that are associated with these events.

f. The attenuation of these motions from source to site.

g. The effects of site characteristics (soil, rock, topography, field conditions, etc.) on the resultant motions.

In this way ground motion parameters are selected that are appropriate for any given site. Such investigations are usually greatly involved and costly. Thus, the procedures are followed only where major engineering works, such as dams or nuclear power plants, are being considered or where safety-related aspects of a structure are critical for special reasons.

Geologic studies

Included in geological studies are interpretations of plate tectonics and satellite imagery. They are nearly always too grand to provide specifics that are of importance in evaluating a site. Thus, they can be treated briefly, with very little investment in time or money. Their benefit is that they enable one to give a fuller account of the setting. Airphoto interpretation and overflights are more meaningful. Their objective is to help locate faults and to judge whether the faults are active or inactive. Slemmons and Glass (1978) provide a useful summary of guidance for the utilization of imagery. Generally, no fault can be accepted from imagery or overflights until it is located on the ground or "ground-truthed."

A fault that is shown to be active must also be determined to be capable, that is capable of generating earthquakes.

The larger the capable fault, the greater the potential earthquake. Thus, relationships have been developed between dimensions of faults and magnitude of earthquakes. Dimensions include length of fault rupture, displacement during movement of the fault, whether the movement is on a primary fault or a branch fault or an accessory fault. Compilations have also included the types of faults, whether strike-slip, thrust or normal, and estimates of seismic moment. The latter may be calculated from the area of a fault plane involved in movement and the rigidity of the rock. The area may be evaluated from the spectral displacement amplitude of long period surface waves.

A useful summary relating faults to earthquake magnitude is provided by Slemmons (1977). Use of the data is ultimately a matter of responsible judgment on the part of the investigator.

Seismic history

Historic earthquakes should be tabulated and plotted geographically along with the geology. The area should be large enough to identify any geotectonic patterns that may be relevant to a site. The tabulation of historic earthquake events, though they are obtained from authoritative sources, should be examined critically. The epicentral intensity of earthquakes are sometimes overstated. Epicenters may also be shifted on the basis of reinterpreting the available data. If the site is important, the historic records should be examined and the interpretations should be checked. The records include newspaper accounts, diaries, early scientific and historical works, etc. A certain caution is in order: no earthquake should be related to a fault unless there is evidence that the fault actually moved during that earthquake.

The seismic history, in combination with the geology and the evidence of fault activity, if any, can be combined to identify earthquake zones. The earthquake zone is an inclusive area over which an earthquake of a determined maximum magnitude, the floating earthquake, may occur anywhere. It is a seismo-tectonic zone and it need not coincide with tectonic provinces.

Historic earthquake data are almost entirely in terms of intensity. Modern instrumental records provide data for magnitudes. There is frequently a need to relate one to another and, of course, there are uncertainties in doing so because of vagaries in the data for each category. However, some general relationship is possible. Figure 1 (Krinitzsky and Chang, 1977) provides such a guide.

Determining Peak Motions

There is no standard procedure for assigning the peak ground motions appropriate for a site. Whatever procedures that are used must consider certain basic problems:

a. The paucity of strong motion records for large earthquakes.

b. The limited data near causative faults.

c. The spread in the available data.

There are two principal approaches as follows:

1. Motions for earthquake intensity. Intensity is a reliable means of earthquake assessment. The intensity scales allow for differences in types of construction and resulting damage. Investigators generally come up with the same intensity for any given site.

Intensities can be attenuated from a source to a site by any of a number of intensity-attenuation charts. Figure 2, from Krinitzsky and Chang (1977) shows a comparison for intensity attenuations in Western United States and Eastern United States. Attenuation differences in these two cases are greatly pronounced.

Figure 3 shows the range in acceleration for Modified Mercalli (MM) intensities obtained from representative worldwide data. One may note also the deficiency of data for MM VIII and greater. It is obvious from the dispersion of the values for acceleration that curves based on the mean or average do not reflect the spread in the data. However, such curves have been widely used for design.

Krinitzsky and Chang (1977) presented charts that show an important difference in peak motions for the Near Field and Far Field. In the Near Field there is much focussing of waves from their source and there is reflection and refraction. There is a buildup of motions from resonance effects and there may be cancellation of motions. There is more high-frequency components of motion. Thus, there is a large spread in the peak motions for any given intensity. In the Far Field, the motions are less diverse, thus they are more orderly and predictable; their peaks are also more subdued. Figure 4 shows a comparison of Near Field and Far Field for acceleration and intensity (Krinitzsky and Chang, 1977). Krinitzsky and Chang have also developed pairs of curves for velocity and displacement. In Figure 4, a mean is shown, a mean plus one standard deviation and a limit of observed data. Also shown is the spread of the data in ten-percent increments. This spread is extended into the areas of higher intensity where there are no data. In this way suitable peak values may be obtained that are relatable to the spread in the data and are projected where there are no data. A mean plus one standard deviation puts one in a good position. If there is no hazard to life and there is a cost-risk benefit from a lesser design, lesser values can be taken. If a structure is on a major fault or is in an area with a high danger to life, such as a dam above an urban area, then it may be desirable to select the very worst motions.

2. Motions for magnitude and distance. Trifunac and Brady (1975) showed a comparison of acceleration versus distance curves for magnitude 6.5 earthquakes prepared from 10 published sources. They found a range of acceleration values of one to two orders of magnitude. Further, their lines do not show the real range in observed acceleration values but are comparisons of means or other groupings of various sorts. Though it is not stated, the distance functions are those of Western United States and might be unsuitable for other areas because of differences in attenuations. Thus, it can be seen that enormous errors are possible in the use of such curves.

The now classic work that established present-day levels of peak motions for earthquakes in relation to magnitude and distance is that of Page and others for the Trans-Alaska Pipeline System (1972). Their work had the benefit of the strong motion records derived from the San Fernando earthquake of February 9, 1971 in which accelerations greater than 1 g were recorded. A caution in using the tables of motions that they specified for various magnitudes of earthquakes are that they are for a frequency range of 1 to 9 Hz suitable for the pipeline. Their filtering of the Pacoima record to eliminate high-frequency components of motion removed about 25 percent of the range in acceleration. Also, their specified motions are for the worst-case situations where the pipeline is directly over active faults. Thus, their tabulated values need to be assessed carefully for use in any other situation.

Donovan (1973) showed acceleration values with distance for worldwide earthquakes and for the San Fernando earthquake. The spreads are shown by the mean, mean plus one standard deviation and mean plus two standard deviations. The total spread of the worldwide data is seen to be several orders of magnitude.

Algermissen and Perkins (1976) adjusted the Schnabel and Seed curves (1973) using attenuations for Central United States developed by Nuttli so that the Schnabel and Seed curves for acceleration could be used for any part of the United States. The curves, however, present problems in accommodating the range that exists in acceleration values and they do not provide guidance for specifying other critical components of motion such as velocity, displacement and duration.

Nuttli and Hermann (1978) provided curves for Central United States (Figure 5) that give acceleration and velocity for magnitude and distance from source. These are useful curves but a few cautions are in order:

a. The magnitude lines, or any equations for such lines, do not describe the dispersion of the data that is possible for each magnitude.

b. Close to the source, the motions may be subject to very large errors because of the great variability of motions in the Near Field and at a causative fault.

c. It is impossible to say whether the magnitude lines are means, or mean plus some deviation or maximum values because their relationships are not specified in these terms and are not consistent throughout the graph.

Under some circumstances, a precise knowledge of a capable fault, and its mechanics of rupture, can be used in two ways to specify motions: (1) by obtaining analogous strong motion records for scaling; and (2) by modifying peak motions to accord with the geometry of wave focussing of the fault rupture (see Bolt, 1981, and Singh, 1981).

Duration

Several investigators have proposed methods of measuring the duration of strong motion shaking. An important approach is an integration of the acceleration peaks with a duration that encompasses the inflection of the curve at the beginning of shaking and at the end. (See Arias, 1970; Trifunac and Brady, 1975; Vanmarcke, 1979).

Probably, the method most widely used in engineering is that of Bolt (1973), called bracketed duration. It is the inclusive time in which the acceleration level equals or exceeds the amplitude threshold of 0.05 g, or 0.10 g, according to the selection. A comparison is made in figure 6 of bracketed durations for soil and for rock by Krinitzsky and Chang (1977), those of Page, et al (1972) and Bolt (1973). A significant difference, roughly a hundred percent, is indicated between soil and rock.

Duration will always provide the greatest uncertainty in specifying earthquake motions. Very simply, a large earthquake may result from ruptures on several planes with these motions fused together in their effects at any one point so that they have the appearance of one earthquake rather than the sum of several. Thus, the Caucete, Argentina, earthquake of 1979 had a magnitude of 7.1 and a bracketed duration (>.05 g) of 48 seconds at a distance of 70 km. Caucete may be three earthquakes. With more data, even more extreme variances should be expected.

Soil versus Rock

Thus far, no differences were indicated between motions for soil sites and rock sites except for duration. Durations show pronounced differences, as stated. For other components of motion, the distinction between soil and rock presents very severe problems because of a large interdispersion of data.

In specifying the design motions at a site, it is customary to specify them as representing motion on level ground surfaces, rock outcrops or bedrock surfaces. This specification is obviously site specific. It is best to specify the motions as surface motions since almost all of our existing accelerograms were obtained from instruments located on the surface. Bedrock motions at depth can be obtained analytically if the soil profile is known.

Seed, et al (1976), provides a chart that is widely used for adjusting peak accelerations for ground conditions. If an acceleration is reduced for soil at a site, increases are probably desired for the velocity and duration.

Spectral Properties

The spectral composition of strong motion records are likely to be affected by site conditions and by distance from earthquake source. The appropriate spectral composition will be obtained by selecting records for scaling from earthquakes that are as analogous as possible to the specified type of faulting, distance from source, attenuation and site conditions. Synthetic accelerograms are likely to contain appropriate spectra but may be somewhat conservative.

Seed, et al (1976), presents a statistical analysis of response spectral shapes that show differences of soil and rock in Western United States. Chang and Krinitzsky (1977) present predominant period characteristics that are related to magnitude and distance together with local geological conditions.

Chang (1981) developed nonsite specific spectra based on geology of the sites and expressed as power density. He found close relationships among peak acceleration, duration and root mean square (rms) accelerations.

Anderson and Trifunac (1978) describe "uniform risk functionals" that have the same probability of exceedance at each frequency, when all the seismicity of a region is evaluated. The uniform risk functional then does not necessarily reflect the shaking from any single earthquake but will provide an inclusive coverage of the motions to be expected at a site.

Site specific earthquake ground motions can be developed without first obtaining peak values for acceleration, velocity, etc. The site response spectrum and the duration can be estimated as a first step. With the spectrum defined, historical accelerograms, scaled or unscaled or artificial ones can be selected according to how they match the prescribed site response. Response spectra generated for rock can be modified for soil at a site by performing a one- or two-dimensional wave propagation or finite element analysis using computer programs such as SHAKE, QUAD-4, LUSH, etc.

Scaling of Accelerograms

Chang (1978) provides a catalogue of earthquakes of Western United States arranged by fault type, magnitude, soil and rock, epicentral distance and peak acceleration, velocity, and displacement. Tabulations also list the duration, predominant period, and focal depth. This source, or the selection of representative earthquakes listed by Hays (1980) in his Table 16 to show appropriate earthquakes for soil and rock sites, may be used to select appropriate strong motion records either to use as they are or for scaling. Vanmarcke (1979) indicates that scaling must be restricted to 2X or less in order to avoid distortion of the spectral properties of the records. The time scale should not be altered unless there are definite spectral values that are desired. The time scale can be repeated or deleted in portions in order to obtain a desired time interval.

The peak accelerations for scaled accelerograms are not suitable as acceleration values for entering smoothed response spectra such as those of the Nuclear Regulatory Guide 1.60. Vanmarcke (1979) proposed a methodology for developing site-specific design response spectra based on appropriate accelerograms from past earthquakes. A rule of thumb for relating the maximum acceleration of a time history to entering an acceleration for a smoothed response spectrum is to reduce the maximum to 60 percent.

UNCERTAINTY IN SPECIFIED MOTIONS

One may accommodate the variance in data by bracketing its spread and selecting safe, encompassing parameters. One may project values into areas of a chart where there are no data. One may apply data from one geographic region to another. The objective is to utilize available data and projections of data in such a way that, should earthquakes occur, there will be no surprises. Thus, the spread in the data and the uncertainties in the extrapolation of data can be accommodated in a reasonably safe manner. Less certain are some of the problems associated with requirements in the methods of analysis and with the use of probabilities.

Method of Analysis

It was pointed out that there are two general approaches for engineering analysis, pseudostatic and dynamic. The dynamic analyses may be in either of two sorts: site specific or non-site specific. Each requires its own input motions. The motions that are specified for these differing analyses are not the same even for identical sites. A single site may require a coefficient of 0.1 for a pseudostatic analysis, 0.42 g for the acceleration peak in a time history and 0.25 g to enter smoothed response spectra of NRC Guide 1.60. A lesser or greater number does not mean that one is more or less conservative than the other. In fact, the reverse may be the case depending on other input parameters used. At present, the relation between motion requirements is a gray area in which satisfactory equivalents have not been entirely worked out. A guide for producing acceleration values appropriate to smoothed response spectra from accelerograms is provided by Vanmarcke (1979).

Deterministic versus Probabilistic Characterization

A deterministic characterization of peak motions is a statement of the appropriate values that may be used in an analysis for a site. The values are obtained from a combination of empirical knowledge, theoretical conceptualization and professional judgment.

Probabilistic characterization makes the assumption that no structure is absolutely safe, no motion is absolutely the maximum. Therefore, it is argued, a probabilistic analysis is needed to estimate the recurrence of whatever motions are assigned, and by projection, to estimate what larger motions there will be and how often such motions will occur. The motions may get to be very severe when they are projected over long periods of time, up to thousands of years. But, the probabilities experts reason that the recurrence of very severe events is extremely low. One-in-a-million is argued to be less risk than driving an automobile. Thus, the public will be expected to accept the risk and live with it.

That there is always something worse that can happen is an argument hard to deny. Can a meteor smash into a dam and demolish it? Yes, it can. Can it happen coincidentally when the reservoir is at its highest? Yes, that can happen too. What then is the probability? It does not follow that someone can put a number on the likelihood of such an event and that the number will have any meaning. The number would have such an enormous range of error that generating it would be indistinguishable from creating pure fiction. It is this lack of reality that occurs to varying degrees in estimates of probabilities in formulating seismic risk.

Recurrence is equated with a rate of increase for peak motions such as acceleration, etc. A difficulty is that the recurrence rate may not reflect peak motions in any satisfactory way. The motions invariably include a large element of interpretation: near field versus far field, spectral content--dependence on distances from source, effects of site conditions, etc. Also, data are very commonly utilized from areas other than those in which they are to be applied. Thus, the procedures in specifying motions are themselves arbitrary and deterministic. Probabilities are merely appended to them, often in a dubious association.

Probabilities have a usefulness if they are restricted to noncritical structures and for time intervals that are within the local data base. For insurance purposes it is desirable to estimate thresholds of damage on a probability basis. Yegian (1979) provides a useful review of methods for probabilistic analysis. A theoretical review of errors in probabilities, especially those that occur in projections of the data base, is given by Veneziano (1982). For a major structure, such as a dam or a nuclear power plant, where the design must be safe, probabilities may be inappropriate. Additionally, engineers are not trained or prepared to accept the fact that there is some finite number (probability) that a dam can fail and many people die. Thus, the concept may not be accepted by those who must assume responsibility.

CONCLUSIONS

There is no standard procedure. There are many decision levels. The decision levels differ from project to project. The values that are appropriate for design may further be influenced by the type of analysis to be made. It is best to review one's results and check them through several approaches and to allow for a consensus. The art has been developing rapidly. One should make allowance for new methods and for additions to the data base. The safest general approach is to base one's selection of ground motions on a large selection of observed data. The data should be appropriate for the situation. The trends in the data should be projected where the data is insufficient and the values should be bracketed or encompassed in such a way that there will be no surprises were an earthquake to occur. And the peak motions should be adjusted so that they are appropriate for the analyses in which they are used.

REFERENCES

Algermissen, S. T., and Perkins, D. M., (1976), A probabilistic estimate of maximum acceleration in rock in the contiguous United States: U.S. Geol. Survey Open-File Rept. 76-416, 45 p.

Anderson, John G. and Trifunac, M. D., (1978), Uniform risk functionals for characterization of strong earthquake ground motion. Bull. Seism. Soc. Am. Vol. 68, No. 1, pp. 205-218.

Applied Technology Council (1978), Tentative provisions for the development of seismic regulations for buildings, National Science Foundation, National Bureau of Standards, ATC 3-06, NBS SP 510, NSF 78-8, 514 pp.

Arias, A. (1970), A measure of earthquake intensity in Seismic Design of Nuclear Power Plants, R. J. Hanon, Editor, the Mass. Inst. Tech. Press.

Bolt, Bruce A., (1973), Duration of strong ground motion: World Conf. on Earthquake Eng., 5th Rome, Proc., v. 2, no. 292, 10 p.

Bolt, Bruce A., (1981), State-of-the-art for assessing earthquake hazards in the United States, Interpretation of strong motion records, Corps of Engineers, Waterways Experiment Station, Misc. Paper S-73-1, Rept. 17, 215 pp.

Chang, Frank K., (1978), State-of-the-art for assessing earthquake hazards in the United States, Catalogue of strong motion earthquake records: Vol. 1, Western United States, 1933-1971, Corps of Engineers, Waterways Experiment Station, Misc. Paper S-73-1, Rept. 9, 28 pp. and 2 appen.

Chang, Frank K., (1981), Site effects on power spectral densities and scaling factors, Corps of Engineers, Waterways Experiment Station, Misc. Paper GL-81-2, 57 pp, 8 tables.

Chang, Frank K., and E. L. Krinitzsky, (1977), State-of-the-art for assessing earthquake hazards in the United States, Duration, spectral content, and predominant period of strong motion earthquake records from Western United States, Corps of Engineers, Waterways Experiment Station, Misc. Paper S-73-1, Rept. 8, 58 pp 1 appen.

Donovan, N. C., (1973), A statistical evaluation of strong motion data including the February 9, 1971 San Fernando earthquake: World Conf. on Earthquake Engineering, 5th Rome, Proc., v. 2, paper 155, 10 p.

Hays, Walter W., (1980), Procedures for estimating earthquake ground motions, Geol. Surv. Prof. Paper 1114, Washington, DC, 77 pp.

Krinitzsky, E. L., and Chang, Frank K. (1977), State-of-the-art for assessing earthquake hazards in the United States, Specifying peak motions for design earthquakes, Corps of Engineers, Waterways Experiment Station, Misc. Paper S-73-1, Rept. 7, 34 pp.

Nuttli, O. W., and Herrmann, R. B., (1978), State-of-the-art for assessing earthquake hazards in the United States, Credible earthquakes for the Central United States, Corps of Engineers, Waterways Experiment Station, Misc. Paper S-73-1, Rept. 12, 99 pp. and 1 appen.

Page, R. A., Boore, D. M., Joyner, W. B., and Coulter, H. W., (1972), Ground motion values for use in the seismic design of the trans-Alaska pipeline system: U. S. Geol. Survey Circ. 672, 23 p.

Schnabel, P. B., and Seed, H. B., (1973), Accelerations in rock for earthquakes in the Western United States: Seismol. Soc. America Bull., v. 62, p. 501-516.

Seed, H. B., Murarka, R., Lysmer, John, and Idriss, I. M., (1976), Relationships of maximum acceleration, maximum velocity, distance from source, and local site conditions for moderately strong earthquakes: Seismol. Soc. America Bull., v. 66, p. 1323-1342.

Singh, J. P., (1981), The influence of seismic source directivity on strong ground motions, Ph.D. dissertation in Grad. Div., Univ. Calif., Berkeley, 183 pp.

Slemmons, David B., (1977), State-of-the-art for assessing earthquake hazards in the United States, Faults and earthquake magnitude, Corps of Engineers, Waterways Experiment Station, Misc. Paper S-73-1, Rept. 6, 129 pp., and 1 appen.

Slemmons, David B., and Charles E. Glass, (1978), State-of-the-art for assessing earthquake hazards in the United States, Imagery in earthquake analysis, Corps of Engineers, Waterways Experiment Station, Misc. Paper S-73-1, Rept. 11, 221 pp. and 2 appen.

Trifunac, M. D., and Brady, A. G., (1975), On the correlation of seismic intensity scales with peaks of recorded strong ground motion, Seismol. Soc. America Bull., v. 65, pp. 139-162.

Vanmarcke, Erik H., (1979), State-of-the-art for assessing earthquake hazards in the United States, Representation of earthquake ground motion: Scaled accelerograms and equivalent response spectra, Corps of Engineers, Waterways Experiment Station, Misc. Paper S-73-1, Rept. 14, 83 pp.

Veneziano, Daniele, (1982), State-of-the-art for assessing earthquake hazards in the United States, Errors in probabilistic seismic hazard analysis, Corps of Engineers, Waterways Experiment Station, Misc. Paper S-73-1, Rept. 18, 132 pp.

Yegian, M. K., (1979), State-of-the-art for assessing earthquake hazards in the United States, Probabilistic seismic hazard analysis, Corps of Engineers, Waterways Experiment Station, Misc. Paper S-73-1, Rept. 13, 130 pp., and 1 appen.

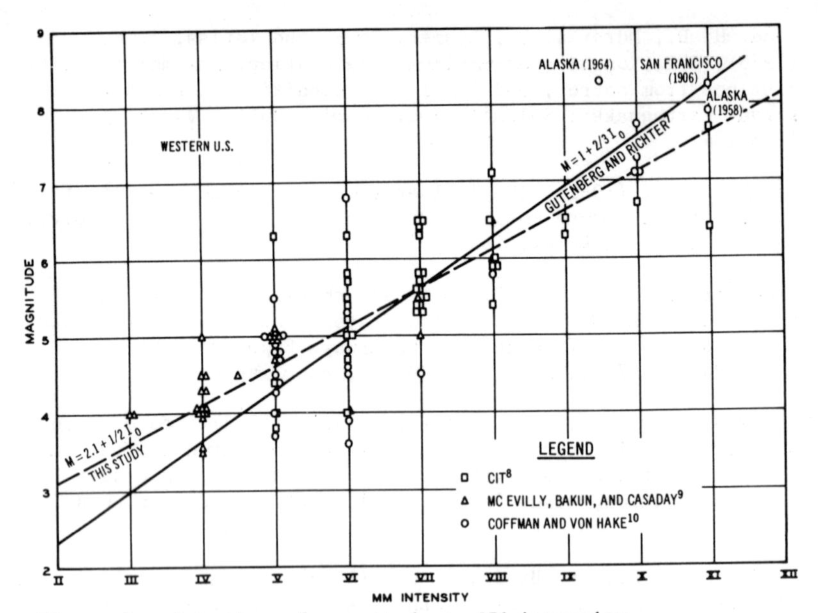

Figure 1. Relation of magnitude to MM intensity. Krinitzsky and Chang (1977)

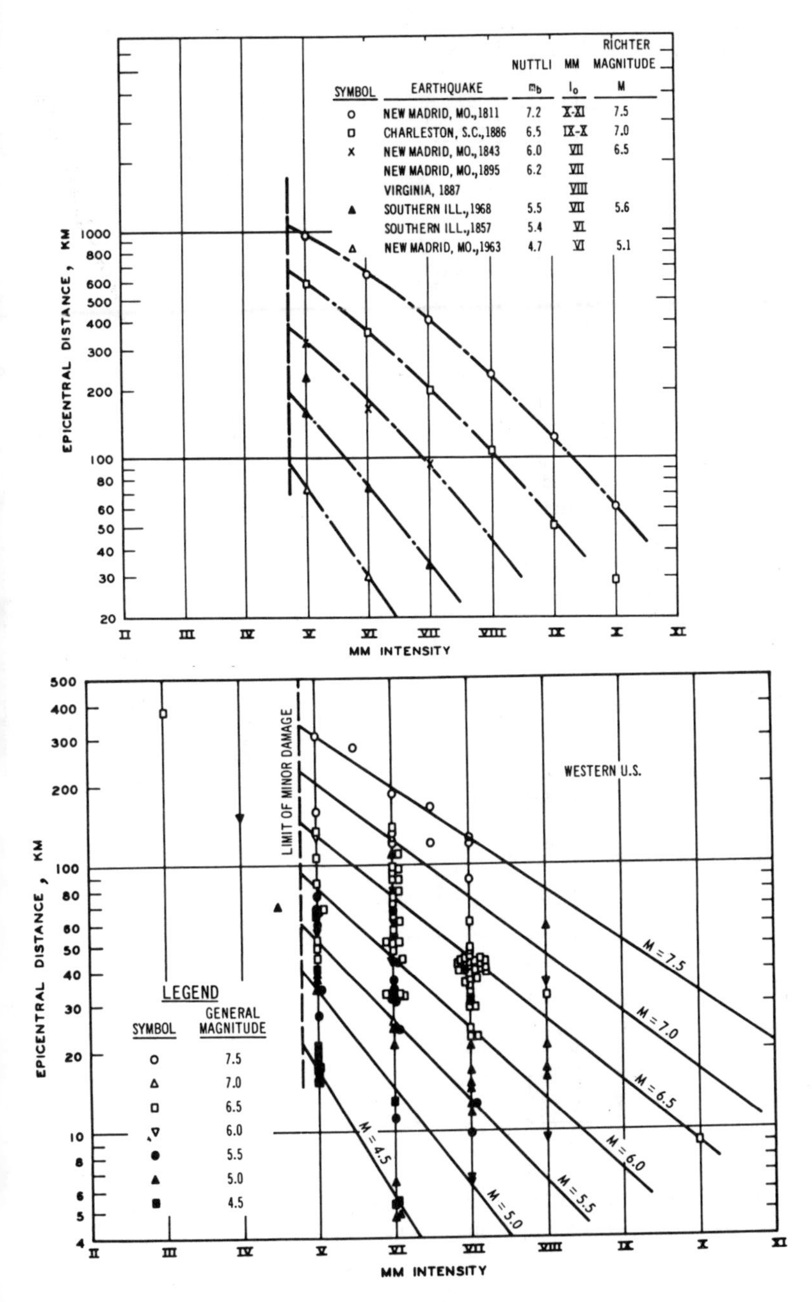

Figure 2. Intensity with distance.
Krinitzsky and Chang (1977).

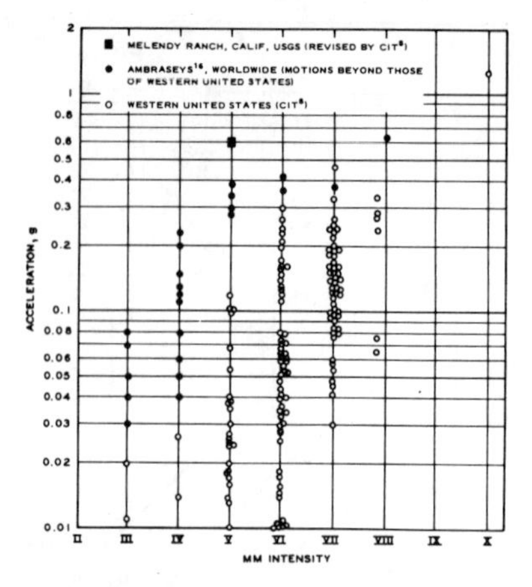

Figure 3. Worldwide range in acceleration values. Krinitzsky and Chang (1977)

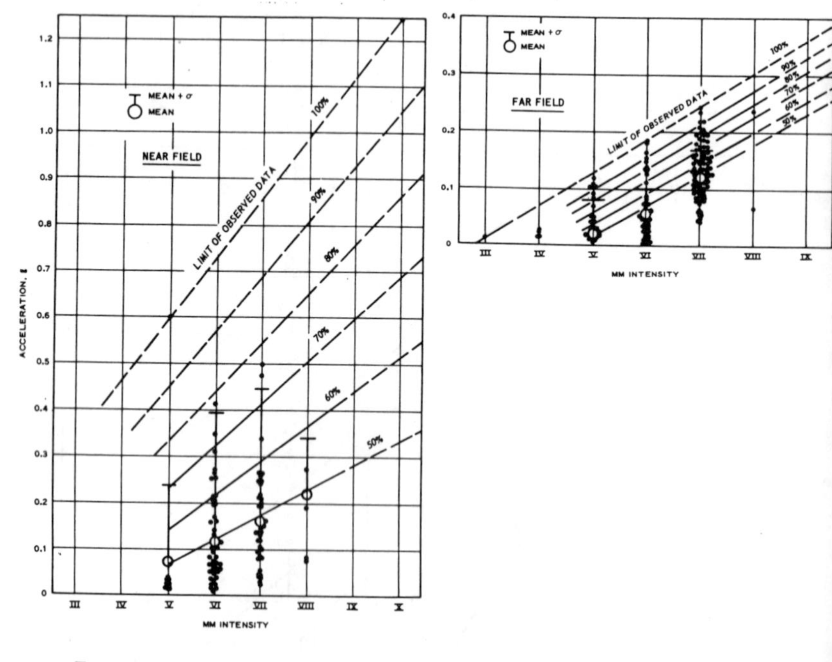

Figure 4. Near Field versus Far Field accelerations. Krinitzsky and Chang (1977)

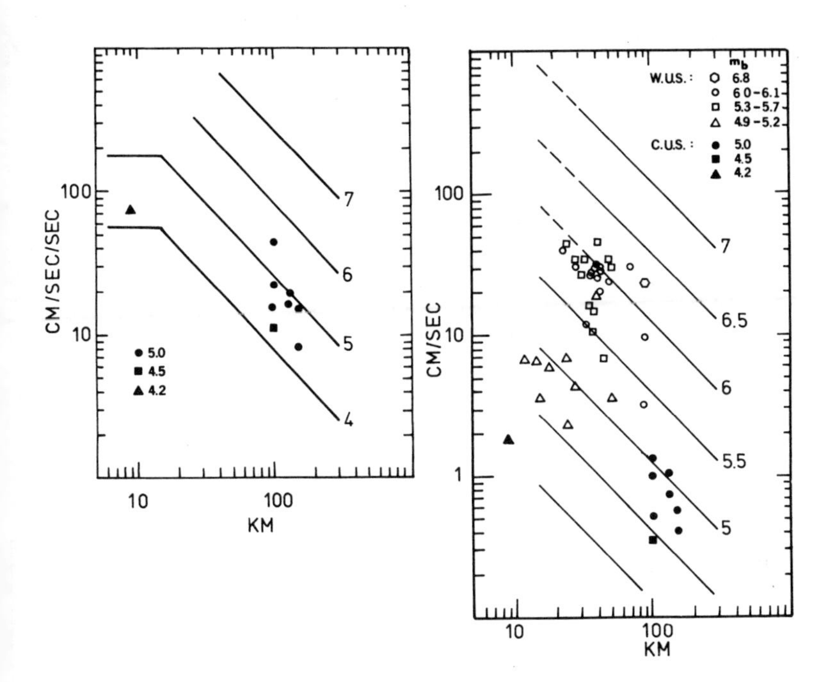

Figure 5. Acceleration and velocity for magnitude-distance, Central United States. Nuttli and Herrmann (1978)

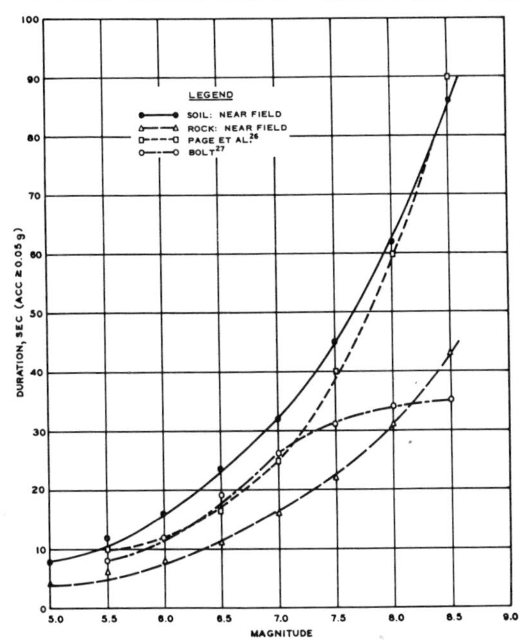

Figure 6. Interpretation of bracketed duration >.05g. Krinitzsky and Chang (1977)

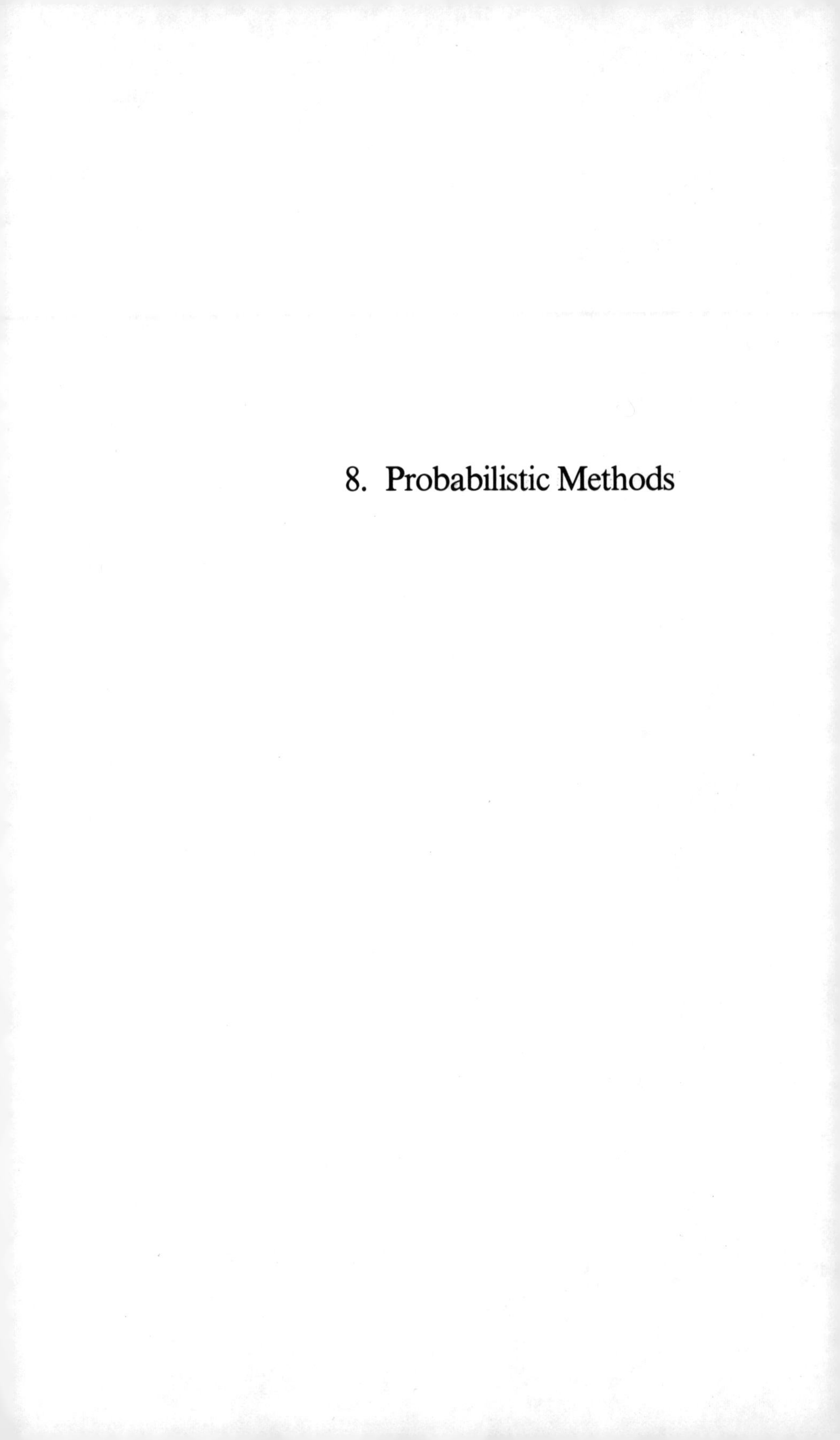

8. Probabilistic Methods

Soil Dynamics & Earthquake Engineering Conference / Southampton / 1982.07.13-15

Probabilistic Evaluation of Liquefaction in a 3-D Soil Deposit

ACHINTYA HALDAR & FRANK J.MILLER
Georgia Institute of Technology, Atlanta, USA

INTRODUCTION

Earthquake-induced liquefaction needs better understanding, since liquefaction can result in damage to property and the environment and can cause suffering and loss of human life. The devastating damage due to soil liquefaction in Anchorage, Alaska and Niigata, Japan in 1964 are but two recent reminders of the critical nature of such events. An enormous amount of research effort has been exerted into this problem since then. Experimental, analytical, and field investigations have led to several methods to evaluate the liquefaction potential of a soil deposit (Christian and Swiger (1975), Donovan (1971), Faccioli (1973), Fardis and Veneziano (1979), Finn et al. (1977), Haldar (1976, 1980, 1981), Haldar and Tang (1979, 1981), Haldar and Miller (1982), Kishida (1969), Liou et al. (1977), Vanmarcke (1970, 1977) Whitman (1971), Yegian and Whitman (1976). These studies also led to better understanding of the factors that influence liquefaction potential. Quite a few factors influence the liquefaction potential. Moreover, each factor influences the evaluation to a different degree, and there is considerable uncertainty in the estimation of each factor. This led to the development of probabilistic approaches in lieu of deterministic approaches to evaluate liquefaction potential. In view of the unpredictability of the earthquake loading, lack of information on the soil deposit, and the unavoidable errors in the modeling of the soil behavior and in the estimation of the design parameters, a systematic probabilistic model has definite advantages over the so-called deterministic methods (Christian and Swiger (1975), Fardis and Veneziano (1979), Haldar (1967, 1980, 1981), Yegian and Whitman (1976)).

The currently available probabilistic methods can be

broadly separated into two groups: Methods based on the case histories of liquefaction available in the literature (Christian and Swiger (1975), Yegian and Whitman (1976)) and methods relating laboratory experimental results and dynamic analysis of the deposit (Donovan (1971), Faccioli (1973), Fardis and Veneziano (1979), Haldar and Tang (1979)). A detailed discussion on this topic can be found in Reference 7. All the aforementioned methods will estimate the probability of liquefaction at a point in the deposit. Generally, the liquefaction risk is evaluated considering the weakest point in the deposit. In order to produce a noticeable amount of damage to a structure, a volume of sand has to undergo a considerable amount of strain. It is very important to identify this critical volume and to model the soil properties in this volume appropriately. Local variations as well as long-distance fluctuations of the soil properties in this liquefiable volume have to be modeled in three dimensions. These can only be modeled effectively using probability theory.

This study systematically identifies all the appropriate parameters affecting the liquefaction potential of a saturated soil deposit during an earthquake, and evaluates the uncertainties therein. The resulting statistical description is extended to include the three-dimensional physical nonhomogeneity of the soil properties in the critical volume. The parameters and their associated uncertainties are then combined into a probabilistic model to evaluate the risk of liquefaction of the critical volume of soil. The objective is to develop a simple but practical and efficient probabilistic method which will estimate the risk of liquefaction in three dimensions.

DETERMINISTIC LIQUEFACTION MODEL

The risk of liquefaction of a soil deposit can be estimated by comparing the in situ shear resistance of a soil element τ_R and and the average equivalent uniform shear stress of intensity τ_A that will act on the soil element for N_{eq} number of cycles during an earthquake. Liquefaction will occur when τ_A is greater is greater than τ_R. τ_A and τ_R can be estimated using a concept similar to that of Seed and Idriss (1971). Haldar and Miller (1982) discussed this concept in great detail.

τ_A can be estimated from the following relationship:

$$\tau_A = S_L \, r_d \, \gamma_s \, h \, \frac{a_{max}}{g} \tag{1}$$

in which S_L = stress ratio to convert the maximum applied shear stress to the uniform shear stress, τ_A, and can be considered as 0.75 (Haldar and Tang (1981)); r_d = the stress reduction coefficient to consider the flexibility of the soil column under consideration; γ_s = the saturated unit weight of the soil; g = the acceleration due to gravity; and a_{max} = the design maximum ground acceleration.

Haldar and Miller (1982) showed that the shear resistance, τ_R, mobilized by an element of soil during earthquake shaking can be modeled effectively by introducing the shear strength parameter R such that

$$R = \frac{\tau_R}{\sigma'_m D_r} \tag{2}$$

in which σ'_m = the average effective normal stress and D_r = the relative density. The parameter R was introduced by Haldar (1976, 1979). σ'_m can be estimated as:

$$\sigma'_m = \frac{\sigma'_1 + \sigma'_2 + \sigma'_3}{3} \tag{3}$$

in which σ'_1, σ'_2 , and σ'_3 are the effective stresses in three directions at a point in the deposit. The in situ value for R can be inferred from laboratory test results if the value for R measured in the laboratory is modified by a corrective factor, C_r, i.e.

$$R_{field} = C_r \cdot R_{lab} \tag{4}$$

Combining Equations 2 and 4, the following expression can be obtained:

$$\tau_R = C_r \cdot R \cdot \sigma'_m \cdot D_r \tag{5}$$

In Equation 5, R is estimated under laboratory conditions, as will be discussed later.

THREE-DIMENSIONAL PROBABILISTIC LIQUEFACTION MODEL

The liquefaction potential of a soil deposit can be evaluated if information on the various parameters in Equations 1 through 5 is available. However, considerable error can be incurred in estimating these parameters. All the parameters mentioned are uncertain to some degree and should therefore be modeled as random variables.

In addition, soil parameters typically exhibit local variations about their average values or about major trends (horizontally and vertically). Thus, an evaluation of the liquefaction potential of the deposit at the so-called "weakest point" may be misleading. As discussed earlier, a sufficient volume of sand has to undergo a considerable amount of strain in order to produce a noticeable amount of damage at the referenced location. This critical volume may contain pockets of very loose as well as very dense sand. Evaluation of the liquefaction potential of this volume of sand considering either the loose or dense pocket will obviously be incomplete. The soil property averaged over the volume would be more representative than the point estimate. Thus, τ_A and τ_R in Equations 1 and 2, respectively, need to be modeled in terms of spatial averages. Taking the spatial average of τ_R over a volume Δv yields:

$$\tau_{R_{\Delta v}} = \frac{1}{\Delta v} \int_{\Delta x} \int_{\Delta y} \int_{\Delta z} \tau_R \,(x,y,z)\, dx\, dy\, dz \qquad (6)$$

in which $\tau_{R_{\Delta v}}$ = the spatial average over the volume Δv, $\Delta v = \Delta x \cdot \Delta y \cdot \Delta z$, and Δx, Δy, Δz are the lengths of the soil volume in the x, y and z directions, respectively. However, $\tau_{R_{\Delta v}}$ is random. Using first-order approximate analysis and assuming a statistically homogeneous soil deposit, Haldar and Miller (1982) showed that the mean value of τ_R can be estimated as

$$\overline{\tau}_{R_{\Delta v}} = \mu_{\tau_R} = \mu_{C_r} \cdot \mu_R \cdot \mu_{\sigma'_m} \cdot \mu_{D_r} \qquad (7)$$

where μ represents the point mean. The three-dimensional characteristics of the relative density in a deposit are expected to be more important than the other parameters. Modeling only D_r in three dimensions (other parameters can be modeled similarly) and assuming statistical independence among parameters, Haldar and Miller (1982) estimated the variance of τ_{R_v} as:

$$\mathrm{Var}\,(\tau_{R_{\Delta v}}) = \mu^2_{\tau_R} \Big[\Omega^2_{N_{\tau_R}} + \Omega^2_{C_r} + \Omega^2_R + \Omega^2_{\sigma'_m} + \Gamma^2_{D_{r_x}} (\Delta x) \cdot \Gamma^2_{D_{r_y}} (\Delta y) \cdot \Gamma^2_{D_{r_z}} (\Delta z) \cdot \Omega^2_{D_r} \Big] \qquad (8)$$

in which Ω represents the point coefficient of variation (COV) and $\Omega_{N_{\tau_R}}$ is the modeling error. $\Gamma^2_{D_{r_x}}(\Delta x)$, $\Gamma^2_{D_{r_y}}(\Delta y)$, and $\Gamma^2_{D_{r_z}}(\Delta z)$ are called the variance functions of D_r in the x, y, and z directions, respectively. They describe the decay of the variance of the spatial average of D_r as the averaging dimensions increase (Vanmarcke (1970, 1977, 1977)).

For all practical purposes, the variance function in the x direction can be estimated as (Vanmarcke (1970, 1977)):

$$\Gamma^2_{D_{r_x}}(\Delta x) = \begin{cases} 1.0; & \Delta x \le \theta_{D_{r_x}} \\ \dfrac{\theta_{D_{r_x}}}{\Delta x}; & \Delta x > \theta_{D_{r_x}} \end{cases} \tag{9}$$

where $\theta_{D_{r_x}}$ is the scale of fluctuation in the x direction. It is a parameter which measures the distance in the x direction within which D_r shows relatively strong correlation from point to point. Intuitively, small values of $\theta_{D_{r_x}}$ imply rapid fluctuations about the average, while large values of $\theta_{D_{r_x}}$ suggest that a slowly varying component is superimposed on the average value. The variance functions in the y and z directions can be estimated similarly knowing $\theta_{D_{r_y}}$ and $\theta_{D_{r_z}}$.

In a similar manner, the spatial average of τ_A over the volume Δv is given by

$$\tau_{A_{\Delta v}} = \frac{1}{\Delta v} \int_{\Delta x} \int_{\Delta y} \int_{\Delta z} \tau_{A_{\Delta v}} (x,y,z)\ dx\ dy\ dz \tag{10}$$

The predictive model of τ_A is given by Equation 1. Though a_{max} and r_d are random variables, their spatial variability can be considered negligible. Directly, the spatial variability of γ_s will not be considered here; however, it is considered indirectly in the spatial variability of D_r. Thus, for a liquifiable volume Δv, the spatial mean and variance of $\tau_{A_{\Delta v}}$ becomes

$$\overline{\tau}_{A_{\Delta v}} = \mu_{\tau_A} = S_L\ \mu_{r_d}\ \mu_{\gamma_s} \cdot h\ \frac{\mu_{a_{max}}}{g} \tag{11}$$

$$\mathrm{Var}\ (\tau_{A_{\Delta v}}) = \mu^2_{\tau_A}\ \Omega^2_{\tau_A} = \mu^2_{\tau_A} \left[\Omega^2_{r_d} + \Omega^2_{\gamma_s} + \Omega^2_{a_{max}} \right] \tag{12}$$

The values of g, S_L and h are assumed to be known.

The probability that the soil volume Δv will liquefy is given by the probability of the event $\tau_{R_{\Delta v}} \le \tau_{A_{\Delta v}}$. Since the statistics of $\tau_{A_{\Delta v}}$ and $\tau_{R_{\Delta v}}$ cannot be adequately defined beyond the first two moments, for simplicity,lognormal distributions can be prescribed for $\tau_{A_{\Delta v}}$ and $\tau_{R_{\Delta v}}$ in estimating the probability of liquefaction. The validity of these probability estimates was studied by Haldar (1976). The probability of liquefaction is thus given by the following:

$$P_{f_{\Delta v}} = 1 - \Phi\left(\frac{\ln\left[\frac{\mu_{\tau_{R_{\Delta v}}}}{\mu_{\tau_{A_{\Delta v}}}} \sqrt{\frac{1+\Omega_{\tau_{A_{\Delta v}}}^2}{1+\Omega_{\tau_{R_{\Delta v}}}^2}} \right]}{\sqrt{\ln\left[(1+\Omega_{\tau_{R_{\Delta v}}}^2)(1+\Omega_{\tau_{A_{\Delta v}}}^2) \right]}} \right) \qquad (13)$$

where $\Phi(\;)$ is the standard normal cumulative distribution function.

Equation 13 can be used to estimate the risk of liquefaction in a volume of sand when both the maximum ground acceleration and the earthquake magnitude at a site are known. Basically, Equation 13 can be used to investigate past case studies.

A GENERAL THREE-DIMENSIONAL PROBABILISTIC LIQUEFACTION MODEL

The procedures described in the previous section would be inadequate if a site has to be designed for future earthquakes where a_{max} and N_{eq} or M are unknown. The risk of liquefaction can still be estimated if the uncertainties in these parameters are incorporated into the model appropriately. Using the theorem of total probability, Haldar (1976) and Haldar and Tang (1979) showed that the probability of liquefaction of a soil volume can be calculated as

$$P_f = \int_{(n_{eq})_o}^{(n_{eq})_u} \int_{0}^{a_{max}} (P_{f_{\Delta v}} \mid a_{max}, n_{eq}) \, f_{A_{max}}(a_{max}) \, f_{N_{eq}}(n_{eq}) \, da_{max} \, dn_{eq} \qquad (14)$$

in which $f_{A_{max}}(a_{max})$ and $f_{N_{eq}}(n_{eq})$ are the density functions of A_{max} and N_{eq}, respectively. $(n_{eq})_u$ and $(n_{eq})_o$ are the values of N_{eq} corresponding to the upper and lower bound magnitudes of the earthquakes. The conditional probability, $(P_{f_{\Delta v}} \mid a_{max}, n_{eq})$ can be estimated using Equation 13.

UNCERTAINTY ANALYSIS OF PARAMETERS

The probabilistic characteristics of all the parameters in Equations 13 and 14 need to be evaluated to estimate the risk of liquefaction. Haldar and Miller (1982) studied all the parameters in great detail; however, they can not be presented here due to lack of space. They will be discussed qualitatively here.

The uncertainty associated with the estimation of the in situ relative density contributes significantly to the overall uncertainty in τ_R. The amount of uncertainty in D_r depends on how it was determined; directly using the information on the maximum, minimum and inplace dry densities, or indirectly using a correlation between the Standard Penetration Test (SPT) value and D_r values. Haldar and Tang (1979) observed that using the direct method the uncertainties in D_r in terms of COV could be of the order of 0.11 to 0.36. When the indirect method is used, Haldar and Miller (1982) observed that the uncertainty could be of the order of 0.20 to 0.35. The uncertainty associated with the shear strength parameter R is also considerable. Using large-scale shaking table test results on liquefaction and considering factors such as system compliance, methods of sample preparation, mean grain size, multidirectional shaking and other secondary factors, Haldar and Miller (1982) evaluated the probabilistic characteristics of R. The uncertainty associated with the prediction of the in situ value of R could be of the order of 0.20 to 0.30.

The uncertainty associated with the load parameters is also considerable. The magnitude and duration of the future earthquake, as well as the maximum ground acceleration at a particular site within a specified time period need to be considered probabilistically. Available geological, seismological and observed records at or near the region concerned need to be considered in estimating the seismic risk of the region. The uncertainty associated with the attenuation equations themselves could be of the order of 0.90 in terms of COV (Haldar (1976)). Haldar and Tang (1979) discussed the uncertainty associated with N_{eq}, the equivalent number of uniform stress cycles corresponding to a design earthquake magnitude. Considering all the relevant information, the seismic risk of a site in terms of design acceleration versus return period can be developed (Haldar (1976)).

RISK OF LIQUEFACTION

Probability of liquefaction, a_{max} and M are known

This method can be used to investigate the case studies of liquefaction or no liquefaction where both the maximum ground acceleration and the earthquake magnitude at a site are known.

A site in Niigata, Japan which liquefied during the 1964 earthquake is considered here. The magnitude of the earthquake was 7.5 and the site experienced 0.16 g maximum ground acceleration. The SPT value of 6 was measured at the critical depth 25 ft. The depth of the water table was 3 ft. from the ground surface. The saturated unit weight and the mean grain size, D_{50}, were considered to be 120 pcf and 0.26 mm, respectively. The maximum and minimum dry densities of the deposit are considered to be 102.7 pcf and 81.5 pcf, respectively, from indirect information. The scales of fluctuation, θ_x, θ_y, and θ_z can not be estimated for the site. The scales of fluctuation in the two horizontal directions are assumed to be the same. θ_x, θ_y, and θ_z are considered to be 120 ft., 120 ft., and 7 ft.,

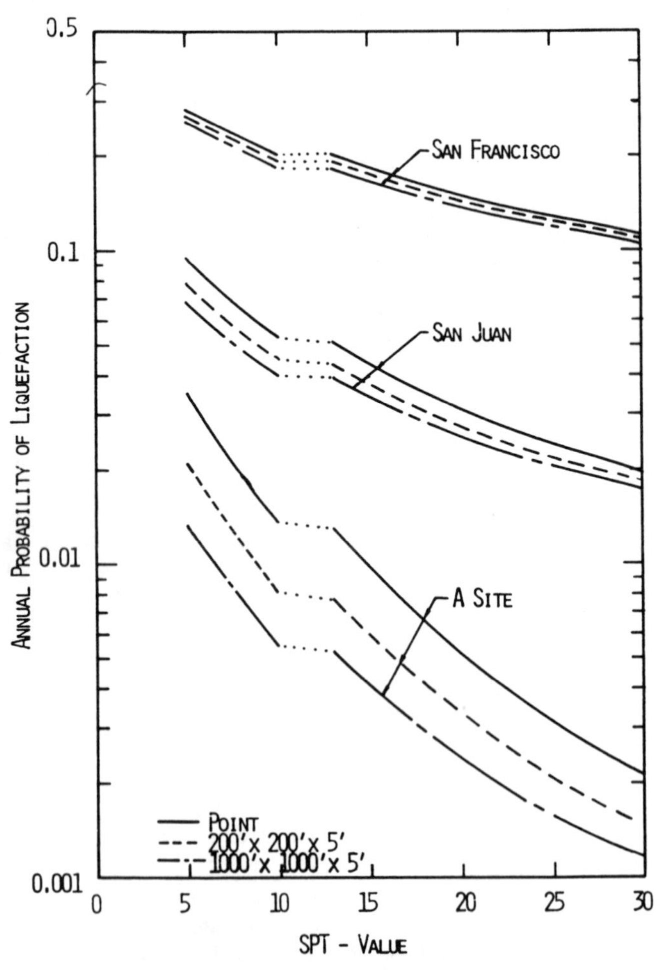

Figure 1 Estimated Annual Probability of Liquefaction

respectively. The volume of the liquefied sand is assumed to be 200 ft. x 200 ft. x 5 ft. The COV of γ_s and h_{WT} can be taken as 0.01 and 0.20, respectively. Using Equation 13, Haldar and Miller (1982) showed that the risk of liquefaction of the soil volume would be around 0.98. The corresponding point estimation of liquefaction is around 0.95.

To show the application of the general three-dimensional probabilistic liquefaction model, the aforementioned site conditions can be assumed to exist in deposits in the San Francisco Bay area and in San Juan, Puerto Rico. These sites are considered here since the seismic risk of the areas is available in the literature. In Figure 1, the annual risks of liquefaction using Equation 14 versus the Standard Penetration Test (SPT) values are plotted for various soil volumes. As expected, some difference is noticeable. When a site is designed against liquefaction, the annual risk of liquefaction would be much smaller. In that case, the differences between different volumes of sand would be considerable. This is also shown in Figure 1.

CONCLUSIONS

A procedure for estimating the probability of liquefaction of a volume of sand is developed here. All the methods presently available estimate the risk of liquefaction at a point in the deposit. In order to produce a noticeable amount of damage at a referenced location, a considerable volume has to liquefy. Nonhomogeneous soil properties in this critical volume are modeled probabilistically in this model. A considerable amount of difference is expected between the risk of liquefaction in a volume and at a point. The seismic activity of the region, as well as the attenuation characteristics should also be given proper consideration. The proposed three-dimensional probabilistic model could complement the deterministic procedures by providing information on the relative risk of liquefaction between various alternatives.

ACKNOWLEDGMENT

This material is based upon work partly supported by the National Science Foundation under Grant No. PFR-8006348. Any opinions, findings, and conclusions or recommendations expressed in this publication are those of the writers and do not necessarily reflect the views of the National Science Foundation.

REFERENCES

Christian, J. T. and Swiger, W. F. (1975) Statistics of Liquefaction and SPT Results. J. of the Geotechnical Engineering Division, ASCE, 101, GT11:1135-1150.

Donovan, N. C. (1971) A Stochastic Approach to the Seismic Liquefaction Problem. Proc. Conference on Statistics and Probability in Soil and Structural Engineering, Hong Kong, 513-535.

Faccioli, E. (1973) A Stochastic Model for Predicting Seismic Failure in a Soil Deposit. Earthquake Engineering and Structural Dynamics, 1, 293-307.

Fardis, M. N. and Veneziano, D. (1979) Probabilistic Liquefaction of Sands During Earthquakes. Report No. R79-14, Massachusetts Institute of Technology, Cambridge, Mass.

Finn, W. D. L., Lee, K. W. and Martin, G. R. (1977) An Effective Stress Model for Liquefaction. J. of the Geotechnical Engineering Division, ASCE, 103, GT6:517-533.

Haldar, A. (1976) Probabilistic Evaluation of Liquefaction of Sand Under Earthquake Motions. Ph.D. Thesis, University of Illinois, Urbana, Illinois.

Haldar, A. (1980) Liquefaction Study - A Decision Analysis Framework. J. of the Geotechnical Engineering Division, ASCE, 106, GT12:1297-1312.

Haldar, A. (1981) Uniform Cycles in Earthquakes: A Statistical Study. Proc. International Conference on Recent Advances in Geotechnical Earthquake Engineering and Soil Dynamics, St. Louis, 1, 195-198.

Haldar, A. (1981) Statistical and Probabilistic Methods in Geomechanics. NATO - Advanced Study Institute, Lisbon.

Haldar, A. and Tang, W.H. (1979) Probabilistic Evaluation of Liquefaction Potential. J. of the Geotechnical Engineering Division, ASCE, 105, GT2:145-163.

Haldar, A. and Tang, W.H. (1981) Statistical Study of Uniform Cycles in Earthquake Motions. J. of the Geotechnical Engineering Division, ASCE, 107, GT5:577-589.

Haldar, A. and Miller, F. J. (1982) Probabilistic Evaluation of Damage Potential in Earthquake-Induced Liquefaction in a 3-D Soil Deposit. Report No. SCEGIT-82-101, School of Civil Engineering, Georgia Institute of Technology, Atlanta, Georgia.

Kishida, H. (1969) Characteristics of Liquefied Sands During Mino-Owari, Tohnankai, and Fukui Earthquake. Soils and Foundations, 9, 1:76-92.

Liou, C. P., Streeter, V. L. and Richart, F. E. Jr. (1977) Numerical Model for Liquefaction. J. of the Geotechnical Division, ASCE, 103, GT6:589-606.

Ohsaki, Y. (1970) Effects of Sand Compaction on Liquefaction During the Tokachioki Earthquake. Soils and Foundations, 10, 2:112-128.

Seed, H. B. and Idriss, I. M. (1967) Soil Liquefaction in the Niigata Earthquake. J. of the Soil Mechanics and Foundation Division, ASCE, 93, SM3:83-108.

Seed, H. B. and Idriss, I. M. (1971) Simplified Procedure for Evaluating Soil Liquefaction Potential. J. of the Soil Mechanics and Foundations Division, ASCE, 97, SM9:1249-1273.

Seed, H. B., Mori, K., and Chan, C. K. (1977) Influence of Seismic History on Liquefaction of Sands. J. of the Geotechnical Engineering Division, ASCE, 103, GT4:246-270.

Vanmarcke, E. H. (1970) Working Paper on Variance Functions. M.I.T. Department of Civil Engineering, Cambridge, Mass.

Vanmarcke, E. H. (1977) Probabilistic Modeling of Soil Profiles. J. of the Geotechnical Engineering Division, ASCE, 103, GT11:1227-1246.

Vanmarcke, E. H. (1977) On the Scale of Fluctuation of Random Functions. Report No. R79-19, M.I.T. Department of Civil Engineering, Cambridge, Mass.

Whitman, R. V. (1971) Resistance to Soil Liquefaction and Settlement. Soils and Foundations, 11, 4:59-68.

Yegian, M. K. and Whitman, R. V. (1976) Risk Analysis for Earthquake Induced Ground Failure by Liquefaction. M.I.T. Seismic Design Decision Analysis Report No. 26, Cambridge, Mass.

Soil Dynamics & Earthquake Engineering Conference / Southampton / 1982.07.13-15

Fragility Curves and Uncertainty of the Seismic Action

F.CASCIATI, L.FARAVELLI & A.GOBETTI
Università di Pavia, Italy

ABSTRACT

The methodology for determining the fragility curve of a seismic-resistant r.c. frame for given details of the ground motion is discussed, particular care being devoted to the definition of the probabilistic failure criterion for the single structural element. Different (either simulated or recorded) ground motion histories are considered and the relevant fragility curves are computed for an example-frame making use of simulation techniques. The possibility of deriving a global fragility curve accounting for the input-detail randomness is investigated and the use of fragility curves in earthquake engineering for reliability analysis purposes is discussed from an operative point of view.

INTRODUCTION

Reliability analysis of seismic-resistant frames has been generally performed considering that a stochastic input acts on a deterministic structure for which a deterministic failure criterion holds (Garetas (1976) and Wen (1979) among others). This policy leads to very accurate results whenever the limit state being analysed is defined by the first yielding: the uncertainties on the elastic constants and on the yield stress of the materials, in fact, are generally negligible in comparison with the randomness of the action. Conversely, if the inelastic behaviour has to be investigated, the uncertainty on the structure parameters and on the failure criterion may have a significant influence on the structural response.

In recent studies (Banon(1980),Banon-Veneziano(1981)), a probabilistic resistance model has been proposed for reinforced concrete sections under cyclic loading. Using that model and a criterion of system failure (e.g. first section failure) one can construct a fragility curve (a plot of the probability of failu

re versus peak ground acceleration or any other seismic intensity measure) for each given ground motion. In this case, the ground motion preserves its shape and is simply scaled according to the intensity parameter. Naturally, different fragility curves will results from different scaled accelerograms.
The present paper investigates the influence of uncertainty on the shape of the ground motion into the calculation of seismic fragilities. This is a first step towards obtaining fragility curves not conditional on the shape of the ground motion. The latter curves are needed for the calculation of seismic reliability: for example, the probability of structural failure due to a generic earthquake results directly from the convolution of the unconditional fragility curve with the probability density function of the seismic intensity parameter at the site of the facility.

BASIC MODELS

Input model

The dynamic analyses of the frame being investigated will be performed for different input accelerograms. They are obtained either by simulation or simply scaling the maximum peak acceleration of recorded ground motions. Two historical records are considered: the N-S accelerogram of duration 40 sec. of 1940 El Centro earthquake and the E-W ground acceleration record of Tolmezzo, relevant to Friuli (Italy) earthquake (May 6, 1976); the last record has been truncated at 15 sec.

Different ground motions have been simulated by sampling realizations of a filtered white-noise. The white-noise, which is made nonstationary by the introduction of an appropriate intensity function (Penzien(1978)), is filtered twice. The first filter cuts high frequencies and its transfer function is selected to have the same analytical form of the transfer function governing the "foundation displacement input/displacement output" system of given damping ratio ξ and undamped natural frequency ν_o. The second filter cuts off very low frequencies and its transfer function is selected to have the form governing the "force input/acceleration output" system of given ξ and ν_o. The numerical values of ξ and ν_o for the two systems have been selected according to the literature (Penzien, 1978), as well as the shape of the intensity function. The duration of simulated ground motions is assumed to be 15 sec. long.

Mechanical behaviour model

Non-linear dynamic analyses are developed by using the finite element code DRAIN-2D (Kanaan-Powell, 1973). The version adopted for the calculations includes a modified-Takeda constitutive law(Fig.1b)) in the plastic hinges and a pair of inelastic springs at each end of the beam-element accounting for pinching of the hysteretic loops from high shear loads and from slippage of the longitudinal reinforcement (Banon, 1980). The last device has not been introduced in the examples, but remains a potential option of the code the authors implemented for running on a SEL

32A computer. All calculations have been performed on this machine.

The DRAIN-2D code makes use of a model assuming that each element of the plane frame being investigated behaves like an elastic beam between inelastic hinges. The constitutive law of inelastic hinges must be selected "a priori" by assigning the hardening as a percentage of the elastic stiffness. Several common (although not explicitly required) simplifying hypotheses are also introduced: the base of the frame is assumed infinitely rigid; only traslational masses are considered and they are lumped at the nodes; P-Δ effects are taken into account by assuming the axial load on each column to be constant and equal to that due to dead load; axial deformation of girders is neglected. Moreover, the single beam element is assumed to deform antisymmetrically in bending. The last assumption is justified if the horizontal beams are subjected to negligible gravity loads, as when columns are designed (according to common seismic codes) to behave elastically under seismic excitation.

Further (sometimes crude) idealizations are finally introduced in the numerical integration of the incremental equations of motion. For this purpose, the code makes use of the simplest integration algorithm, assuming the acceleration of the single lumped mass to be constant during each step. No device is adopted to avoid yielding overshooting (Casciati-Faravelli-Gobetti (1980)), consisting in an overestimation of the yielding properties during the single integration step,and no device is introduced to avoid kinematic overshooting (Faravelli-Gobetti(1980)), consisting in overestimations of displacements and velocities during the first integration steps.

Fatigue damage model

Reinforced concrete elements undergo progressive deterioration when subjected to intense cyclic loading and some may break down suddenly (Fig. 1a). The relevant physical process, so-called "low-cycle fatigue", is not well known from a theoretical point of view, while a large dispersion of member resistance has been observed in experimental investigations.

Recent work (Banon (1980), Veneziano-Banon (1981), Perrault (1980), Veneziano (1981)) has demonstrated the opportunity of treating low-cycle fatigue in terms of aggregate force-deformation parameters. The most important steps in deriving the resistance model are described next.

Two damage parameters $D_1(t)$ and $D_2(t)$ are used, which depend on the deformation history of each critical section. The flexural damage ratio $D_1(t)$ is defined by the ratio between the initial flexural stiffness K_f and the reduced secant stiffness K_r (Fig. 1c)

$$D_1(t) = K_f/K_r(t) = 24\,EI/L^3\,K_r(t) \qquad (01)$$

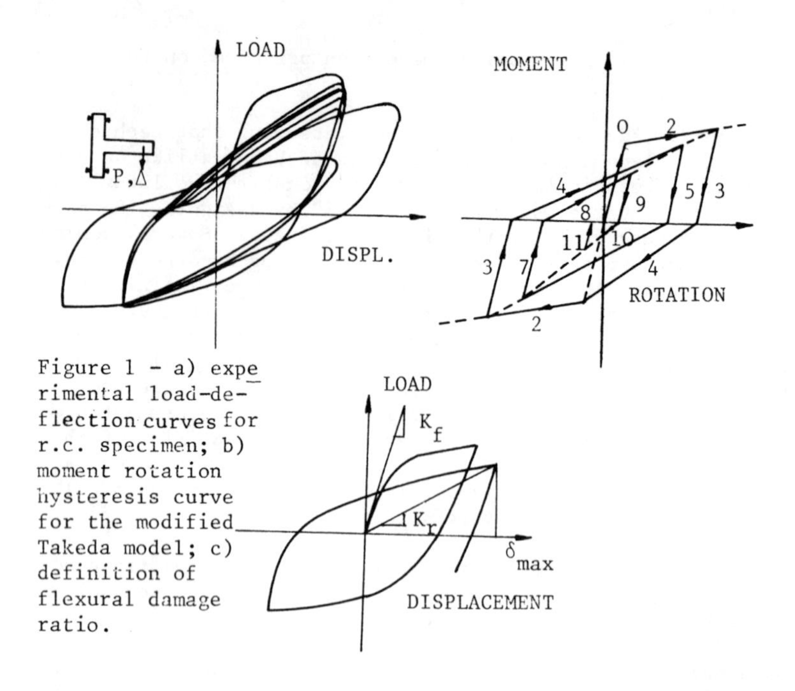

Figure 1 - a) experimental load-deflection curves for r.c. specimen; b) moment rotation hysteresis curve for the modified Takeda model; c) definition of flexural damage ratio.

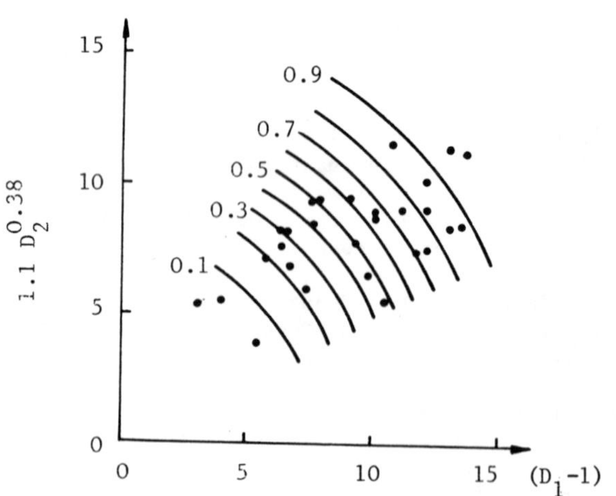

Figure 2 - Probabilistic failure criterion for the single critical section.

The second equality holds in the case of antisymmetric bending of the element of length L. This parameter D_1 is chosen to represent damage due to large displacement, while the other parameter, the normalized dissipated energy $D_2(t)$, is taken to represent the damage due to many cycles at relatively low displacement. D_2 is defined as the ratio between the energy dissipated by inelastic rotation at one end of the member and the maximum elastic energy stored in the member in antisymmetric bending:

$$D_2(t) = \int_0^t M(\tau)\theta(d\tau)/\frac{1}{2} M_y\theta_y = \int_0^t M(\tau)\theta d(\tau)/(M_y^2 L/12\,EI) \qquad (02)$$

In Eq. (02) t is the time since the beginning of loading and $\theta(d\tau)$ is the increment of the inelastic spring rotation at one end of the element during the time interval from τ to $\tau + d\tau$. Again the last equality holds in the case of antisymmetric bending.
If a single damage parameter R is to be defined, then it has been found (Hasofer-Veneziano-Banon (1981)) that an appropriate choice is

$$R = \left[(D_1 - 1)^2 + (b\, D_2^{\gamma})^2\right]^{\frac{1}{2}} \qquad (03)$$

in which $\gamma = 0.38$ and $b = 1.10$. Laboratory data have been used to fit a Weibull distribution to R of the type:

$$P_R(r) = 1 - \exp\left(-\frac{\phi_\circ}{\beta+1}(r)^{\beta+1}\right) \qquad (04)$$

where $\phi_\circ = 1.13\ 10^{-4}$ and $\beta = 3.05$. The associated values of the mean and variance are:

$$\mu_R = 12.1 \quad ; \quad Var[R] = s_R^2 = 11.2 \qquad (05)$$

Model (04) (see Figure 2) has been employed in this paper as a probabilistic criterion for local failure, $P_R(r)$ giving the probability of failure of the single section characterized by the value r of the damage parameter R.

For the joint distribution of resistances R_i of all components in the structure, one can define a Weibull defined random variable $\mu°$(with mean $\mu_{\mu°} = \mu_R$ and variance $s^2_{\mu°}$) representing the variability common to all components of the structure and then take the conditional distribution of the resistances $P(R_i|\mu°)$ to be Weibull distributed with mean $\mu°$ and variance $s^2_{R_i|\mu°}$. The resulting unconditional joint distribution has the following characteristics

$$E[R_i] = \mu_R$$

$$Var[R_i] = s^2_{\mu°} + s^2_{R_i|\mu°} \qquad (06)$$

$$\rho = \rho_{R_i,R_j} = Cov[R_i,R_j]/(Var[R_i]Var[R_j])^{\frac{1}{2}} = s^2_{\mu°}/(s^2_{\mu°} + s^2_{R_i|\mu°})$$

Then, once $Var[R_i]$ and ρ have been selected, the joint resistance distribution is fully defined (Veneziano (1981)).
Nevertheless, for sake of simplicity, all calculation have been developed in this paper for independent resistances R_i with mean $\mu_{R_i}=\mu_R=12.1$, according to Eq. (05), and variance $s^2_{R_i}=s^2_R/2=5.6$. The latter assumption accounts for the smaller dispersion that characterizes the resistances of an actual frame in comparison with the variability of the results of nonhomogeneous laboratory tests.

Limit-state and reliability calculations

The ultimate limit state of a r.c. frame may be defined either in a brittle or in a ductile form.
In the first case the structure is assumed to collapse as soon as a section of the frame fails, therefore the reliability of the system $\mathbb{R}$ is given by (Banon-Veneziano (1981))

$$\mathbb{R} = \int_0^{\infty} \prod_{i=1}^{n} (1 - P(R_i|\zeta))p_{\mu}(\zeta)d\zeta \tag{07}$$

where n is the number of structural members, $P(R_i|\zeta)$ is the conditional distribution of the resistances given $\mu^{\circ}=\zeta$ and $p_{\mu^{\circ}}(\zeta)$ is the probability density function of μ°.
By contrast, the analysis must proceed after the first section failure if the static redundancy of the structure is accounted for; for this purpose the failed element is removed from the structure and the global stiffness properties modified. After successive failures of different sections and corresponding modifications of structure geometry, then, the frame may either become no longer statically determinate or undergo large displacements, so that the equations of motion should be changed. For both cases, in order to define a common limit state, one can assign a limit value u_L to displacements: its attainment will be denoted as *progressive failure* (Perrault (1981), Veneziano (1981)). All reliability calculations under this assumption have been developed for u_L = 76.2 cm (30") following the guide-lines outlined in next Sections.

FRAGILITY CURVES

For the purpose of reliability calculations the basic result of dynamic analysis is a time history, for each critical section, of the damage parameter R in Eq. (03).
Making use of the distribution of Eq. (04), the probability of failure of each section in the structure is determined and subsequently the reliability of the frame itself, dependent on the amount of correlation assumed between independent member resistances.

Weakest-link failure

If the frame fails when the first section fails the probability of failure of the structure for a given accelerogram and a given value of related intensity is given by Eq. (07). Then, scaling the accelerogram, several reliability values are obtained and a (conditional) fragility curve of the type in Figure 3 can be drawn for the earthquake under investigation. Varying the accelerogram, different fragility curves are finally derived.

Rendundant system

Even if the simplification that failed members are no longer active components of the structure is accepted, calculations of reliability against progressive failure cannot be performed analytically because of randomness in the response due to the uncertainty on which members fail and at what times. However, it is possible to simulate the resistances R_i' for each section and trace the behaviour in time of this simulated system: if at the end of a given loading history the maximum displacement is lower than u_L, one obtains a successful trial, otherwise a failure. By repeating this simulation for different samples R_i', the probability of failure of the whole system for that frame and for that loading history (conditional fragility) can be estimated as the fraction of simulations for which the maximum displacement during the motion exceeds u_L.
In order to obtain the relevant fragility curve (Figure 3), the previous procedure must be repeated for several scaled loading histories.

Actual probability of failure

Convolution of the fragility curve with the probability density function of the seismic intensity parameter at the site of the facility (for instance the PDF of the 50-year peak acceleration (Figure 3)) gives the probability of failure of the structure under the condition that next ground motion will have the same time history one employed in the derivation of the fragility curve. The relation between this result and the actual probability of failure of the structure might be estimated by analysis of several ground motions and by statistics of the relevant probabilities of failure. Thus, for many ground motions one should construct the fragility curves to introduce in the convolution calculations: this task appears to be bearable for the brittle limit state (Casciati-Faravelli-Veneziano (1982)), but it is absolutely unthinkable to do with reference to the ultimate limit state defined by progressive failure.

In the absence of clear theoretical guidelines, a first step towards obtaining unconditional fragility curves is to develop the calculation of different fragility curves for a given structure and, from the analysis of the results, to outline general aspects and trends.

Two remarks

Before the results of developed calculations are presented, it is suitable to state in advance two remarks, operative the first and theoretical the second.
Everywhen reliability assessments are performed by simulation, the results are often unreliable because of the low number of experiments that can be born. That is the case of progressive failure, where for each value of the seismic intensity one can perform 50 or 100 analyses, achieving a direct frequency estimation of failure of few percents; in any case, one cannot estimate the left tail of the fragility curve that is of interest for

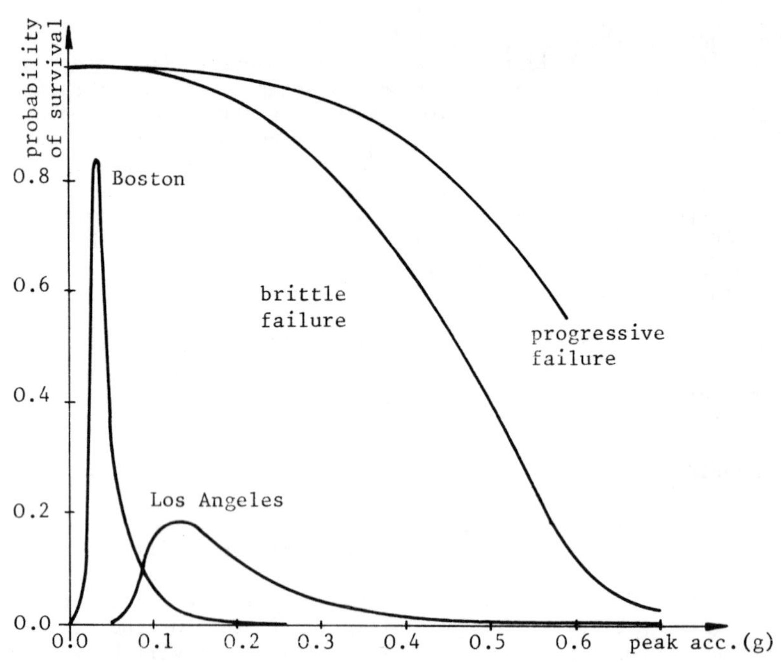

Figure 3 - Fragility curves and examples of 50-year PDF's of peak accelerations

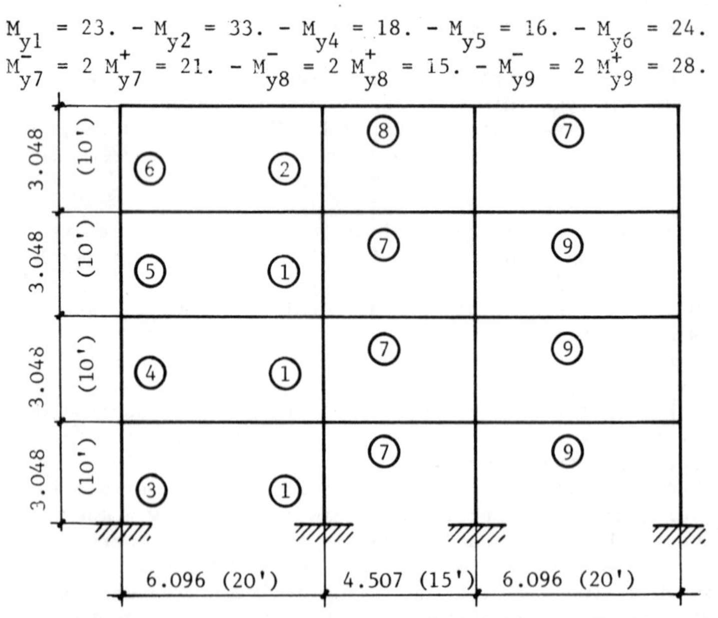

Figure 4 - 4-storey building frame for which calculations are developed (M_y in t.m)

the successive convolutions. It follows the necessity of introducing for the fragility curve an analytical model whose parameters may be determined by simulation ("*hybrid simulation method*" (Casciati (1981)). Having in mind that the fragility curve coincides with the cumulative distribution function (CDF) of the peak acceleration a_{p_R} causing failure of the structure, the choice of an analytical model reduces to the selection of an appropriate function for this CDF. For this purpose, a lognormal distribution is often adopted: next section is also devoted to verify this assumption.

The theoretical remark concerns with the meaning of the peak acceleration a_{p_R} causing failure. It appears to be a resistance parameter, but if W is a resistance parameter, i.e. a loading causing failure, a given deterministic structure that fails when subjected to the load W"> W', does not fail under the load W'. Conversely, in the present dynamic case, a given deterministic structure that does not fail for $a''_{p_R} > a'_{p_R}$ may fail when the maximum peak acceleration of the considered accelerogram assumes the value a'_{p_R}, as it will be shown in the following example.
Therefore the fragility curve provides a global averaged information (the conditional probability of failure increases as the value of the seismic intensity increases, i.e. more structures are expected to fail for higher intensity values), but it is quite different from the "conditional probability of failure" (CPF) often introduced in statical calculations, because the parameter on the abscissa does not have the necessary meaning of resistance.

NUMERICAL EXAMPLE

The frame

Numerical calculations have been developed with reference to the 4-storey frame of Figure 4 designed according to the U.S. Uniform Building Code (1973 version) (Banon (1980)). All girders have 30.48cm x 50.80 cm (12"x20") sections and all columns have 30.48 cm x 45.72 cm (12"x18") sections.
For dynamic analysis, 100 per cent of dead load (assumed to be 332 kg/l (2400 p/f)) and 25 per cent of live load (110.7 kg/m (800 p/f)) are applied to each floor.
All beams and columns are assumed to have equal stiffnesses in the positive and negative loading directions, the effective elastic stiffness of each beam and column being taken to be 45% and 60% of the elastic stiffnesses respectively. The associated natural period is 0.86 sec.; for the relevant natural mode a damping ratio of 3%, is aimed and hence the structural damping matrix is taken to be 0.0083 times the global elastic stiffness matrix.
The yield moment capacities of beams and columns are shown in Figure 4, and correspond to a concrete strength of 281 kg/cm^2 (4000 psi). For the constitutive law in every critical section, one introduced a hardening coefficient of 0.03, while an unloading

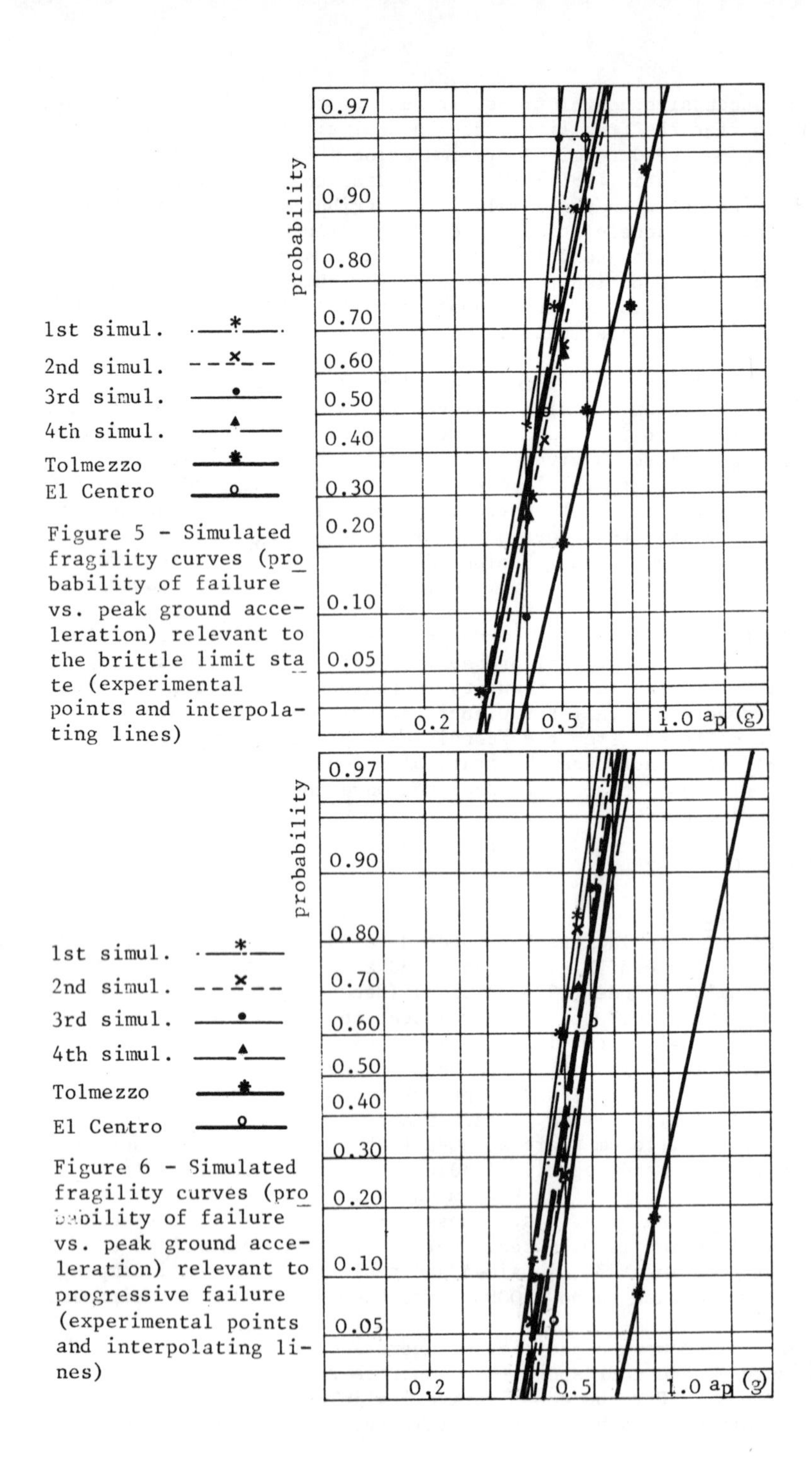

Figure 5 - Simulated fragility curves (probability of failure vs. peak ground acceleration) relevant to the brittle limit state (experimental points and interpolating lines)

Figure 6 - Simulated fragility curves (probability of failure vs. peak ground acceleration) relevant to progressive failure (experimental points and interpolating lines)

parameter α = 0.3 modifies the unloading stiffness (Litton(1975)). For each value of seismic intensity, 50 numerical experiments were carried out and the relevant probabilities of failure were estimated as the fraction of simulations providing failure. The results are drawn, in lognormal paper, in Figs 5 and 6 for the brittle and progressive failure. The hypothesis of lognormal distribution for the failure acceleration appears to be acceptable.

Four simulated ground motions have been considered and their fragility curves settle down on the left of the two relevant to recorded accelerograms introduced for comparison. In particular, the peak ground acceleration appears a very bad definition of seismic intensity for the Friuli earthquake. However, even if a sophisticated definition of intensity, making use of the average of the motion pseudo-velocity (spectral ordinates) over a given frequency interval, is adopted, all fragility curves drawn near in the brittle case, whereas, for progressive failure, the fragility curves obtained for simulated accelerograms maintain ordinates higher than the ones relevant to recorded accelerograms. Such a circumstance may depend on the long continuous time interval (≃10 sec.) in which simulated accelerograms are considered to be stationary, so that large positive values of ground acceleration may alternate to large negative values without any attenuation form of the type shown by real records. If this statement is true in general, an upper bound of the actual unconditional probability of failure can be derived by simulating a ground motion with a long stationary interval and making use of the relevant fragility curve for successive reliability calculations. Possibly, one can generate more ground motions and derive an *average fragility curve*. This average curve is shown as a dashed line in Fig. 6. Table I summarizes the parameters of the lognormal distributions obtained after graphical interpolation in the progressive failure case.

Table I - Parameters of the lognormal distributions interpolating the experimental results (peak ground acceleration in g)

Ground Motion	Median	Standard deviation
El Centro	0.58	0.15
Tólmezzo	1.10	0.22
1st simul.	0.49	0.15
2nd simul.	0.52	0.12
3rd simul.	0.48	0.14
4th simul.	0.56	0.17
average	0.51	0.15

Finally, in order to emphasize that the collapse acceleration a_{pR} has not the meaning of resistance, note that the 19th simulated structure, when excited by the first simulated accelerogram, does not fail for the scale factor a_p equal to 0.5 g and 0.7 g, but attains the limit displacement limit state for a_p = 0.6 g. Other similar cases were encountered in performing numerical analyses (Belotti-Bissolati (1982)).

CONCLUSIONS

With reference to a particular r.c. frame, the relevant fragility curves for different ground motions have been obtained in the form of lognormal distribution functions. Moreover, the curves obtained for simulated accelerograms lead to compute risks of failure greater than the risk associated with recorded ground motions.
If further numerical experiments will confirm this result for different frames and ground motions, the fragility curve might be easily computed making use of the lognormal assumption, and a safe estimation of the actual probability of failure of the structure might be obtained by the knowledge of the curve associated with any simulated ground motion. Further investigation, however, appears to be necessary in order to confirm the generality of such a result and to prove the accuracy of the mechanical and fatigue models on which all calculation are founded.

ACKNOWLEDGMENT

The authors are very indebted to prof. Daniele Veneziano for many useful discussions and appropriate suggestions. Funds have been provided by the Italian National Research Council (C.N.R.).

REFERENCES

Banon H. (1980), Prediction of Seismic Damage on Reinforced Concrete Frames, Ph. D. Thesis, Department of Civil Engineering M.I.T., Cambridge

Banon H. and Veneziano D. (1981), Seismic Safety of Reinforced Concrete Members and Structures, to be published on Earthquake Engineering and Structural Dynamics

Belotti F. and Bissolati E. (1982), Comportamento di strutture intelaiate in cemento armato in zona sismica (in Italian), SM Thesis, Dep. of Structur. Mechanics, University of Pavia

Casciati F., Faravelli L. and Gobetti A. (1980) - Elasto-Plastic Response Analysis of Seismic-Resistant Frames Via Mathematical Programming, Proc. of WCEE, Istanbul, Vol. 5, 521-528

Faravelli L. and Gobetti A.(1980), Analisi dinamica di telai elasto-plastici: programmazione lineare parametrica e "metodi α"(in Italian),Proc. 5th Congr. AIMETA, Palermo

Garetas G. (1976), Random Vibration Analysis of Inelastic Multi-Degree-of-Freedom Systems Subjected to Earthquake Ground Motions, Res. Report R 76-39, Dep. of Civ. Eng., M.I.T.

Hasofer A.M., Veneziano D. and Banon H. (1981), Risk Analysis in More than One Dimension, to be published on Journal of Applied Probability

Kanaan A.E., Powell G.H. (1973), General Purpose Computer Program for Dynamic Analysis of Inelastic Plane Structures, EERC, Univ. of California, Berkeley

Litton R.W. (1975), A Contribution to the Analysis of Concrete Structures under Cyclic Loading, Ph. D. Thesis, Department of Civil Engineering, University of California, Berkeley

Penzien J. (1978), Non-Linear Stochastic Response of Structures to Strong Earthquake Ground Motions, First Int. Seminar on Advances in Vibrations, London

Perrault S.M. (1980), Probabilistic Model of Low-Cycle Fatigue and Progressive Failure for Reinforced Concrete Frames, S.M. Thesis, Dep. of Civil Eng., M.I.T., Cambridge

Veneziano D. (1981), Probabilistic Seismic Resistance of R.C. Frames, 3rd Int. Conf. on Structural Safety and Reliability (ICOSSAR), Trondheim, 241-258

Wen Y.K. (1979), Stochastic Seismic Response Analysis of Histeretic Multi-Degree-of-Freedom Structures, Earthquake Engineering and Structural Dynamics, Vol. 7, 181-191

Casciati F. (1981), Probabilistic Analysis of Inelastic Structures, Proc. 3rd Seminar on "Reliability Nuclear Power Plants", Paris, August

Casciati F., Faravelli L. and Veneziano D. (1982), The Seismic Fragility of R.C. Frames Including Progressive Failure, to be presented at 7th ECEE, Athens

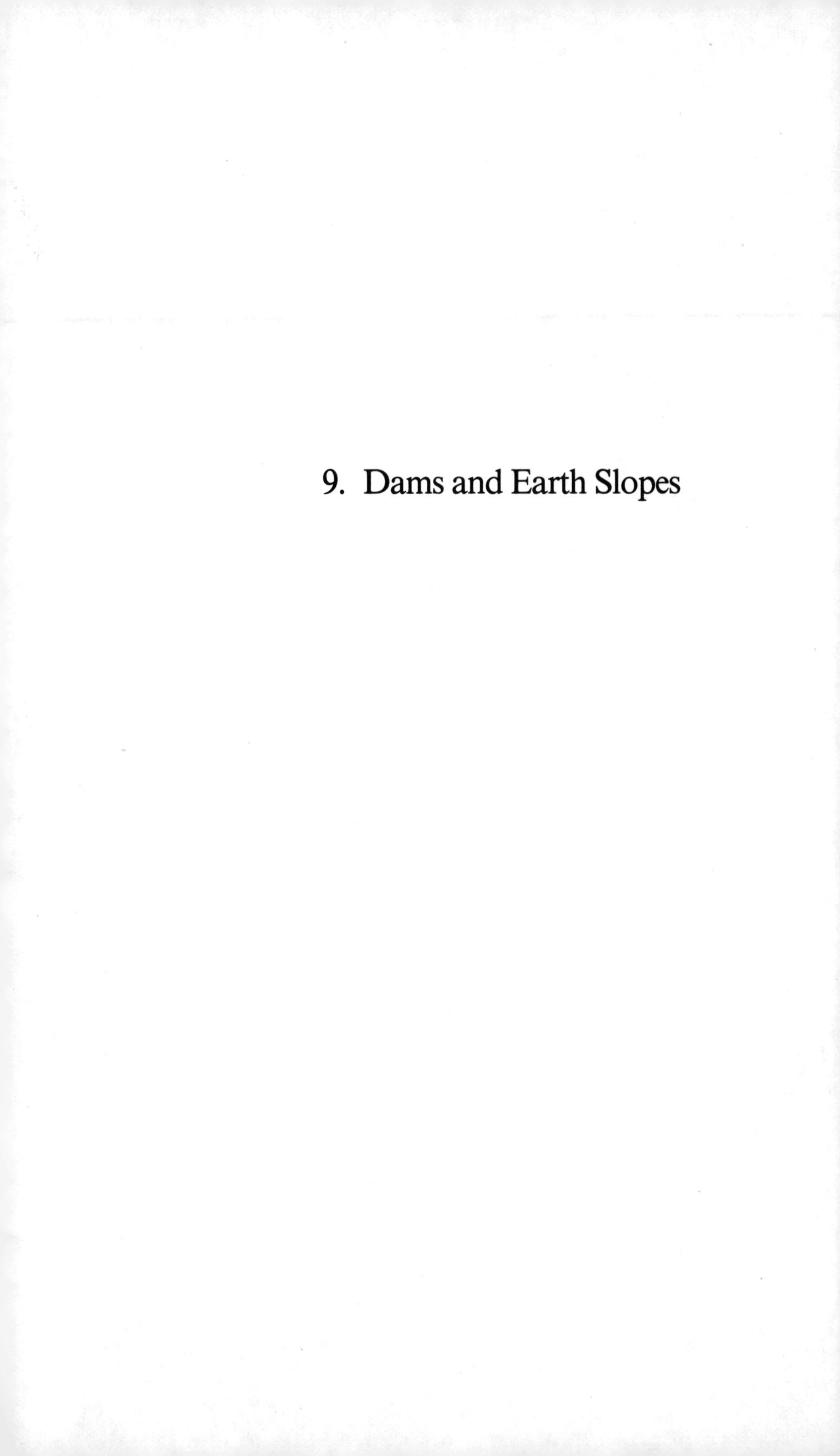

9. Dams and Earth Slopes

Soil Dynamics & Earthquake Engineering Conference / Southampton / 1982.07.13-15

Random Vibration Analysis of Finite Element Models of Earth Dams

D.A.GASPARINI & WEI-JOE SUN
Case Western Reserve University, Cleveland, OH, USA

INTRODUCTION

Safe designs for dams subject to earthquakes depend on dynamic analyses using engineering models of dams and of the excitation. In design, it is necessary to recognize the uncertainties inherent in such models and in the corresponding dynamic responses. Increasing the sophistication of only one of the models may not decrease such uncertainties and may not lead to more economical designs but merely increased computational costs.

The complexity of the geometry, support conditions and material behavior of actual earth and rockfill dams is formidable. Nevertheless, elastic, one or two dimensional dam models have provided useful insights on behavior, and dynamic responses of such models have been utilized in design. Shear beams, either homogeneous or inhomogeneous are the simplest models. Linear elastic plane strain finite element models are more sophisticated but concomitantly, more expensive. Finite element models can naturally represent the inhomogeneity of actual dams but refined failure criteria, design procedures, and an assessment of the variability in the excitation are also required in conjunction with finite element models to achieve safer, more economic designs.

Acceleration time histories or response spectra are common engineering models of earthquakes which can be used as input for deterministic dynamic analysis procedures. Design response spectra may be smoothed "envelope" versions which conservatively represent a broader, fuller frequency content and a longer duration than those of any one real earthquake. In response spectrum dynamic analyses, modal responses or the effects of multiple-component excitation must be superposed in an approximate way. Modal time history analysis with multiple excitation components, on the other hand, yields responses to a specific frequency content and correlation between components.

For random vibration or stochastic analyses the excitation may have any broad, full frequency content desired (as represented by "envelope" response spectra). Importantly, any correlation between excitation components can be explicitly considered. Moreover, modal responses can be superposed exactly, accounting for the time-varying correlation between modal responses. RMS responses obtained from random vibration analyses reflect the uncertainty in the phasing of an excitation. Random vibration analyses are very economical, even if finite element models are used, because RMS responses evolve smoothly in time.

This paper presents results from random vibration analyses of finite element models of dams. The objectives are: one, to quantify the effects of correlation between vertical and horizontal components on dam responses; two, to compare RMS responses of finite element models with those of shear beam models; and, three, to evaluate the significance of non-stationarity in the excitation and responses. A brief explanation of the random vibration formulation is given first. Then the dynamic properties of the model used are presented, because interpretation of the random vibration results relies on such properties. Finally, the excitation is described and the analytical results are presented and discussed.

RANDOM VIBRATION MODAL TIME HISTORY (R.V.M.T.H.) ANALYSIS

The random vibration approach used to compute responses is a time domain formulation. It is described in detail in Ref.[3]. Its capabilities/limitations are as follows:

Excitation - The excitation may be non-stationary and vector valued. The time variation of intensity and also the duration of each component are defined by a strength function; a strong motion duration is not specified. Each pair of components can be correlated differently and the correlation may vary in time. The excitation may also be non-white, with the frequency content of each component defined by an associated two-parameter filter.

System - Any linear multi-degree-of-freedom system which has been decoupled into modes can be analyzed. The system may be a continuous one such as a beam or a discrete one such as a finite element model.

Analysis - The computations are analogous to those of a modal time history analysis. Analytical expressions are used for the transition matrix [3] of each modal state vector. The evolutionary mean and covariance matrices of each modal state vector as well as the cross covariance matrix between each pair of modal state vectors are computed analytically.

Output - The time histories of the mean and variance of any response which is expressed as a linear function of the modal state coordinates are computed by modal superposition.

Contributions from the cross correlation between modal responses are included. Non-stationary probabilities of exceedance of very high thresholds are computed using evolutionary variances of a response and its time derivative.

In summary, the approach given in Ref. [3] may be called a random vibration modal time history analysis. The response statistics reflect the fact that the phasing of an excitation having a specific frequency content and intensity is random. Any additional variability in responses arising from uncertainty in the dynamic properties of the system or from uncertainties in the frequency content and intensity of the excitation may be quantified by repeated random vibration analyses.

The formulation has been used to study linear, inhomogeneous shear beam models of dams [4]. It has also been used in a piece-wise-linear way to study nonlinear behavior [5].

DAM MODELS ANALYZED

The dam model of Clough and Chopra [1] was analyzed. Its parameters and the finite element grid used are shown in Fig. 1. The model is homogeneous, isotropic, linear elastic and consists of plane strain elements. The dynamic properties of the model are given in Ref. [1]. Since they are important for interpretation of the random vibration analyses, the first ten natural frequencies are given in Table 1. Also given are modal participation factors for excitation in the horizontal direction, Γ_x, and for the vertical direction, Γ_y. It is important to note that $\Gamma_x = 0.0$ for modes 2, 5, 6, 9, 10; i.e. those modes, which are symmetric, are not excited by horizontal excitation. On the other hand, $\Gamma_y = 0.0$ for modes 1, 3, 4, 7, 8; i.e., those modes, which are antisymmetric, are not excited by vertical excitation.

Fig. 2 shows the σ_x, σ_y and τ_{xy} stress mode shapes for modes 1 and 2. Such contours are obtained by imposing displacement vectors proportional to the mode shapes. The static gravity stress contours, which may be interpreted as mean stresses about which earthquake-induced stresses fluctuate, are shown in Fig. 3 . They were deduced from the plots given in Ref. [2]. The data in Figs. 2 and 3 are relevant for interpreting the output of the random vibration analyses.

Also given in Table 1b are the dynamic properties of a shear beam model derived from the parameters given in Fig. 1.

EXCITATION

The excitation consisted of horizontal and/or vertical ground acceleration components applied at the support nodes. Components are defined by parameters which control the frequency content, correlation (functions) with other components and strength functions. The latter define the variation of the earthquake intensity with time. For stationary excitation, the maximum intensity value is proportional to the power spectral density function used in frequency domain random vibration analyses.

Two "filter" parameters: ω_f, which determines a predominant frequency, and ζ_f, which affects the frequency bandwidth define the frequency content of each component. $\omega_f = 4\pi$ and $\zeta_f = 0.4$ were the values used for both the vertical and horizontal excitation components. Studies [6] have shown that such values are representative of earthquakes.

Two excitation strength functions were used, a stationary one and another non-stationary one, the latter was meant to represent the intensity variation of an actual earthquake. They are shown in Fig. 4 . The maximum intensity for the horizontal component was assumed as $Q_{max} = 746.5\ cm^2/sec^3$. Such an intensity can be approximately related to other, deterministic measures [4]; the assumed value corresponds to a strong motion earthquake with an RMS acceleration of about 0.1g. The maximum intensity of the vertical component was taken as 2/3 that of the horizontal component, this is consistent with some current practice (e.g., U.S.-N.R.C.).

The correlation between components was assumed to be constant in time and equal to the extreme values of 0.0 and 1.0. The extremes were used to indicate the potential effects of correlation on responses, although any other intermediate (time varying) value may be used.

STUDIES OF STATIONARY RESPONSES

Horizontal Excitation - Fig. 5 shows contours of stationary RMS horizontal and vertical relative displacements, absolute accelerations, normal stresses and shear stresses. Only half the model is shown because stationary responses are symmetric. At the axis of symmetry, RMS responses of the finite element model are compared with those of a shear beam model.

The RMS horizontal relative displacement and absolute acceleration responses are quite uniform along horizontal sections. Corresponding shear beam responses correlate very well. The RMS vertical relative displacement and absolute acceleration generally have lower values than corresponding horizontal responses. Peak values occur near the face at about 0.3H from the crest of the dam. Since the horizontal earthquake component excites only the antisymmetric modes, the RMS vertical relative displacement and absolute acceleration are zero at the axis of symmetry.

The RMS horizontal normal stress is zero at the axis of symmetry and maximum near the face at midheight. The contours are very similar to those of the first stress mode shape (see Fig. 2) implying that the response is largely due to the first mode. Clough and Chopra [1] also show deterministic horizontal normal stress contours (at a particular instant of time) which indicate that the first mode is dominant. Chopra used the N-S component of the 1940 El Centro earthquake (a_{max} = 0.33g)and computed a maximum horizontal stress near the face of about 35 psi (2.49 Kg/cm^2); correspondingly, the horizontal component of excitation used herein had an RMS acceleration of about 0.1g and the stationary RMS horizontal normal stress near the face was about 0.9 Kg/cm^2. It is important to compare such magnitudes with the gravity load horizontal normal stresses given in Fig. 3 . Then the known risk of seismically-induced shallow slope failures becomes apparent.

The stationary RMS vertical normal stresses are smaller than the horizontal stresses. They are zero at the axis of symmetry and the contours again imply the dominance of the first mode (see Fig. 2). Similarly, the RMS shear stresses are largely determined by the first mode; at the axis of symmetry the predictions of the finite element model and the shear beam are quite close.

Vertical Excitation - Fig. 6 shows stationary RMS responses to vertical excitation having an intensity 2/3 that of the horizontal excitation. Because the vertical component excites only the symmetric modes, the shear stresses and the horizontal relative displacements and absolute accelerations at the axis of symmetry are zero. The horizontal and vertical normal stresses are maximum at the base on the axis of symmetry where high confining stresses due to gravity loads also occur (see Fig. 3). The contours of the shear stress and the vertical normal stress both imply that the second mode is dominant in determining those responses.

Simplistically then, the horizontal excitation causes higher horizontal responses which are primarily determined by the first mode; the vertical excitation causes larger vertical responses which are largely determined by the second mode. However, higher modes may be significant for earthquakes having other frequency contents or for models having different geometries.

Horizontal + Vertical Excitation - Fig. 7 shows stationary RMS responses to combined horizontal and vertical components which are either perfectly correlated or perfectly uncorrelated. Of course for uncorrelated components, the response variances are additive. At the axis of symmetry, responses are entirely determined either by the horizontal or by the vertical component therefore correlation does not affect those responses. Moreover, since responses throughout the section are not equally excited by both components but rather one component dominates, the effects of correlation are small throughout the dam. Fig. 8 illustrates this fact by showing stationary RMS responses at section A-A of Fig. 1 for the four loading conditions used. Only the normal stresses at that section have significant contributions from both components and so the effect of correlation is larger.

NON-STATIONARY RESPONSES

Thus far, only stationary RMS responses have been discussed. However, the formulation can compute time histories of the covariance matrix of any response vector. This capability is important because, in reality, both the excitation and the response may remain non-stationary. This, in turn, implies lower RMS responses and lower probabilities of exceeding response thresholds. Such probabilities can, in fact, be easily calculated within the formulation since evolutionary values of the variances of a response and its time derivative are computed.

The non-stationary output is extensive, so only some representative responses are presented. Fig. 9 shows the evolutionary RMS horizontal relative displacement and absolute acceleration responses at 0.2H and the RMS shear stress at 0.9H on the axis of symmetry of the finite element model. The non-stationary excitation defined by Fig. 4 was used. Also shown in Fig. 9 are the stationary RMS values and evolutionary RMS values from the shear beam model. Responses very close to stationary were attained, even for such a short duration motion, primarily because the modal damping values are relatively high. The near stationary value is not maintained for a significant amount of time, thus lowering probabilities of exceeding given responses. The shear beam yields very similar evolutionary responses. There is no effect of correlation between excitation components because the vertical component does not affect the responses shown.

Fig. 10 shows evolutionary RMS stresses at element 27. The correlation between components does have a small effect because both components excite those responses. Near stationary responses are attained. Of course, dams having lower fundamental frequencies, are less likely to achieve stationary responses.

CONCLUSIONS/FURTHER WORK

A finite element model of an earth dam was subjected to excitation having a specific frequency content and an intensity which approximately corresponds to 0.1g RMS ground acceleration. One set of responses was largely determiend by the horizontal excitation component and the first mode; another set was largely determined by the vertical component and the second mode. Thus the effects of correlation between components were not large. The responses of the shear beam model were very close to corresponding ones of the finite element model. However, the finite element model clearly indicates the risk of slope failure near the faces of the dam.

Continuing work will examine non-homogeneous and/or zoned dams. Probabilities of exceeding thresholds will be computed and the concept of "principal directions" as given by Penzien and Watabe [7] will be examined. The overall objective is to formulate failure criteria and a design method based on random vibration analytical results.

ACKNOWLEDGEMENTS

The financial support of the U.S. National Science Foundation (Grant PFR 80-17684) is kindly acknowledged. Discussions with Prof. Gary Bianchini were very helpful in the analysis and presentation of the results.

REFERENCES

1. Clough, R.W. and Chopra, A.K. "Earthquake Stress Analysis in Earth Dams" ASCE Journal of the Engineering Mechanics Division, Vol. 92, No. EM2, April, 1966

2. Clough, R.W. and Woodward, R.J. "Analysis of Embankment Stresses and Deformations" ASCE Journal of the Soil Mechanics & Foundation Division, Vol. 93, No. SM 4, July 1967

3. DebChaudhury, A. and Gasparini, D.A. "Response of MDOF Systems to Vector Random Excitation" ASCE Journal of the Engineering Mechanics Division, Vol. 108, No. EM2, April 1982

4. Gazetas, G., DebChaudhury, A. and Gasparini, D. "Random Vibration Analysis for the Seismic Response of Earth Dams" Geotechnique, 31, No. 2, 261-277, 1981

5. Gazetas, G., DebChaudhury,A. and Gasparini, D. "Stochastic Estimation of the Nonlinear Response of Earth Dams to Strong Earthquakes" Soil Dynamics and Earthquake Engineering, 1982, Vol. 1, No. 1 pp 39-46

6. Lai, S-S., P. "Overall Safety Assessment of Multistory Steel Buildings Subjected to Earthquake Loads" MIT Department of Civil Engineering Research Report R80-26, June 1980

7. Penzien, J. and Watabe, M. "Characteristics of 3-Dimensional Earthquake Ground Motions" Earthquake Engineering and Structural Dynamics, Vol. 3, 3-9,(1974)

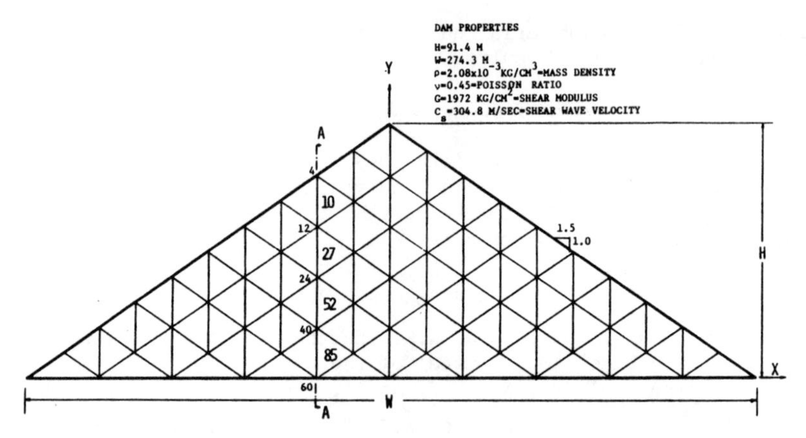

FINITE ELEMENT DISCRETIZATION WITH 100 CONSTANT STRAIN TRIANGULAR ELEMENTS AND 66 NODES

FIG. 1-DAM PROPERTIES

Mode	Frequency ω_i	Participation Factors Γ_x	Γ_y
1	7.715	-50.13	0
2	12.52	0	-27.33
3	14.60	- 7.016	0
4	19.31	-19.97	0
5	20.12	0	7.989
6	23.10	0	- 7.418
7	23.75	4.463	0
8	25.95	0.471	0
9	26.76	0	29.75
10	28.77	0	19.91
ζ_i = 0.2 in all modes			

a) Finite Element Model

Mode	Frequency ω_i	Participation Factors Γ_x
1	8.016	1.602
2	18.40	-1.065
3	28.85	.8514
4	39.30	- .7296
5	49.77	.6485
6	60.24	-.5895
ζ_i = 0.2 in all modes		

b) Shear Beam Model

TABLE 1 - DYNAMIC PROPERTIES OF DAM MODELS

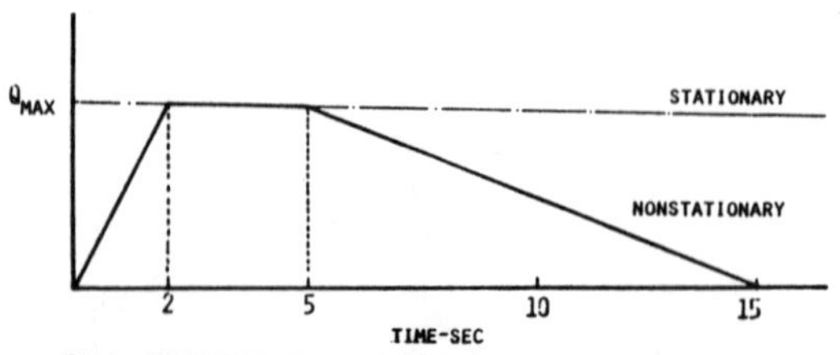

FIG.4 -STATIONARY AND NON-STATIONARY STRENGTH FUNCTIONS FOR THE EXCITATION COMPONENTS

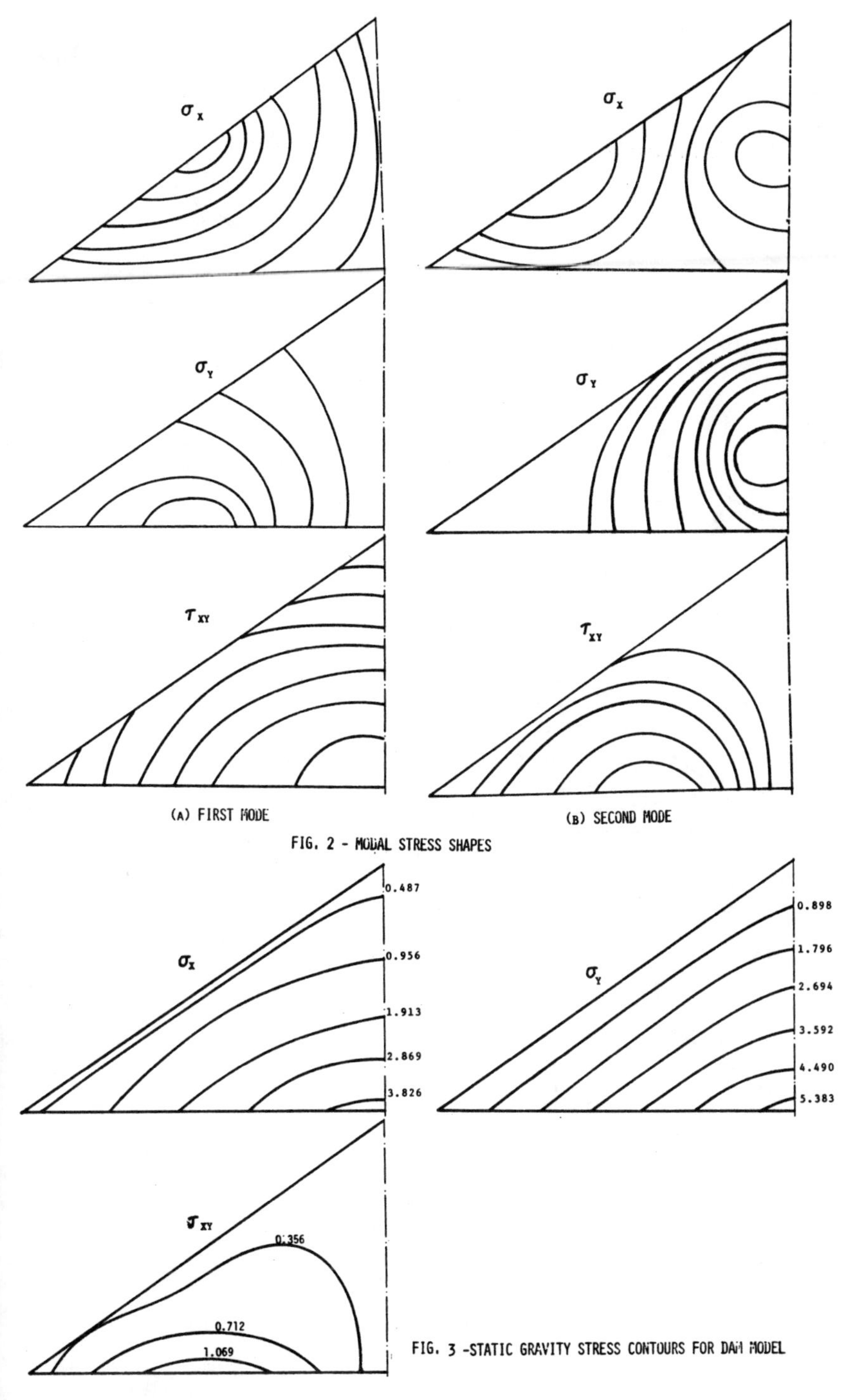

FIG. 2 - MODAL STRESS SHAPES

FIG. 3 -STATIC GRAVITY STRESS CONTOURS FOR DAM MODEL

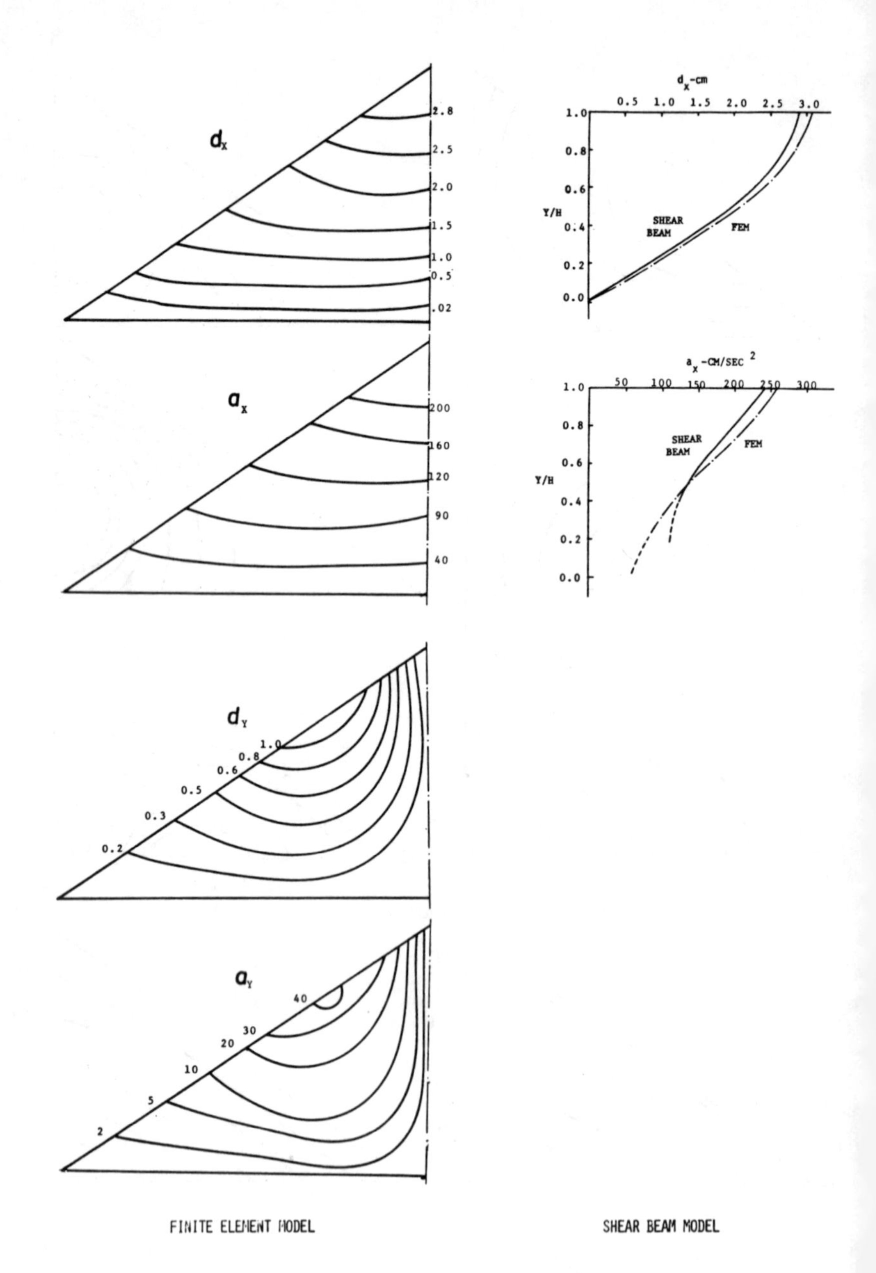

FIG. 5-STATIONARY RMS RESPONSES TO HORIZONTAL EXCITATION

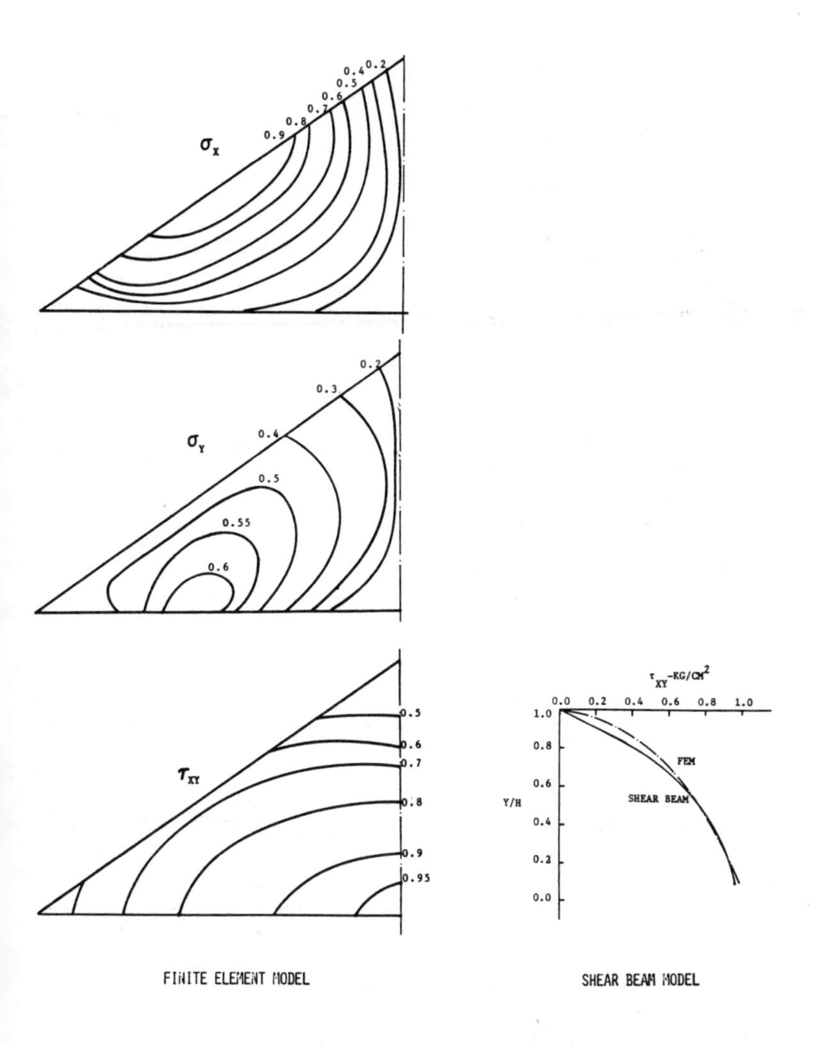

FIG. 5-STATIONARY RMS RESPONSES TO HORIZONTAL EXCITATION

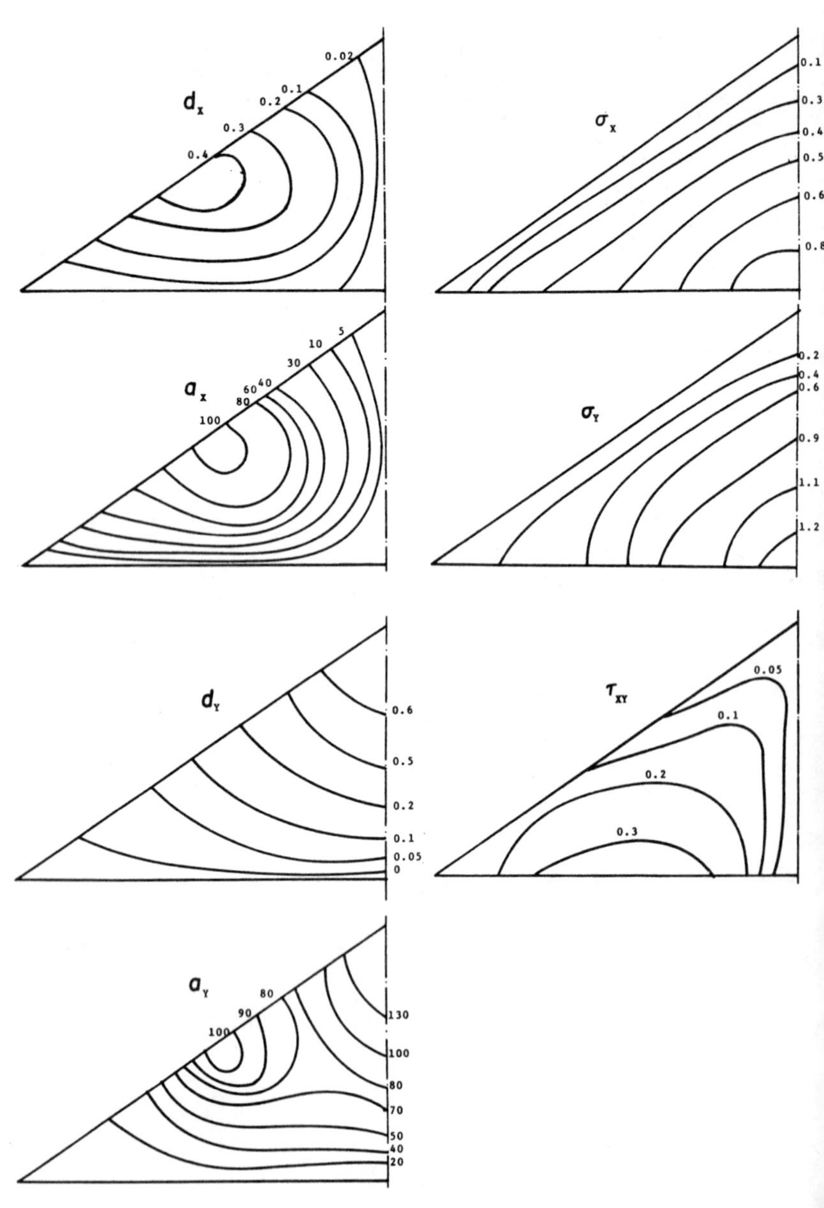

FIG. 6-STATIONARY RMS RESPONSES TO VERTICAL EXCITATION

FIG. 7-STATIONARY RMS RESPONSES TO COMBINED VERTICAL AND HORIZONTAL EXCITATION

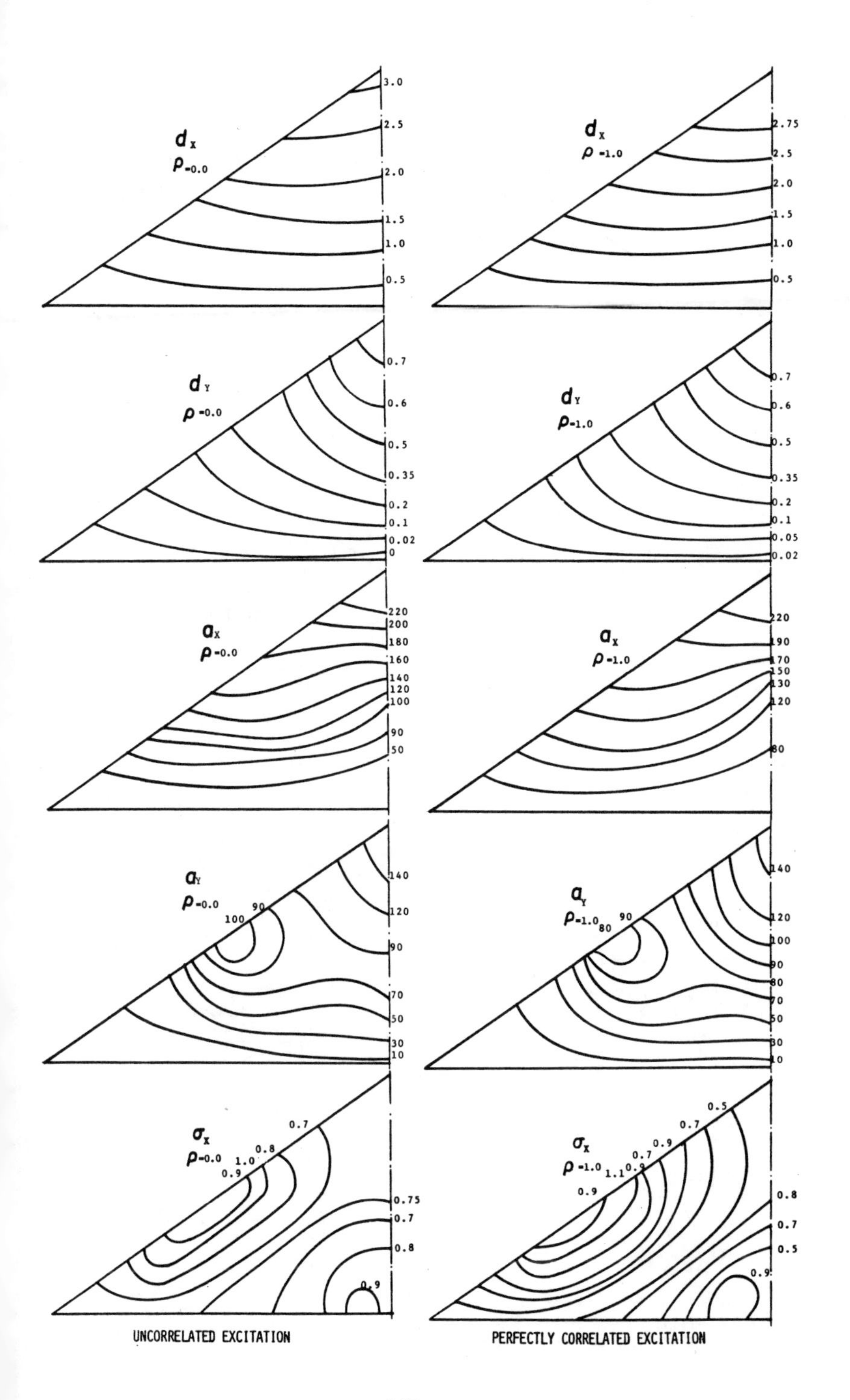

d_X ρ=0.0
3.0
2.5
2.0
1.5
1.0
0.5
d_X ρ=1.0
2.75
2.5
2.0
1.5
1.0
0.5
d_Y ρ=0.0
0.7
0.6
0.5
0.35
0.2
0.1
0.02
0
d_Y ρ=1.0
0.7
0.6
0.5
0.35
0.2
0.1
0.05
0.02
a_X ρ=0.0
220
200
180
160
140
120
100
90
50
a_X ρ=1.0
220
190
170
150
130
120
80
a_Y ρ=0.0
100
90
140
120
90
70
50
30
10
a_Y ρ=1.0
80
90
140
120
100
90
80
70
50
30
10
σ_X ρ=0.0
0.7
0.8
1.0
0.9
0.75
0.7
0.8
0.9
σ_X ρ=1.0
0.5
0.7
0.9
0.7
0.9
1.1
0.9
0.8
0.7
0.5
0.9
UNCORRELATED EXCITATION
PERFECTLY CORRELATED EXCITATION

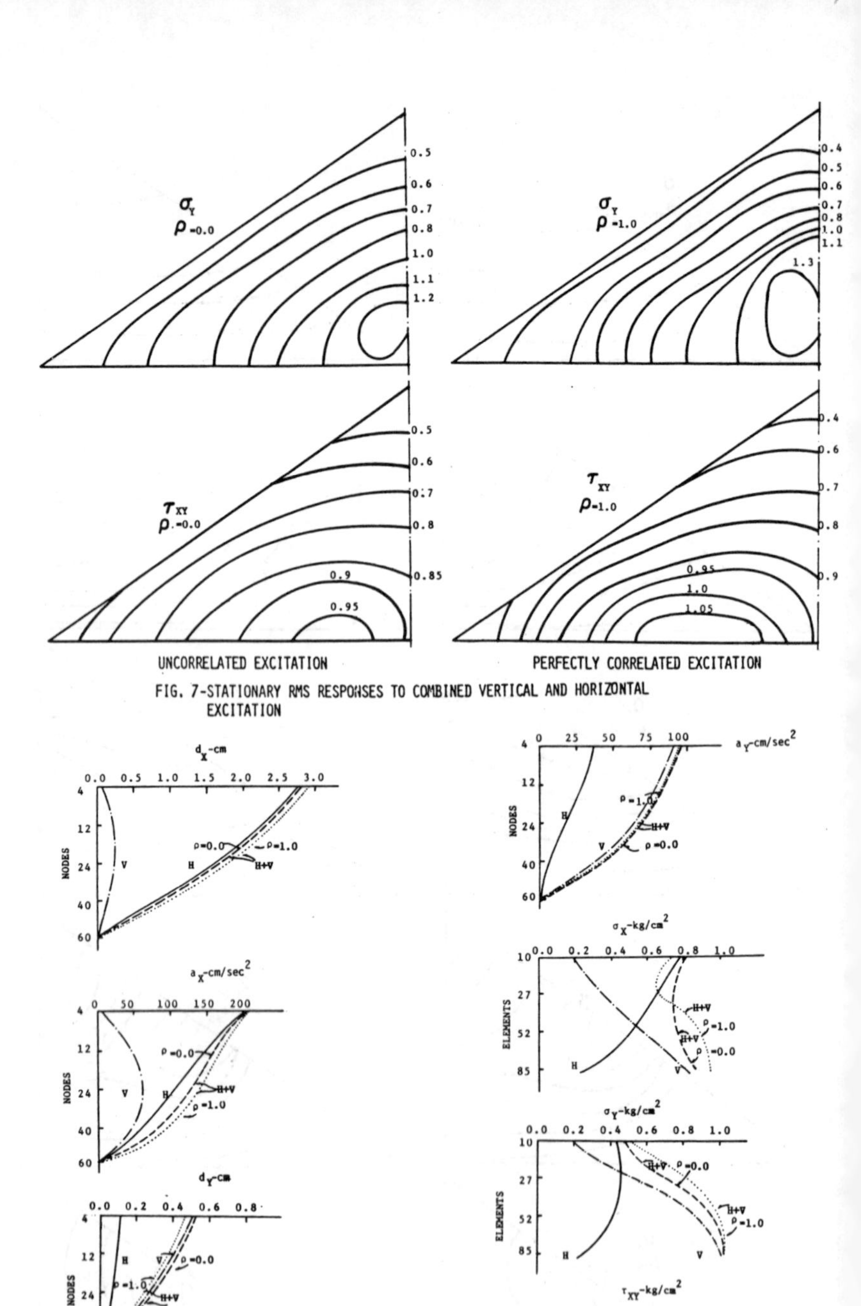

FIG. 7-STATIONARY RMS RESPONSES TO COMBINED VERTICAL AND HORIZONTAL EXCITATION

FIG. 8-EFFECTS OF CORRELATION BETWEEN EXCITATION COMPONENTS AND RESPONSES AT SECTION A-A (SEE FIG.1)

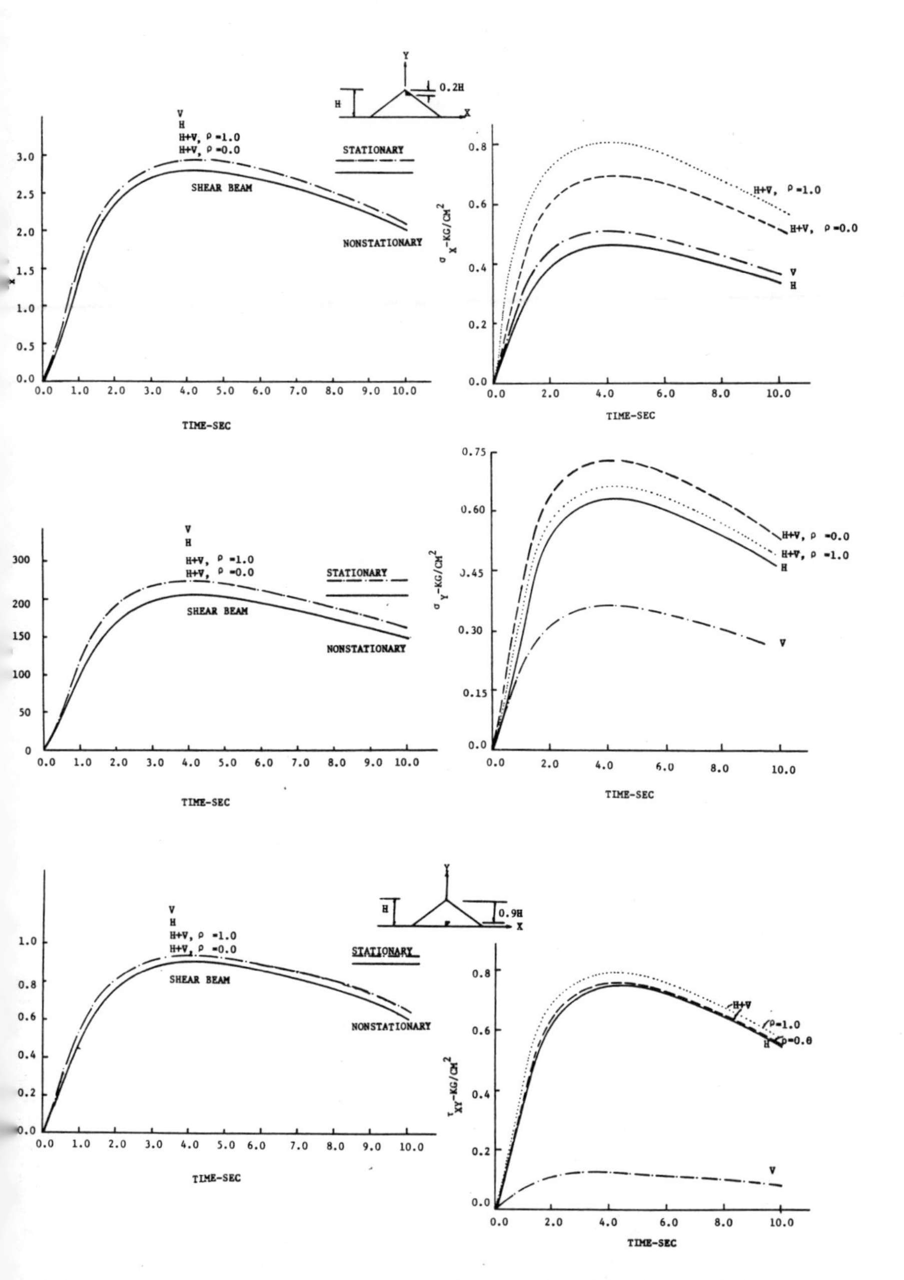

FIG. 9-EVOLUTIONARY RMS RESPONSES FROM F.E. AND SHEAR BEAM MODELS

FIG. 10-EFFECTS OF CORRELATION ON NONSTATIONARY RMS STRESSES AT ELEMENT 27

Soil Dynamics & Earthquake Engineering Conference / Southampton / 1982.07.13-15

Probabilistic Seismic Safety of Earth Slopes

M.SEMIH YÜCEMEN
Middle East Technical University, Ankara, Turkey

INTRODUCTION

The design and analysis of earth slopes are characterized by numerous uncertainties resulting from insufficient information and inadequate knowledge of the subsoil conditions. Properties of soil generally vary from location to location and also may change with time. Statistical methods for estimating the reliability of earth slopes were employed by various authors. Earlier studies, such as those by Biernatowski (1969), Wu and Kraft (1970a), Barboteu (1972), Yücemen and Tang (1975), Morla Catalan and Cornell (1976) and Matsuo (1976) were based on the plane strain assumption. In the more recent studies, probabilistic three-dimensional models of slope stability were proposed by Vanmarcke (1977a, 1980) and Veneziano et al. (1977). In these models, finite failure widths are considered and the spatial variation of shear resistance in all three directions is accounted for. Results of these studies as well as some other deterministic analyses (Baligh and Azzouz, 1975) indicate the importance of the three-dimensional effects.

In all of the previous studies, except the one by Wu and Kraft (1970b), the load effects are restricted to the dead weight of the soil lying above the potential failure surface. Because the variability in the soil density is negligible compared to that in the shear resistance, the driving moment is assumed to be deterministic. The aim of this paper is to extend the earlier work on the probabilistic three-dimensional models of slope stability to the case where the load effect is also random. For this purpose earthquake loads are taken into consideration and the reliability functions are given for an earth slope in seismic environment. The method accounts for the spatial variability of shear strength as well as the spatial and temporal variability of the ground acceleration.

In the probabilistic formulation, both the resistance to sliding and earthquake loading are taken as multiparameter random functions (random field). A detailed discussion of random fields are given by Cramer and Leadbetter (1967), Panchev (1971),

Pugachev (1965) and Yaglom (1973). Vanmarcke (1979a, 1979b) has developed an approach based on the concept of variance function and the scale of fluctuation for the analysis of homogeneous random fields.

This paper attempts only to formulate the three-dimensional seismic stability of earth slopes in the time domain and based on the theory of random fields. The implementation, using actual data will be the subject of a future paper.

PROBABILISTIC MODELS FOR RESISTANCE AND SEISMIC LOADING

In checking the seismic safety of earth slopes, besides shear stresses required for equilibrium one has to consider the shear stresses resulting from the ground acceleration. In a severe earthquake, it is expected that the moment attempting to cause the earth mass to slide down will exceed the resisting moment a few times for short periods during the earthquake. Successive earthquake pulses large enough to cause instability will cause successive slides in the downhill direction. For example, in the case of dams, the cumulative downhill slip may cause a significant lowering of the crest and hence to overtopping. In this case the problem will be the extent of sliding which can be safely tolerated by the dam. Two types of "unsafe" events will be considered here: (i) Slip: the initiation of a slide and (ii) Total failure: the cumulative slide length exceeding the allowable permanent displacement.

The deterministic seismic stability evaluations of earth slopes or embankments are generally carried out in terms of a seismic coefficient, A (Makdisi and Seed, 1978). The quantity Ag corresponds to the acceleration, where g is the acceleration of gravity. In the probabilistic formulation that follows, both the resistance to sliding and the earthquake loading will be expressed in terms of seismic coefficients.

Resistance to sliding motion

It will be assumed that an earth mass will start sliding when the overturning moment due to gravitational and seismic forces exceeds the resisting moment. While the sliding progresses, the resistance (and the gravitational force) will continue to act whereas the seismic forces will be gradually reduced to zero and then reversed, so that the sliding of the earth mass is decelerated and then stopped. When earthquake acceleration is again increased in the downhill direction the sliding commences again and the process is repeated. With each repetition the accumulated displacement will increase and when the earthquake ends a certain amount of permanent displacement will remain.

The resistance to sliding will be described by the yield seismic coefficient A_r, and its exceedence will initiate the slide. From a deterministic point of view, A_r is the pseudo static seismic coefficient that reduces the factor of safety to unity. The yield seismic coefficient can be expressed in terms of the resisting and driving moments as follows:

Consider the sliding cross sectional element of the earth

slope shown in Figure 1, where a circular arc of radius r defines the sliding surface. It is assumed that the entire slope is in cohesive soil and it is appropriate to use undrained strength throughout the slope. The weight of the element is denoted by W and it has a lever arm d' about the center of rotation. Let $A_r(x)$ be the yield seismic coefficient (acting at the center of gravity of the cross section) at any location x along the axis of the slope, that the slope can resist without any slide occurring. Consideration of dynamic equilibrium of driving and resisting moments at a typical cross section $x=x_o$ yields to:

$$A_r(x_o)W(x_o)d'' = M_R(x_o) - M'_R(x_o) \tag{1}$$

Here, d" is the moment arm of the seismic force, $M_R(x_o)$ is the total resisting moment under dynamic conditions and $M'_R(x_o)$ denotes the resisting moment under static conditions and equals to $M'_D(x_o) = W(x_o)d'$, where $M'_D(x_o)$ is the driving moment due to the weight of the soil element. Therefore,

$$A_r(x_o) = \frac{1}{W(x_o)d''} [M_R(x_o) - W(x_o)d'] \tag{2}$$

In Equation (2), A_r is expressed in terms of other variables. Among these variables, M_R, which depends on the shearing strength of soil mobilized during an earthquake, is the main source of uncertainty. Since the variability in the geometric parameters and soil density is small, $W(x_o)$, d' and d" will be treated as deterministic quantities.

In a probabilistic approach to the three-dimensional slope stability problem, it is assumed that M_R may vary depending on the location of the cross sectional element. However, we shall assume that the statistical character of the random deviations about the mean will not change throughout the slope and also the soil properties show no significant dependence on depth. In the light of these assumptions the soil mass will be treated as a statistically homogeneous medium. Accordingly, A_r, is taken as a weakly stationary random process whose statistical parameters (mean $\bar{A}_r$, standard deviation $\tilde{A}_r$ and coefficient of variation δ_r) are to be estimated from those of M_R using Equation (2).

The three-dimensional formulation of slope stability requires consideration of "end effects" which will depend on the location of the cylindrical failure surface (Vanmarcke, 1980). For the surface located between $x_1 = x_o - b/2$ and $x_2 = x_o + b/2$, the resulting yield seismic coefficient may be expressed as follows:

$$A_{r,b}(x_o) = \frac{1}{M''_{D,b}(x_o)} [M_{R,b}(x_o) - M'_{D,b}(x_o) + R_e] \tag{3}$$

where, b = width of the failure area; x_o= center of the failure area; $M''_{D,b}(x_o)$ = total driving moment due to the weight of soil mass above the failure surface acted upon by unit seismic coefficient, for uniform cross sections between x_1 and x_2 it equals to $W(x_o)bd'' = M''_D b$; $M'_{D,b}(x_o)$ = total driving moment due to the weight of soil mass above the failure surface, for uniform cross sections between x_1 and x_2 it equals to $W(x_o)bd' = M'_D b$; R_e = contribution of the end sections of the failure surface to the resisting moment; $M_{R,b}(x_o)$ = integral of the resisting moments $M_R(x)$ for the cross sections within the failure area located between x_1 and x_2, i.e.

$$M_{R,b}(x_o) = \int_{x_1}^{x_2} M_R(x)\, dx \tag{4}$$

If we assume that the cross section of the slope does not vary along the axis of the slope, then by using the results obtained by Vanmarcke (1980) and based on the first-order approximation we can show that

$$\bar{A}_{r,b} = \frac{1}{\bar{M}''_D} [\bar{M}_R(1 + d/b) - \bar{M}'_D] \tag{5}$$

$$\tilde{A}_{r,b} = \frac{1}{\bar{M}''_D} \tilde{M}_R \, \Gamma_R(b) \tag{6}$$

$$\delta_{r,b} = \frac{\tilde{M}_R \, \Gamma_R\,(b)}{\bar{M}_R(1 + d/b) - \bar{M}'_D} \tag{7}$$

where, a "bar" and a "tilde" over a random variable denote the mean and standard deviation of the random variables; M_R, M'_D and M''_D are the cross sectional (plane strain) values of the resisting and driving moments; $d = R_e/\bar{M}_R$, has the dimension of length and is obtainable from a plane strain deterministic analysis; $\Gamma_R(b)$ is the dimensionless reduction factor to be applied to the standard deviation, $\tilde{M}_R$, due to the smoothing effect of the integration operation given by Equation (4), and δ denotes the coefficient of variation.

In general application, through the use of a variance function (square of the reduction factor for the standard deviation), the variance of a spatially averaged soil property can be obtained directly from the corresponding point variance of the same property. The decrease in the standard deviation due to one-dimensional spatial averaging is represented by the reduction factor $\Gamma(x)$ and the decrease in the variance by $\Gamma^2(x)$. Here, the averaging is done along the axis of the slope and over a finite length of b. Vanmarcke (1979b) has tabulated the variance functions for some representative correlation structures along with the corresponding scales of fluctuation. The scale of fluctuation can simply be interpreted as the "effective correlation distance". An approximate relation for $\Gamma_R^2(b)$ in terms of the scale of fluctuation, θ, is given as follows (Vanmarcke 1980):

$$\Gamma_R^2(b) \simeq \begin{cases} 1 & b \leq \theta \\ \theta/b & b \geq \theta \end{cases} \quad . \tag{8}$$

where θ is the scale of fluctuation associated with M_R.

According to Equation (8), for averaging widths that are less than the effective correlation distance, there will be no reduction in the variance due to spatial averaging. For averaging distances exceeding θ, the decrease in variance is inversely proportional to the averaging width, with scale of fluctuation being the proportionality factor. Equation (8) is exact for $b=0$ and $b=\infty$. A detailed discussion on the assessment of the scale of fluctuation as well as more refined approximations for the variance functions are given by Vanmarcke (1979a, 1979b).

Earthquake induced loading

The earthquake loading will be described by the earthquake induced seismic coefficient $A_s(\underset{\sim}{u},t)$ operating at any point of the potential sliding mass. In the most general case $A_s(\underset{\sim}{u},t)$ is a space-time process with $\underset{\sim}{u}$ being the three-dimensional space vector ($\underset{\sim}{u}'$:{x,y,z}) and t represents time. In other words $A_s(\underset{\sim}{u},t)$ is a four-dimensional random field. However, by making some simplifying assumptions the dimensionality of this most general case may be reduced.

In our case, the main spatial variability is along the depth (z-axis) and along the axis of the slope (x-axis). The variability of earthquake induced acceleration along the x and z-axes are of different character and will exhibit a different correlation structure. It is expected that the earthquake induced acceleration will vary along the depth consistent approximately with the first mode of vibration, with a maximum value at the top of the slope (Makdisi and Seed, 1978). Therefore, almost a perfect correlation will exist along the z-axis. Here, our interest lies in the seismic coefficient acting at the center of gravity of the sliding cross sectional element, averaged over the full depth. This quantity will be denoted by $\bar{A}_z$ and can be interpreted as the mean amplification factor, since it quantifies the mean change in the base excitation over the total height of the slope. The mean amplification factor is assumed to remain the same during the earthquake and along the axis of the slope (i.e. it is independent of t and x). The value of $\bar{A}_z$ is to be determined from a dynamic analysis. Such a method is described by Makdisi and Seed (1978) for dams. The uncertainty associated with the dynamic analysis model will be quantified by the coefficient of variation, δ_z. Tang (1981) estimated the coefficient of variation in the dynamic amplification factor to be greater than 0.35 for an embankment using the results of a study made for the response of reinforced concrete structures.

In accordance with the above discussion, the earthquake induced seismic coefficient is expressed as the product of three factors as follows:

$$A_s(\underset{\sim}{u},t) = \bar{A}_z . A(x,t) . a_p \tag{9}$$

The last term in Equation (9) is a deterministic scaling factor (peak ground acceleration corresponding to a specified return period) to be obtained from seismic hazard curves. Failure probabilities computed for a given value of a_p will be a conditional probability. The conditionality may be removed by integrating the conditional probabilities over the distribution of a_p. $A(x,t)$ is a space-time process describing the behavior of base acceleration (as a fraction of g) at time t and at a point x along the axis of the slope due to the strong motion record of an earthquake normalized to a maximum base acceleration of 1g. No realization of $A(x,t)$ is available, however, $A(x,t)$ may be inferred from $A(t)$ as will be explained in the next section. Therefore, in selecting the strong motion record and in adjusting the peak acceleration, $A(t)$ will form the basis.

The total driving moment due to earthquake excitation on the cross sectional elements within a failure surface located between x_1 and x_2 can be written as follows:

$$M''_{D,b}(x_o) = \int_{x_1}^{x_2} W(x_o) d'' \bar{A}_z A(x,t) a_p \, dx$$

$$= M''_{D,b}(x_o) \bar{A}_z a_p Y_b(x_o,t) \quad (10)$$

In Equation (10), $M''_{D,b}(x_o)$ denotes the driving moment due to the sliding soil mass of width b acted upon by unit seismic coefficient, and

$$Y_b(x_o,t) = \int_{x_1}^{x_2} A(x,t) dx \quad (11)$$

The total driving moment computed from Equation (10) is conditional on the given value of a_p, and accordingly the future computations based on this driving moment will also be conditional on a_p.

As seen, the earthquake induced seismic coefficient, $A_{s,b}(x_o,t)$ to be compared with the yield seismic coefficient $A_{r,b}(x_o)$ (given by Equation (3)) is

$$A_{s,b}(x_o,t) = \bar{A}_z a_p Y_b(x_o,t) \quad (12)$$

With these assumptions, the statistical analysis for the earthquake loading is reduced to the consideration of the one-dimensional (spatial) integral of the random field $A(x,t)$.

The complete time history of ground motions, $A(t)$, during an earthquake will have nonstationary features, especially during the buildup and decay periods of the earthquake. As a result, $A(x,t)$ is a nonhomogeneous random field. However, in this study we shall consider only the simple model based on the assumption that any change in the spectral composition during the whole earthquake may be neglected. Bolotin (1969) took this as a first approximation considering that the high frequency jolts which ordinarily terminate an earthquake exert very little influence on the strength, as compared with the powerful low frequency impulses. If this assumption is made, the components of the acceleration of the ground may be represented as the realization of a stationary stochastic process modulated with the aid of a deterministic envelope function of time. Chang et al. (1979) have analyzed four major California earthquake records and indicated that the principal source of nonstationarity of the earthquake acceleration data lies on the time dependence of variance of the driving noise process rather than the filtering parameters. The earthquake records were subdivided into nonoverlapping time windows in order to eliminate the nonstationarity of the variance. Over these short segments of time the statistical characteristics of the earthquake excitations were approximately constant; in other words local stationarity was achieved. The nonstationarity was problematic only for the initial 0-5 secs., during which the variance always builds up from essentially zero to some finite (perhaps even maximal) value. Here, as a first approximation we shall assume that A(t) could be transformed into a stationary process, and thus we shall take A(x,t) to be a weakly homogeneous random field. For weakly homogeneous random fields, the mean value is a constant (in our case it is zero) and the covariance function depends only on the difference of the instants of time and the difference of the locations along the x-axis at which the ordinates of the random field are taken. Expressed in mathematical terms, the covariance function for this two-dimensional random field may be written as follows:

$$B_A(x,t,x',t') = B_A(x-x',\ t-t') = B_A(\xi,\tau) \tag{13}$$

where, $\xi = x-x'$ is the distance between two cross sections and $\tau = t-t'$ is the time lag.

As discussed earlier, A(x,t) is assumed to be a zero-mean weakly homogeneous random field. The point variance of A(x,t) will be denoted by $\tilde{A}^2$ and it equals to $B_A(0,0)$. For notational convenience, the reference to x_o will be dropped and the spatial integral process will be denoted by $Y_b(t)$. Note that for homogeneous random fields the statistics of $Y_b(t)$ do not depend on x_o. The mean of $Y_b(t)$ will be zero, since A(x,t) is a zero-mean random field. The variance of $Y_b(t)$ will be expressed as a function of the point variance $\tilde{A}^2$ and the one-dimensional (spatial) reduction factor $\Gamma_A^2(b)$ (Vanmarcke 1979b):

$$\tilde{Y}_b^2 = \tilde{A}^2\ b^2\ \Gamma_A^2(b) \tag{14}$$

In order to compute $\Gamma_A^2(b)$, one needs information on the spatial correlation of the ground acceleration along the x-axis. Direct data on the spatial variation of earthquake induced motions are generally not available. However, if we assume that seismic waves propagate along the x-axis with a constant velocity, c, then the space-time covariance function $B_A(\xi,\tau)$ may be expressed solely based on the temporal covariance function

$B_o(\tau)$ as follows:

Let $A_o(t)$ be the ground motion (as a fraction of g) at a point along the x-axis with the temporal covariance function $B_o(\tau)$. Here the subscript "o" is used to denote that the functions under consideration are one-dimensional. Since this motion travels at a constant velocity, c,

$$A_o(t) = A(x, t + x/c) \tag{15}$$

Replacing t by t-x/c in Equation (15) we obtain

$$A_o(t - x/c) = A(x,t) \tag{16}$$

Then, since A(x,t) is a weakly stationary zero-mean random field

$$B_A(\xi,\tau) = E[A(x,t)A(x',t')] = E[A_o(t-x/c)A_o(t'-x'/c)]$$

$$= B_o[(t-x/c)-(t'-x'/c)] = B_o(\tau - \xi/c) \tag{17}$$

where E(.) denotes expectation. Hence, the space-time covariance function $B(\xi,\tau)$ is expressed by the one-dimensional covariance function $B_o(\tau)$. Phsically it means that the heterogeneities of the A-field move along the x direction with a constant speed c without any changes, as if it is a "frozen" random field.

Due to high velocities of seismic shear waves, at a given instant of time, all cross sections of the slope over a width of b will be subjected practically to the same earthquake pulse, unless a very large b value is under consideration. Hence, due to nearly perfect spatial correlation, $\Gamma_A^2(b)$ will be close to 1.0 and $\tilde{Y}_b \simeq b\tilde{A}$.

The mean of $A_{s,b}(x_o,t)$ is zero. Assuming A_z and Y_b to be statistically independent and using the first-order approximation, the standard deviation, $\tilde{A}_{s,b}$, becomes:

$$\tilde{A}_{s,b} \simeq (\tilde{A}_z^2 + b^2\,\tilde{A}^2)^{\frac{1}{2}}\,a_p \tag{18}$$

RISK OF FAILURE AT A SPECIFIC LOCATION

Failure in this section is defined as the occurrence of the event, whereby the cumulative downward displacement experienced during an earthquake by a soil mass of width b, centered at a specified point x_o along the axis of the slope, exceeds the allowable displacement. During an earthquake excitation, whenever the earthquake induced seismic coefficient $A_{s,b}(x_o,t)$ exceeds the yield seismic coefficient, $A_{r,b}(x_o)$, a permanent deformation (downward slip) will take place, and these deformations will accumulate. Here, it is assumed that the deformations will take place only in the downward direction. We also assume that the yield seismic coefficient remains the same during the earthquake. For cohesive soils the dynamic shearing strength will be close to the static shearing strength (Makdisi and Seed, 1978). We shall assume that the factors affecting the resistance to sliding remain the same after each slip during an earthquake.

The expected number of slips taking place during an earthquake of duration T will be obtained by using the level crossing approach. This information coupled with the information on the expected magnitude of permanent deformation per slip, will give the expected value of the total displacement resulting from a single earthquake.

A soil mass of width b centered at location x_o will slide downwards whenever $A_{s,b}(t) > A_{r,b}$ at any time between t=0 and t=T. Therefore, the "slip" along the potential failure surface can be defined as

$$[\text{Slip}] \equiv [A_{s,b}(t) > A_{r,b} \; ; \; t \in (0,T)] \tag{19}$$

$A_{s,b}(t)$ is a zero-mean random process, and here we are interested in the events where this process passes from below to above the level $A_{r,b}$, which will be referred to "upcrossings". We seek the mean rate, ν, of occurrence of such events. Although, $A_{r,b}$ is assumed to be constant with time (i.e. factors influencing resistance do not change during an earthquake and with the number of slips previously occurred), it is a random variable due to the variability of the shearing resistance along the potential failure surface. The problem of upcrossing a random threshold can be solved by conditioning the mean upcrossing rate on the specific values of $A_{r,b}$. The total solution is then found by integrating the conditional upcrossing rate over the probability density function of $A_{r,b}$.

To compute the mean number of upcrossings of a specified level $A_{r,b}=a_r$, it is necessary to know the distribution of $A_{s,b}(t)$. The total solution requires also the probability distribution of $A_{r,b}$. $A_{s,b}(t)$ and $A_{r,b}$ are assumed to be a Gaussian process and a Gaussian variable, respectively, on the basis of the central limit theorem, since both quantities result essentially from the summation (integration) of the contributions of the cross sectional elements lying between x_1 and x_2. The mean number of upcrossings per unit time, conditional on the specified level $A_{r,b}=a_r$, is denoted by $\nu(a_r)$ and will be computed from the following expression:

$$\nu(a_r) = \frac{\tilde{\dot{A}}_{s,b}}{2\pi \, \tilde{A}_{s,b}} \exp(-a_r^2/2\tilde{A}_{s,b}^2) \tag{20}$$

where $\dot{A}_{s,b}$ represents the derivative of $A_{s,b}$ and

$$\tilde{\dot{A}}_{s,b}^2 = - \frac{d^2 \, B_o(\tau)}{d\tau^2} \bigg|_{\tau=0} \tag{21}$$

In the above equations, $B_o(\tau)$ is the covariance function of $A_o(t)$ and τ is the time lag. Here it is assumed that $A_{s,b}(t)$ is a differentiable random process in continuous time with continuous derivative.

The overall mean rate of upcrossings per unit time, ν is

$$\nu = \int_{-\infty}^{\infty} \nu(a_r)\, f_{A_{r,b}}(a_r)\, da_r \tag{22}$$

in which $f_{A_{r,b}}$ is the probability density function of $A_{r,b}$ and given as

$$f_{A_{r,b}}(a_r) = \frac{1}{\sqrt{2\pi}\,\tilde{A}_{r,b}} \exp\left[-\frac{1}{2}\left(\frac{a_r - \bar{A}_{r,b}}{\tilde{A}_{r,b}}\right)^2\right] \tag{23}$$

The value of ν forms the basis for predicting the probability distribution of extreme values of the Gaussian process $A_{s,b}(t)$. For slopes of engineering interest the threshold will be high. The upcrossings of a high level by a smooth Gaussian process are asymptotically Poisson. If the safety considerations require that no slip of the soil mass under consideration occurs, then the failure probability will have the following form for a time interval of (0,T):

$$P_f = 1 - \exp(-\nu\, T) \tag{24}$$

Here, we assumed that the process starts below the threshold with probability 1.0 and the upcrossings occur randomly in accordance with the Poisson process.

On the other hand, if some permanent deformation is permitted, then the cumulative permanent displacement Δ_t developed during the earthquake excitation becomes to be the quantity of concern. Δ_t will be the sum of the contributions Δ_i, each resulting from an upcrossing of the yield level, i.e.

$$\Delta_t = \sum_{i=1}^{N} \Delta_i \tag{25}$$

where, N is the random number of upcrossings during the interval (0,T). (The duration of the earthquake will also be a random quantity; however, here T is assumed to be known). The expected value of N is

$$E(N) = \bar{N} = \nu\, T \tag{26}$$

The distribution of N is approximately Poisson for high-reliability slopes which imply high yield levels.

The amount of permanent deformation per slip, Δ, can be estimated from empirical data (Newmark (1965), Makdisi and Seed (1978)) or from theoretical considerations. For example, Karnopp and Scharton (1966) derived an approximate expression for the average amount of permanent deformation, Δ, resulting from an isolated crossing of the yield level in terms of the variance of the response and the yield level. If the magnitude of the permanent deformation per slip is assumed to be the same at every slip and equals to Δ, then the expected value of the total displacement Δ_t, becomes:

$$E(\Delta_t) = \nu_t \, T \, \Delta \tag{27}$$

Consistent with the above assumptions and notation, the event of total failure will be

$$[\text{Total Failure}] = [\Delta_t > \Delta_a] \tag{28}$$

where Δ_a denotes the allowable displacement. To compute the probability of total failure, a probability distribution should be assumed for Δ_t.

RISK OF SLOPE FAILURE

The survival of the slope as a system requires that no failure occurs along the axis of the slope over a total length of L and during the time interval of (0,T). In this case, failure will mean the occurrence of a single slip within any portion of the slope with a width of b. Failure, in terms of cumulative deformations will not be pursued here.

Let, U(x,t) be a "composite" random field defined as follows:

$$U(x,t) = A_{s,b}(x,t) - A_{r,b}(x) \tag{29}$$

We shall assume the failure width b to be fixed. A slope failure due to a single slip will occur if U(x,t) crosses into the unsafe region, $\{U(x,t) > 0\}$, at any location between b/2 and L-b/2 and during any instant between 0 and T, i.e.

$$[\text{Slope Failure}] = [U(x,t) > 0;\ x\ \varepsilon(b/2, L-b/2) \text{ and } t\ \varepsilon(0,T)] \tag{30}$$

$A_{s,b}(x,t)$ and $A_{r,b}(x)$ are both assumed to be Gaussian, based on the central limit theorem. Therefore, U(x,t) will be a Gaussian random field. We define

$$V(x,t) = U(x,t) - \overline{U} \tag{31}$$

with $\overline{U}$ being the mean of U(x,t), so that V(x,t) becomes a zero-mean Gaussian field. Since $A_{s,b}(x,t)$ is a zero-mean random field, from Equation (29) we get $\overline{U} = -\overline{A}_{r,b}$. Therefore, the new unsafe region becomes: $\{V(x,t) > \overline{A}_{r,b}\}$.

For the case of random fields, such as V(x,t) there is no obvious generalization of the concept of number of upcrossings, while the number of maxima above a level generalizes readily (Hasofer, 1976). For high thresholds, the upcrossings are rare Poisson events and the slope failure probability, P_F, can be approximated as follows (Morla Catalan and Cornell, 1976):

$$P_F = Pr(N \geq 1) \simeq Pr(N^* \geq 1) = 1-\exp(-\overline{N}^*) \tag{32}$$

where, N = the number of excursions of the level $\overline{A}_{r,b}$; N* = the number of maxima above level $\overline{A}_{r,b}$ and $\overline{N}^*$ = the mean (expected) value of N*. Therefore, to compute the failure probability only the expected value of N* is required.

An asymptotic formula for the mean number of maxima of a random field above high levels in stationary Gaussian random fields has been obtained by Nosko (1969), and a proof has been given by Hasofer (1976). Based on this formula, the expected number of maxima above level $\overline{A}_{r,b}$, per unit "area" (unit distance by unit time), λ, is as follows:

$$\lambda = |K|^{\frac{1}{2}} \frac{\overline{A}_{r,b}}{(2\pi)^{3/2}\,\tilde{V}^3} \exp(-\overline{A}^2_{r,b}/2\tilde{V}^2) \tag{33}$$

Accordingly, for the region under consideration: $\{x\ \varepsilon(b/2,L-b/2)$ and $t\ \varepsilon(0,T)\}$

$$E(N^*) = \overline{N}^* = (L-b)T\ \lambda \tag{34}$$

In Equation (33), $\tilde{V}$ is the standard deviation of V(x,t) and $|K|$ is the determinant of the covariance matrix of the first partial derivatives of V(x,t). The covariance matrix [K] is a 2x2 symmetric matrix with elements

$$k_{11} = \mathrm{Cov}(\frac{\partial V}{\partial x}, \frac{\partial V}{\partial x}) = - \frac{\partial^2 B_V(\xi,\tau)}{\partial \xi^2}\Big|_{\xi=\tau=0}$$

$$k_{22} = \mathrm{Cov}(\frac{\partial V}{\partial t}, \frac{\partial V}{\partial t}) = - \frac{\partial^2 B_V(\xi,\tau)}{\partial \tau^2}\Big|_{\xi=\tau=0} \tag{35}$$

$$k_{12} = \mathrm{Cov}(\frac{\partial V}{\partial x}, \frac{\partial V}{\partial t}) = - \frac{\partial^2 B_V(\xi,\tau)}{\partial \xi\, \partial \tau}\Big|_{\xi=\tau=0}$$

where, Cov(.) denotes the covariance and $B_V(\xi,\tau)$ is the covariance function of V(x,t) which is to be computed from the covariance function of U(x,t). The random field U(x,t) is a "composite" random field, since it is based on two component random fields (Equation (29)). Assuming the two component random fields to be independent, the correlation properties of U(x,t) may be computed as a weighted average of the basic properties of the component random fields. The weighting factors are taken inversely proportional to the variances of the component fields (Vanmarcke, 1979a).

Vanmarcke (1979b) has also derived a general expression for the asymptotic mean rate of occurrence of maxima of local averages of an n-dimensional random field above a specified high level and showed that it converges to the result attributed to Nosko (1969). According to Vanmarcke's approach, by fixing t and x, the two-dimensional random field V(x,t) is described in terms of the one-dimensional (conditional) random processes $V_{x|t}(x)$ and $V_{t|x}(t)$, respectively. Then,by using the one-dimensional theory of extremes, the expected value of the random lengths of the excursions of these one-dimensional random processes are computed. These may be viewed as the "dimensions" of the area of the regions where the original two-dimensional random field equals or exceeds the critical level. If these random lengths are assumed to be statistically independent, then the product of their expectations will give the expected value of the "excursion area". From the expected value of the "excursion area" and assuming a Poisson process for high level exceedences, the mean rate of excursions per unit area is obtained. This quantity also approximates the expected number of maxima above the critical level, per unit area of the random field. For the details of this approach the reader is referred to Vanmarcke (1979b).

Once the expected value of the number of maxima above $\bar{A}_{r,b}$ is obtained, either from Equations (33) and (34) or by following Vanmarcke's method, then the slope failure probability can be computed directly from Equation (32). It should be kept in mind that this failure probability is based on the assumption that the slope fails as a system, when any soil mass of width b along the axis of slope undergoes at least a single downward slip. Therefore, this will form a conservative estimate of the slope failure probability based on the exceedence of the allowable displacement by the cumulative permanent displacement.

SUMMARY AND CONCLUSIONS

This paper attempts to formulate probabilistically the seismic safety of earth slopes in the space-time domain. Soil properties are assumed to be statistically homogeneous throughout the earth slope and remain unchanged with time. The resistance to sliding is modeled as a spatial random process and the earthquake induced loading as a space-time random field. The resistance to sliding and the earthquake excitation are both expressed in terms of "equivalent" seismic coefficients. It is shown that spatial correlation of the ground acceleration along the axis of the slope can be estimated using temporal correlation data.

Failure models, based on the occurrence of a single slip and based on the magnitude of accumulated permanent displacement are considered. In the proposed formulation, the risk of failure at a specific location is evaluated along with the risk that a failure will occur anywhere along the axis of the slope. The corresponding failure probabilities are obtained by using the one-dimensional level crossing approach and its extension to the two-dimensional random fields. The effect of seismic activity during the lifetime of an earth slope is reflected into the analysis by specifying the value of maximum ground acceleration, a_p, which is to be obtained from a seismic hazard analysis.

In the application of the probabilistic framework outlined herein, the major difficulty will be the assessment of the various covariance functions from actual data. Computational difficulties may also arise in removing the conditionality on the failure probabilities resulting from the assumptions on certain input parameters (for example, a_p and Δ). A realistic evaluation of the reliability of an earth slope should also include the modeling uncertainties as well as uncertainties associated with soil testing.

ACKNOWLEDGEMENTS

This paper is based in part on the research conducted during the author's stay at Massachusetts Institute of Technology, which was made possible by a Fulbright advanced research fellowship. The author wishes to express his sincere appreciation to Prof. E. H. Vanmarcke, with whom he had many useful discussions. Thanks are due Miss F. Yeşiladalı who has expertly typed the manuscript.

REFERENCES

Baligh, M.M. and Azzouz, A.S. (1975) End Effects on Stability of Cohesive Slopes. Proc. ASCE, J. Geotech. Eng. Div., 101, GT 11.

Barboteu, G. (1972) Reliability of Earth Slopes. M.Sc. Thesis, Dept. Civil Eng., M.I.T., Cambridge, Mass.

Biernatowski, K. (1969) Stability of Slopes in Probabilistic Solution. Proc. 7th Intern. Conf. Soil Mech. Foundation Eng., Mexico, 2:527-530.

Bolotin, V.V. (1969) Statistical Methods in Structural Mechanics. Holden Day Inc., San Francisco.

Chang, M.K., Kwiatkowski, J.W., Nau, R.F., Oliver, R.M. and Pister, K.S. (1979) ARMA Models for Earthquake Ground Motions. Earthquake Eng. Research Center, University of California, Rep. 79/19.

Cramer, H. and Leadbetter, M.R. (1967) Stationary and Related Stochastic Processes. J. Wiley Inc., New York.

Hasofer, A.M. (1976) The Mean Number of Maxima Above High Levels in Gaussian Random Fields. J. Appl. Probability, 13:377-379.

Karnopp, D. and Scharton, T.D. (1966) Plastic Deformation in Random Vibration. J. Acoust. Soc. Am., 39:1154-1161.

Makdisi, F.I. and Seed, H.B. (1978) Simplified Procedure for Estimating Dam and Embankment Earthquake-Induced Deformations. Proc. ASCE, J. Geotech. Eng. Div., 104:849-867.

Matsuo, M. (1976) Reliability in Embankment Design. Dept. Civil Eng., M.I.T., Cambridge, Mass., Rep. R76-33.

Morla Catalan, J. and Cornell, C.A. (1976) Earth Slope Reliabili[illegible] by a Level-Crossing Approach. Proc. ASCE, J. Geotech. Eng. Div[illegible] 104:849-867.

Newmark. N.M. (1965) Effects of Earthquakes on Dams and Embankments. Geotechnique, London, 5:139-159.

Nosko, V.P. (1969) The Characteristics of Excursions of Gaussian Homogeneous Fields Above a High Level. Proc. USSR-Japan Symposium on Probability, Novosibirsk.

Panchev, S. (1971) Random Functions and Turbulence. Pergamon Press.
Pugachev, V.S. (1965) Theory of Random Functions and Its Application to Control Problems, Pergamon Press.
Tang, W.H. (1981) Probabilistic Evaluation of Loads. Proc. ASCE, J. Geotech. Eng. Div., 107:287-304.
Vanmarcke, E.H. (1977a) Reliability of Earth Slopes. Proc. ASCE, J. Geotech. Eng. Div., 103:1247-1265.
Vanmarcke, E.H. (1977b) Probabilistic Modeling of Soil Profiles. Proc. ASCE, J. Geotech. Eng. Div., 103:1227-1246.
Vanmarcke, E.H. (1979a) On the Scale of Fluctuation of Random Functions. Dept. Civil Eng., M.I.T., Cambridge, Mass., Rep. R79-19.
Vanmarcke, E.H. (1979b) Averages and Extremes of Random Fields. Dept. Civil Eng., M.I.T., Cambridge, Mass., Rep. R79-43.
Vanmarcke, E.H. (1980) Probabilistic Stability Analysis of Earth Slopes. Eng. Geology, Elsevier, Amsterdam, 16:29-50.
Veneziano, D., Camacho, D. and Antoniano, J. (1977) Three-Dimensional Models of Slope Reliability. Dept. Civil Eng., M.I.T., Cambridge, Mass., Rep. R77-17.
Wu, T.H. and Kraft, L.M. (1970a) Safety Analysis of Slopes. Proc. ASCE, J. Soil Mech. Foundations Div., 96:609-630.
Wu, T.H. and Kraft, L.M. (1970b) Seismic Safety of Earth Dams. Proc. ASCE, J. Soil Mech. Foundations Div., 96:1987-2006.
Yaglom, A.M. (1973) An Introduction to the Theory of Stationary Random Functions. Dover Publications Inc., New York.
Yücemen, M.S. and Tang, W.H. (1975) Long Term Stability of Soil Slopes: A Reliability Approach. Proc. 2nd ICASP, Aachen, 215-230.

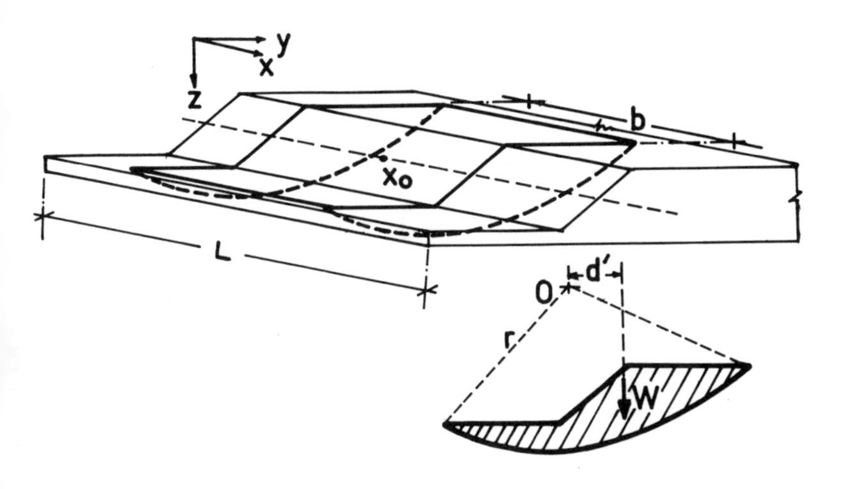

Figure 1 Sliding soil mass and the cross sectional element at $x=x_o$

Soil Dynamics & Earthquake Engineering Conference / Southampton / 1982.07.13-15

Seismic Design Stresses and Stability of Soil Media

M.P.SINGH & M.GHAFORY-ASHTIANY
Virginia Polytechnic Institute & State University, USA

INTRODUCTION

To predict the behavior of soil media like foundation, slope, earthdam, etc., subjected to earthquake induced ground motion, often dynamic analyses are performed. As soil behaves non-linearly under vibratory loads, this nonlinearity must be considered in these analyses. For a given ground motion time history, such analyses could be step-by-step nonlinear in which a nonlinear stress-strain law such as shown in Fig. 1 is followed or equivalent linear iterative approach in which each iteration some equivalent linear soil properties such as shown in Fig. 2 are used. These approaches could be used with finite element formulation or for the solution of wave propagation equation (Martin, 1975, Schnabel, et. al., 1972) of the soil media.

For seismic design of important structural and soil systems, the generalized seismic inputs are often defined in terms of smoothed response spectra (e.g., Housner, 1970). These spectra are presumably equivalent to an ensemble of earthquake time histories that can possibly occur on a site. Other generalized forms of seismic inputs which are collectively equivalent to an ensemble of earthquake motions are usually defined stochastically, such as by a spectral density function. To evaluate a design of a soil system for such seismic design inputs, the nonlinear or equivalent linear approaches mentioned above can be employed if an ensemble of ground motion time histories that are collectively equivalent to the design input are obtained and used in the analysis. However, analyses for several time histories can become cumbersome and expensive. Therefore to avoid use of any time history and to use directly the inputs like response spectra or spectral density function, an equivalent linear approach was proposed earlier by Singh and Khatua (1978). In this approach, the nonlinear behavior of soil is defined in terms of equivalent strain dependent shear modulus and damping curves

(Seed and Idriss, 1970). This approach uses input design spectra directly without any intervening steps of generating ground motion time histories, and therefore it will often be referred to as "direct" approach. In this paper , further details of this approach and the results of a numerical simulation study adopted for its validation are presented.

ANALYTICAL FORMULATION OF DIRECT APPROACH

The approach has been developed for finite element discretization of a soil media for which the equations of motion can be written as

$$[M]\{\ddot{x}\} + [D(\varepsilon)]\{\ddot{x}\} + [S(\varepsilon)]\{x\} = -[M]\{r\}\ddot{x}_g(t) \quad (1)$$

in which [M], [D] and [S] are the mass, damping and stiffness matrices; {x} = relative displacement vector; $x_g(t)$ = the base acceleration at time t; {r} = displacement influence coefficient vector. The matrices [D] and [S] depend upon the level of strain ε in the finite elements.

In the proposed approach, the strain dependent matrices [D] and [S] in Eq. 1 are replaced by some suitable constant matrices [C] and [K] as

$$[M]\{\ddot{x}\} + [C]\{\dot{x}\} + [K]\{x\} = -[M]\{r\}\ddot{x}_g(t) \quad (2)$$

The elements of matrices [M] and [K] are obtained by the technique of stochastic linearization (e.g., Iwan and Yang, 1970) in terms of the elemental damping ratios and shear moduli as follows:

$$\overline{\beta}_q = E[\beta_q(\varepsilon)] \quad (3)$$

$$\overline{G}_q = \frac{E[G_q(\varepsilon)\{x_q\}'[\overline{K}_q]'[\overline{K}_q]\{x_q\}]}{E[\{x_q\}'[\overline{K}_q]'[\overline{K}_q]\{x_q\}]} \quad (4)$$

where, for element q, $\overline{\beta}_q$ = equivalent linear damping ratio; $\beta_q(\varepsilon)$ strain dependent damping ratio; $\overline{G}_q$ = equivalent shear modulus normalized by a value $\overline{G}_m$; $G_q(\varepsilon)$ = normalized strain dependent shear modulus; $[\overline{K}_q]$ = the stiffness matrix obtained for the element with a shear modulus of $\overline{G}_m$; and E(•) denotes the expected value of (•). Various assumptions made in the development of Eqs. 3 and 4 and their effect on final solution are discussed in more detais in a report by Singh and Ashtiany (1980).

It is noted that the solution of Eq. 2 is required to obtain $\overline{\beta}_q$ and $\overline{G}_q$. For this an iterative approach is used, in which one starts with some predecided values of $\overline{\beta}_q$ and $\overline{G}_q$ and solves Eq. 2 to further refine these values in the next iteration. The iterations are continued till a conversion in the final values of $\overline{\beta}_q$ and $\overline{G}_q$ is obtained. In this respect, this iterative methodology is similar to that being used in other equivalent linear methods.

The solution of equivalent linear Eq. 2, in an iteration, can be obtained by any suitable technique. For an input earthquake motion defined by an acceleration time history, step-by-step numerical integration scheme can be used. For the motion characterized by a random process, the statistics of the response quantity can be obtained by random vibration procedures (Singh and Chu, 1976). For earthquake input motions defined by response spectrum curves, one has to use the modal analysis approach as the spectra provide the maximum response in each mode directly. For such seismic inputs, various expected values required in Eq. 3 can be written in terms of given spectrum values and peak factor F_j (implicitly built in the response spectra definition) as follows:

$$E[R_q^2] = \sum_{j=1}^{N} \frac{1}{F_j^2} \frac{\gamma_j^2 \phi_j^2(q) R_a^2(\omega_j)}{\omega_j^2} + 2 \sum_{j=1}^{N} \sum_{k=j+1}^{N} \gamma_j \gamma_k \phi_j(q) \phi_k(q) CTX \tag{5}$$

$$E[x_r x_s] = \sum_{j=1}^{N} \gamma_j^2 \phi_j(r) \phi_j(s) R_a^2(\omega_j) / \omega_j^4 F_j^2 + \sum_{j=1}^{N} \sum_{\substack{k=1 \\ j \neq k}}^{N} \gamma_j \gamma_k \{\phi_j(r) \phi_k(s) + \phi_j(s) \phi_k(r)\} CTX \tag{6}$$

$$E[\varepsilon_q x_r] = \sum_{j=1}^{N} \gamma_j^2 \phi_j(r) \zeta_j(q) R_a^2(\omega_j) / \omega_j^4 F_j^2 + \sum_{j=1}^{N} \sum_{k=1}^{N} \gamma_j \gamma_k \{\phi_j(r) \zeta_k(q) + \phi_k(r) \zeta_j(q)\} CTX \tag{7}$$

in which CTX is defined as:

$$CTX = \{A_{jk} R_a^2(\omega_j) + B_{jk} \omega_j^2 r^2 R_v^2(\omega_j)\} / (F_j^2 \omega_j^4) + \{C_{jk} R_a^2(\omega_k) + D_{jk} \omega_k^2 R_v^2(\omega_k)\} / (F_k^2 \omega_k^4) \tag{8}$$

and $R_a(\omega_j)$, $R_v(\omega_j)$ = pseudo-acceleration and relative velocity spectra values at modal frequency ω_j and damping β_j, respec tively; F_j = the peak factor for response spectrum value at frequency ω_j and damping β_j; $r = \omega_j/\omega_k$; and the factors A_{jk}, B_{jk} etc. are obtained from the solution of the following simultaneous equations:

$$\begin{bmatrix} 0 & 1 & 0 & 1 \\ 1 & v_k & 1 & v_j r^2 \\ v_k & 1 & v_j r^2 & r^4 \\ 1 & 0 & r^4 & 0 \end{bmatrix} \begin{Bmatrix} A_{jk} \\ B_{jk} \\ C_{jk} \\ D_{jk} \end{Bmatrix} = \begin{Bmatrix} 0 \\ 1 \\ u \\ r^2 \end{Bmatrix} \qquad (9)$$

in which $v_k = -2(1-2\beta_k^2)$, $v_j = -2(1-2\beta_j^2)$ and $u = -(1+r^2-4\beta_j\beta_k r)$.

F_j, generally depends upon the frequency and damping ratio of the mode. However, not much error is introduced if only a single value, probably corresonding to the most dominant mode, is used. In this investigation a single value, F, corresponding to the fundamental mode of the system obtained from time history simulation study (Singh and Ashtiany, 1980) has been used. Also the same peak factor is used for pseudo acceleration, relative velocity and relative acceleration response spectra values. The relative velocity response spectra required in these expressions are different from the pseudo velocity spectra, especially in the high frequency range. If these spectra are not available, the approximate expression suggested by Singh (1980) can be used to obtain $R_v(\omega_j)$ from the acceleration spectrum values. In the simulation study, however, the spectra were generated for there use in these expressions. Eq. 4 also requires the evaluation of the expected values of expressions like $[G_q(\varepsilon)x_r x_s]$ and $[G_q(\varepsilon)x_r^2]$. These are obtained by Singh and Ashtiany (1980).

MEASURES OF STABILITY UNDER EARTHQUAKE GROUND MOTIONS

Earthquake motions induce inertia forces as well as gradually degrade the soil strength by successive accumulation of deformation (in cohesive soils) or pore pressure (in sandy soils). In an earthdam or a slope, these two effects together may cause slip circle type of soil failures. In a foundation media, this degradation may be responsible for a structural failure by subsidence.

This degradation is often measured in terms of cyclic damage (Annaki and Lee, 1976) and is evaluated according to the well known Palmgren-Miner hypothesis. Using the probability distribution of stress peaks, and the S-N (stress versus number of cycles-to-failure) curve for the soil, the expected value of cumulative damage for a stationary stress response can be obtained from the following expression (Singh and Ashtiany, 1980):

$$E[D] = \frac{\overline{T}M}{C}\left[\frac{\overline{\alpha}(\sigma_s\sqrt{2})^b}{2}\left\{\Gamma\left(\frac{b+2}{2}\right) + I_1\right\} + I_2\right] \qquad (10)$$

in which

$$I_1 = \int_0^\infty t^{b/2}\,\mathrm{erf}\{\overline{\alpha}\sqrt{t}/\sqrt{1-\overline{\alpha}^2}\}\exp(-t)dt \qquad (11)$$

and

$$I_2 = \frac{\sqrt{1-\overline{\alpha}^2}}{\sigma_s\sqrt{2\pi}}\,\frac{\{2\sigma_s^2(1-\overline{\alpha}^2)\}^{\frac{b+1}{2}}}{2}\,\Gamma\left(\frac{b+1}{2}\right) \qquad (12)$$

in which $\overline{T}$ = equivalent stationary duration of time history; $\Gamma(\cdot)$ = the complete Gamma function; M = expected number of stress peaks per unit time; σ_s = standard deviation of stress; α = the band width parameter defined as = $\nu_0/M = \sigma_{\dot{s}}^2/\sigma_s\sigma_{\ddot{s}}$; ν_0 = zero crossing rate of stress response; $\sigma_{\dot{s}}$ and $\sigma_{\ddot{s}}$ = standard deviations of first and second time derivates of stress s; and C and b are the parameters of S-N curve defined as $Ns^b = C$. To obtain the σ_s, $\sigma_{\dot{s}}$ and $\sigma_{\ddot{s}}$ expressions similar to Eq. 5 can be used with appropriate values for the mode shapes of stress.

Eq. 10 provides only the mean value of damage. However, as the soil systems will tend to have a large effective damping ratio at the stress levels of any practical consequence the damage variability, according to Crandall and Mark (1963) and also substantiated by the numerical results by Singh and Ashtiany (1980b), may not be significant.

Using the concept of equivalent uniform stress proposed by Seed, et al (1975), the extent of damage sustained by soil can be also be quantified in terms of available factor of safety. As shown by Singh and Khatua (1978), the factor of safety for a log-log linear S-N curve is just equal to $(D)^{-1/b}$. For nonlinear S-N curve, some weighted measure of factor of safety (Singh and Khatua, 1978) can be used.

To obtain cyclic damage and factor of safety for a stress time history obtained in the time history analysis of the simulation study, a methodical procedure as described by Singh (1981) was employed. This procedure is entirely consistent with the probabilistic method used in the direct approach as it employs the similar peak response characteristics which are obtained in the simulation study.

SIMULATION STUDY

To check the validity of the proposed approach, a numerical simulation study has been conducted in which the results obtained by the proposed direct approach are compared with the results obtained by a nonlinear approach which considers the hysteretic behavior of soils under cyclic loads. The nonlinear approach developed by Streeter, et. al. (1974) was used. This approach probably provides the most accurate nonlinear response for a one-dimensional wave propagation problem. Since the proposed and nonlinear approaches are quite different from each other in as much as they use different forms of seismic input, model the soil system differently and even use different soil parameters, it is essential that the problem parameters be chosen consistently to obtain comparable results. This consistency in various parameters chosen in the two approaches is achieved as follows:

Seismic Input The proposed direct approach is meant to be used in design evaluations and therefore has been formulated for its use with seismic input defined by ground response spectra. The nonlinear approach on the other hand uses earthquake time histories as seismic input. To make these two inputs consistent, earthquake motion time histories (with certain frequency and intensity characteristics) were synthetically generated for their use in the time history approach, and the very same time histories were used to generate a set of average spectrum curves for their use in the proposed approach.

The peak factor value required in the direct approach could be approximately assessed analytically. However, to avoid possible discrepancy in the results obtained by the two approaches due to approximate value of peak factor, its value as implied in the analysis was obtained by simulation. For this the time history response of the oscillators at several spectral frequencies were processed to obtain the mean square response time history. The peak factor was then defined as the response spectrum value divided by the maximum root mean squared response. The calculated peak factor values ranged 2.6 at high frequency to 1.4 at very low frequencies. A value of 2. obtained at the dominant period and damping of the system was selected for use in the direct approach. To examine the sensitivity of the results with respect to the choice of the peak factor, the results at other values are also reported.

Shear Modulus and Damping A typical soil strata with 10 layers, 60 ft. deep and with different low strain shear moduli and S-N curve parameters as shown in Table 1, has been analyzed by the two methods. For nonlinear analysis, the stress-strain behavior of each layer is characterized by Ramberg-Osgood curve. The equivalent constitutive properties for use in the direct approach are defined in terms of strain dependent shear modulus and damping curves which are consistent with such a hysteretic stress strain behavior. The secant modulus is used as the equivalent shear modulus and the equivalent damping ratio is related to the energy loss in a hysteresis cycle (Seed and Idriss, 1970 and Singh and Ashtiany, 1980). The variations of normalized equivalent shear modulus, G, and damping ratio, β with strain are shown in Fig. 2.

CONSTRUCTION OF SYSTEM DAMPING MATRIX

In this section some methods which have been used be used to define the element damping matrix $[C_q]$ and then the system damping matrix [C] for finite element analyisis are described.

Method 1: In absence of any better rational procedure, the Raleigh's damping matrix has often been used to define the damping matrix for a system. That is,

$$[C] = a[M] + b[k] \tag{13}$$

This is a proportional damping matrix. For this matrix the damping ratio in any mode i could be written as

$$2\beta_i = \frac{a}{\omega_i} + b\omega_i \tag{14}$$

However, the damping ratio in a soil system changes from element to element depending upon the level of strain. To incorporate this feature of soil systems, Idriss et al (1974) proposed the use of Eq. 13 at the element level, i.e.,

$$[C_q] = a[m_q] + b[k_q] \tag{15}$$

To define a and b in this equation, they used Eq. 14 with $a = \beta\omega_1$, $b = \beta/\omega_1$ in which ω_1 = fundamental frequency of the system. The element damping matrices, $[C_q]$, are then assembled to obtain the system damping matrix [C].

Method 2: The above procedure relies heavily on the first mode to determine a and b. This generally damps out the higher modes. To avoid this, two different frequencies were used with Eq. 14 to obtain $a = 2\beta\omega_i\omega_j/(\omega_i+\omega_j)$ and $b = 2\beta/(\omega_i+\omega_j)$ This assumes that modes i and j have same modal damping β. Again the choice of ω_i and ω_j is arbitrary. For the soil system examined in this investigation, more uniform modal damping ratios were obtained when the first and the last frequencies were used.

Method 3: In this procedure, all frequencies are used in Eq. 14 to estimate a and b by minimizing the mean square error between the desired and calculated values of β_i. It is desired to have the same damping ratio in all modes. Different weights can be assigned to the errors in various modes. Thus, if the weightage assigned are W_i, then the total weighted mean square error is:

$$e_d = \sum_{i=1}^{N} W_i \left(\frac{a}{\omega_i} + b\omega_i - 2\beta_i\right)^2 \qquad (16)$$

The minimization of e_d, for parameters a and b, requires that:

$$\begin{bmatrix} A_{11} & A_{12} \\ A_{21} & A_{22} \end{bmatrix} \begin{Bmatrix} a \\ b \end{Bmatrix} = \begin{Bmatrix} C_1 \\ C_2 \end{Bmatrix} \qquad (17)$$

in which

$$A_{11} = \sum_{i=1}^{N} W_i/\omega_i^2, \quad A_{12} = \sum_{i=1}^{N} W_i, \quad A_{21} = A_{12},$$

$$A_{22} = \sum_{i=1}^{N} W_i\omega_i^2, \quad C_1 = 2\sum_{i=1}^{N} W_i\beta_i/\omega_i \text{ and } C_2 = 2\sum_{i=1}^{N} W_i\beta_i\omega_i. \qquad (18)$$

Eq. 17 provides a and b for the definition of element damping matrices by Eq. 15. Various weighting schemes were considered. The scheme with equal weights gave more uniform distribution of modal damping ratio which were somewhat similar to the ones obtained in Method 2, discussed earlier. Also these two methods give the closest response and safety factor values when compared with the time history anlaysis results.

NUMERICAL RESULTS

The numerical results have been obtained by the nonlinear and proposed approaches for a 10-layered strata with maximum shear modulus, mass density and S-N curve parameters for layers given in Table 1. For nonlinear analysis, the hysteretic stress-strain law for the soil in various layers is assumed to be defined by Ramberg-Osgood relation shown in Fig. 1. Also shown in Fig. 2 are the strain dependent normalized shear modulus and damping ratio curves, consistent with this stress-strain law. The cyclic failure rule is assumed to be described by the S-N curves defiend as $Ns^b = C$.

Table 1: Shear Modulus and S-N Curve Parameters for the Layered Media Model

Layer No.	Depth ft.	Maximum Shear Modulus	Effective Unit wt., lbs/cft	S-N Curve Parameters b	$C^* = C/S_e^b$
1	6	755	70.4	5.454	0.05558
2	6	1400	72.2	5.454	0.05558
3	6	1858	73.0	5.924	0.09700
4	6	2317	74.5	5.924	0.09700
5	6	2786	76.2	5.924	0.09700
6	6	3124	76.7	5.924	0.09700
7	6	3444	77.1	5.924	0.09700
8	6	3700	77.1	5.764	0.23563
9	6	3995	77.5	5.764	0.23563
10	6	4284	78.0	5.764	0.23563

S_e = effective overburden pressure

The bench-mark numerical results were obtained by nonlinear time history analyses for 54 synthetic time histories. Response quantities like maximum shear stress, zero crossing rate, peak statistics, cyclic damage and factor of safety were obtained at various locations in the strata. The averages of these values are compared with the similar results obtained in the direct approach. The results for three different acceleration levels of 0.1, 0.2 and 0.4 were obtained. Here the results for only 0.2G acceleration are presented. The results for the other two levels are similar in trend and are available elsewhere (Singh and Ashtiany, 1980). The results with all three methods of damping matrix construction, described in the preceeding section, are designated as "direct approach-1" when Method 1 is used to construct an element damping matrix and "direct approach-2" and "direct approach-3" when the other two methods are used.

Fig. 3 compares the variation of maximum shear stress with depth obtained by various methods. The stress trend obtained by the direct approaches is similar to that obtained by the nonlinear approach and the best comparison is obtained when the elemental damping matrices are formed by the least squares procedure, that is Method 3. The percent difference between stresses obtained by the nonlinear and direct approaches is largest at top. However it is well recognized that near the surface the stress and safety predictions are generally uncertain and also of little practical significance.

Fig. 4 shows the factor of safety against liquefication type of failure in various layers obtained by the nonlinear and direct approaches. Here again, Method 3 of direct approach gives a better prediction. The direct approach values are obtained for an equivalent stationary duration of 2 secs., which is half of the strong motion phase. A better corresondence was obtained when the equivalent stationary duration was assumed equal to 1.25 secs. Although, equivalent duration depends on many complex factors, it appears that a value one fourth of the strong motion phase of the input motion will provide a good estimate of the duration for safety evaluation purposes. However, more research is required to establish better and rational guidelines.

Fig. 5 shows the maximum stresses obtained for various assumed values of the peak factor. The differences in various curves show that the use of a correct peak factor value is rather important in the prediction of response. Similar differences were observed in the factor of safety values. Various analytical procedures are available to obtain peak factor values. Whether these methods could provide a consistent value for the analysis performed here was not investigated further. Rather, a value which was consistent with the seismic input used in the investigation was obtained in simulation study and used in the analyses.

As quantities like number of peaks/sec., M, and band width parameter, α, are used in safety evaluation in the direct approach, these were also obtained in the time history analysis and compared with those obtained in the direct approach. Again it was seen that approach-3 predicted these better, with a much better correspondence in the results of the peak rate (Singh and Ashtiany, 1980).

CONCLUSIONS

A good comparison of the response results obtained by the nonlinear and direct approaches indicates that the latter provides a rational analytical method for evaluation of maximum stresses and stability of an earth media. Seismic input in the form of response spectra can be used in the proposed approach. This makes it especially suitable for evaluation of a design.

Since the results of a one-dimensional problem only are examined and compared, the observations made here, in a strict sense, are applicable to such earth structure problems only. However, the proposed methodology is quite general and can be used for two or three dimensional situations as well.

The response results obtained in the direct approach are observed to depend upon the values of peak factor and equivalent earthquake duration used. Some analytical procedures are available to define the inherent values of peak factors built in the response spectra curves. Still some further research effort is warranted to obtain these in a simple usuable form. More research effort is also required to obtain an equivalent duration which would include the nonstationarity of input. The numerical results obtained for a typical horizontal strata analyzed here indicate that the equivalent earthquake duration should be much smaller than the strong motion phase of the input motion.

Given the correct values of peak factor and equivalent duration, the direct approach will provide a conservative estimate of response and seismic safety. That is, the calculated stresses will be a little higher and the factors of safety a little lower than the values obtained by a rigourous nonlinear approach. The observed differences in results, though not large and crucial, can be attributed to: (1) inadequate modeling of energy dissipation characteristics in the analysis; (2) choice of peak factor values used in the analysis; (3) discretizaion error in the finite element formulation used in the direct approach; and (4) certain soil layer property averaging procedures employed in the algorithm used in the nonlinear approach. Out of these factors, the energy dissipation modeling procedure is probably the most important. Inadequate modeling of energy dissipation could be due to two reasons: (1) representation of hysteretic behavior by strain dependent damping ratio, (2) the method of formation of an element damping matrix. The definition of a proper damping matrix of a system has always been elusive. Thus a need for a continued development to define a better energy dissipation model in a system by damping matrix is identified.

ACKNOWLEDGEMENT

This work was sponsored by the United States National Science Foundation under Grant No. PFR-7823095. This support is gratefully acknowledged.

REFERENCES

Annaki, M. and Lee, K. L. (1976), Equivalent Uniform Cycle Concept for Soil Dynamics, Preprint ASCE Convention, Philadelphia, PA, Sept.-Oct., 227-254.

Crandall, S. H. and Mark, W. D. (1963), Random Vibration, Academic Press, New York.

Housner, G. (1970), Design Spectrum, Chap. 5 in Earthquake Engineering, Ed. R. L. Weigel, Prentice Hall, Englewood Cliffs, NJ.

Idriss, I. M., Seed, H. B. and Seriff, N. (1974), Seismic Response by Variable Damping Finite Elements, J. of Geo. Eng. Div., ASCE, 100, No. 1.

Iwan, W. D. and Yang, I. M. (1970), Application of Statistical Linearization Techiques to Nonlinear Multi-Degree-of-Freedom Systems, J. App. Mech., 39, 545-550.

Martin, P. P. (1975), Non-linear Methods for Dynamic Analysis of Ground Response, Ph.D., Diss. Univ. of Cal., Berkeley.

Schnabel, P. B., Lysmer, J. and Seed, H. B. (1972), SHAKE - A Computer Program for Earthquake Response Analysis of Horizontally Layered Sites, EERC, Rep. No. 72-12, Univ. of Cal., Berkeley.

Seed, H. B. and Idriss, I. M. (1969), Influence of Soil Conditions on Ground Motions During Earthquakes, J. of Soil Mech. & Fdn. Div., ASCE, 95, SM1.

Seed, H. B. andd Idriss, I. M. (1970), Soil Moduli and Damping Factors for Dynamic Response Analysis, EERC 70-10, Univ. of Cal., Berkeley.

Seed, H. B. Idriss, I. M., Lee, K. L. and Makdisi, F. (1975), Representation of Irregular Stress Time Histories by Equivalent Uniform Stresss Series in Liquefaction Analysis, EERC 75-79, Univ. of Cal., Berkeley.

Singh, M. P. and Chu, S. L. (1976), Stochastic Considerations in Seismic Analysis of Structures, Int. J. of Earth. Eng. and Str. Dyn., 4, 295-307.

Singh, M. P. and Khatua, T. P. (1978), Stochastic Seismic Stability Prediction of Earth Dams, Spec. Conf. in Earth. Eng. and Soil Dyn., Pasadena, CA.

Singh, M. P. (1980a), Seismic Design Input for Secondary Systems, J. of St. Div., ASCE, 160, ST2, 505-517.

Singh, M. P. and Ashtiany, M. G. (1980b), Seismic Stability Evaluation of Earth Structures, Rep. No. VPI-E-80.30, Virginia Polytechnic Institute & State Univ., Blacksburg, VA, USA.

Streeter, V. L., Wylie, E. B. and Richart, F. E. (1974), Soil Motion Computations by Characteristics Methods, J. of Geo. Eng. Div., ASCE, 100, GT3.

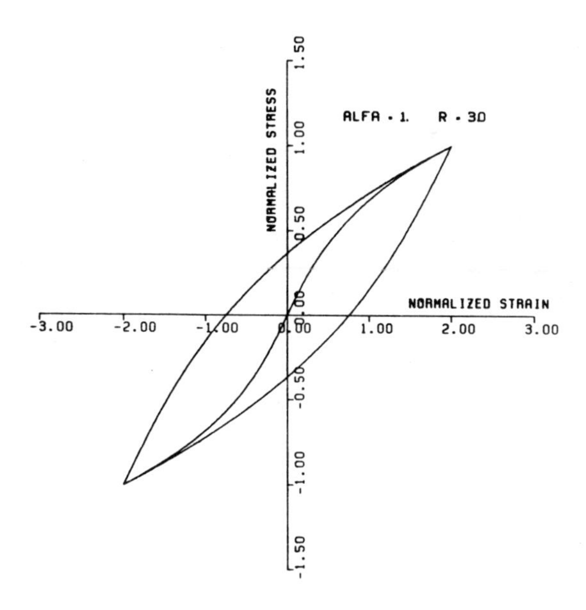

Fig. 1 Ramberg-Osgood Stress-Strain Curves

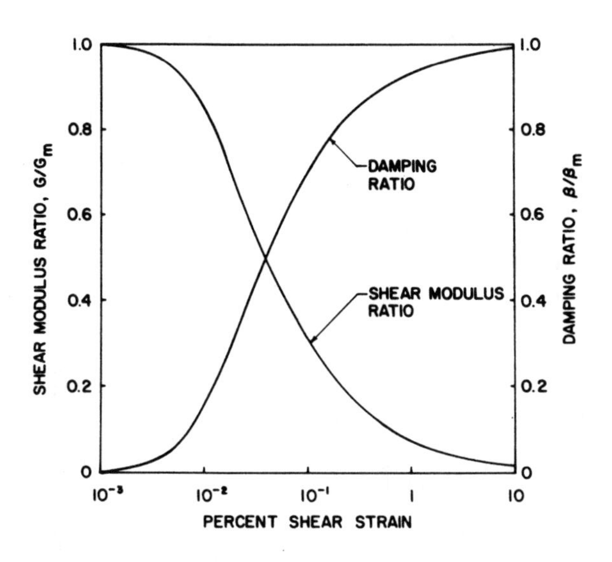

Fig. 2 Strain Dependent Shear Modulus and Damping Ratio

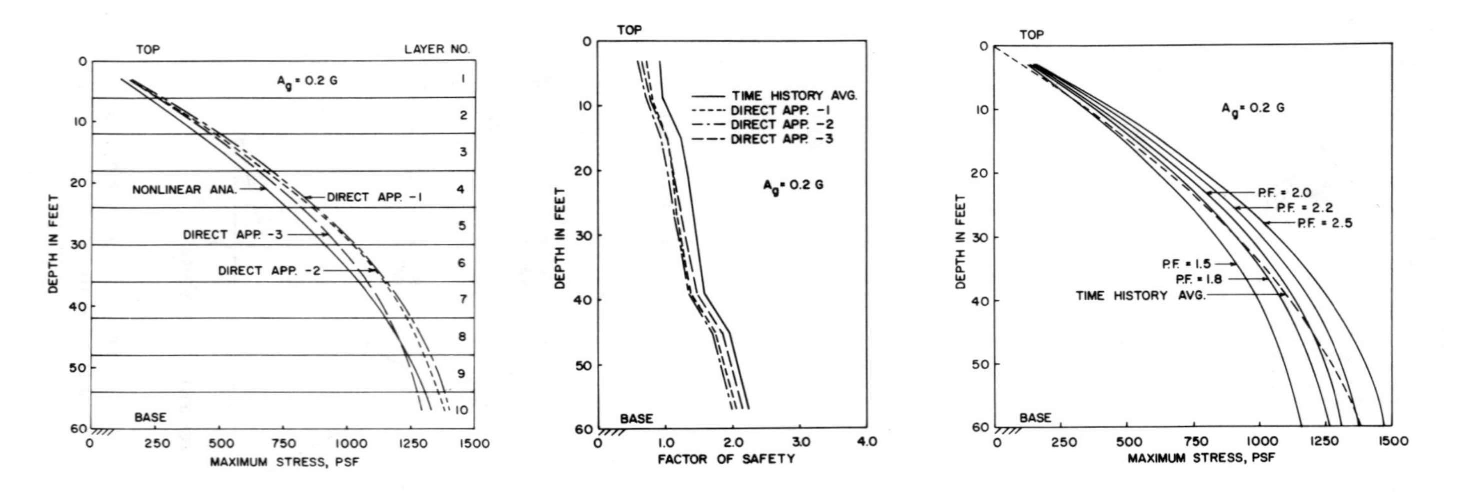

Fig. 3

Variation of Maximum Stress Calculated by Nonlinear and Direct Approaches

Fig. 4

Factor of Safety Calculated by Nonlinear and Direct Approaches

Fig. 5

Effect of Peak Factor on Maximum Stress Calculated by Direct Approach

Soil Dynamics & Earthquake Engineering Conference / Southampton / 1982.07.13-15

Three-Dimensional Dynamic Analysis of Nonhomogeneous Earth Dams

AHMED M. ABDEL-GHAFFAR & AIK-SIONG KOH
Princeton University, NJ, USA

ABSTRACT

A three-dimensional dynamic analysis of nonhomogeneous earth dams is presented. The analysis is based on the Rayleigh-Ritz method and on the assumption that the dam materials are isotropic and linearly elastic but have properties which can vary as functions of position within the dams. The dams are further assumed to sit in rigid canyons which can have sloping sides and/or floors. The base and sides of each dam are fixed to the canyon while the remaining faces of the dam are free. The free vibration modes and frequencies of the Santa Felicia Dam, in California, are computed and compared with the measured results from full-scale tests and earthquake response records. In addition, the Fourier transform of the computed displacement response of the dam (subjected to the 1971 San Fernando earthquake as measured by an accelerogram at the base of the dam) is presented and this computed response is compared with the recorded response at the crest. The computed results agree reasonably well with the measured ones.

INTRODUCTION

In analyzing the seismic behavior of earth dams, past investigators have used many different models, each with many simplifying assumptions. The materials have always been assumed to be linear with homogeneous properties or with properties varying as functions of depth from the crest of the dams. The dams, also, have always been assumed to be symmetrical prismatic wedges sitting in rigid rectangular canyons. Upstream-downstream vibrations have been decoupled from the longitudinal vibrations and the two problems have been solved separately. The two-dimensional-Shear Beam models have been used most often (Refs. 4, 8, 9,10, 12) although bending effects, which are small for most earth dams, can be incorporated in the upstream-downstream cases (Refs. 4,10,11). In the shear beam models, the strains and stresses on the faces of any horizontal upstream-downstream

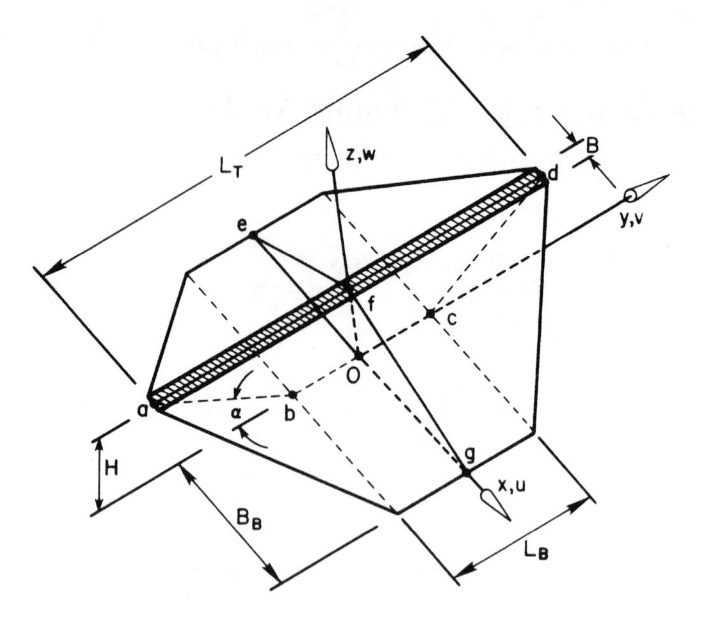

Figure 1.

GEOMETRY OF THE THREE-DIMENSIONAL DAM MODEL

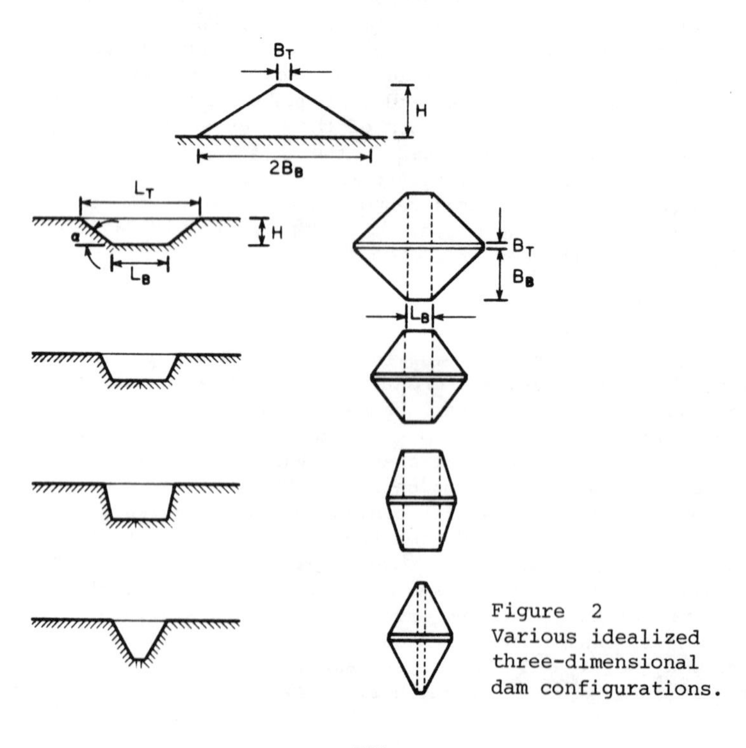

Figure 2
Various idealized three-dimensional dam configurations.

element strip have been assumed to be uniformly distributed. These models have provided useful information but they lack the ability to incorporate the three-dimensional effects which have been found to be significant in field tests (Refs. 2,6).

Recently, a few efforts have been made to extend these analytical models into three-dimensional ones. Martinez and Bielak (13) used a two-dimensional finite element discretization of the dam's cross-section together with a Fourier expansion in the longitudinal direction. They concluded that three-dimensional deformation behavior yields higher stiffness characteristics than does the two-dimensional idealization. Ohmachi (14) also extended the shear-wedge model into a simplified finite element model in which cross-sectional triangular slices were chosen as elements. He concluded that three-dimensionality has an important effect on the earthquake response of dams located in narrow canyons.

In this paper, the Rayleigh-Ritz method is used to analyze the three-dimensional earth dams sitting on rigid rectangular, trapezoidal, or triangular canyons. The dam materials are assumed to be linearly elastic and isotropic but to have properties which change as functions of position within the dams. A comparison is made between the measured dynamic properties of an existing earth dam (determined from real earthquake records (Ref. 3), and from full-scale tests (Ref. 2)), and those computed using the proposed technique. It is found that the computed results agree reasonably well with the measured ones.

EQUATION OF MOTION

A variational energy approach utilizing both Hamilton's principle and the Rayleigh-Ritz method was used to obtain the equation of motion for the three-dimensional free vibration of non-homogeneous earth dams. Figures 1 and 2 show the idealized three-dimensional dam configurations that can be analyzed by the proposed simplified procedures. For the isotropic, linearly elastic dam materials in a general three-dimensional state of stress the strain energy density is given by

$$U_0 = \frac{1}{2}\lambda(\varepsilon_x + \varepsilon_y + \varepsilon_z)^2 + G(\varepsilon_x^2 + \varepsilon_y^2 + \varepsilon_z^2) + \frac{1}{2}G(\gamma_{xy}^2 + \gamma_{yz}^2 + \gamma_{xz}^2). \tag{1}$$

The total strain energy in the vibrating dam is given by the integral of U_0 throughout the dam's domain Ω.

$$U = \int dU = \iiint U_0 dxdydz = \int_\Omega U_0 d\Omega. \tag{2}$$

In the above equations, λ is Lame's constant which is represented by the expression: $\lambda = \frac{\nu E}{(1+\nu)(1-2\nu)} = \frac{2\nu G}{1-2\nu}$; E is the

Young's Modulus, ν is the Poisson's ratio, and G is the Shear Modulus. The normal strains are $\varepsilon_x = \partial u/\partial x$, $\varepsilon_y = \partial v/\partial y$ and $\varepsilon_z = \partial w/\partial z$, and the shear strains are $\gamma_{xy} = \partial u/\partial y + \partial v/\partial x$, $\gamma_{yz} = \partial v/\partial z + \partial w/\partial y$ and $\gamma_{zx} = \partial w/\partial x + \partial u/\partial z$ where u, v, and w are the displacements in the x, y, and z directions, respectively (see Fig. 1).

The total kinetic energy for zero ground motion is given by

$$T = \frac{1}{2}\int_{\Omega} \rho(\dot{u}^2 + \dot{v}^2 + \dot{w}^2)\,d\Omega \quad , \tag{3}$$

where ρ is the mass density of the material and the dot denotes the time derivatives of the displacements.

The Rayleigh-Ritz method is now used, and it assumes that the displacement fields are summations of basis functions:

$$\{u\} = \begin{Bmatrix} u(x,y,z;t) \\ v(x,y,z;t) \\ w(x,y,z;t) \end{Bmatrix} = \sum_{n=1}^{N} \begin{Bmatrix} \bar{u}_n\Phi_n(x,y,z) \\ \bar{v}_n\Psi_n(x,y,z) \\ \bar{w}_n\Theta_n(x,y,z) \end{Bmatrix} \sin \omega t \ , \tag{4}$$

where Φ_n, Ψ_n, and Θ_n belong to a set of basis functions which satisfy the essential boundary conditions of the dam, while $\bar{u}_n$, $\bar{v}_n$ and $\bar{w}_n$ are the corresponding generalized coordinates.

Equation 4 is now substituted in the Hamilton principle:

$$\int_{t_1}^{t_2} \delta(T - U)\,dt = 0 \quad ,$$

and an integration by parts is performed.

A set of $3 \times N$ equations can be obtained to give the eigenvalue problem:

$$[K]\{\bar{u}\} - \omega^2[M]\{\bar{u}\} = 0 \quad , \tag{5}$$

where ω_i, $i = 1,2,3,\ldots,N$, determine the modal frequencies, and the eigenvectors $\{\bar{u}_i\}$, $i = 1,2,3,\ldots,N$, give the corresponding mode shapes of the dam's free vibrations. The displacement vector is given by

$$\{\bar{u}\} = \{u_1,v_1,w_1,u_2,v_2,w_2 \ \ldots \ u_N,v_N,w_N\}^T \tag{6}$$

The mass matrix is given by

$$[M] = \begin{bmatrix} [m_{11}] & [m_{12}] & --- & [m_{1N}] \\ [m_{21}] & [m_{22}] & --- & [m_{2N}] \\ --- & --- & & --- \\ [m_{N1}] & --- & --- & [m_{NN}] \end{bmatrix} \tag{7}$$

in which

$$[m_{mn}] = \int_{\Omega} \rho \begin{bmatrix} \Phi_m \Phi_n & 0 & 0 \\ 0 & \Psi_m \Psi_n & 0 \\ 0 & 0 & \Theta_n \Theta_m \end{bmatrix} d\Omega. \tag{8}$$

And finally, the stiffness matrix is expressed as

$$[K] = \begin{bmatrix} [K_{11}] & [K_{12}] & --- & [K_{1N}] \\ [K_{21}] & [K_{22}] & --- & [K_{2N}] \\ --- & --- & --- & --- \\ [K_{N1}] & --- & --- & [K_{NN}] \end{bmatrix}, \tag{9}$$

in which

$$[K_{mn}] = \int_{\Omega} \begin{bmatrix} \begin{pmatrix} \lambda\Phi_{m,x}\Phi_{n,x} + \\ 2G\Phi_{m,x}\Phi_{n,x} + \\ G\Phi_{m,y}\Phi_{n,y} + \\ G\Phi_{m,z}\Phi_{n,z} \end{pmatrix} & \begin{pmatrix} \lambda\Phi_{m,x}\Psi_{n,y} + \\ G\Phi_{m,y}\Psi_{n,x} \end{pmatrix} & \begin{pmatrix} \lambda\Phi_{m,x}\Theta_{n,z} + \\ G\Phi_{m,z}\Theta_{n,x} \end{pmatrix} \\ \begin{pmatrix} \lambda\Psi_{m,y}\Phi_{n,x} + \\ G\Psi_{m,x}\Phi_{n,y} \end{pmatrix} & \begin{pmatrix} \lambda\Psi_{m,y}\Psi_{n,y} + \\ 2G\Psi_{m,y}\Psi_{n,y} + \\ G\Psi_{m,x}\Psi_{n,x} + \\ G\Psi_{m,z}\Psi_{n,z} \end{pmatrix} & \begin{pmatrix} \lambda\Psi_{m,y}\Theta_{n,z} + \\ G\Psi_{m,z}\Theta_{n,y} \end{pmatrix} \\ \begin{pmatrix} \lambda\Theta_{m,z}\Phi_{n,x} \\ G\Theta_{m,x}\Phi_{n,z} \end{pmatrix} & \begin{pmatrix} \lambda\Theta_{m,z}\Psi_{n,y} + \\ G\Theta_{m,y}\Psi_{n,z} \end{pmatrix} & \begin{pmatrix} \lambda\Theta_{m,z}\Theta_{n,z} + \\ 2G\Theta_{m,z}\Theta_{n,z} + \\ G\Theta_{m,x}\Theta_{n,x} + \\ G\Theta_{m,y}\Theta_{n,y} \end{pmatrix} \end{bmatrix} d\Omega. \tag{10}$$

(Note: $\Phi_{n,x} = \frac{\partial \Phi_n}{\partial x}$)

COORDINATE TRANSFORMATION

In order to evaluate the integrals which make up the elements of [M] and [K], a coordinate transformation is made. The dam in x, y, z coordinates is transformed into a cuboid in the ξ, η, ζ coordinates using linear shape functions N1, N2,... N8. (Fig.3).

$$\left.\begin{aligned} x &= N^1 x_1 + N^2 x_2 + \ldots\ldots + N^8 x_8 \\ y &= N^1 y_1 + N^2 y_2 + \ldots\ldots + N^8 y_8 \\ z &= N^1 z_1 + N^2 z_2 + \ldots\ldots + N^8 z_8 \end{aligned}\right\} , \tag{11}$$

where

$$\left.\begin{aligned} N^1 &= \frac{1}{4}(1-\xi)(1-\eta)(1-\zeta) , & N^2 &= \frac{1}{4}(1+\xi)(1-\eta)(1-\zeta) \\ N^3 &= \frac{1}{4}(1+\xi)(1+\eta)(1-\zeta) , & N^4 &= \frac{1}{4}(1-\xi)(1+\eta)(1-\zeta) \\ N^5 &= \frac{1}{4}(1-\xi)(1-\eta)\zeta , & N^6 &= \frac{1}{4}(1+\xi)(1-\eta)\zeta \\ N^7 &= \frac{1}{4}(1+\xi)(1+\eta)\zeta , & N^8 &= \frac{1}{4}(1-\xi)(1+\eta)\zeta \end{aligned}\right\} \tag{12}$$

Consider the evaluation of the integral, I_1, within an element of the [M] matrix:

$$I_1 = \iiint_{\Omega} \rho(x,y,z)\Phi_m(x,y,z)\Phi_n(x,y,z)\,dxdydz \quad . \tag{13}$$

In the ξ, η, ζ coordinates this integral can be written as

$$I_1 = \int_0^1\int_{-1}^1\int_{-1}^1 \rho(\xi,\eta,\zeta)\bar{\Phi}_m(\xi,\eta,\zeta)\bar{\Phi}_n(\xi,\eta,\zeta)J(\xi,\eta,\zeta)\,d\xi d\eta d\zeta, \tag{14}$$

where J is the Jacobian given by

$$J = \begin{vmatrix} \frac{\partial x}{\partial \xi} & \frac{\partial x}{\partial \eta} & \frac{\partial x}{\partial \zeta} \\ \frac{\partial y}{\partial \xi} & \frac{\partial y}{\partial \eta} & \frac{\partial y}{\partial \zeta} \\ \frac{\partial z}{\partial \xi} & \frac{\partial z}{\partial \eta} & \frac{\partial z}{\partial \zeta} \end{vmatrix} \tag{15}$$

Over the ξ, η, ζ domain, Gaussian Quadrature numerical integration can be used conveniently

$$I_1 \simeq \sum_{k=1}^{N} \sum_{\ell=1}^{N} \sum_{m=1}^{N} \rho(\xi^k,\eta^\ell,\zeta^m)\,\bar{\Phi}_m(\xi^k,\eta^\ell,\zeta^m)\,\bar{\Phi}_n(\xi^k,\eta^\ell,\zeta^m)\,J(\xi^k,\eta^\ell,\zeta^m)\,w_1^k w_2^\ell w_3^m, \tag{16}$$

where N is the order of Gaussian Quadrature integration; $(\xi^k,\eta^\ell,\zeta^m)$ are the coordinates of the Gauss points, and w_1^k, w_2^ℓ, w_3^m are weights at ξ^k,η^ℓ,ζ^m, respectively.

Now consider an integral within an element of the stiffness matrix

$$I_2 = \iiint_\Omega \lambda(x,y,z)\,\Phi_{m,x}(x,y,z)\,\Phi_{n,x}(x,y,z)\,dxdydz \quad . \tag{17}$$

Using the same transformation as previously, one can write

$$I_2 = \int_0^1\int_{-1}^1\int_{-1}^1 \lambda(\xi,\eta,\zeta)\,\hat{\Phi}_{m,x}(\xi,\eta,\zeta)\,\hat{\Phi}_{n,x}(\xi,\eta,\zeta)\,J(\xi,\eta,\zeta)\,d\xi d\eta d\zeta, \tag{18}$$

where

$$\hat{\Phi}_{m,x} = \bar{\Phi}_{m,\xi}\frac{\partial\xi}{\partial x} + \bar{\Phi}_{m,\eta}\frac{\partial\eta}{\partial x} + \bar{\Phi}_{m,\zeta}\frac{\partial\zeta}{\partial x} \quad ,$$

or more generally

$$[\hat{\Phi}_{m,x} \;\; \hat{\Phi}_{m,y} \;\; \hat{\Phi}_{m,z}] = [\bar{\Phi}_{m,\xi} \;\; \bar{\Phi}_{m,\eta} \;\; \bar{\Phi}_{m,\zeta}] \begin{bmatrix} \frac{\partial\xi}{\partial x} & \frac{\partial\xi}{\partial y} & \frac{\partial\xi}{\partial z} \\ \frac{\partial\eta}{\partial x} & \frac{\partial\eta}{\partial y} & \frac{\partial\eta}{\partial z} \\ \frac{\partial\zeta}{\partial x} & \frac{\partial\zeta}{\partial y} & \frac{\partial\zeta}{\partial z} \end{bmatrix} . \tag{19}$$

To evaluate the matrix $[\partial\xi/\partial x]$ we use the identity

$$\begin{bmatrix} \frac{\partial x}{\partial\xi} & \frac{\partial x}{\partial\eta} & \frac{\partial x}{\partial\zeta} \\ \frac{\partial y}{\partial\xi} & \frac{\partial y}{\partial\eta} & \frac{\partial y}{\partial\zeta} \\ \frac{\partial z}{\partial\xi} & \frac{\partial z}{\partial\eta} & \frac{\partial z}{\partial\zeta} \end{bmatrix} \begin{bmatrix} \frac{\partial\xi}{\partial x} & \frac{\partial\xi}{\partial y} & \frac{\partial\xi}{\partial z} \\ \frac{\partial\eta}{\partial x} & \frac{\partial\eta}{\partial y} & \frac{\partial\eta}{\partial z} \\ \frac{\partial\zeta}{\partial x} & \frac{\partial\zeta}{\partial y} & \frac{\partial\zeta}{\partial z} \end{bmatrix} = \begin{bmatrix} 1 & 0 & 0 \\ 0 & 1 & 0 \\ 0 & 0 & 1 \end{bmatrix} . \tag{20}$$

Hence

$$\left[\frac{\partial\xi}{\partial x}\right] = \left[\frac{\partial x}{\partial\xi}\right]^{-1} = \mathrm{Adj}\left[\frac{\partial x}{\partial\xi}\right] \Big/ J \quad .$$

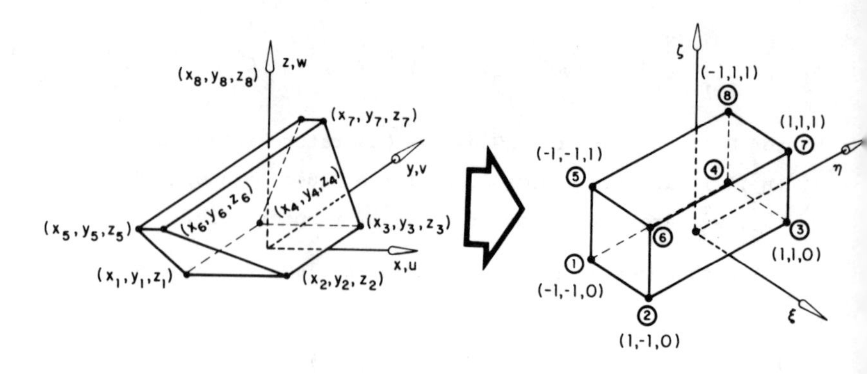

Figure 3
COORDINATE - TRANSFORMATION OF THE THREE-DIMENSIONAL ANALYSIS OF EARTH DAMS

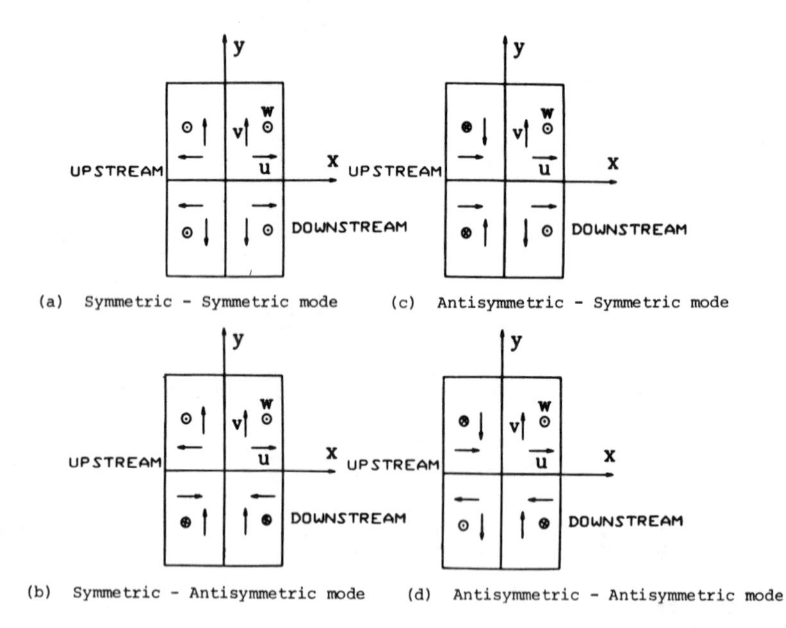

(a) Symmetric - Symmetric mode

(c) Antisymmetric - Symmetric mode

(b) Symmetric - Antisymmetric mode

(d) Antisymmetric - Antisymmetric mode

Figure 4 The reduction of the problem into four simple problems.

The integral I_2 can now be evaluated using the Gaussian Quadrature as in Eq. 16. This applies for all the integrals in the matrices [K] and [M].

In the ξ, η, ζ domain the displacement components can be represented in general by

$$u = \sum_{k=1}^{\infty} \sum_{\ell=1}^{\infty} \sum_{m=1}^{\infty} \left(A^k \sin \frac{k\xi\pi}{2} + B^k \cos \frac{(k-1)\xi\pi}{2} \right) \times \left(C^\ell \sin \ell\eta\pi + D^\ell \cos \frac{\eta\pi}{2}(2\ell - 1) \right) \left(E^m \sin \frac{m\zeta\pi}{2} \right) \qquad \text{(21-a)}$$

$$v = \sum_{k=1}^{\infty} \sum_{\ell=1}^{\infty} \sum_{m=1}^{\infty} \left(\bar{A}^k \sin \frac{k\xi\pi}{2} + \bar{B}^k \cos \frac{(k-1)\xi\pi}{2} \right) \times \left(\bar{C}^\ell \sin \ell\eta\pi + \bar{D}^\ell \cos \frac{\eta\pi}{2}(2\ell - 1) \right) \left(\bar{E}^m \sin \frac{m\zeta\pi}{2} \right) \qquad \text{(21-b)}$$

$$w = \sum_{k=1}^{\infty} \sum_{\ell=1}^{\infty} \sum_{m=1}^{\infty} \left(\hat{A}^k \sin \frac{k\xi\pi}{2} + \hat{B}^k \cos \frac{(k-1)\xi\pi}{2} \right) \times \left(\hat{C}^\ell \sin \ell\eta\pi + \hat{D}^\ell \cos \frac{\eta\pi}{2}(2\ell - 1) \right) \left(\hat{E}^m \sin \frac{m\zeta\pi}{2} \right) \qquad \text{(21-c)}$$

The above displacements automatically satisfy the geometric boundary conditions which state that displacements are zero at the canyon floor and on the canyon faces. Hence, any dam geometry that can be linearly transformed into the cuboid in the ξ, η, ζ domain can be adequately analyzed. It can be seen that ρ, G, and λ can be functions of x, y, z; i.e., the dams can be nonhomogeneous. The important requirement is that the dam material be considered isotropic and linearly elastic.

COMPUTATIONAL REDUCTION

Tremendous computational advantages can be obtained by exploiting the symmetries of the dam geometry and material properties. If the dam is assumed to be symmetric about a vertical plane through the dam center oriented in the upstream-downstream direction (i.e., y = constant) and another vertical plane oriented along the crest of the dam (i.e., x = constant), then the analysis can be done in four small runs rather than one large expensive run. The first run will calculate the modes symmetric with respect to both the above mentioned planes, and the displacement components can be represented by: (see Fig. 4a)

$$\left.\begin{aligned}
u &= \sum_{p=1}^{\infty} \sum_{q=1}^{\infty} \sum_{r=1}^{\infty} A^{pqr} \sin \frac{p\xi\pi}{2} \cos(2q - 1)\frac{\pi\eta}{2} \sin \frac{r\zeta\pi}{2} \\
v &= \sum_{p=1}^{\infty} \sum_{q=1}^{\infty} \sum_{r=1}^{\infty} \bar{A}^{pqr} \cos \frac{(p - 1)\xi\pi}{2} \sin q\pi\eta \sin \frac{r\zeta\pi}{2} \\
w &= \sum_{p=1}^{\infty} \sum_{q=1}^{\infty} \sum_{r=1}^{\infty} \hat{A}^{pqr} \cos \frac{(p - 1)\xi\pi}{2} \cos(2q - 1)\frac{\eta\pi}{2} \sin \frac{r\zeta\pi}{2}
\end{aligned}\right\} \quad (22)$$

The second run will calculate the modes symmetric with respect to the upstream-downstream plane but anti-symmetric with respect to the crest plane. The displacement components are: (see Fig. 4b)

$$\left.\begin{aligned}
u &= \sum_{p=1}^{\infty} \sum_{q=1}^{\infty} \sum_{r=1}^{\infty} B^{pqr} \cos \frac{(p - 1)\xi\pi}{2} \cos(2q - 1)\frac{\eta\pi}{2} \sin \frac{r\zeta\pi}{2} \\
v &= \sum_{p=1}^{\infty} \sum_{q=1}^{\infty} \sum_{r=1}^{\infty} \bar{B}^{pqr} \sin \frac{p\xi\pi}{2} \sin q\eta\pi \sin \frac{r\zeta\pi}{2} \\
w &= \sum_{p=1}^{\infty} \sum_{q=1}^{\infty} \sum_{r=1}^{\infty} \hat{B}^{pqr} \sin \frac{p\xi\pi}{2} \cos(2q - 1)\frac{\eta\pi}{2} \sin \frac{r\zeta\pi}{2}
\end{aligned}\right\} \quad (23)$$

The third run will calculate the modes anti-symmetric with respect to the upstream-downstream plane and symmetric about the crest plane (see Fig. 4c)

$$\left.\begin{aligned}
u &= \sum_{p=1}^{\infty} \sum_{q=1}^{\infty} \sum_{r=1}^{\infty} C^{pqr} \sin \frac{p\xi\pi}{2} \sin q\eta\pi \sin \frac{r\zeta\pi}{2} \\
v &= \sum_{p=1}^{\infty} \sum_{q=1}^{\infty} \sum_{r=1}^{\infty} \bar{C}^{pqr} \cos \frac{(p - 1)\xi\pi}{2} \cos(2q - 1)\frac{\eta\pi}{2} \sin \frac{r\zeta\pi}{2} \\
w &= \sum_{p=1}^{\infty} \sum_{q=1}^{\infty} \sum_{r=1}^{\infty} \hat{C}^{pqr} \cos \frac{(p - 1)\xi\pi}{2} \sin q\eta\pi \sin \frac{r\zeta\pi}{2}
\end{aligned}\right\} \quad (24)$$

The fourth and final run will calculate the modes that are anti-symmetric with respect to both the upstream-downstream and crest planes (Fig. 4d)

$$u = \sum_{p=1}^{\infty} \sum_{q=1}^{\infty} \sum_{r=1}^{\infty} D^{pqr} \cos \frac{(p - 1)\xi\pi}{2} \sin q\eta\pi \sin \frac{r\zeta\pi}{2}$$

$$v = \sum_{p=1}^{\infty} \sum_{q=1}^{\infty} \sum_{r=1}^{\infty} \bar{D}^{pqr} \sin \frac{p\xi\pi}{2} \cos(2q - 1)\frac{\eta\pi}{2} \sin \frac{r\zeta\pi}{2}$$

$$w = \sum_{p=1}^{\infty} \sum_{q=1}^{\infty} \sum_{r=1}^{\infty} \hat{D}^{pqr} \sin \frac{p\xi\pi}{2} \sin q\eta\pi \sin \frac{r\zeta\pi}{2} \qquad (25)$$

In addition to splitting the problem into four smaller parts, each of these parts needs to be integrated over only one quarter of the dam, further improving the accuracy for a given integration order. The details of this computer program can be found in Ref. 1.

It should be noted that the basis functions chosen here are not the only ones possible. They are, however, a good choice because from the analysis of the two-dimensional shear beam models (Refs. 4, 8, 9, 10) the mode shapes are found to be sinusoidal, although a better choice might be to use Bessel Functions for the mode shape in the vertical direction as these functions are the analytical solutions of some shear beam models.

COMPUTATION OF THE NATURAL FREQUENCIES AND MODE SHAPES OF SANTA FELICIA DAM

As an example, the following case, which approximates the Santa Felicia Dam in California, is analyzed (the structural description of the dam can be found in Refs. 2 and 3). The model dam has the shape shown in Fig. 1 which is a good approximation of the real dam whose dimensions are: $H = 236.5$ ft, $B_T = 30.0$ ft, $B_B = 535.3$ ft, $L_T = 1275.0$ ft, and $L_B = 450.0$ ft. The dam is assumed to be made of a nonhomogeneous, linearly elastic and isotropic material. To account for the effects of confining pressures the following distribution of the shear modulus is used

$$G(x,z) = G_0\left(\frac{H - z}{H}\right)^{1/3}\left[1 - \left(\frac{x}{B_B - 2.22}\right)^2\right] , \qquad (26)$$

where G_0 is the shear modulus at the base of the dam. The above relationship was based on the in-situ wave-velocity measurements (3) from which the shear wave velocity in the dam varied from about 600 ft/sec to 1200 ft/sec and the Poisson's ratio was a consistent 0.45. The mass density was found to be rather constant throughout the dam and had a value of about 4.04 slugs.

For the modulus distribution used in the computation, the maximum shear modulus (G_0) is adjusted so that the first upstream-downstream model frequency matches that measured in the actual dam during full-scale forced and ambient vibration tests (see Refs. 2 and 5). The first upstream-downstream frequency

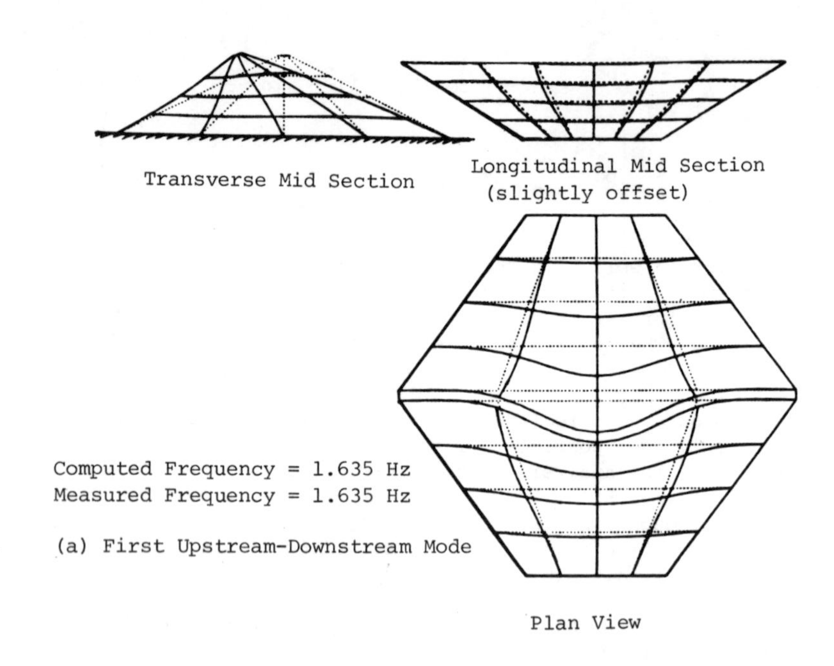

(a) First Upstream-Downstream Mode

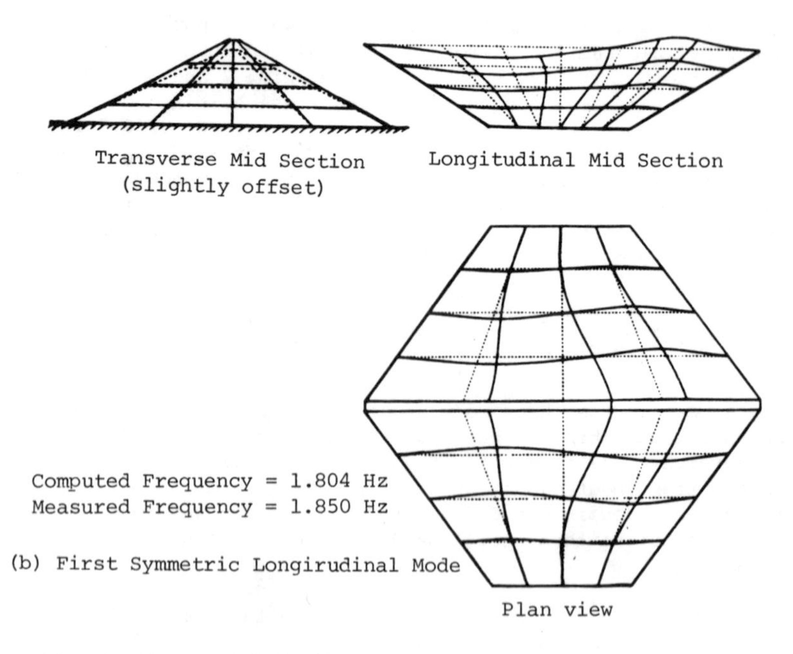

(b) First Symmetric Longirudinal Mode

Fig. 5 Computed Mode Shapes

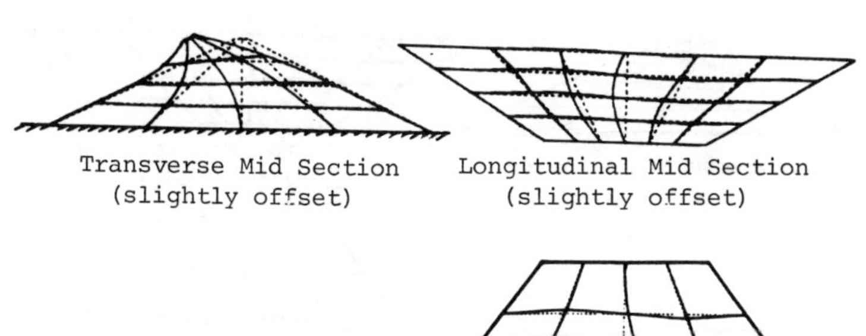

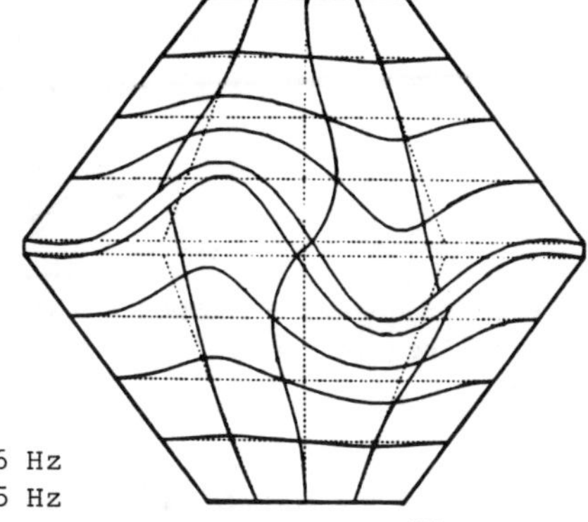

Computed Frequency = 1.816 Hz
Measured Frequency = 1.875 Hz

Plan View

(c) First Antisymmetric Upstream-Downstream Mode

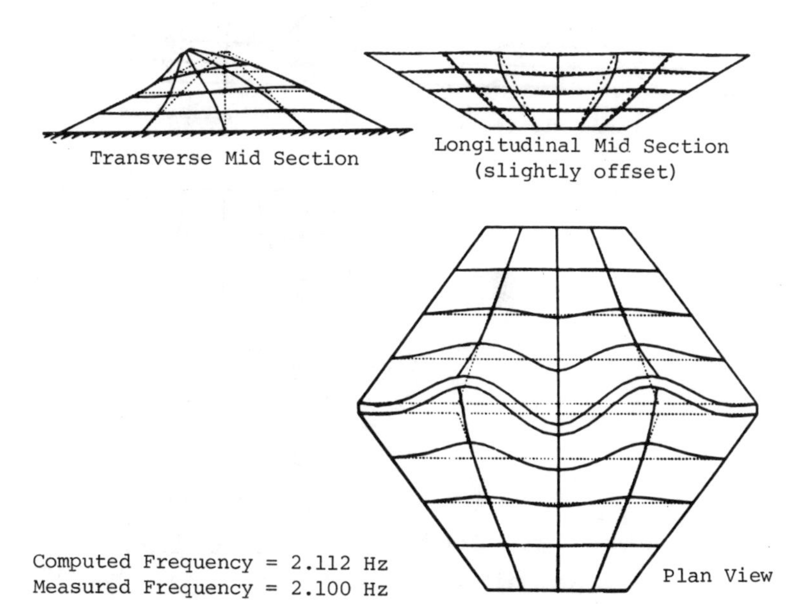

(d) Second Symmetric Upstream-Downstream Mode

Figure 6 Computed MOde Shapes

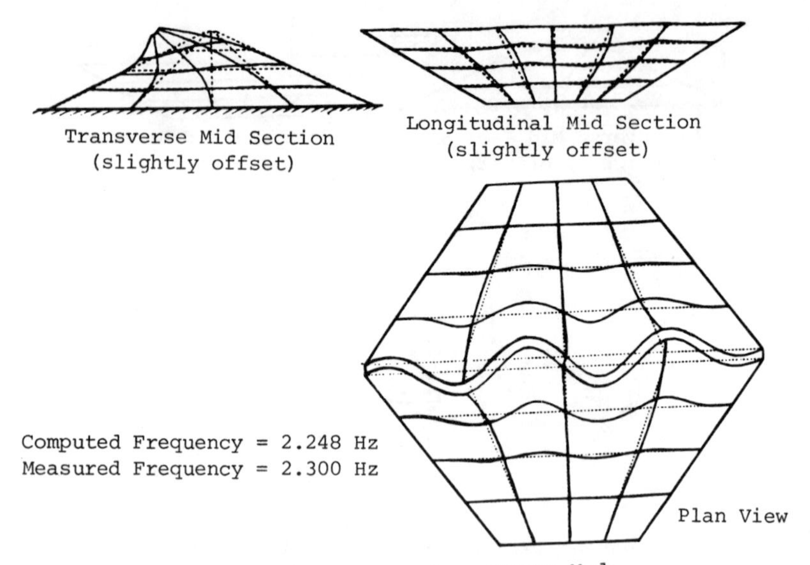

(e) Second Antisymmetric Upstream-Downstream Mode

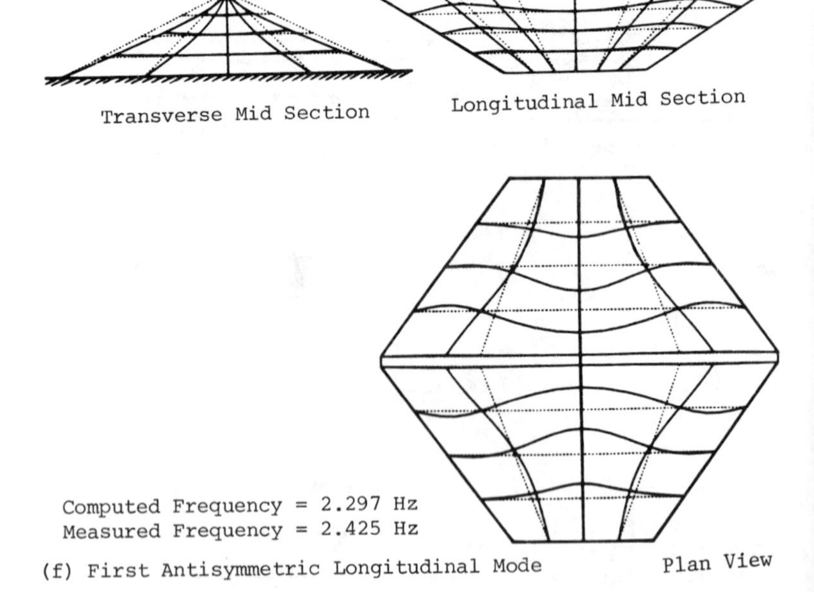

(f) First Antisymmetric Longitudinal Mode

Figure 7 Computed Mode Shapes

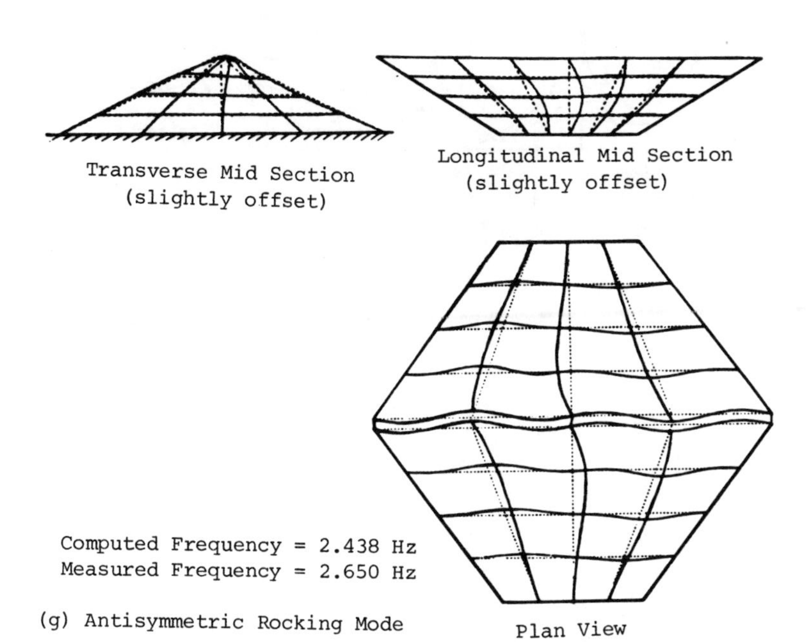

(g) Antisymmetric Rocking Mode

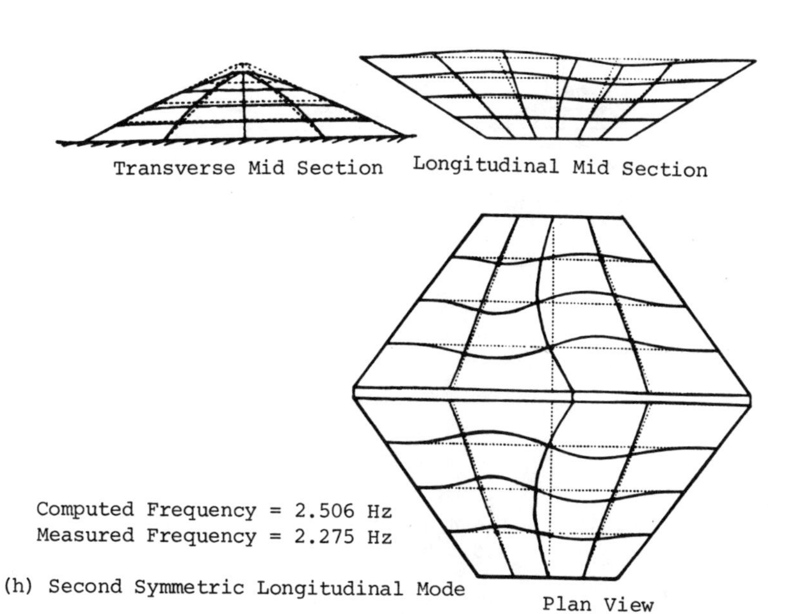

(h) Second Symmetric Longitudinal Mode

Figure 8 Computed Mode Shapes

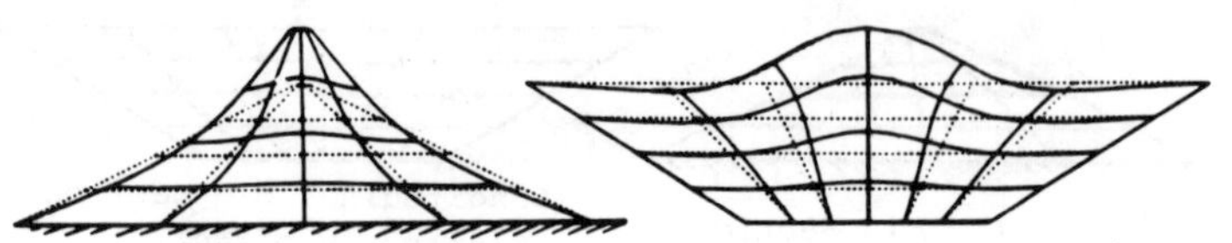

Transverse Mid Section Longitudinal Mid Section

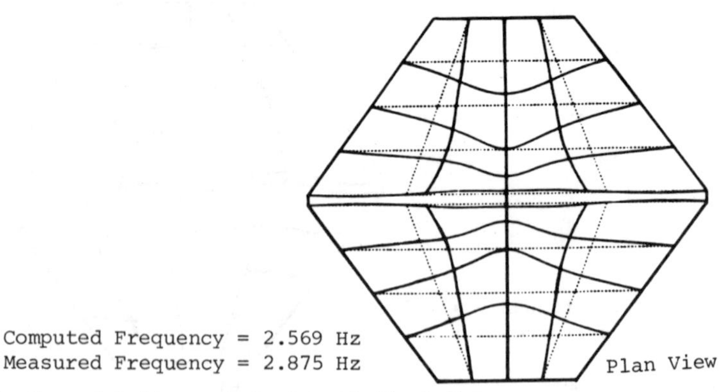

(i) Second Antisymmetric Longitudinal Mode

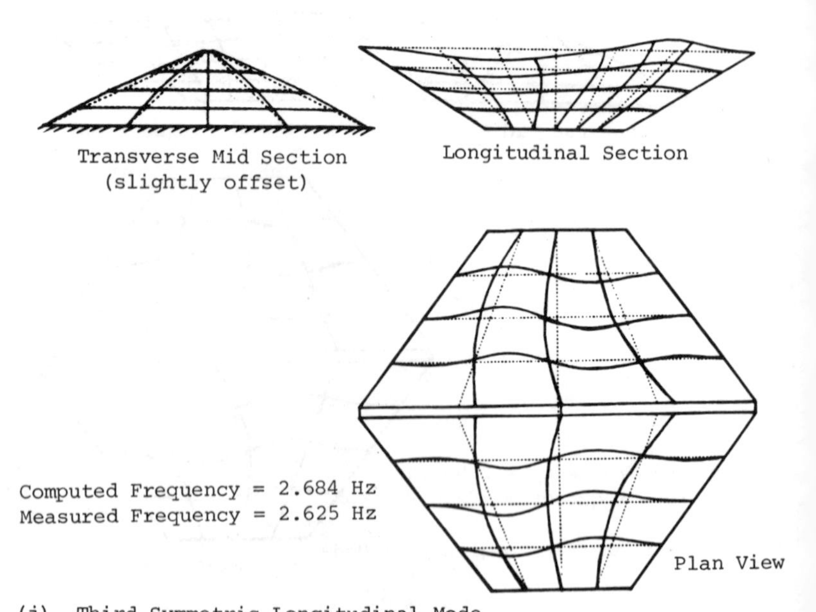

(j) Third Symmetric Longitudinal Mode

Figure 9 Computed Mode Shapes

is used for the match because it is the most reliable measurement made during the field test.

Figures 5 to 9 show the computed mode shapes of the longitudinal mid-section (afdboc of Fig. 1), the transverse mid-section (efgo of Fig. 1) and a plan view; they also show the comparison between the measured resonant frequencies (Ref. 2) and the computed ones. The maximum displacements and components of displacements are shown for scaling and for the comparison between the plots of different sections. It was found that certain plots of the mid-section will show zero displacements in the plane of the section. In order to provide more meaningful plots for these sections, exaggerated plots of displacements of a section slightly offset from the true mid-sections are given instead. The caption 'slightly offset' marks these plots in Figs. 5 to 9. Other, higher modes and associated frequencies can be found in Ref. 1.

The results obtained (both the modal configurations and the associated frequencies) agree well with those from the field tests. The model has revealed more than the two-dimensional shear beam models by showing the effects of coupling between the three orthogonal displacement components as well as more accurately taking into account the canyon shape. The computation for the natural frequencies and the mode shapes was inexpensive because the basis functions used are close approximations of the dam modes as derived from the two-dimensional models. This utilization of some previous information on the mode shapes reduces the number of basis functions needed for reasonable accuracy - unlike in the finite element methods where such prior knowledge cannot be utilized readily. The mass and stiffness matrices are full as opposed to the more easily solved banded matrices in the finite element methods but the dimensions are usually an order of magnitude or two smaller. Besides, for three-dimensional domains it is difficult to have node numbering arrangements that give small bandwidths in the finite element methods.

MAXIMUM STRAINS AND STRESSES

The modal participation functions of the maximum principal dynamic strains and stresses, maximum shear strains and stresses and the volumetric strains and stresses can be plotted to give a reasonable idea of the state of strains and stresses within the dam for each mode. The maximum principal strain and stress will give an idea where cracking will occur, the maximum shear strain and stress will show where local yielding of the dam material will occur and the volumetric strain and stress will identify the areas where liquefactions are possible.

Figures 10 and 11 show the distributions of the maximum principal strain, maximum shear strain, and volumetric strain for the first upstream-downstream and longitudinal mode shapes

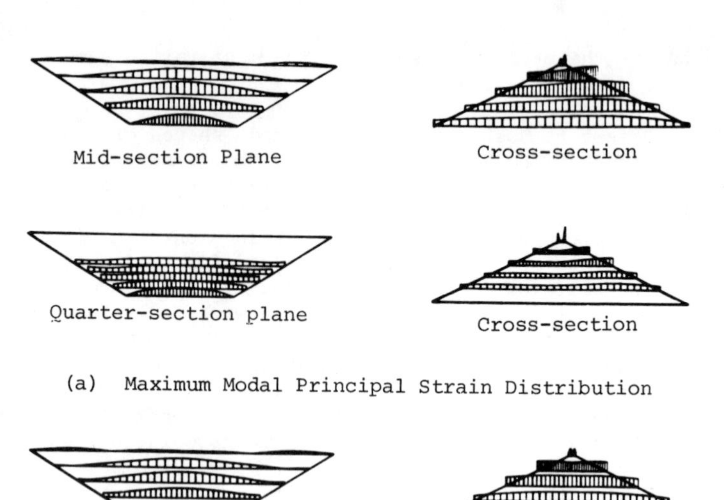

(a) Maximum Modal Principal Strain Distribution

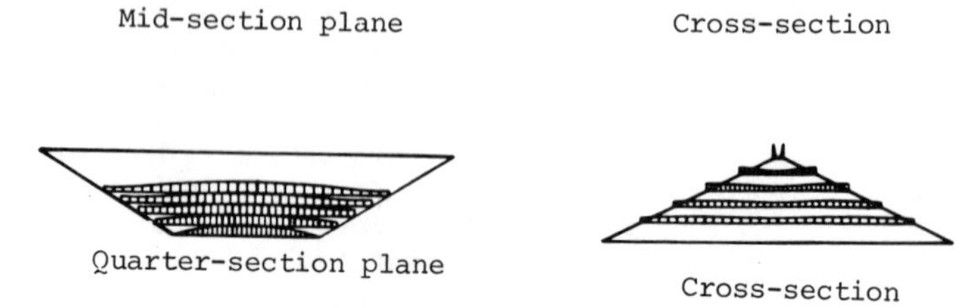

(b) Maximum Modal Shear Strain Distribution

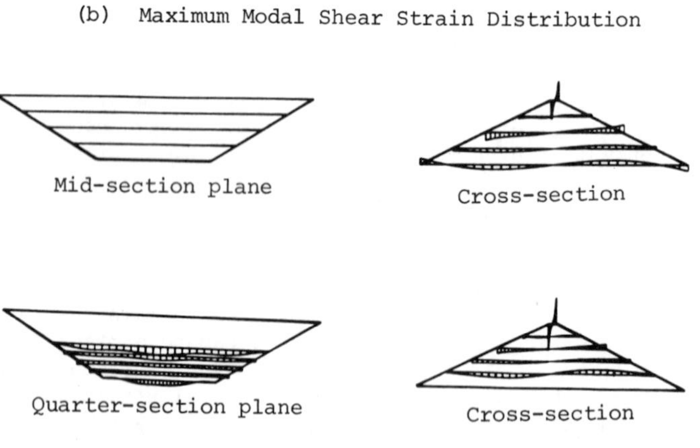

(c) Maximum Modal Spherical Strain Distribution

Figure 10 Modal participation functions of strains for the first upstream-downstream mode.

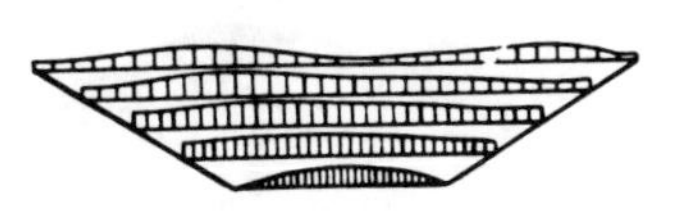

Mid-section Plane

Cross section

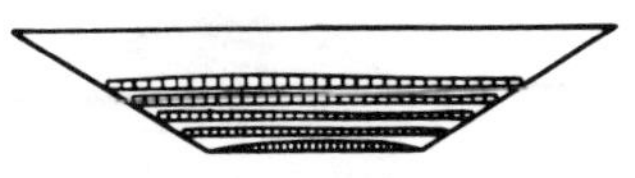

Quarter-section Plane

Cross section

(a) Maximum Modal Principal Strain Distribution

Mid-section Plane

Cross section

Quarter section Plane

Cross section

(b) Maximum Modal Shear Strain Distribution

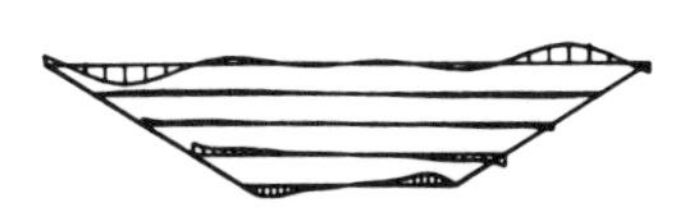

Mid-section Plane

Cross section

Quarter section Plane

Cross section

(c) Modal Spherical Strain Distribution

Figure 11 Modal participation functions of strains for the first longitudinal mode.

Mid Section Plane

Cross Section

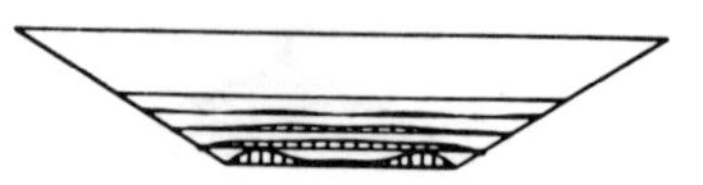

Quarter-Section Plane

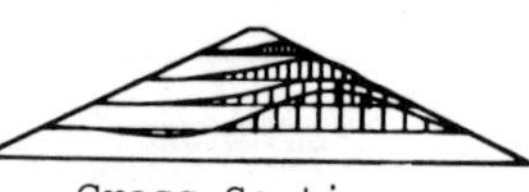

Cross Section

(a) Maximum Modal Principal Stress

Mid-Section Plane

Cross Section

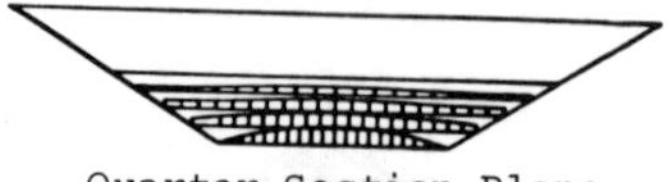

Quarter Section Plane

Cross Section

(b) Maximum Modal Shear Strain Distribution

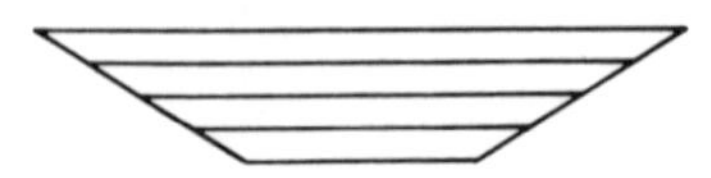

Mid-Section Plane

Cross Section

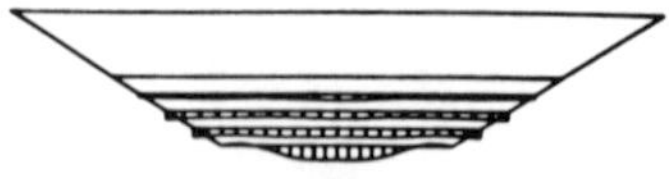

Quarter Section Plane

Cross Section

(c) Modal Spherical Strain Distribution

Figure 12 Modal participation functions of stresses for first upstream-downstream mode.

at four different sections (two mid-sections, afdcb and efgo of Fig. 1, and two quarter-sections parallel to the mid-sections). The corresponding maximum stresses for the upstream-downstream fundamental mode are shown in Fig. 12.

EARTHQUAKE RESPONSE ANALYSIS

For a dam on a rigid foundation subjected to ground accelerations in the three orthogonal directions, the earthquake response can be obtained routinely by utilizing the modal solution:

$$u_k = \sum_{m=1}^{\infty} \phi_k^m(x,y,z)T^m(t)$$

where $\phi_k^m(x,y,z)$ are the mode shapes of the dam. The details of this routine analysis can be found in Ref. 1.

On February 9, 1971 the San Fernando Earthquake shook the Santa Felicia Dam. Two accelerograms on and in the vicinity of the dam made records of the vibrations at the base and on the crest. (See Refs. 3 and 5) These records are used to check the three-dimensional model presented in this paper. The base absolute acceleration record is assumed to be the uniform ground motion input and the participation of fifteen modes are used to compute the relative displacements of the dam at the position where the crest accelerogram is sited. It should be noted that the fundamental frequency of the model is matched with that observed from the recorded response of the dam (Ref. 3). This fundamental frequency is 1.45 Hz.

The computed time-history of displacements, velocities and accelerations in the three orthogonal directions at the crest location correlate very well with the recorded (and processed) records. Because of space limitations, these time-history results are not shown here (but are presented in Ref. 1); however, the Fourier amplitude spectra of the computed displacement responses of the three components are plotted in Fig. 13 to compare with the amplification spectra of the dam as calculated from the field data. The amplification spectra are calculated by dividing the Fourier amplitude spectra of the recorded accelerations at the crest by the corresponding spectra at the base to indicate the natural frequencies and the relative contribution of the different modes of vibration.

For the frequency range covered by the computed spectra of Fig. 13 (only fifteen modes are considered) the main response characteristics are found to be the same as those of the recorded motion (Refs. 3,7). In the upstream-downstream direction, a single dominant mode (the fundamental) is evident, while in the longitudinal direction a significant contribution from a few higher modes was observed; finally, there is a scatter

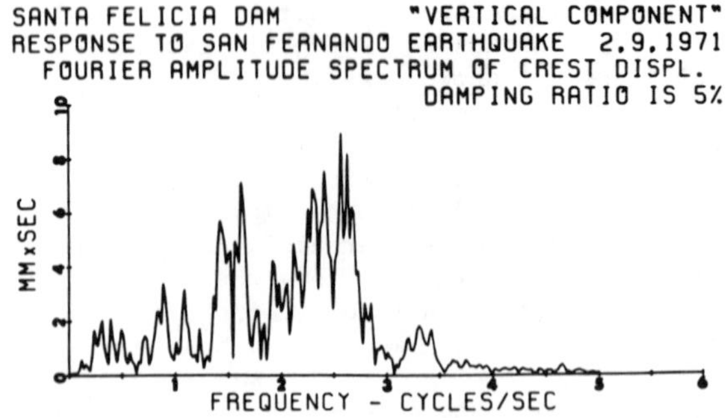

SANTA FELICIA DAM "UPSTREAM-DOWNSTREAM COMP."
RESPONSE TO SAN FERNANDO EARTHQUAKE 2.9.1971
FOURIER AMPLITUDE SPECTRUM OF CREST DISPL.
DAMPING RATIO IS 5%

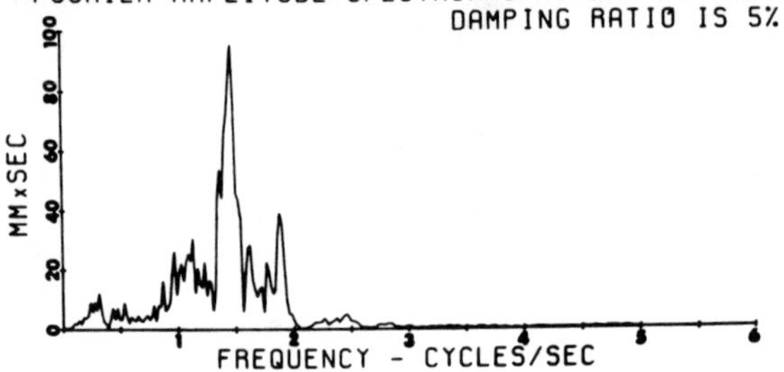

SANTA FELICIA DAM "LONGITUDINAL COMPONENT"
RESPONSE TO SAN FERNANDO EARTHQUAKE 2.9.1971
FOURIER AMPLITUDE SPECTRUM OF CREST DISPL.
DAMPING RATIO IS 5%

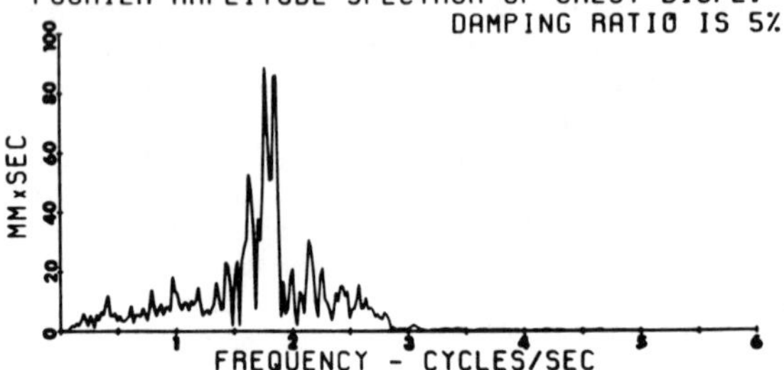

SANTA FELICIA DAM "VERTICAL COMPONENT"
RESPONSE TO SAN FERNANDO EARTHQUAKE 2.9.1971
FOURIER AMPLITUDE SPECTRUM OF CREST DISPL.
DAMPING RATIO IS 5%

MM x SEC
0 2 4 6 8 10
1 2 3 4 5 6
FREQUENCY - CYCLES/SEC

Figure 13 Fourier amplitude spectra of the computed displacement response (three components).

of the sharp modal peaks over the whole frequency range in the vertical direction.

CONCLUSIONS

The Rayleigh-Ritz method is used in the three-dimensional analysis of nonhomogeneous earth dams. Any dam geometry that can be linearly transformed into a cuboid in a new set of orthogonal coordinates can be adequately analyzed. The dam material is assumed isotropic and linearly elastic, but the elastic moduli can be functions of space. Tremendous computational advantages can be obtained by exploiting the symmetries of the dam geometry and material properties.

The proposed model gives results that are in relatively good agreement with those measured on a real dam in the field. The variation of the moduli within the dam has been shown to be important to the dynamic behavior of dams, and more intensive study of this variation will prove necessary for better predictions.

The results of modal displacements, strains and stresses can be used in the design of earthquake resistant dams, and to evaluate the safety of existing ones. The model has added advantage of being relatively economical in computational cost. Generally, it will prove suitable for preliminary design studies before more detailed computations are attempted.

REFERENCES

1. Abdel-Ghaffar, Ahmed M. and Koh, Aik-Siong (1981) "Three-Dimensional Dynamic Analysis of Nonhomogeneous Earth Dams," Department of Civil Engineering, Report No. 81-SM-26, Princeton University, Princeton, New Jersey.

2. Abdel-Ghaffar, Ahmed and Scott, Ronald F. (1981) "Vibration Tests of Full-Scale Earth Dam," Journal of the Geotechnical Engineering Division, ASCE, Vol. 107, No. GT3, Proc. Paper 16096, pp. 241-269.

3. Abdel-Ghaffar, A. M., and Scott, R. F. (1979) "Analysis of Earth Dam Response to Earthquakes," Journal of the Geotechnical Engineering Division, ASCE, Vol. 105, No. GT12, Proc. Paper 15033, pp. 1379-1404.

4. Abdel-Ghaffar, A. M., and Koh, Aik-Siong (1981) "Longitudinal Vibration of Nonhomogeneous Earth Dams," International Journal of Earthquake Engineering and Structural Dynamics, Vol. 9, No. 3, pp. 274-305.

5. Abdel-Ghaffar, A. M., and Scott, R. F. (1979) "Shear Moduli and Damping factors of Earth Dams," Journal of the Geotechnical Engineering Division, ASCE, Vol. 105, No. GT12, Proc. Paper 15034, pp. 1405-1426.

6. Abdel-Ghaffar, A. M., and Scott, R. F. (1981) "Comparative Study of Dynamic Response of an Earth Dam," Journal of the Geotechnical Engineering Division, ASCE, Vol. 107, No. GT3.

7. Abdel-Ghaffar, A. M. and Koh, Aik-Siong (1981) "Earthquake Induced Longitudinal Strains and Stresses in Nonhomogeneous Earth Dams," International Journal of Earthquake Engineering and Structural Dynamics, Vol. 9, pp. 521-542.

8. Ambraseys, N. N., (1960) "On the Shear Response of a Two-Dimensional Wedge Subjected to an Arbitrary Disturbance," Bulletin of the Seismological Society of America, Vol. 50, No. 1, pp. 45-56.

9. Gazetas, G. (1981) "Shear Vibrations of Vertically Inhomogeneous Earth Dams," Int. Jnl. Num. Anal. Meth. Geomech., Vol. 6, No. 4.

10. Gazetas, G. (1981) "Longitudinal Vibrations of Embankment Dams," Journal of the Geotechnical Division, ASCE, Vol. 107, No. GT1, Proc. Paper, 15980, pp. 21-40.

11. Hatanaka, M. (1955) "Fundamental Consideration on the Earthquake Resistant Properties of Earth Dams," Bulletin No. 11 Disaster Prevention Research Institute, Kyoto University, Japan.

12. Makdisi, F. I. and Seed, H. B. (1978) "Simplified Procedure for Estimating Dam and Embankment Earthquake Induced Deformations," Journal of Geotechnical Engineering Division, ASCE, Vol. 104, No. GT7, pp. 850-867.

13. Martinez, B. and Bielak, J. (1980) "On the Three-Dimensional Seismic Response of Earth Structures," Proceedings of the Seventh World Conference on Earthquake Engineering, September 8-13, 1980, Istanbul, Turkey, pp. 523-530.

14. Ohmachi, T. (1981) "Analysis of Dynamic Shear Strain Distributed in Three-Dimensional Earth Dam Models," Proceedings of the International Conference on Recent Advances in Geotechnical Earthquake Engineering and Soil Dynamics, St. Louis, Missouri, Vol. 1, pp. 459-464.

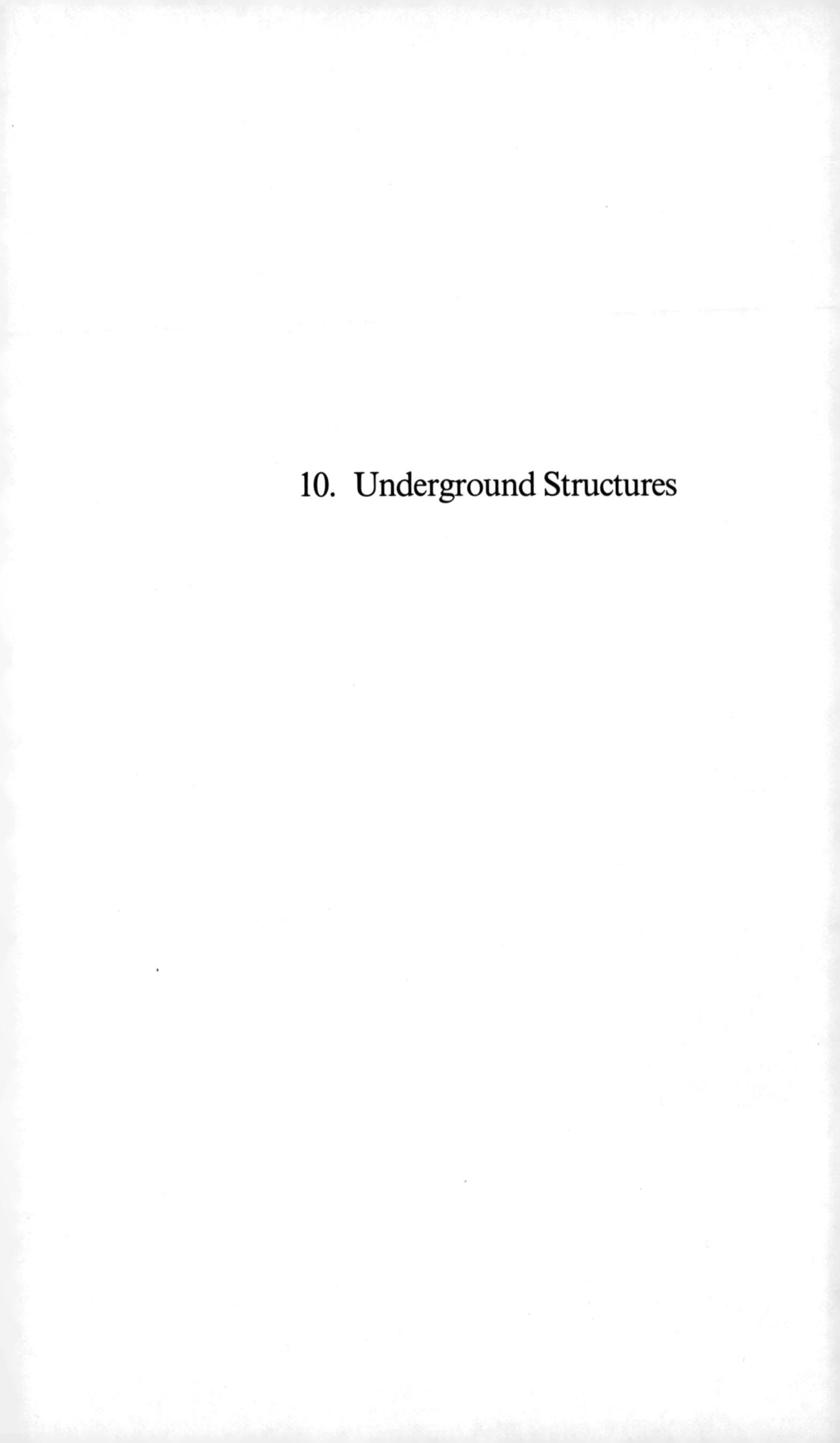

10. Underground Structures

Soil Dynamics & Earthquake Engineering Conference / Southampton / 1982.07.13-15

Study of Ground Strains From Strong Motion Array Data for Lifeline Application

L.R.L.WANG & Y.YEH
University of Oklahoma, Norman, USA

K.KAWASHIMA & K.AIZAWA
Ministry of Construction, Tsukuba, Japan

ABSTRACT

The seismic behavior of lifeline systems is predominately controlled by the ground displacement/strain characteristics. This study investigates the ground displacements and ground strains from the latest strong-motion arrays data recorded at Public Works Research Institute and Ashitaka sites, Japan.

At PWRI site, which have both horizontal and vertical arrays, separated by approximately 50m, it is found that the ground displacements seem to be dominated by the shear wave. Depth effects on the amplification of ground displacements have been observed.

At Ashitaka site, which has only surface arrays, the strains calculated with a distance ranging from 400m to 1300m could not represent the peak ground strain condition. To be useful to lifeline application, more strong motion accelerographs apart no more than 50 meters must be installed.

I. INTRODUCTION

The seismic behavior of lifeline systems, such as water, sewer, oil or gas pipelines, etc. is predominately controlled by the ground displacement or ground strain characteristics. In application, the free field motion on the surface of the ground would be sufficient to represent the ground motion input, because for buried lifelines, the buried depth in general is rather shallow, say 1m to 2m deep as compared to subsurface soil layer.

Although it is quite important, the direct measurement of ground strains is very difficult, if not impossible. However, with the installation of strong motion dense arrays, (1, 2, 6) it is now possible to calculate the ground strains from the earthquake records with some degree of accuracy.

Table 1 **Maximum Accelerations (gal), Maximum Velocities (kine) and Maximum Displacements (cm) of EQ 13 at PWRI Site**

Point	Comp.	Max. Accel.	Max. Vel.	Max. Disp.
1	NS	39.47	2.40	-0.299
	EW	40.08	-2.48	0.342
	UD	-12.64	-0.79	-0.112
2	NS	13.05	-1.19	0.242
	EW	-14.94	1.03	0.189
	UD	-5.94	0.54	-0.081
3	NS	15.20	-1.22	0.233
	EW	-13.95	1.00	0.180
4	NS	12.61	-0.99	-0.212
	EW	14.48	1.29	0.208
	UD	-5.86	0.56	0.085
5	NS	11.81	-1.04	-0.206
	EW	14.59	1.32	0.224
6	NS	-14.41	-1.03	-0.243
	EW	-13.58	-1.11	-0.206
	UD	6.81	0.46	-0.074
7	NS	43.04	-2.43	-0.282
	EW	44.15	3.52	0.473
	UD	-11.61	-0.77	0.102
8	NS	25.29	-1.54	-0.202
	EW	23.76	1.57	0.247
	UD	7.35	-0.40	0.061
9	NS	32.58	2.24	0.328
	EW	35.49	2.78	0.446
13	NS	-22.51	-1.20	0.152
	EW	28.00	-1.81	0.242
	UD	-6.75	-0.43	-0.052

Table 2 **Max. Acc. (gal), Max. Vel. (kine) and Max. Disp. (cm) at Ashitaka site after coordinate transformation**

Segment	Station	Comp.	Max. Acceleration [gal]	Max. Velocity [cm/sec]	Max. Displacement [cm]
1 ∿ 2	1	NS	-73.10	3.33	1.69
		EW	122.00	-3.31	0.90
		UD	73.80	2.13	-0.66
	2	NS	-181.00	-6.84	0.79
		EW	169.00	-9.61	0.93
		UD	68.80	2.62	-0.54
2 ∿ 3	2	NS	-192.00	-7.11	0.80
		EW	156.00	-9.49	-0.86
		UD	68.80	2.62	-0.55
	3	NS	-163.00	-7.48	0.92
		EW	-120.00	-9.12	-0.74
		UD	-36.80	-1.82	-0.57
3 ∿ 5	3	NS	-151.00	-7.96	-0.77
		EW	135.00	-8.90	-0.90
		UD	-36.80	-1.82	-0.57
	5	NS	99.40	-3.82	0.73
		EW	-111.00	2.82	0.82
		UD	46.00	1.51	0.44

Recently, the Public Works Research Institute, Tsukuba Science City, Japan has deployed a local laboratory array consisting of 20 accelerometers at PWRI site and a simple extended array consisting of 5 accelerographs at Ashitaka area in Shizuoka Prefecture. Since their installations, there have been a number of earthquake data recorded from these sites. Preliminary results on subsurface strains calculated by using some earlier data have been reported. (4, 5)

To supplement the earlier report (5), this study is to investigate the surface ground strains from the latest data recorded at PWRI and Ashitaka sites for future application to lifelines. Also the subsurface ground strains from PWRI site, where data from vertical arrays are available, were investigated in order to determine the depth effects.

II. BACKGROUND DATA

(A) PWRI Site

The strong motion arrays, designated as A, B and C fields, at PWRI are shown in Figure 1. The geological condition around the Institute is more or less uniform. Shear wave velocities of the upper and lower diluvial deposits are approximately 250m/sec and 400m/sec, respectively. The number of accelerometers, the number of channels of strong-motion arrays at these fields and the details of the triggering, recording and processing of the array systems can be found in the earlier reports (4, 5) and thus will not be repeated.

Since last report (5) after December 1980, three earthquakes, occurred in the middle of Chiba Prefecture, and Southwest of Ibaraki Prefecture, have been recorded at PWRI site, denoted as EQ-13, 14 and 15. Due to the fact that the magnitudes of EQ 14 and 15 are too small to give meaningful result, this study uses EQ 13 data only. Table 1 gives the maximum accelerations of the channels that were functioned properly during the earthquake. Figure A1 shows the time histories of the corrected accelerations of a typical Point (Point 1) at PWRI Site.

The details of the geological and seismic profiles, time histories of all corrected accelerations, calculated ground velocities and displacements at the site are given in the original report (8).

(B) Ashitaka Site

The Ashitaka site is located between Numazu City and Fuji City in Shizuoka Prefecture facing Suruga Bay. This is a simple extended array on the surface of the ground crossing Ukishimagahara in almost NS direction. The layout of strong motion accelerographs is shown in Figure 2. Note from this figure that Point 4 had no data due to mal-function of the accelerographs. Thus, there are only 3 segment routes, i. e.,

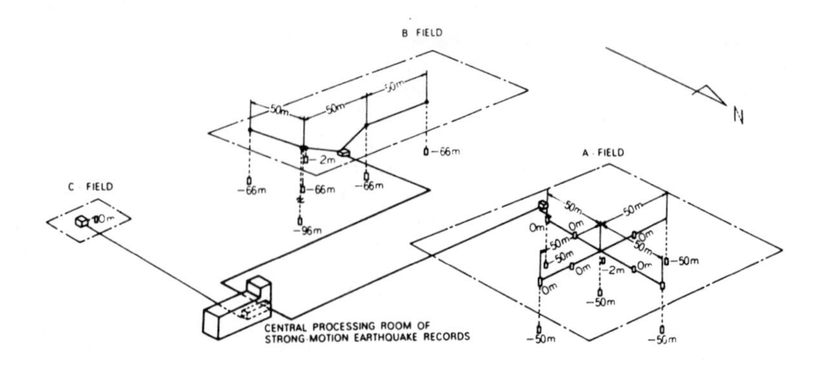

Fig.1 LOCAL LABORATORY ARRAYS AT PWRI

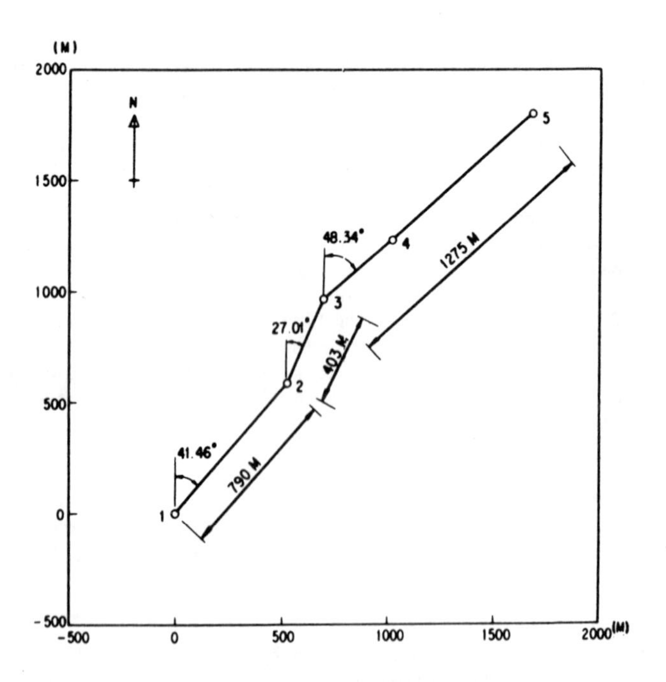

Fig. 2 Layout of strong motion accelerometers installed in Ashitaka area, Japan

(Note: No date as at point 4 due to malfunction of the instruments)

1-2, 2-3 and 3-5, at the site.

Shear wave velocity of subsurface ground of Ashitaka site is in the range between 70m/sec to 200m/sec. Tuff breccia exists at the depth of approximately 35m below the surface and has a shear wave velocity of approximately 650m/sec.

Due to the fact the accelerographs, which are in NS and EW directions, do not coincide with the segment route directions of the array as shown in Figure 2, a coordinate transformation is needed in order to calculate the ground strains along the route of the arrays, as follows:

$$\begin{cases} X = x_0\cos\theta + y_0\sin\theta \\ Y = -x_0\sin\theta + y_0\cos\theta \end{cases} \quad (1)$$

where X, Y, are the coordinates along the segment route, x_0 and y_0 are EW and NS direction, θ is the angle of rotation.

After the coordinate transformation, the time histories of the corrected accelerations for 3 segments in Ashitaka site were calculated (8). Figure A3 gives a sample corrected acceleration time history for segment 1-2. The maximum acceleration along three segment routes are given in Table 2.

III. CALCULATION OF GROUND VELOCITIES AND GROUND DISPLACEMENTS

The determination of ground velocity and ground displacement (3) can be obtained by integrating the originally recorded acceleration data, $\tilde{a}(t)$, once and twice through Fourier transformations. The procedure first requires the transformation from time domain to frequency domain as:

$$\tilde{F}(f) = \int_{-\infty}^{\infty}\tilde{a}(t)e^{-2\pi ift}dt \quad (2)$$

To remove the undesirable errors (noise) in low and high frequency ranges, a band-pass filter must be applied to the acceleration records:

$$F(f) = F_c(f)\cdot\tilde{F}(f) \quad (3)$$

In this study, a trapezoid type filter is used. Mathematically, using lowest filtering frequency, $f_{\ell\ell}$; lower to upper frequency, $f_{\ell u}$; upper to lower frequency, $f_{u\ell}$ and highest filtering frequency, f_{uu} designations, $F_c(f)$ is expressed as follows:

$$F_c(f) = \begin{cases} 0 \ \dots\dots\dots\dots\dots\dots & f < f_{\ell\ell} \\ \dfrac{f - f_{\ell\ell}}{f_{\ell u} - f_{\ell\ell}} \dfrac{e^{i\phi(f)}}{R(f)} \ \dots\dots\dots & f_{\ell\ell} < f < f_{\ell u} \\ \dfrac{e^{i\phi(f)}}{R(f)} \ \dots\dots\dots\dots\dots & f_{\ell u} < f < f_{u\ell} \\ \dfrac{f_{uu} - f}{f_{uu} - f_{u\ell}} \quad \dfrac{e^{i\phi(f)}}{R(f)} \ \dots\dots\dots & f_{u\ell} < f < f_{uu} \\ 0 \ \dots\dots\dots\dots\dots\dots & f > f_{uu} \end{cases} \tag{4}$$

in which R(f) and $\phi(f)$ represent amplification and phase of sensors, respectively.

For broader applications, three different filters, called F1, F2 and F3, with different low frequency limits were used.

After the filtering operation, the modified or corrected acceleration a(t) can be obtained as:

$$a(t) = \int_{-\infty}^{\infty} F(f)\, e^{2\pi i f t} df \tag{5}$$

and the velocity v(t) and displacement d(t) associated with a(t) are calculated in the forms:

$$v(t) = \int_{-\infty}^{\infty} \frac{F(f)}{2\pi i f}\, e^{2\pi i f t} df \tag{6}$$

$$d(t) = \int_{-\infty}^{\infty} \frac{F(f)}{-4\pi^2 f^2}\, e^{2\pi i f t} df \tag{7}$$

Based on the characteristics of sensors used at the PWRI site, the amplification R(f) and phase $\phi(f)$ in Eq. (4) were taken as:

$$R(f) = \sqrt{\left[\frac{1 - (f/f_0)^2}{2hf/f_0}\right]^2 + 1}$$

$$\phi(f) = \tan^{-1}\left[\frac{2hf/f_0}{1 - (f/f_0)^2}\right] \tag{8}$$

in which f_0 and h, representing natural frequency and damping ratio of sensors in a type of velocity feed-back servo accelerometer.

By applying this procedure with F3 filter to the strong motion data at PWRI site, the time histories of velocities and displacements for PWRI site were obtained. The displacement time history for Point 1 is shown in Figure A1.

Note that at the beginning of this study, all three

filters have been used in the analysis, but it was found that F3 was most satisfactory.

Similarly using F3 filter, the time histories of velocities and displacements for Ashitaka site were calculated. The displacement time histories for Points 1 and 2 of Segment 1 are given in Figure A3. In this case, the amplification R(f) and phase $\phi(f)$ were taken as unity and zero, respectively, because the sensors used at the site are displacement feed back servo accelerometers having a flat sensitivity between DC and 50 Hz.

After integrations, the maximum values of velocities and displacements for PWRI site and Ashitaka site are given in Tables 1 and 2, respectively.

IV. ANALYSIS OF GROUND STRAINS

(A) Finite Element Strains At PWRI Site

Because PWRI site consists both vertical and horizontal arrays, it is possible to construct a tetrahedron, shown in Figure 3, for strain analysis (9). At each nodal point, the displacements u(t), v(t) and w(t) in the tetrahedron are assumed to be linear in x-, y- and z-direction. Mathematically, they can be expressed as:

$$\begin{cases} u = \alpha_1 + \alpha_2 x + \alpha_3 y + \alpha_4 z \\ v = \beta_1 + \beta_2 x + \beta_3 y + \beta_4 z \\ w = q_1 + q_2 x + q_3 y + q_4 z \end{cases} \tag{9}$$

where the coefficients α_m, β_n and q_k can be determined by prescribing the nodal coordinates. Following Ref. 9, Eq. (9) is rewritten in matrix form as:

$$\begin{Bmatrix} u \\ v \\ w \end{Bmatrix} = \frac{1}{6V} \begin{bmatrix} u_i & u_j & u_m & u_p \\ v_i & v_j & v_m & v_p \\ w_i & w_j & w_m & w_p \end{bmatrix} \begin{Bmatrix} a_i + b_i x + c_i y + d_i z \\ a_j + b_j x + c_j y + d_j z \\ a_m + b_m x + c_m y + d_m z \\ a_p + b_p x + c_p y + d_p z \end{Bmatrix} \tag{10}$$

where

$$a_i = \det \begin{vmatrix} x_i & y_i & z_i \\ x_m & y_m & z_m \\ x_p & y_p & z_p \end{vmatrix} ; \quad b_i = -\det \begin{vmatrix} 1 & y_i & z_i \\ 1 & y_m & z_m \\ 1 & y_p & z_p \end{vmatrix}$$

$$c_i = -\det \begin{vmatrix} x_i & 1 & z_i \\ x_m & 1 & z_m \\ x_p & 1 & z_p \end{vmatrix} ; \quad d_i = -\det \begin{vmatrix} x_i & y_i & 1 \\ x_m & y_m & 1 \\ x_p & y_p & 1 \end{vmatrix} \tag{11}$$

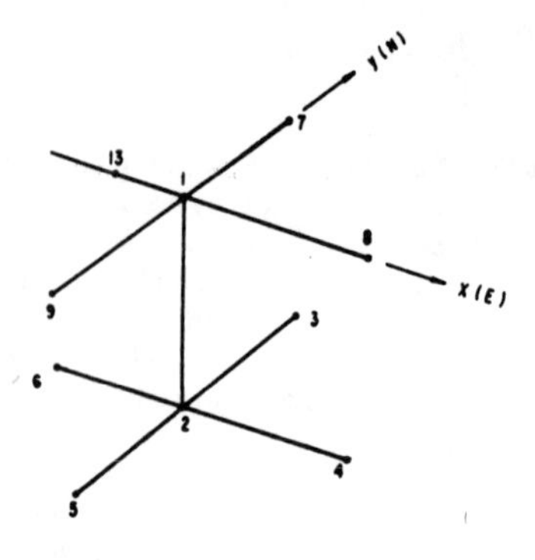

(a) The Deployment of Both Vertical and Horizontal Arrays

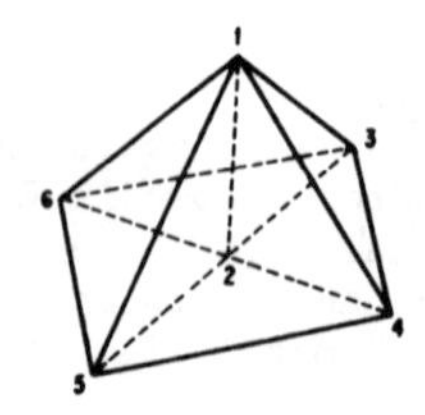

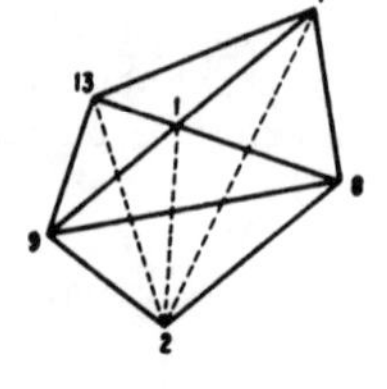

(b) Configuration of Lower Pyramid

(c) Configuration of Upper (inverse) Pyramid

Figure 3

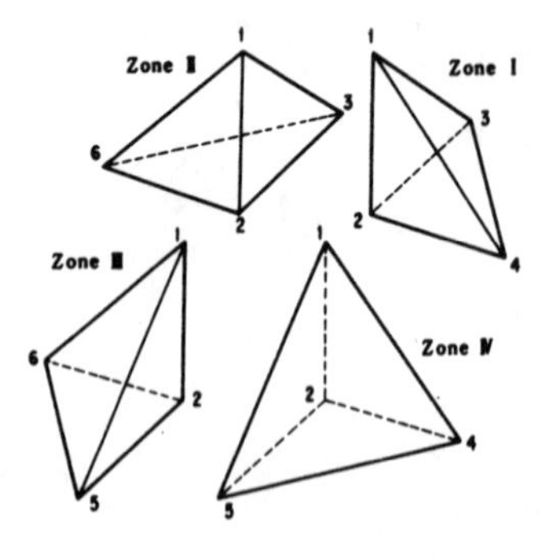

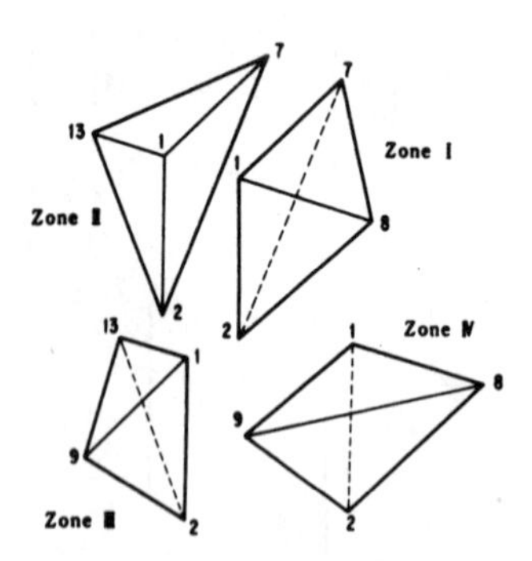

(a) Tetrahedrons for Calculating Lower Level Strains

(b) Tetrahedrons for Calculating Upper Level Strains

Fig. 4

$$6V = \det \begin{vmatrix} 1 & x_i & y_i & z_i \\ 1 & x_j & y_j & z_j \\ 1 & x_m & y_m & z_m \\ 1 & x_p & y_p & z_p \end{vmatrix} \tag{12}$$

and x_k, y_k, z_k are coordinates of point k (k=i, j, m, p) and other constants, a_k, b_k, c_k and d_k (k=i, j, m, p) can be obtained in the same way by changing the subscript in the order of i, j, m and p.

The determination of strains (7) of a ground element surface, $\{\varepsilon\}$, can be done as:

$$\{\varepsilon\} = \begin{Bmatrix} \varepsilon_x \\ \varepsilon_y \\ \varepsilon_z \\ \gamma_{xy} \\ \gamma_{yz} \\ \gamma_{zx} \end{Bmatrix} = \begin{Bmatrix} \partial u/\partial x \\ \partial v/\partial y \\ \partial w/\partial z \\ \dfrac{\partial u}{\partial y} + \dfrac{\partial v}{\partial x} \\ \dfrac{\partial v}{\partial z} + \dfrac{\partial w}{\partial y} \\ \dfrac{\partial w}{\partial x} + \dfrac{\partial u}{\partial z} \end{Bmatrix} \tag{13}$$

By substituting Eq. (9) into Eq. (13), $\{\varepsilon\}$ can be written in the form:

$$\{\varepsilon\} = [B]\,\{\delta\} = [B_i\ B_j\ B_m\ B_p]\,\{\delta\} \tag{14}$$

where

$$[B_k] = \frac{1}{6V} \begin{bmatrix} b_k & 0 & 0 \\ 0 & c_k & 0 \\ 0 & c_k & d_k \\ c_k & b_k & 0 \\ 0 & d_k & c_k \\ d_k & 0 & b_k \end{bmatrix}, (k = i, j, m, p) \tag{15}$$

$$\{\delta\} = \{\delta_i\ \delta_j\ \delta_m\ \delta_p\}^T \tag{16}$$

$$\{\delta_k\} = \{u_k\ v_k\ w_k\}^T \tag{17}$$

According to the above procedure, a computer program has

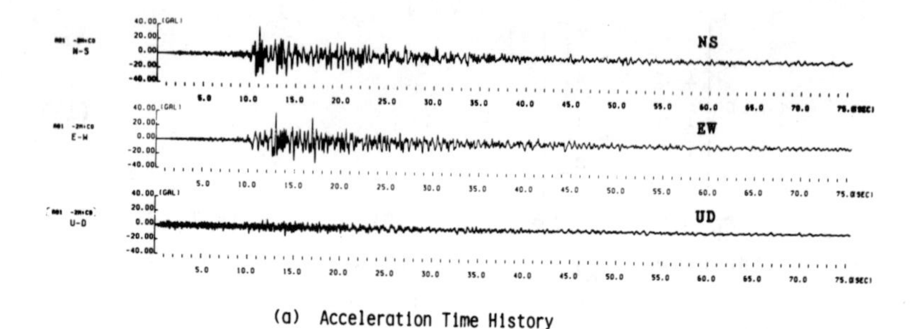

(a) Acceleration Time History

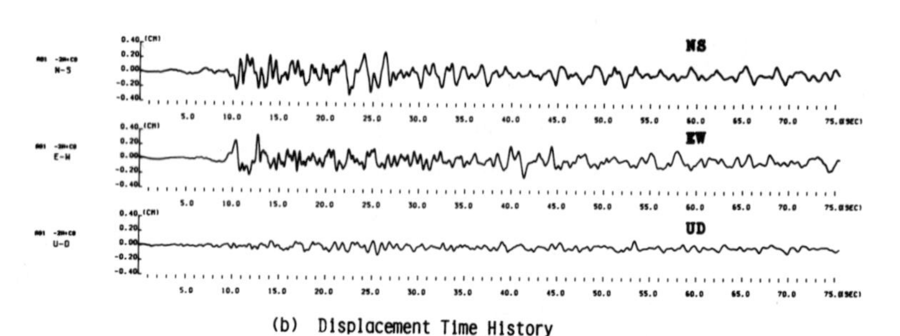

(b) Displacement Time History

Figure A1 Ground Responses of Point 1 at PWRI site by E013

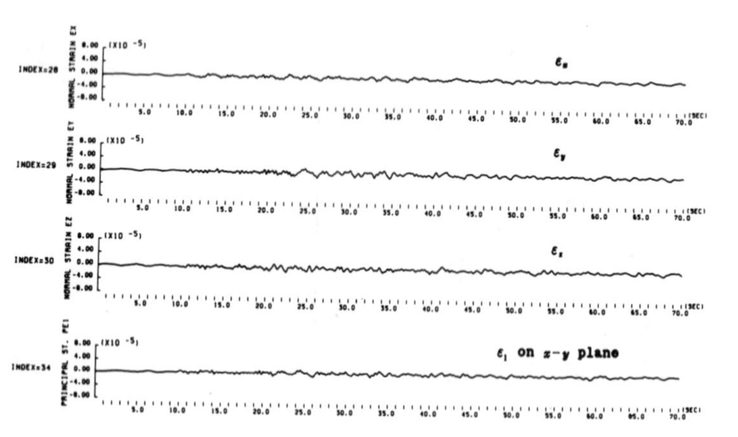

Figure A2

Ground Strains for Zone I Lower Level at PWRI Site by E013

been written and is used in this study.

In consideration to evaluate the depth effects on ground strains, 10 observation stations at PWRI were selected to form 8 different tetrahedrons as shown in Figure 4 for the strain analysis.

For lifeline applications, the principal strains, which dominate the system response behavior, must be determined. Since most lifeline systems are constructed in a horizontal plane (i. e., x-y plane) with respect to the ground, only ε_x, ε_y and γ_{xy} are necessary to calculate the principal strains in x-y plane. Thus, the principal strains (7) in x-y plane are:

$$\varepsilon_{p1} = \frac{\varepsilon_x + \varepsilon_y}{2} + \sqrt{\left[\frac{\varepsilon_x - \varepsilon_y}{2}\right]^2 + \frac{1}{4}\gamma^2_{xy}}$$

$$\varepsilon_{p2} = \frac{\varepsilon_x + \varepsilon_y}{2} - \sqrt{\left[\frac{\varepsilon_x - \varepsilon_y}{2}\right]^2 + \frac{1}{4}\gamma^2_{xy}} \qquad (18)$$

and the principal angle is

$$\tan 2\theta = \frac{\gamma_{xy}}{\varepsilon_x - \varepsilon_y} \qquad (19)$$

Based on above described procedure, the ground strains, ε_x, ε_y, γ_{xy} as well as the principal strains and principal angle for the xy plane of the 8 tetrahedrons, 4 for surface or upper level strains and 4 for below ground or lower level (50m deep) strains were calculated. Sample strains for Zone I of Lower Level are shown in Figure A2. The maximum strains are listed in Table 3.

(B) Normal Surface Strains At Ashitaka Site

Due to the fact that Ashitaka site has accelerographs on the ground surface only, it is not possible to obtain the finite element strains described earlier. Upon transformation of coordinates, the normal ground strain ε_{ij}, along the array segment route can easily be determined as:

$$\varepsilon_{ij} = \frac{u_x^i - u_x^j}{L_{ij}} \qquad (20)$$

where ε_{ij} represents the average strain between nodes i and j and u_x^i and u_x^j are ground displacements in ij direction at node i and j, respectively; L_{ij} is the distance between i and j. Based on above procedure, the strain time histories for the

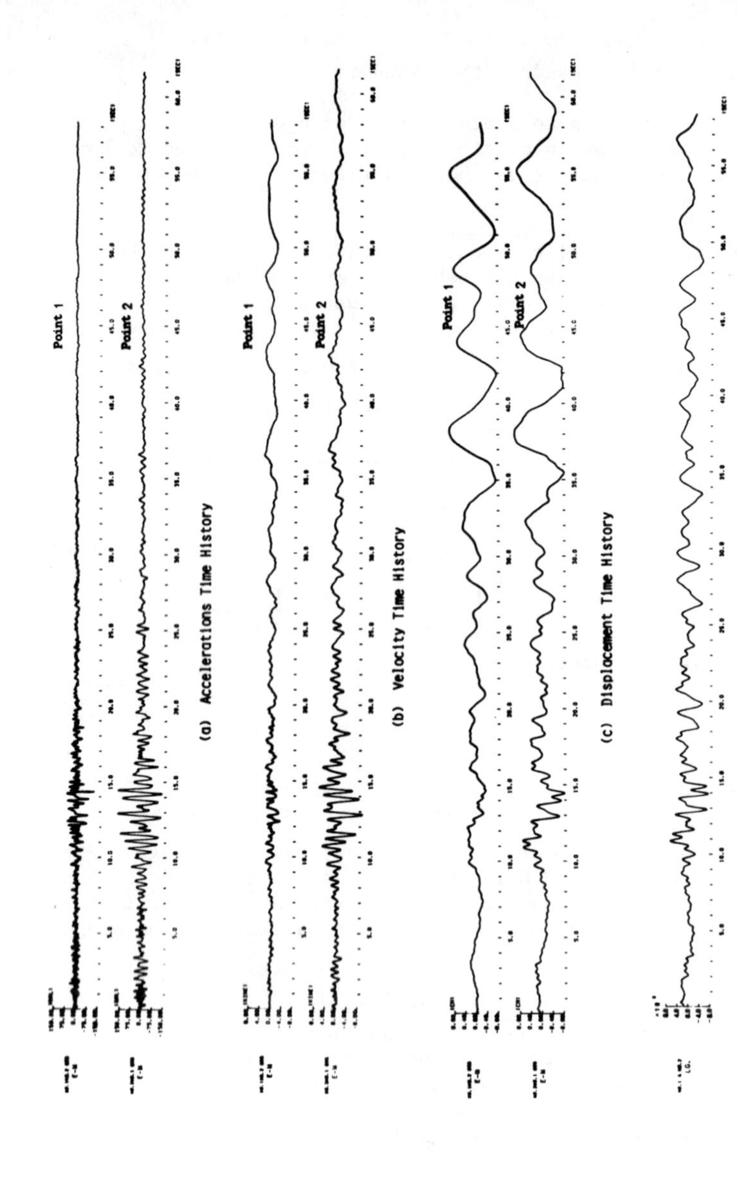

Ground Response Time Histories of Segment 1-2 at Ashitaka Site

Figure A3

Table 3 Maximum Ground Strains [$\times 10^{-6}$] at PWRI Site for EQ 13

Zone	Strain	Lower Level	Upper Level
I	ε_x	9.4	-27.2
	ε_y	-14.5	15.4
	ε_z	-12.4	-12.4
	γ_{xy}	-13.4	-46.4
	γ_{yz}	-	-54.2
	γ_{zx}	59.7	68.2
II	ε_x	12.1	52.4
	ε_y	-14.5	15.4
	ε_z	-12.4	-12.4
	γ_{xy}	15.8	66.4
	γ_{yz}	-	-54.2
	γ_{zx}	60.4	47.3
III	ε_x	12.1	52.4
	ε_y	17.3	14.2
	ε_z	-12.4	-12.4
	γ_{xy}	20.1	81.4
	γ_{yz}	-	-
	γ_{zx}	60.4	47.3
IV	ε_x	9.4	-27.2
	ε_y	17.3	14.2
	ε_z	-12.4	-12.4
	γ_{xy}	-19.1	32.2
	γ_{yz}	-	-
	γ_{zx}	59.7	68.2

Table 4 Maximum Normal Ground Strains [$\times 10^{-6}$] along the Route of Array at Ashitaka Area

Segment	Strain ε_x
1 ∿ 2 (ℓ = 790m)	-7.6
2 ∿ 3 (ℓ = 403m)	-22.4
3 ∿ 5 (ℓ = 1275m)	6.7

Ashitaka site were calculated (8). The ground strain time history for Segment 1-2 is shown in Figure A3. The maximum strains at Ashitaka site are given in Table 4.

V. DISCUSSIONS

In examining the maximum ground displacements at PWRI site from Table 1, one can find that the displacements on the surface in general are larger than those underground as expected. It is interesting to note, however, that the amplification due to the depth effect in the EW direction is more than in the NS direction. Since the epicenter of EQ 13 was located at 71km south of PWRI, such amplification in the direction normal to seismic wave propagation direction may be attributed to the shear wave.

The amplification on ground strains can be seen in Table 3 and similar conclusions are observed. To quantify the amplification due to depth effect, the magnification ratios, which is defined as the maximum upper level strain divided by the lower level strain, for PWRI site were studied (8). It is found that the amplification of axial strain in EW direction ranged between 3 to 4. For NS direction, however, there was little amplification (magnification ratio = 1.07 for Zone I and II) or even reversed amplification (0.82 for Zone III and IV). Such reversed amplification, perhaps, is due to the variation of local geology/soil conditions at the site. For future application to lifeline systems, earthquake input (acceleration and/or velocity) must be correlated ground strains along with site environments (geological and/or soil conditions). More data must be obtained.

For Ashitaka site, the strains, which were calculated over a distance ranged from 400m to 1275m, were too low to represent the peak ground strains for lifeline application. Note that the original purpose of the strong motion array deployment at Ashitaka site was to study the wave propagation characteristics. Thus, to be useful to lifeline application, more strong motion accelerographs apart no more than 50m must be installed.

ACKNOWLEDGEMENT

The writing of this report was during the period when the first author was working on an NSF project and was a Visiting Researcher at Public Works Research Institute. The financial supports from the National Science Foundation and Japan Science and Technology Agency are greatly appreciated.

The authors wish to thank Dr. T. Okubo, Dr. S.C. Liu, Dr. E. Kuribayashi, Dr. T. Iwasaki and Mr. O. Ueda for their cooperation and encouragement.

VI. REFERENCES

1) Iwan, W. D. (Ed.) (1978) International Association for Earthquake Engineering: Strong-Motion Earthquake Instruments Arrays, Proceedings of the International Workshop on Strong-Motion Earthquake Instruments Arrays, Honolulu, Hawaii, U. S. A.

2) Iwan, W. D. (1979) The Deployment of Strong-Motion Earthquake Instrument Arrays, Earthquake Engineering and Structural Dynamics, 7.

3) Iwasaki, T. (1981) Free Field and Design Motions During Earthquakes, International Conference on Recent Advances in Geotechnical Earthquake Engineering and Soil Dynamics, St. Louis, Missouri, U. S. A.

4) Okubo, T., Iwasaki, T. and Kawashima, K. (1980) Dense Instrument Array Program of the Public Works Research Institute for Observing Strong Earthquake Motion, Some Recent Earthquake Engineering Research and Practice in Japan, The Japanese National Committee of the International Association for Earthquake Engineering.

5) Okubo, T., Iwasaki, T. and Kawashima, K. (1981) Dense Instrument Array Program of the Public Works Research Institute and Preliminary Analysis of Some Records, 13th Joint Meeting, U. S.-Japan Panel on Wind and Seismic Effect, UJNR, Tsukuba, Japan.

6) Sakagami, Y., Okubo, T., Iwasaki, T. and Kawashima, K. (1980) Dense Instrument Array Program of the Public Works Research Institute for Observing Strong Earthquake-Motion, 12th Joint Meeting, U. S.-Japan Panel on Wind and Seismic Effect, UJNR, Washington, D. C., U. S. A.

7) Timoshenko, S. and Goodier, J. N. (1951) Theory of Elasticity, McGraw-Hill.

8) Wang, L. R. L., Kawashima, K., Yeh, Y. H. and Aizawa, K. (1981) Study of Ground Strains from Strong Motion Array Data for Lifeline Application. Technical Report LEE002, School of Civil Engineering and Environmental Science, University of Oklahoma, Also Tech. Note ,Public Works Research Institute, 39.

9) Zienkiewics, O. C. (1971) The Finite Element Method in Engineering Science, McGraw-Hill.

Soil Dynamics & Earthquake Engineering Conference / Southampton / 1982.07.13-15

Earthquake Observation on Two Submerged Tunnels at Tokyo Port

M.HAMADA, Y.SHIBA & O.ISHIDA
Taisei Corp., Tokyo, Japan

INTRODUCTION

In order to examine the antiseismic safety of submerged tunnels the dynamic characteristics of the deformation and the strain were studied by the experimental and the analytical approach. Dynamic strain and acceleration were measured simultaneously in two adjacent submerged tunnels, which had differences to each other on the dimension, the flexibility of the joints and the soil condition. By the qualitative and quantitative inspection of the observed records during two typical earthquakes, in one of which long period vibration was predominant due to surface waves and in the other short period vibration dominated due to body waves, the followings were clarified.
The strain caused in the submerged tunnels during earthquakes had mostly linear and static relationship with the dynamic strain of the surrounding ground and the flexible joints between tunnel elements had much effectiveness on the reduction of the tunnel strain.

OBSERVATION OF DYNAMIC RESPONSE OF TWO SUBMERGED TUNNELS

Outline of tunnels and geophysical condition

Two submerged tunnels, where the dynamic response was observed, are located neighboring to each other in Tokyo Port as shown in Figure-1(a). The length of two tunnels (call as A and B tunnel) are 1,035m and 744m respectively and the distance between two tunnels is about 3.0km. These tunnels consist of 9 (Tunnel A) and 6 (Tunnel B) reinforced concrete elements with flat rectangular section shown in Fig-2 and each element is connected by flexible joints in Fig-3 for the consideration of the earthquake resistance of the tunnels. Tunnel B has much more flexible joints which are by only P.C. cables compared with Tunnel A where the joints are made by steel shear keys and mortar concrete.
Tunnel A was buried in soft alluvial clay layer with S-wave

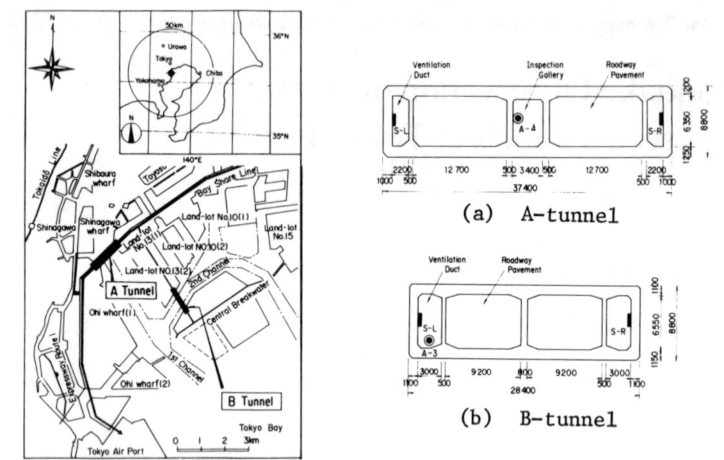

(a) Location of two Submerged Tunnels

Fig. 2 Section of Tunnel

(a) A-tunnel

(b) B-tunnel

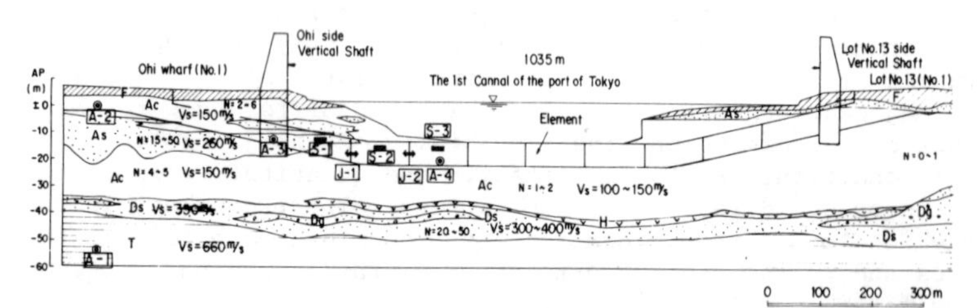

(b) Soil Profile and Measuring Instruments of A-Tunnel

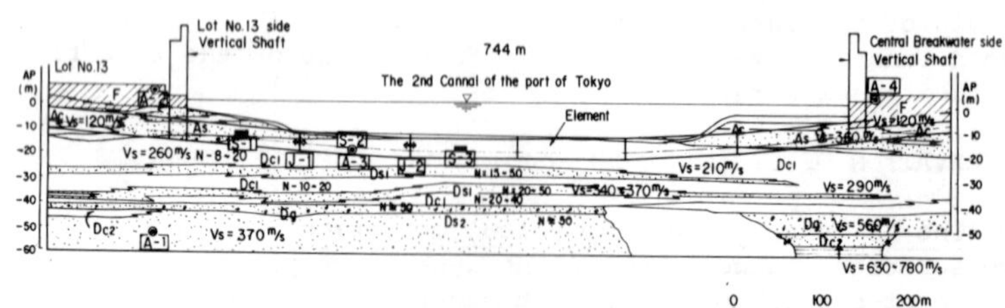

(c) Soil Profile and Measuring Instruments of B-Tunnel

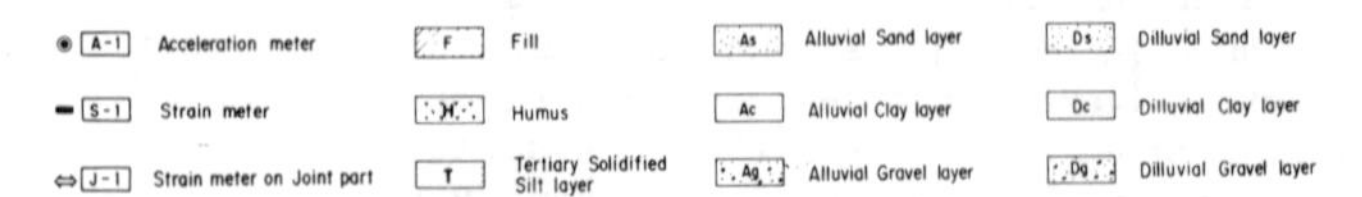

Fig. 1 Location and General View of two Submerged Tunnels

velocity of 100 ∿ 150m/s, while Tunnel B mostly in dilluvial clay layer with about 210m/s S-wave velocity as shown in Fig-1 (b), (c). Tertiary silt layer with the S-wave velocity of above 600m/s can be found at 50m ∿ 60m depth under the ground surface.
Two shafts for the ventilation are located on the both banks and are separated completely from the tunnels to move freely to each other during earthquakes.

Earthquake observation

Acceleration in horizontal and vertical direction was measured in the tertiary layer, on the ground surface, in the ventilation shafts and in the tunnel, while dynamic strains in the direction of the tunnel axis on the inner surface of the side wall as shown in Fig-1, 2. From two strains recorded on the both side wall the strain due to push-pull deformation of the tunnel in the axial direction (call as axial strain) and the one due to bending deformation around the vertical axis (bending strain) can be obtained separately. The strain gauges used for the measurement have enough sensitivity of $1/10 \times 10^{-6}$ which are made of a 1.0m length steel rod and a differential meter.

Observed earthquakes

More above ten earthquakes with the small ∿ medium intensity, the acceleration was larger than 10 cm/sec^2, were observed. In this paper the dynamic response of the tunnel and the surrounding ground during two typical earthquakes will be studied.
The first earthquake was caused by a shallow fault with the depth of about 10km and had a long epicentral distance of 90km (later as Izu-earthquake), while the second one was by a deep fault with 80km depth and had a medium epicentral distance of 40km (Chiba-earthquake). The magnitude of these two earthquakes is 6.7 and 6.1, respectively.
During the former earthquake long period vibration by surface waves was dominant because of the shallow fault and the long epicentral distance. On the contrary during the latter one short period vibration by body waves was remarkable due to the deep fault.
Fig-5 shows the acceleration and the dynamic strain of B-tunnel recorded during Izu-earthquake. From the inspection of these records the following qualitative characteristics can be clarified;

(a) Acceleration is large during 0 ∿ 40 sec of the records and rather high frequency vibration is dominant in this term. During the term after 40 sec long period component of 7.0 sec ∿ 8.0 sec is notable, especially in the axial direction (A1X and A2X). These long period vibration can be considered due to surface waves caused by the deep structure of the ground around Tokyo area reported by Seo.[2] They can be judged to be caused by Love wave since the epicentral direction of Izu-earthquake; shown in Fig-4, is nearly perpendicular to the B-tunnel's axis.

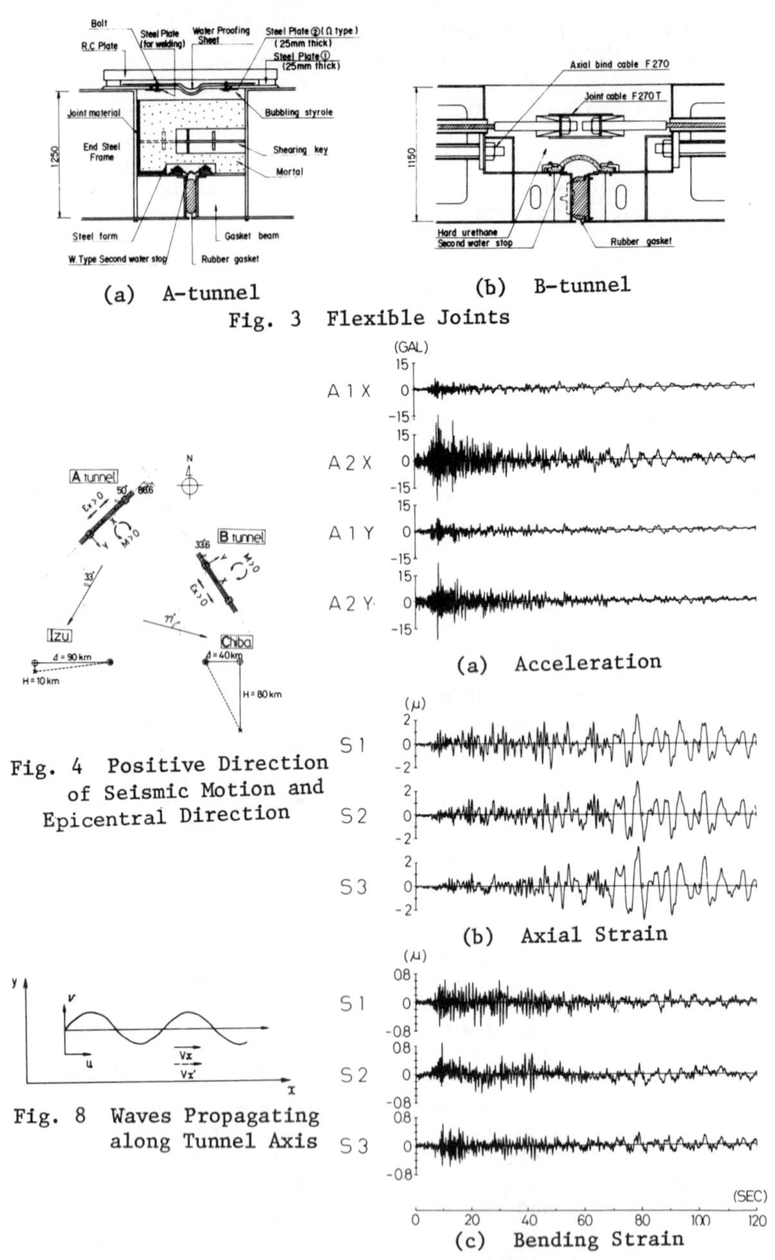

(a) A-tunnel (b) B-tunnel

Fig. 3 Flexible Joints

Fig. 4 Positive Direction of Seismic Motion and Epicentral Direction

Fig. 8 Waves Propagating along Tunnel Axis

(a) Acceleration

(b) Axial Strain

(c) Bending Strain

Fig. 5 Acceleration and Dynamic Strain of B-tunnel during Izu-Earthquake

(b) The axial strain is about $4 \sim 5$ times larger than the bending strain and dominant during the latter term of the record, after 40 sec, where the long period is remarkable in the acceleration, while the bending strain during the first term where the acceleration is large and the short period is prominent.
(c) The tunnel can be considered to be deformed uniformly along the axis particularly in the latter term of the records because of the good similarity of the axial strains at three observation points (S1, S2 and S3). The bending strains at these three points have much differences to each other. It can be concluded that the bending deformation mostly depends on the local condition of the dynamic characteristics of the surrounding ground compared with the push-pull deformation.

Fig-6 and 7 show the acceleration and the strain recorded in A and B tunnel during Chiba-earthquake. The followings can be indicated qualitatively;
(a) During Chiba-earthquake the short period vibration by body waves was much more dominant compared with Izu-earthquake mentioned above. The acceleration has large amplitude within only first 5 sec of the record, but the axial and the bending strains are maintaining the comparably large value after the main term of the acceleration.
(b) The axial strain is two times larger than the bending one in B-tunnel (Fig-6), while about same magnitude in A-tunnel (Fig-7) having flatter cross-section which can be considered to cause the bending strain more easily.
As a general conclusion, it can be pointed out that the axial strain will be larger than the bending strain and should be important on the earthquake resistant design of submerged tunnels, but it is necessary to pay some attention to the bending strain in the case of a tunnel with much flat section.

Consideration of dynamic deformation characteristics of submerged tunnel

By numerous researches on the dynamic behavior on the underground structures like the buried pipe-lines and inground tanks, it has been clarified that the deformation and the stress of underground structures due to earthquakes do not depend on the inertia force but on the relative displacement of surrounding grounds, namely on the dynamic ground strain.
Now, the waves travelling apparently along the tunnel axis are written as

$$u = f\left(t - \frac{x}{V_x}\right)$$
$$v = g\left(t - \frac{x}{V'_x}\right) \qquad \cdots\cdots\cdots\cdots \quad (1)$$

where u and v are each component of the seismic motion in the axial and the lateral directions, respectively. V_x and V'_x are the apparent phase velocities in each direction which are

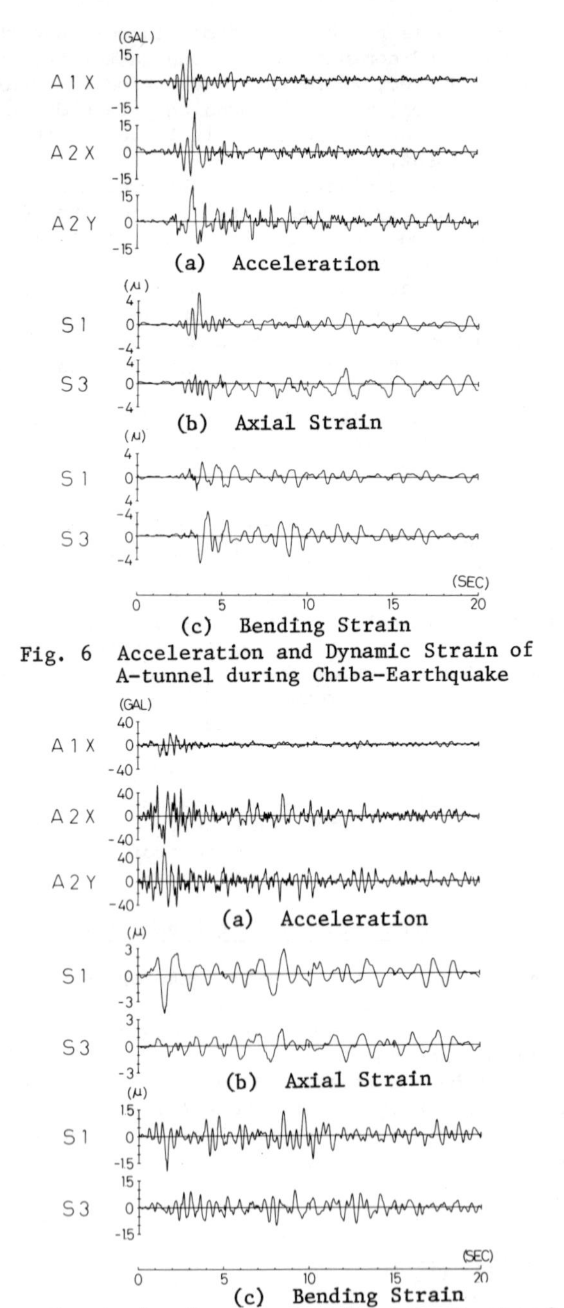

Fig. 6 Acceleration and Dynamic Strain of A-tunnel during Chiba-Earthquake

Fig. 7 Acceleration and Dynamic Strain of B-tunnel during Chiba-Earthquake

assumed to be constant without respect to the frequency.
The normal strain ε_n of the ground in the axial direction is

$$\varepsilon_n = \frac{\partial f}{\partial x} = \frac{1}{V_x} \frac{\partial f}{\partial t} \quad \cdots\cdots\cdots \quad (2)$$

and the curvature of the ground deformation $1/\rho = \frac{\partial^2 g}{\partial x^2}$ is

$$\frac{1}{\rho} = \frac{\partial^2 g}{\partial x^2} = \frac{1}{V'_x{}^2} \frac{\partial^2 g}{\partial t^2} \quad \cdots\cdots\cdots \quad (3)$$

$\frac{\partial f}{\partial t}$ and $\frac{\partial^2 g}{\partial t^2}$ are the particle velocity and the acceleration.
Fig-9, 10 show the comparison of the axial strain with the particle velocity in the axial direction as well as of the bending strain with the particle acceleration in the lateral direction. The positive direction of acceleration, velocity and strain as well as the epicentral directions of two earthquakes are shown in Fig-4.
From the results shown in Fig-9, 10 the followings can be obtained about the relationship between the dynamic strain and the seismic motion.

(a) The axial strains have very similar wave forms and Fourie spectra to the particle velocities of the seismic motion in the axial direction, while the bending strains to the accelerations in the lateral direction. For an example, the Fourie spectra in Fig-9 shows that the predominant periods of the axial velocity V4X and the lateral acceleration A4Y are 1.2 sec, 0.4 sec and 0.9 sec which coincide well with those of the axial strain S1A and the bending S1B, respectively. Similarly a good coincidence can be also found in the spectra shown in Fig-10.

(b) By considering the relation between the bending strain and the lateral acceleration shown in Equation (3) and the positive direction of the local coordinates in both tunnels shown in Fig-4, it can be found that the bending strain should have same phase to the lateral acceleration in A-tunnel and opposite phase in B-tunnel without respect to the incident angle of the seismic motion to the tunnels. Furthermore, the axial strain should have same phase with the axial velocity in the both tunnels during Chiba-earthquake, because the apparent phase velocities of the incident seismic motion along the axes are negative, but opposite phase in A-tunnel during Izu-earthquake because of the positive apparent phase velocity as shown in Fig-4.
The relationship of the wave phases shown in Fig-9 and 10 is well consistent with the above-mentioned consideration. For example, the axial strains of both tunnels have same phase with the axial particle velocities during Chiba-earthquake shown in Fig-9(a) and Fig-10(a), while the bending strains have same phase in A-tunnel in Fig-9(b) and opposite in B-tunnel in Fig-10(b). And also, the phase of the axial strain of A-tunnel during Izu-earthquake is opposite to one of the velocity as shown in Fig-9(c).

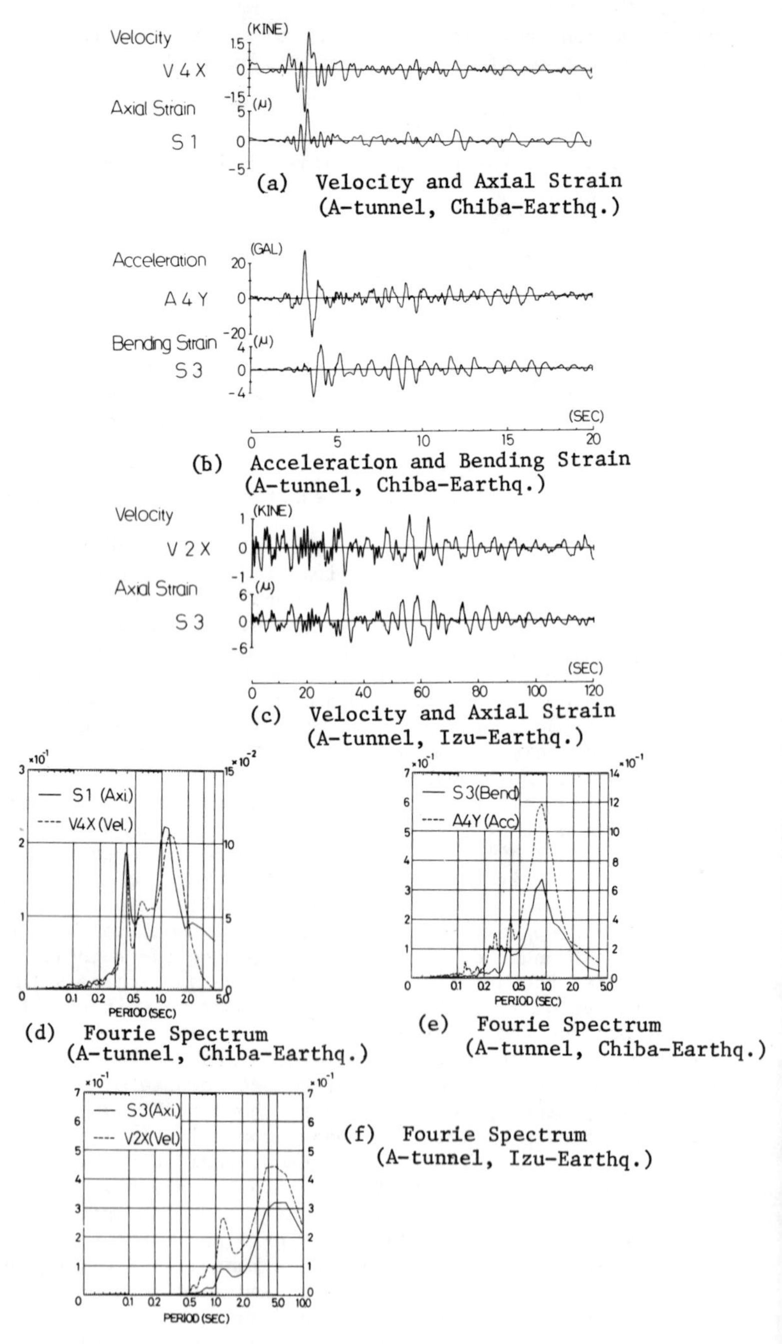

Fig. 9 Comparison of Velocity, Acceleration and Strain

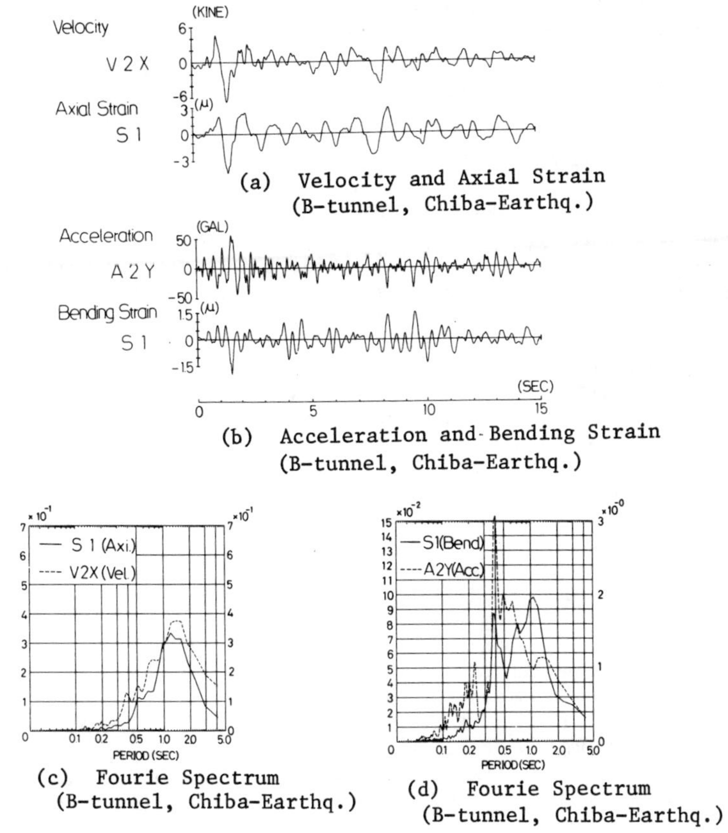

(a) Velocity and Axial Strain (B-tunnel, Chiba-Earthq.)

(b) Acceleration and Bending Strain (B-tunnel, Chiba-Earthq.)

(c) Fourie Spectrum (B-tunnel, Chiba-Earthq.)

(d) Fourie Spectrum (B-tunnel, Chiba-Earthq.)

Fig. 10 Comparison of Velocity, Acceleration and Strain

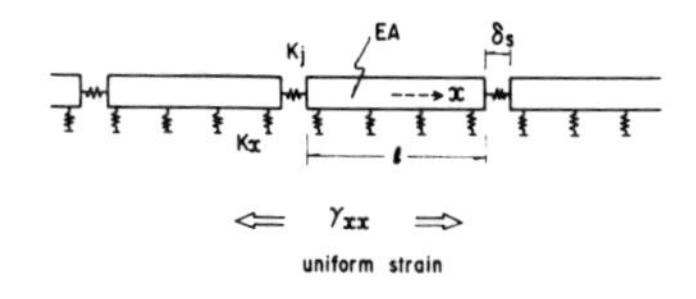

Fig. 11 Simplified Numerical Model

NUMERICAL CONSIDERATION ON DYNAMIC STRAIN OF TUNNEL AND RELATIVE DISPLACEMENT OF FLEXIBLE JOINT

Simplified numerical model

As mentioned in the previous section the dynamic strain due to earthquakes has a linear and static relationship with the strain of the surrounding ground. However, it is tough work to estimate the exact values of the tunnel strain quantitatively because of the difficulties on the exact evaluation of the ground strain which depends on the complicated geophysical condition in the wide area along the tunnel axis as well as on characteristics of incident seismic waves. And also, there is some difficulty on the evaluation of the real flexibility of joints between tunnel elements and the exact coefficient of the subgrade reaction.

In this section the probable magnitude range of the dynamic strain of submerged tunnels will be examined on the following simplified assumptions;

(a) The tunnel is consisted of a infinite chain of tunnel elements and joints as shown in Fig-11. This assumption can be considered to give larger magnitude of the tunnel strain, which is on safe side on the design when the tunnel is not connected with the ventilation building or is connected with more flexible joints than the ones between elements.

(b) Only axial strain is calculated, since it is generally larger than the bending strain and the key point for the earthquake resistant design.

(c) The axial strain of ground is uniform along tunnel axis. This assumption represents the case when the apparent length of the seismic waves along the tunnel axis is much larger than the tunnel's length and also result in the safe side design.

In the simplified numerical model shown in Fig-11 the dynamic strain of the tunnel ε_x and the relative displacement of flexible joint δ_s are obtained as follow:

$$\varepsilon_x = \gamma_{xx} \left\{ 1 - \frac{\cosh \beta x}{\cosh \frac{\beta \ell}{2}} \left(1 - \frac{2 \dfrac{\tanh \frac{\beta \ell}{2}}{\beta \ell}}{\dfrac{EA}{k_j \ell} + \dfrac{2 \tanh \frac{\beta \ell}{2}}{\beta \ell}} \right) \right\} \quad \ldots \quad (4)$$

$$\delta_s / \ell = \gamma_{xx} \frac{\dfrac{2}{\beta \ell} \tanh \dfrac{\beta \ell}{2}}{1 + \dfrac{k_j \ell}{EA} \dfrac{2 \tanh \frac{\beta \ell}{2}}{\beta \ell}} \quad \ldots \quad (5)$$

$$\beta = \sqrt{\frac{k_x}{EA}}$$

where EA and ℓ are the stifness of push-pull deformation and the length of the tunnel element, while k_j and k_x are the spring constant of the flexible joint and the coefficient of the subgrade reaction. γ_{xx} is the dynamic strain of ground in the axial direction.
Fig-12 and 13 show the maximum strain of the tunnel which occures in the center of the element, obtained by substituting $x = 0$ into Equation (4) and the relative displacement of the joint.
According to Fig-12 and 13 the ratio of the tunnel strain to the ground one, shown in the ordinate, is always less than 1.0 and the relative displacement of the joint is smaller than $\gamma_{xx} \cdot \ell$. When the stiffness ratio parameter $\beta\ell$ increases, which means that the stiffness of the tunnel becomes smaller or the coefficient of the subgrade reaction larger, the strain ratio increases toward 1.0 and the relative displacement decreases. On the contrary in the case when the stiffness ratio parameter is small and the joint is much flexible, the relative displacement becomes close to $\gamma_{xx} \cdot \ell$ and the tunnel strain decreases.

Examination of element's stiffness and joint's flexibility of two tunnels

Table-1 shows the strain ratio and the relative displacement of joints recorded at two tunnels during Chiba and Izu-earthquake. The max. relative displacements of joints in this table were picked up from Fig-14. The relative displacement of B-tunnel during Chiba-earthquake is the mean value of J1 and J2 shown in the figure. The apparent phase velocity of the seismic motion along the each tunnel axis was calculated from the epicentral direction angles and the S-wave velocity on the deep crust in Tokyo area, which was evaluated to be about 1,600 m/sec from the time lag at two observation points during Izu-earthquake.
The stiffness ratio parameter $\beta\ell$ shown in Table-1 was estimated on the assumption the K-value, the coefficient of the subgrade reaction, in the axial direction was within 0.1 $kg/cm^2/cm \sim 0.5$", which could be reasonable range by considering that the surrounding fill of the tunnel was generally soft and the coefficient would be smaller for the large dimension structure like submerged tunnels compared with small structures or structural elements like foundation piles. $\beta\ell$ was estimated to be $0.42 \sim 0.94$ for A-tunnel and to be $0.50 \sim 1.12$ for B-tunnel.
The shadowed area in Fig-12 and 13 shows the probable zone of the strain ratio $\varepsilon_x/\gamma_{xx}$ and non-dimensional relative displacement $\delta_x/\gamma_{xx} \cdot \ell$ for two submerged tunnels obtained by the examination in Table-1. From this result it can be understood that the ratio of the joint spring to the stiffness of the tunnel $k_j\ell/EA$ is much larger than in B-tunnel. This is a reasonable result by considering that B-tunnel has much more flexible joint which was made by only P.C. cables than A-tunnel the joint of which was by steel shear keys and mortar concrete.

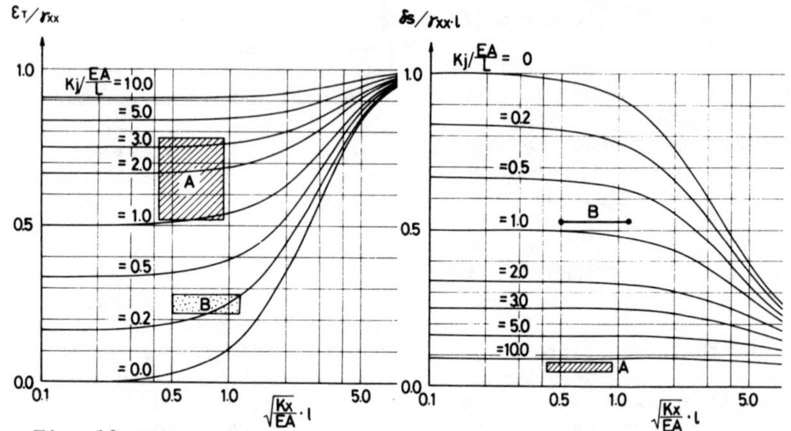

Fig. 12 Maximum Axial Strain of Tunnel

Fig. 13 Maximum Relative Displacement of Joint

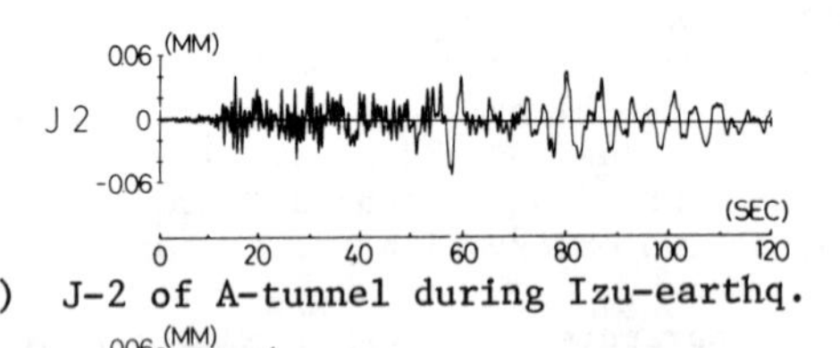

(a) J-2 of A-tunnel during Izu-earthq.

(b) J-1 of A-tunnel during Chiba-earthq.

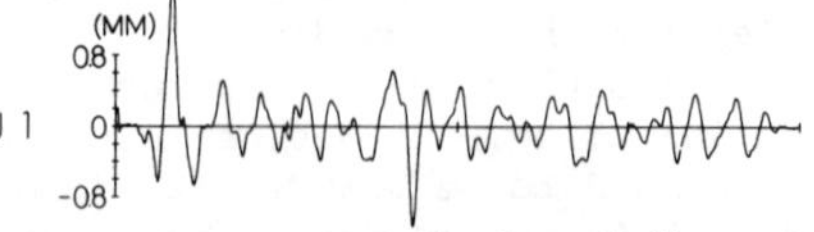

(c) J-1 of B-tunnel during Chiba-earthq.

(d) J-2 of B-tunnel during Chiba-earthq.

Fig. 14 Relative Displacement Records

Table - 1 Examination of Stiffness of Tunnel and Flexibility of Joint

Tunnel	A (ℓ=115 m)		B (ℓ=124 m)	
$\sqrt{\frac{k_x}{EA}} \cdot \ell$ (Assumed)	0.42 ~ 0.94		0.50 ~ 1.12	
Earthquake	Chiba	Izu	Chiba	Izu
Max. Strain of Tunnel ε_T	5.4×10^{-6} (1)	5.0×10^{-6} (1)	4.4×10^{-6} (2)	2.5×10^{-6}
Max. Relative Disp. of Joint δ_s (mm)	0.061	0.053	≒1.04	——
Max. Axial Particle Velocity Vx (cm/sec)	2.9 (1)	1.1 (1)	6.7 (2)	4.8
Max. Apparent Phase Vel. v_x (m/sec)	2800	1720	4200	2200
Max. Ground Strain $\gamma_{xx} = \frac{V_x}{v_x}$	10.4×10^{-6}	6.4×10^{-6}	16.0×10^{-6}	21.6×10^{-6}
Strain Ratio $\frac{\varepsilon_T}{\gamma_{xx}}$	0.52	0.78	0.28	0.22
Relative Disp. Ratio $\frac{\delta_s}{\gamma_{xx} \cdot \ell}$	0.05	0.08	0.53	——

※ (1) Fig-9 (2) Fig-10

CONCLUSION

Earthquake observation was carried out at two adjacent submerged tunnels at Tokyo Port, measuring the strain of the tunnel's side wall and the seismic motion of the surrounding ground during two typical earthquakes.
By the inspection of earthquake records and some numerical consideration on the deformation characteristics of tunnels, the followings were obtained as the main conclusions.

(a) The strain due to the push-pull deformation in the axial direction is larger than that due to the bending deformation around the vertical axis. Particularly, the axial strain is dominant during an earthquake with long period vibration component.

(b) The axial strain has mostly linear and static relationship with the particle velocity of the seismic motion, while the bending strain with the acceleration.

(c) The flexibility of the joints between tunnel elements has much effectiveness on the reduction of the tunnel strain.

References

(1) Tamura, C., Okamoto, S. and Hamada, M. (1975) Dynamic Behavior of A Submerged Tunnel during Earthquakes, Report of The Institute of Industrial Science, The University of Tokyo, Vol.24, No.5, March.

(2) Seo, K. (1978) Earthquake Motions Modulated by Deep Soil Structure, 5th Japan Earthquake Engineering Symposium.

Soil Dynamics & Earthquake Engineering Conference / Southampton / 1982.07.13-15

Dynamic Response of Buried Frameworks

B.BANDYOPADHYAY & G.D.MANOLIS
State University of New York, Buffalo, USA

ABSTRACT

This work addresses the numerical determination of the dynamic response of framed structures buried in a horizontally layered soil medium under the action of vertically propagating shear waves emanating from a distant earthquake source. The methodology consists of applying the Laplace transform with respect to time to the governing equations of motion of the soil and the structure and of subsequently using the finite element methodology to obtain the solution in the transformed domain. A numerical inversion of this solution yields the response in the time domain. Finally, a numerical example to illustrate the method and demonstrate its advantages is presented.

INTRODUCTION

In recent years, the use of underground space has attracted attention as an important factor for achieving goals such as energy conservation, improved environment, industrial development, power generation, and efficient transportation. In addition, underground structures are subjected to less severe earthquake damage because, in general, the effects of shear and surface seismic waves decrease with depth of structure embedment. The dynamic analysis of structures founded on or embedded in geological media has received the attention of investigators since the earlier 1960s. Most of the research work since has focused on surface buildings, and relatively little work on underground structures is presently available. A very comprehensive state of the art review on the dynamic soil-structure interaction phenomenon was recently prepared by Johnson (1981). That review recapitulates that there are currently two major categories of numerical techniques available for soil-structure analyses, namely approximate continuum and discrete (lumped parameter) methods. The most widely employed approximate continuum method at present is the Finite

Element Method (FEM). In principle, the FEM is a versatile technique for soil-structure interaction problems because it can handle complex structure geometry, medium inhomogeneities, and complicated material behavior in both two and three dimensions. In spite of these advantages, a deficiency of that method is that a semi-infinite medium is represented by a finite size model. Therefore, considerable efforts, e.g., Lysmer and Waas (1972) and Kausel et al (1975), have been aimed at obtaining special transmitting boundaries that are placed at the ends of the finite element mesh to allow for energy radiation. Discrete methods, on the other hand, effectively uncouple the structure from the medium with due consideration to the interaction phenomenon and allow for an efficient dynamic analysis of the structure alone. This is achieved by using appropriate values for the mass, stiffness, and damping coefficients (impedance functions) that essentially represent the medium. Lumped parameter models have been successfully used in foundation problems, e.g., Veletsos and Wei (1971) and Luco and Westmann (1971). Extension of these concepts to underground structures is hampered by the fact that the boundary value problem to be solved for the analytic determination of the appropriate impedance functions is very difficult, even after relaxing some of the boundary conditions.

This work attempts the construction of a simple and efficient, yet engineering sound mathematical model to represent the dynamic behavior of framed underground structures embedded in horizontally layered soil deposits under conditions of plane strain. Both the structure and the soil are assumed to exhibit linear elastic or viscoelastic material behavior. Structural and soil viscous type of dampings are also considered in the formulation. The basic step is the application of the Laplace transform with respect to time to the governing equations of motion for the soil layers and the structural elements. This results in changing the form of these equations to ordinary differential equations in the Laplace domain. It then becomes possible, by using closed form solutions to these ordinary differential equations and following the standard procedure described in Manolis and Beskos (1980), to construct dynamic stiffness influence coefficients for both soil and flexural beam elements. These beam and soil models defined in the Laplace transform domain have the advantage of retaining a continuous mass distribution, thus leading to the exact solution of the problem within the framework of the theory used. Dynamic stiffness influence coefficients for flexural elements under harmonic excitations are discussed in Clough and Penzien (1975), and similar coefficients for piles were more recently introduced by Novak and Aboul-Ella (1978). The present method is capable of representing any variation of frame and soil properties by simply combining beam and soil elements so as to reflect this variation. The dynamic excitation, which is prescribed at the bottom of the soil stratum resting on rigid bedrock, can be of any arbitrary time variation. Finally, the solution is obtained in discrete form in the Laplace domain, and a numerical inverse Laplace transform is required to produce the dynamic response in the time domain. This last subject is discussed in detail in Narayanan and Beskos (1982).

FORMULATION AND SOLUTION

The problem under consideration is shown in Figure 1 (a), along with the cartesian coordinate system chosen.

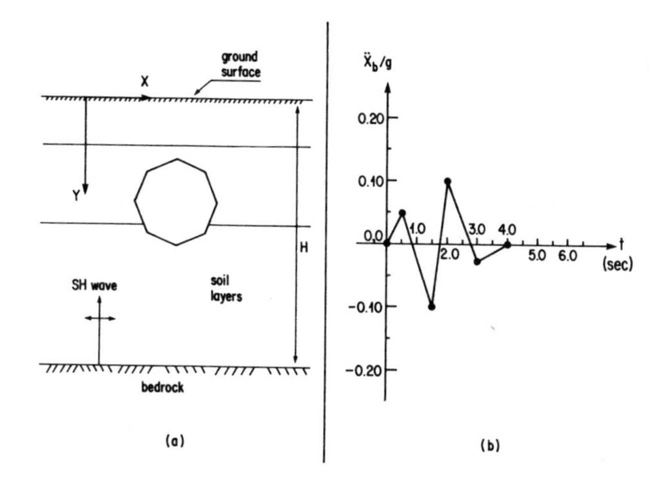

Figure 1. (a) Frame embedded in soil medium under the action of an incoming shear wave.
(b) Simplified accelerogram applied at bedrock.

At first, the soil stratum, which is assumed to be continuous and linear elastic or viscoelastic, is approximated by a vertical shear beam model. Therefore, the equation of lateral motion for a soil element, which is any horizontal layer of soil with constant values of mass density ρ, viscous damping c_s and shear modulus G, is of the form

$$G\,(\partial^2 u_s/\partial y^2) - c_s \dot{u}_s - \rho \ddot{u}_s = \rho \ddot{x}_b \qquad (1)$$

where dots indicate differentiation with respect to time t. Furthermore, $u_s(y,t)$ is the relative displacement of the soil with respect to the bedrock displacement $x_b(y,t)$. The total soil displacement x_s is, of course equal to $x_b + u_s$.

The Laplace transform $\bar{f}(s)$ of a function f(t) which is zero for negative t is defined as

$$\bar{f}(s) = L\{f(t)\} = \int_0^\infty f(t)e^{-st}dt \tag{2}$$

where s is the Laplace transform parameter. An application of the aforementioned transform to equation (1) under zero initial conditions yields

$$G(d^2\bar{u}_s/d_y{}^2) - (\rho s^2 + c_s s)\bar{u}_s = \rho L(\ddot{x}_b) \tag{3}$$

The homogeneous solution of equation (3) is of the form

$$\bar{u}_s(y,s) = Ae^{Ky} + Be^{-Ky} \tag{4}$$

where $K = \sqrt{(\rho s^2 + c_s\ s)/G}$ and A and B are constants.

On the basis of the displacement function, equation (4), the following transformed nodal force-displacement relationship can be constructed using standard finite element techniques :

$$\begin{Bmatrix} \bar{X}_I \\ \bar{X}_J \end{Bmatrix} = \begin{bmatrix} \bar{V}_{11} & \bar{V}_{12} \\ \bar{V}_{21} & \bar{V}_{22} \end{bmatrix} \begin{Bmatrix} \bar{u}_{s_I} \\ \bar{u}_{s_J} \end{Bmatrix} \tag{5}$$

In equation (5), $\bar{V}_{11} = \bar{V}_{22} = (GKa)\coth(Kl)$ and $\bar{V}_{12} = \bar{V}_{21} = -(GKa)\operatorname{csch}(Kl)$, where l is the length of the soil element, a is its shear area taken as equal to unity, and subscripts I and J denote nodal points. The stiffness matrix of the whole soil stratum is obtained by appropriate superposition of the stiffness matrices of the individual soil beam elements. Finally, the imposed nodal loads at the top and bottom of the soil stratum are of the form

$$\begin{aligned} \bar{X}|_{y=0} &= -0.5\ \rho_{av}\ H\ A\ L\ (\ddot{x}_b) \\ \bar{X}|_{y=H} &= 0.5\ \rho_{av}\ H\ A\ L\ (\ddot{x}_b) \end{aligned} \tag{6}$$

respectively, where ρ_{av} is the average mass density and H the total height of the stratum to be discretized. Furthermore, $L(\ddot{x}_b)$ is the direct Laplace transform of the accelerogram prescribed at the bedrock.

Next, the rigid-jointed framework is considered. The equation of lateral motion of a beam element of length L on an elastic foundation is of the form

$$EI(\partial^4 v/\partial y^4) + K_s v + c\dot{v} + m\ddot{v} = -m\ddot{x}_s \qquad (7)$$

where EI is the flexural rigidity, K_s is the soil resistance modulus, c is the viscous damping coefficient, and m is the mass per unit length. Furthermore, v(y,t) is the beam relative to the soil lateral displacement and $\ddot{x}_s(y,t)$ is the absolute soil acceleration. Applying the Laplace transform under zero initial conditions to equation (7) results in

$$EI(d\bar{v}^4/dy^4) + (K_s + cs + ms^2)\,\bar{v} = -m\,L(\ddot{x}_b) - ms^2\bar{u}_s \qquad (8)$$

where $\ddot{x}_s$ has been decomposed to $\ddot{x}_b$ and $\ddot{u}_s$. The same procedure used previously and also elaborated in Manolis and Beskos (1980) is followed for the construction of the transformed stiffness matrix equation for the beam element. As far as the axial behavior of the beam element in question is concerned, coefficients similar to the flexural ones can easily be derived from processing the homogeneous governing equation of axial motion,

$$EA(\partial^2 u/\partial y^2) - K_w u - c\dot{u} - m\ddot{u} = 0 \qquad (9)$$

In the above equation, u(y,t) is the displacement along the axis of the member, K_w is the modulus of soil resistance, A the axial area, and the remaining quantities were defined earlier. The axial stiffness coefficients are appropriately added to the flexural ones to complete the transformed stiffness matrix equation for the beam element, as shown in Table 1. The following quantities need to be defined:

$$\begin{aligned} \beta &= ((ms^2 + cs + K_s)/EI)^{1/4} \; ; \; \gamma = e^{2kL} \\ k &= \beta/\sqrt{2} \; ; \; \Delta = \gamma^2 - 2\gamma(1+2s^2)+1 \\ s &= \sin kL \; ; \; N = 2EIk/\Delta \\ c &= \cos kL \; ; \; \lambda = \sqrt{(ms^2 + cs + K_w)/EA} \end{aligned} \qquad (10)$$

Also, Θ is the rotation at a nodal point and M the corresponding moment. The transformed forces acting on the beam elements involve contributions from both the soil displacements and the bedrock accelerations. Assuming a linear variation of $\bar{u}_s$ along a soil layer, the input forces are of the form

$$\begin{aligned} \bar{F}_I &= 0 \\ \bar{Y}_I &= (7\,\bar{u}_{sI} + 3\,\bar{u}_{sJ})\,ms^2\,L/20 - L(\ddot{x}_b)\,mL/2 \end{aligned} \qquad (11)$$

$$\overline{M}_I = -(3\overline{u}_{sI} + 2\overline{u}_{sJ})\ ms^2\ L^2/60$$

and similarly for node J. It should be noted that for a beam element inclined with respect to the Y-axis, routine coordinate system transformations must be done.

TABLE 1. Transform stiffness matrix for a beam element.

Force	=						Displacement
$\bar{F}_I$	$AE\lambda\coth(\lambda L)$						$\bar{u}_I$
$\bar{Y}_I$	0	$2Nk^2(\gamma^2 + 4\gamma sc - 1)$			symmetric		$\bar{v}_I$
$\bar{M}_I$	0	$Nk(-\gamma^2+2\gamma\cdot(1-2s^2)-1)$	$N(\gamma^2-4\gamma cs^2-1)$				$\bar{\theta}_I$
$\bar{F}_J$	$-AE\lambda\,\mathrm{csch}(\lambda L)$	0	0	$AE\lambda\coth(\lambda L)$			$\bar{u}_J$
$\bar{Y}_J$	0	$(4Nk^2/\sqrt{\gamma})(-\gamma^2\cdot(s^3+sc^2+c)-\gamma(s^3+sc^2-c))$	$(2Nk/\sqrt{\gamma})(\gamma^2-\gamma)\cdot(s^3+sc^2+s)$	0	$2Nk^2(\gamma^2+4\gamma sc-1)$		$\bar{v}_J$
$\bar{M}_J$	0	$(2Nk/\sqrt{\gamma})(\gamma-\gamma^2)\cdot(s^3+sc^2+s)$	$(2N/\sqrt{\gamma})(\gamma^2\cdot(s^3+sc^2-c)+\gamma\cdot(s^3+sc^2+c))$	0	$-Nk(-\gamma^2+2\gamma\cdot(1-2s^2)-1)$	$N(\gamma^2-4\gamma cs^2-1)$	$\bar{\theta}_J$

Also, linear viscoelastic material behavior can be accounted for by invoking the correspondence principle in the Laplace transform domain, e.g., Flugge (1967).

The solution obtained for the dynamic response of an underground structure by using the methodology developed is, of course, in the Laplace transformed domain and in discrete form so that a numerical inverse transformation is necessary to bring the solution to the original time domain. The pertinent inversion integral is of the form

$$f(t) = (1/2\pi i) \int_{\beta-i\infty}^{\beta+i\infty} \overline{f}(s)\ e^{st} ds \tag{12}$$

where $i = \sqrt{-1}$ and $\beta>0$ is arbitrary, but greater than the real part of all the singularities of $\overline{f}(s)$. The integration over the complex plane required in equation (12) is performed numerically through the use of the algorithm described in Durbin (1974), which is essentially a quadrature scheme for the complex integral shown above. The transform parameter s is a complex number of the form $s = \beta + i\,(2\pi/T)\,n$, where T is the total time of interest and n ranges from 1 to N, the total number of steps used in the algorithm.

Finally, it should be mentioned that the soil influence coefficients in equation (5) can be used in a deconvolution process, that is to find the accelerogram at bedrock resulting from a prescribed accelerogram at the surface. This subject is further elaborated in Manolis and Beskos (1982), as is the procedure for numerically performing the direct Laplace transformation required in equation (11).

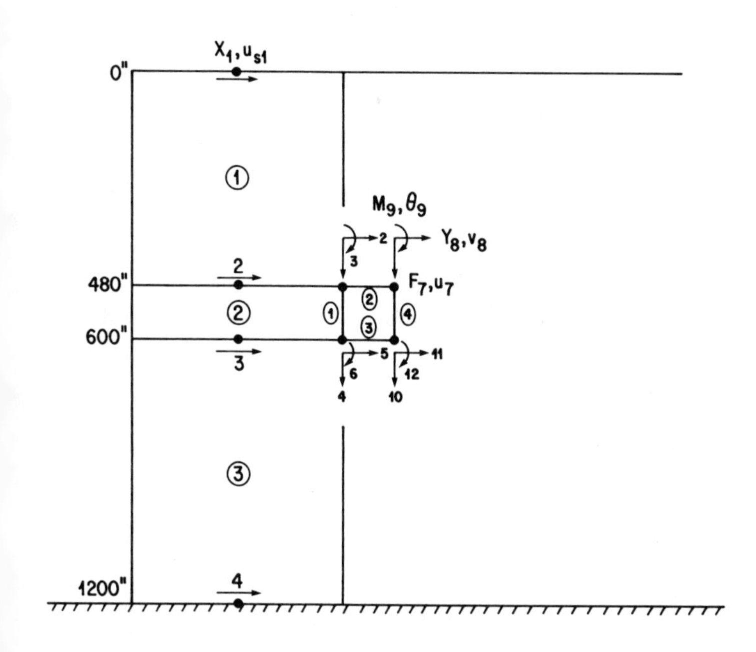

Figure 2. Discretization of the soil-frame system for the present method.

NUMERICAL EXAMPLE

Consider a square steel frame embedded in a homogeneous soil deposit whose material properties are assigned the following numerical values :

$$G = 2400 \text{ lb/in}^2 \qquad \rho = 0.00018 \text{ lb-sec}^2/\text{in}^4 \tag{13}$$

$\lambda = 0.10$

where λ is the fraction of critical damping and $c_s = 2\pi\omega_1\lambda$, with ω_1 being the first fundamental frequency of the soil stratum equal to $(\pi/2H)\sqrt{G/\rho}$. In reference to Figure 2, the soil deposit is subdivided into 3 elements in order to accommodate the structure, which in turn is discretized into four beam elements. Figure 2 also shows the pertinent degrees of freedom for the soil medium (4) and for the frame (12). The soil rests on bedrock where the simplified accelerogram shown in Figure 1(b) is applied at time t = 0 seconds. The frame is designed according to routine structural design procedures, and the structural shapes selected are shown in Table 2 below.

TABLE 2. Member selection for the square frame.

Member	Shape	m ($lb\text{-}sec^2/in^2$)	A (in^2)	I_{xx} (in^4)
Top girder	W 24 x 90	0.02029	27.70	2690.0
Bottom girder	W 8 x 20	0.00432	5.89	69.4
Columns	W 14 x 87	0.01878	25.60	967.0

Also, $E = 29{,}000{,}000\ lb/in^2$, Poisson's ratio $\nu = 0.29$, and the viscous damping coefficient c is taken as equal to zero.

The members of the frame are assumed to be in full contact with the surrounding soil at all times. If this is not the case, then the problem becomes a nonlinear one and is beyond the scope of this work. The parameters K_s and K_w, the soil resistances to horizontal and vertical frame structure motions, respectively, are obtained from Novak et al (1978). These parameters are derived as the reactions of an embedded cylindrical body under conditions of plane strain. Most of the assumptions in Novak et al (1978) are reasonably consistent with the ones made in this work and are as follows: (1) the infinite medium is homogeneous and linear viscoelastic and (2) the embedded cylinder is massless, infinitely long and undergoes small displacements. These parameters are of the general form $G(S_1 + iS_2)$, and their real part is a stiffness term while their imaginary part is a radiation damping term. The S_1 and S_2 coefficients are, in general, functions of the dimensionless frequency $a_o = r_o\omega/V_s$, the ratio G'/G and ν, where V_s is the shear wave velocity of the viscoelastic medium equal to $\sqrt{G/\rho}$ and G' is the imaginary component of the soil complex compliance while G is the real one. For the purpose of this work, r_o is taken as a representative radius of the frame members' cross-section. The frequency ω is the imaginary part of the Laplace transform parameter s , i.e. equal to $(2\pi/T)n$. Therefore, the appropriate numerical values of K_s and K_w are input in the program at each value of the parameter s and the values of K_w are halved to reflect the fact that during axial vibrations only half the frame members' area is in contact with the soil.

In addition, the finite element program SAP IV of Bathe et al (1973) was employed to determine the dynamic response of this example. The mesh used in SAP IV consisted of 189 quadrilateral plane strain elements and 12 flexural beam elements for a total of 236 nodal points. The recommendations of Segol et al (1975) were adopted in the construction of this mesh.

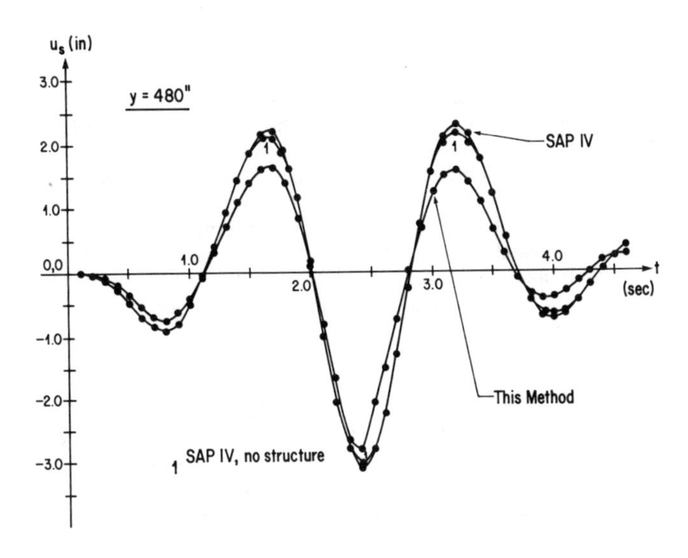

Figure 3. Displacements in the soil stratum at y = 480 inches.

Figure 3 plots the relative soil displacements u_s versus time obtained by the two methods at a depth of 480 inches from the surface of the soil stratum. The two methods produce results that are in good agreement, except for the peaks occuring around 1.5 , 2.5 and 3.5 seconds, where the results on the average may differ by as much as 20%. Also, the FEM predicts negligible vertical displacements, a fact that confirms the validity of treating the soil stratum as a one dimensional shear beam by the present method. Concurrently plotted is the displacement at the same location by the FEM in the absence of the framed structure, but it is almost indistinguishable from the other FEM curve except at the peaks, where the difference is less than 5%. This indicates that the displacements of the frame with respect to the soil are very small when compared to the displacements of the soil with respect to the bedrock. Therefore, the structure essentially conforms to the displacements of the surrounding soil and as a result the dynamic forces developed in the frame should be of small magnitude when compared to the static overburden forces. This last conclusion is confirmed in Figure 4, where the time variation of the shear force developed in the upper left-hand corner of the frame is plotted. In particular, the maximum shear forces developed are 130 lb. at 1.3 seconds and 117 lb. at 1.6 seconds as predicted by the present method and the FEM, respectively. These forces are negligible when compared to the overburden force of 63,000 lb. used in the static design of the frame.

Finally, it should be mentioned that the present method utilized N = 50 sampling values for the parameter s while the FEM utilized 50 time points in a modal analysis solution scheme. The former method required 5.82 seconds of execution time and cost $0.98 while the latter required 33.91 seconds of execution time and cost $31.67. It is, therefore, observed that the present method is more efficient than the FEM for comparable results. The difficulty with the present method is that a good selection of the soil reaction parameters K_s and K_w is problematic.

CONCLUSIONS

An approximate numerical method based on linear theories has been developed for the dynamic analysis of underground framed structures under conditions of plane strain. It is concluded that the proposed method has certain advantages when compared with other numerical methods traditionally used for problems of this kind. In particular, 1) the exact solution is obtained within the scope of the approximations made because the dynamic stiffness influence coefficients retain a continuous distribution of mass for both the soil medium and the structure and 2) data preparation from the part of the user is minimal and the computer implementation of the method requires small amounts of computer time and memory.

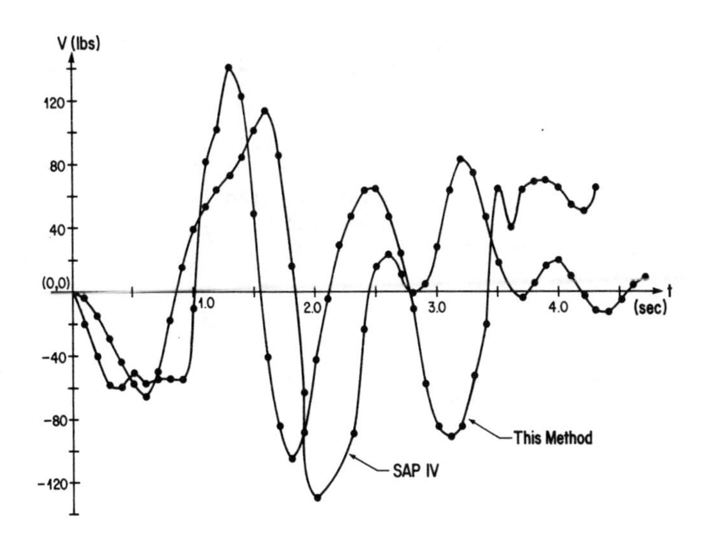

Figure 4. Shear force at the upper left-hand side corner of the frame.

The present method is quite versatile as it combines the advantages of the FEM and of the Laplace transform. The method may be improved to approximately account for large strains developed in the soil by simply adjusting the influence coefficients in an iterative manner according to the level of strains developed.

ACKNOWLEDGEMENT

The University Computing Services of the SUNY at Buffalo are acknowledged for making their facilities available to the authors.

REFERENCES

Bathe, K.J., Wilson, E.L. and Peterson, F.E. (1973) SAP IV: A Structural Analysis Program for Static and Dynamic Response of Linear Systems. Report No. EERC 73-11, Univ. of California, Berkeley.

Clough, R.W. and Penzien, J. (1975) Dynamics of Structures. McGraw-Hill, New York.

Durbin, F. (1974) Numerical Inversion of Laplace Transforms: An Efficient Improvement to Dubner and Abate's Method. Comp. J., 17, 371-376.

Flugge, W. (1967) Viscoelasticity. Blaisdell,Waltham, Mass.

Johnson, J.J. (1981) Soil Structure Interaction: The Status of Current Analysis Methods and Research. U.S. Nuc. Reg. Commission, Washington.

Kausel, E., Roesset, J.M. and Waas, G. (1975) Dynamic Analysis of Footings on Layered Media. Proc. ASCE, 101, EM5: 679-693.

Luco, J.E. and Westman, R.A. (1971) Dynamic Response of Circular Footings. Proc. ASCE, 97, EM5: 1381-1395.

Lysmer, J. and Waas, G. (1972) Shear Waves in Plane Infinite Structures. Proc. ASCE, 98, EM1: 85-105.

Manolis, G.D. and Beskos, D.E. (1980) Thermally Induced Vibrations of Beam Structures. Comp. Meth. Appl. Mech. Engng., 21, 337-355.

Manolis, G.D. and Beskos, D.E. (1982) Dynamic Response of Framed Underground Structures. Comp. and Structures, to appear.

Narayanan, G.V. and Beskos, D.E. (1982) Numerical Operational Methods for Time-Dependent Linear Problems. Int. J. Num. Meth. Engng., to appear.

Novak, M. and Aboul-Ella, F. (1978) Impedance Functions of Piles in Layered Media. Proc. ASCE, 104, EM3: 643-661.

Novak, M., Nogami, T. and Aboul-Ella, F. (1978) Dynamic Soil Reactions for Plane Strain Case. Proc. ASCE, 104, EM4: 953-959.

Segol, G.,Abel, J.F. and Lee, P.C.Y. (1975) Finite Element Mesh Gradation for Surface Waves. Proc. ASCE, 101, GT11: 1177-1181.

Veletsos, A.S. and Wei, Y.T. (1971) Lateral and Rocking Vibrations of Footings. Proc. ASCE, 97, SM9: 1227-1248.

Soil Dynamics & Earthquake Engineering Conference / Southampton / 1982.07.13-15

Seismic Response Analysis of Multi-Span Continuous Bridge with High Piers on Deep Pile Foundations

HIROKAZU TAKEMIYA
Okayama University, Japan

YOSHIKAZU YAMADA
Kyoto University, Japan

ABSTRACT

Seismic response analysis has been carried out for a multi-span continuous bridge with high piers on deep pile foundations in consideration of the soil, foundation and superstructure interaction. The substructure method is applied to advantage for the present system, formulating the soil-foundation and the superstructure separately first, and then integrating them into the complete system. Special interest is placed on the seismic wave propagation effect which, in the vertical direction, is due to the kinematic interaction between soil and foundation, and in the horizontal direction, due to the out-of-phase input motions at the respective foundation.

INTRODUCTION

Present stat-of-the-art for the soil-structure interaction problem is classified as the complete system analysis and the substructure method. The first method, as represented by the common finite element method, deals with a descretized soil region and structural part in a unified way and compute the system response in one step[4]. The soil-structure interaction is thus completely solved provided that the outer boundary condition is properly defined. This method, however, increases the numbers of degrees of freedom tremendously as the structure becomes huge and complex. The two-step approximate analysis is often taken by first adopting a simplified stick model for the superstructure and then giving the foundation top response as an input for the complex superstructur. The accuracy of this solution therefore depends on the simplified superstructural model. There may be the cases when the superstructural behavior is hardly represented by such simplification. The flexible multi-span continuous bridge reported herein is among them.

The substructure method[1,7],on the other hand deals with the superstructure and the soil-foundation system separately first, and then inegrates them with use of their continuity conditions: force equilibrium and displacement compatibility at their interface. The former analysis is devoted to finding the soil impedance functions and the effective seismic input motion associated with the foundation motion at a specific section, which are incorporate into the superstructural analysis. Thus, the different method of analysis may be taken between these subsystems. Another advantage of the substructure method is to decrease the total degrees of freedom for the integrated system by assuming the normal mode decomposition for the superstructure. The formulation of a multipl input system enables the coupling between superstructure and foundations.

The soil-structure interaction phenomenon in seismic environment involves two effects on the structure[2]: One is the soil impedance effect which induces an inertial interaction as to change the dynamic properties of the superstructure, and the other is a modification effect of the seismic input motion from the free field situation due to the kinematic interaction dependin on the local stiffness change along the foundation face. The soil impedance is taken into account in most design process if necessary; however, the kinematic interaction is seldom considered. Instead, the soil surface motion of free field is used as an input It is worth to note that this effect grows significant as the enbedment of the foundation becomes deep.

Most seismic analysis of bridge structures with a multiple foundations are made by assuming a uniform earthquake excitation at their base. However, upon considering the traveling nature of the seismic wave, this formulation may be improper, especially when the foundations are subject to the excitation of around π of-of-phase at the adjacent foundations[9]. There is a possibility that the higher vibration modes of the structure are amplifie in this situation. Althongh the spatial variation of the seismic motion is defined by the phase difference and the amplitude attenuation with the distance it travels, only the former effect is accounted in this investigation.

In what follows, the three-span continuous bridge with high piers on deep piles, as illustrated in Figure 1, is analyzed with the aforementioned things in mind.

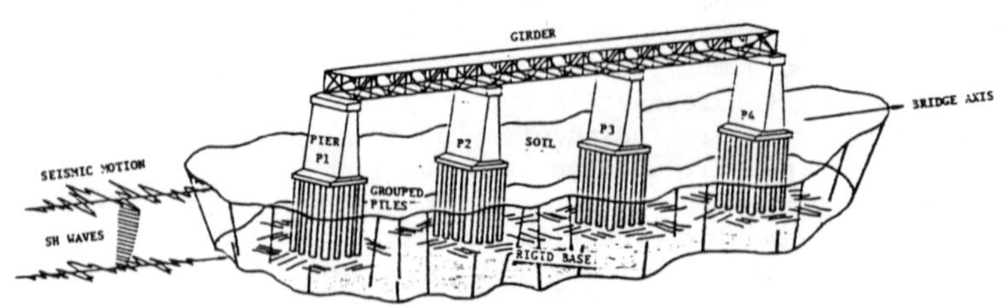

Figure 1. Structural model for analysis

FORMULATION

Soil-foundation system

The modeling is made such that: The soft soil layers overlie a rigid base rock at the top of which the seismic input motion is prescribed. The pile foundation is composed of a group of piles embedded in this surface layers; their top is rigidly connected with a rigid body pier-footing and their tip is built in the base rock. Only the soil reaction along piles is taken into account, neglecting that at the footing face.

The dynamic situation, when subjected to the base input motion, is analyzed by breaking it into two phenomena: the free field motion without piles and the pile vibration due to the loading at pile head in consideration of the soil reaction by the scattering wave from it [8]. The one dimensional SH wave solution is used for the former situation. The modified plane strain solution, which is substantiated from the comparison with the more rigourous three dimensional one, is adopted for the soil reaction in the latter situation.

In view of the fact that the fundamental natural frequency of the surface soil layers h_1 gives the cut-off frequency for the radiating wave, the following plane strain soil reaction [5] may be valid only above this frequency range:

For the lateral reaction in the j-th layer

$$P_T(z_j) = -\pi G_j a_j^2 \frac{4K_1(q_j^*)K_1(s_j^*) + s_j^*K_1(q_j^*)K_0(s_j^*) + q_j^*K_0(q_j^*)K_1(s_j^*)}{q_j^*K_0(q_j^*)K_1(s_j^*) + s_j^*K_1(q_j^*)K_0(s_j^*) + q_j^*s_j^*K_0(q_j^*)K_0(s_j^*)} \qquad (1)$$

in which $K_m(\ )$ signifies the modified Bessel function of the 2nd kind of order m, and the arguments are given by

$$q_j^* = q_j r = \frac{ia_j}{\eta_j\sqrt{1+iD_j}} \qquad s_j^* = s_j r = \frac{ia_j}{\sqrt{1+iD_j}}$$

and G_j=the shear modulus of soil, $\eta_j = V_{pj}/V_{sj}$, V_{pj}, V_{sj}=the longitudinal and shear wave velocity, respectively, D_j=the internal damping factor which yields twice of the viscous damping ratio for the Voigt type soil, $a_j = r\omega/V_{sj}$ denotes the dimensionless frequency, r=the radius of pile, and ω=the driving frequency.

For the longitudinal direction, the soil reaction becomes

$$P_L(z_j) = 2\pi G_j(1+iD_j)\frac{p_j^*K_1(p_j^*)}{K_0(p_j^*)} \qquad (2)$$

in which

$$p_j^* = p_j r = \frac{ia_j}{\sqrt{1+iD_j}}$$

Applying the transfer matrix method for a beam vibration with varying soil reaction [8], we obtain the impedance functions which define the force-displacement relationship at the pile head. Making an analogy of a set of springs and dashpots yields

$$[K^*_{pile}] = \begin{bmatrix} K_{uu} & K_{u\theta} & 0 \\ K_{\theta u} & K_{\theta\theta} & 0 \\ 0 & 0 & K_z \end{bmatrix} + i\omega \begin{bmatrix} C_{uu} & C_{u\theta} & 0 \\ C_{\theta u} & C_{\theta\theta} & 0 \\ 0 & 0 & C_z \end{bmatrix} \quad \text{for } \omega \geq h_1 \qquad (3)$$

in which the suffix denotes the direction associated with the motion.

In the frequency range below h_1, only the internal soil damping exists, so that Equation 3 may be modified as

$$[K^*_{pile}] = \begin{bmatrix} K_{uu}(1+iD_{uu}) & K_{u\theta}(1+iD_{u\theta}) & 0 \\ K_{\theta u}(1+iD_{\theta u}) & K_{\theta\theta}(1+iD_{\theta\theta}) & 0 \\ 0 & 0 & K_z(1+iD_z) \end{bmatrix} \quad \text{for } \omega < h_1 \qquad (4)$$

Since the soil layers exert an inertia force on pile accordin to their vibration, the driving force along the pile is therefore to be transferred to that at the pile head when the pile head impedance functions are used for the superstructural analysis. The former quantity is evaluated as the opposite of fixing forces of the pile head for the base rock input situation. Thus obtained pile head impedance and the effective forces are summed to comply with the rigid pier-footing displacement. However, the so-called pile grouping effect which confines the scattered wave among the gathered piles is not considered herein. Figure 2 illusrates the above consideration. The governing equation of motion for one foundation is therefore expressed, for the coupling of rocking and sway, as

$$\begin{bmatrix} M_F & \\ & J_G \end{bmatrix} \begin{Bmatrix} \ddot{u}_G \\ \ddot{\theta}_G \end{Bmatrix} + \sum_{p=1}^{\text{Piles}} [\alpha]_p^T [K^*_{pile}] [\alpha]_p \begin{Bmatrix} u_G - u_g \\ \theta_G \end{Bmatrix} = [\alpha]_F^T \begin{Bmatrix} Q(t) \\ M(t) \end{Bmatrix} + [\alpha]_p^T \{F\}_s \qquad (5)$$

in which u_G, θ_G=the absolute horizontal displacement and rotation angle, respectively at the footing,s gravity center (G.C.), u_g= the base rock displacement, M_F=the mass of footing, J_G=the mass moment of inertia with respect to the footing's G.C., $M(t)$, $Q(t)$ =the moment and horizontal force at the footing top, $\{F\}_s$=the effective seismic input, and $[\alpha]_p$, $[\alpha]_F$=the displacement transformation matrices from footing's G.C. to each pile head and to its top, respectively. Denoting the mass matrix, $[M]_F$, the equivalent stiffness and damping matrices, $[K]_F$ and $[C]_F$, respectively, we get the abbreviated form for Equation 5 as

$$[M]_F\{\ddot{u}\}_F + [C]_F\{\dot{u}\}_F + [K]_F\{u\}_F = [C]_F\{\dot{u}_g\} + [K]_F\{u_g\} + [\alpha]_F^T\{R\}_{sup} + [\alpha]_p^T\{F\}_s \qquad (6)$$

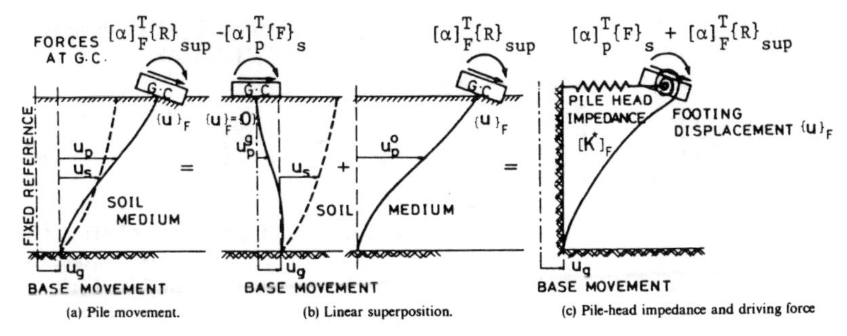

Figure 2. Soil-pile foundation interaction

Superstructure

The matrix formulation with beam elements gives the governing equation of motion

$$[M]_{sup}\{\ddot{u}\}_{sup} + [C]_{sup}\{\dot{u}\}_{sup} + [K]_{sup}\{u\}_{sup} = \{F\}_{sup} \tag{7}$$

for the relased boundary condition; namely, the pier base is subject to have degrees of freedom in connection with the foundation movement. When the boundary nodal displacements, denoted by j, are separated from the free nodal displacements, denoted by i, and the corresponding partitioning is made for the coefficient matrices, Equation 7 becomes

$$\begin{bmatrix} [M_{jj}] & [M_{ji}] \\ [M_{ij}] & [M_{ii}] \end{bmatrix}\begin{Bmatrix} \{\ddot{u}_j\} \\ \{\ddot{u}_i\} \end{Bmatrix} + \begin{bmatrix} [C_{jj}] & [C_{ji}] \\ [C_{ij}] & [C_{ii}] \end{bmatrix}\begin{Bmatrix} \{\dot{u}_j\} \\ \{\dot{u}_i\} \end{Bmatrix} + \begin{bmatrix} [K_{jj}] & [K_{ji}] \\ [K_{ij}] & [K_{ii}] \end{bmatrix}\begin{Bmatrix} \{u_j\} \\ \{u_i\} \end{Bmatrix} = \begin{Bmatrix} \{F_j\} \\ \{0\} \end{Bmatrix} \tag{8}$$

in which $\{F_j\}$ is the internal force vector at the interface with foundations. Expressing the free nodal displacement as the sum of a quasi-static component $\{u_i^u\}$ due to the pier base (or the foundation top) movement, and the dynamic component $\{u_i^c\}$ for the constraint boundary at this section [3], then

$$\{u_i\}_{sup} = \{u_i^u\} + \{u_i^c\} \tag{9}$$

In order to evaluate the former response, the static condensation is carried out for Equation 8, deriving the transfter matrix [β] as

$$[\beta] = -[K_{ii}]^{-1}[K_{ij}] \tag{10}$$

The dynamic response, on the other hand, is assumed to be decomposed into the normal modes, resulting in the following orthogonality conditions of the modal matrix [Φ] with respect to the mass, damping and stiffness matrices.

$$[\Phi]^T[M_{ii}][\Phi]=[I], \quad [\Phi]^T[C_{ii}][\Phi]=\lceil 2\xi_\ell\omega_\ell \rfloor, \quad [\Phi]^T[K_{ii}][\Phi]=\lceil \omega_\ell^2 \rfloor \tag{11}$$

in which ω_ℓ=the natural frequency of ℓ-th mode, ξ_ℓ=the damping factor, [I]=the identity and $\lceil \; \rfloor$ means the diagonal matrix. Hence, the above transformation becomes

$$\begin{Bmatrix} \{u_j\} \\ \{u_i\} \end{Bmatrix}_{sup} = \begin{bmatrix} [I] & [0] \\ [\beta] & [\Phi] \end{bmatrix} \begin{Bmatrix} \{u_j\} \\ \{q\} \end{Bmatrix} = [T] \begin{Bmatrix} \{u_j\} \\ \{q\} \end{Bmatrix} \tag{12}$$

Substituting Equation 12 into Equation 8 and premultiplying the transpose of the transformation matrix [T] to the same equation to maintain the symmetric nature of the relevant coefficient matrices, we get the governing equation expressed in terms of the boundary displacements $\{u_j\}$ and the normal mode co-ordinates {q} as

$$\left[\begin{array}{c|c} [M_{jj}]+[\beta]^T[M_{ij}]+[M_{ji}][\beta]+[\beta]^T[M_{ii}][\beta] & ([M_{ji}]+[\beta]^T[M_{ii}])[\Phi] \\ \hline \text{sym.} & [I] \end{array}\right] \begin{Bmatrix} \{\ddot{u}_j\} \\ \{\ddot{q}\} \end{Bmatrix}$$

$$\left[\begin{array}{c|c} [\beta]^T[C_{ii}][\beta] & [\beta]^T[C_{ii}][\Phi] \\ \hline \text{sym.} & \lceil 2\xi_\ell\omega_\ell \rfloor \end{array}\right] \begin{Bmatrix} \{\dot{u}\} \\ \{\dot{q}\} \end{Bmatrix} + \left[\begin{array}{c|c} [K_{jj}]+[K_{ji}][\beta] & [0] \\ \hline \text{sym.} & \lceil \omega_\ell^2 \rfloor \end{array}\right] \begin{Bmatrix} \{u_j\} \\ \{q\} \end{Bmatrix} = \begin{Bmatrix} \{F_j\} \\ \{0\} \end{Bmatrix} \tag{13}$$

Integrated system

The present structure has a multiple foundations so that Equation 6 holds for each foundation. Expressing these equations in a matrix form gives

$$[\tilde{M}]_F\{\ddot{\tilde{u}}\}_F + [\tilde{C}]_F\{\dot{\tilde{u}}\}_F + [\tilde{K}]_F\{\tilde{u}\}_F = [\tilde{C}]_F\{\dot{\tilde{u}}_g\} + [\tilde{K}]_F\{\tilde{u}_g\} + [\tilde{\alpha}]_F^T\{\tilde{R}\}_{sup} + [\tilde{\alpha}]_p^T\{\tilde{F}\}_s \tag{14}$$

in which the vectors and the coefficients matrices are enlarged for the numbers of foundations.

In order to combine all the foundations in the above and the superstructure in Equation 13, as an integrated coupled system the displacement compatibility

$$\{u_j\} = [\tilde{\alpha}]_F\{\tilde{u}\}_F \tag{15}$$

and the force equilibrium

$$\{F_j\} + \{\tilde{R}\}_{sup} = \{0\} \tag{16}$$

are used at their interface. Since the pile head impedance functions and the effective input motion are frequency dependent, the frequency domain analysis is preferred for the solution. The result is then

$$\begin{Bmatrix} \{\hat{q}(\omega)\} \\ \{\hat{u}(\omega)\}_F \end{Bmatrix} = \begin{bmatrix} [K_{qq}(\omega)] & [K_{qu_F}(\omega)] \\ [K_{qu_F}(\omega)]^T & [K_{u_Fu_F}(\omega)] \end{bmatrix}^{-1} \begin{Bmatrix} \{0\} \\ [P]_F\{\hat{u}_g(\omega)\} \end{Bmatrix} \tag{17}$$

in which $\{\hat{q}(\omega)\}, \{\hat{u}(\omega)\}_F, \{\hat{F}(\omega)\}_s$ and $\{\hat{u}_g(\omega)\}$ are the Fourier Transforms of $\{q(t)\}$, $\{u(t)\}_F$, $\{F(t)\}_s$ and $\{u_g(t)\}$, respectively, and

$$[K_{qq}(\omega)] = \lceil (-\omega^2 + 2\xi_\ell \omega_\ell \omega + \omega_\ell^2) \rfloor$$

$$[K_{qu_F}(\omega)] = -\omega^2 [\Phi]^T ([M_{ii}][\beta] + [M_{ij}])[\tilde{\alpha}]_F$$

$$[K_{u_F u_F}(\omega)] = -\omega^2 \Big\{ [M]_F + [\tilde{\alpha}]_F^T ([\beta]^T [M_{ii}][\beta] + [\beta]^T [M_{ij}] + [M_{ji}][\beta] + [M_{jj}])[\tilde{\alpha}]_F \Big\} + i\omega[\tilde{C}]_F + [\tilde{K}]_F + [\tilde{\alpha}]_F^T ([K_{ji}][\beta] + [K_{jj}])[\tilde{\alpha}]_F$$

$$[P]_F = [K]_F + i\omega[C]_F + [\tilde{\alpha}]_p^T [\{F(\omega)\}_s \mid \{0\}]$$

The frequency response in the original co-ordinates are obtained from the back-transformation of Equation 12. The inverse Fourier transform gives the corresponding response time history. The block diagram of Figure 3 explains the above analysis.

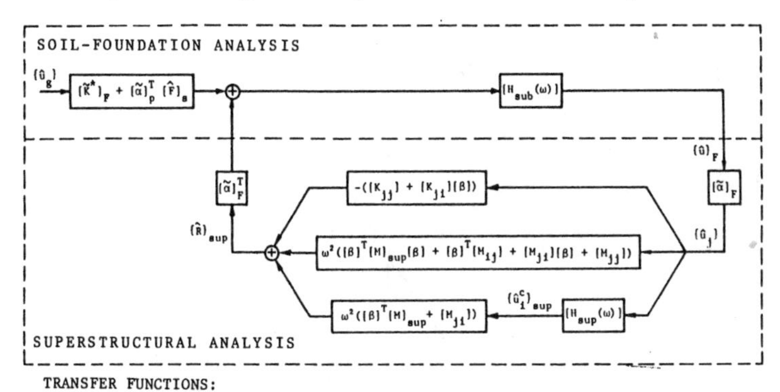

TRANSFER FUNCTIONS:

$[H_{sup}(\omega)] = [\Phi][1/\{(\omega_\ell/\omega)^2 + i2\xi_\ell(\omega_\ell/\omega) - 1\}][\Phi]^T([M_{ii}][\beta] + [M_{ij}])$

$[H_{sub}(\omega)] = (-\omega^2[\tilde{M}]_F + i\omega[\tilde{C}]_F + [\tilde{K}]_F)^{-1}$, $[\tilde{K}^*]_F$ = Complex subgrade impedance

Figure 3. Block diagram for analysis

Wave propagation

Consider an oblique SH wave hitting the top of the base rock with an incident angle α from the vertical axis, as shown in Figure 4. The apparent phase lag

$$\tau = \frac{D \sin\alpha}{V_b} \tag{18}$$

is observed at this situation between two points on the base rock level but apart by D in horizontal distance. V_b denotes the shear wave velocity in the base rock medium. The soil particle motion in the surface layers is then described by discrete Fourier transform

$$u_s(x,z;t) = \sum_{\nu=0}^{N/2} A(z,\omega_\nu) u_g(\omega_\nu) \exp[i\omega_\nu(t - \frac{x}{V_b}\sin\alpha)] \tag{19}$$

in which $A(z,\omega_\nu)$ satisfies the one-dimensional shear wave equation in the vertical direction and is imposed as unity at the base rock level, while $\hat{u}_g(\omega_\nu)\exp[i\omega_\nu(t-x\cdot\sin/V_b)]$ represents the wave propagation in the horizontal direction.

Hence, for a multiple-supported bridge structure, the traveling seismic motion is input at the foundations. For instance, at the i-th foundation

$$u_{gi}(t)= u_g(t-\tau_i),\ \dot{u}_{gi}(t)= \dot{u}_g(t-\tau_i),\ \ddot{u}_{gi}(t)= \ddot{u}_g(t-\tau_i) \quad (20)$$

or in the frequency domain their Fourier transform

$$\hat{u}_{gi}(\omega)=\hat{u}_g(\omega)e^{-i\omega\tau_i},\ \hat{\dot{u}}_{gi}(\omega)=\hat{\dot{u}}_g(\omega)e^{-i\omega\tau_i},\ \hat{\ddot{u}}_{gi}(\omega)=\hat{\ddot{u}}_g(\omega)e^{-i\omega\tau_i} \quad (21)$$

in which $u_g(t)$ and $\hat{u}_g(\omega)$ are for the first foundation and the time lag for the i-th foundation is computed from

$$\tau_i= \frac{\sum_{k=1}^{i} D_k \sin\alpha}{V_b} \quad (22)$$

depending on the span length D_k.

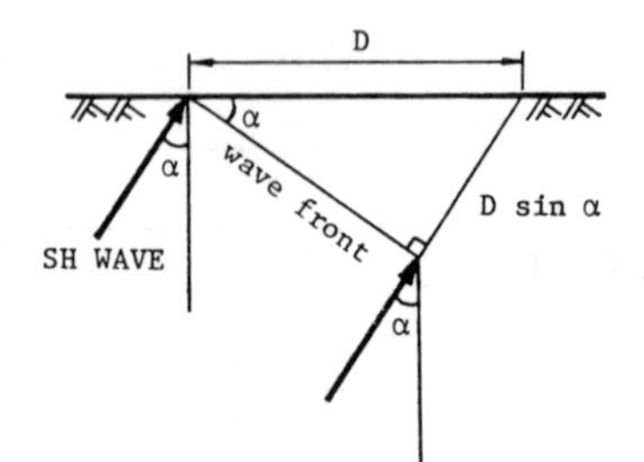

Figure 4. Wave propagation

NUMERICAL RESULTS AND DISCCUSION

Model for analysis

The three-span continuous bridge, as illustrated in Figure 1, and whose structural dimensions are indicated in Table 1, is analyzed. The pile layout is shown in Figure 5. The soil properti at the construction site are listed in Table 2, which are the results after the strain-dependent site response analysis for one dimensional SH wave propagation in the vertical direction[6]. Although a small variation exists in the soil profile at the respective foundation site, the values in Table 2 are used for all the foundation site.

The structural behavior in the direction perpendicular to the bridge axis is of interest in this investigation, so that the out-of-plane degrees of freedom, as indicated in Figure 6, ar taken into account. At each foundation, two degrees of freedom of rocking and sway are imposed. The seismic wave is represented by the SH wave to travel in the direction of bridge axis which causes the out-of-phase input at each foundation, depending on the shear wave velocity and the angle of incidence at the base rock level.

Dynamic characteristics

The fixed base superstructure is decomposed into the normal modes. The first nine modes are depicted in Figure 7 in which the participation factor indicated is based on a uniform excitation. Since the present superstructure is characterized by the flexible girder and the stiff piers,the torsional motion is easily induced at the former structural part. The modes shapes show that the fundamental mode is due to the in-phase pier movement while most of the higher modes are due to the out-of-phase pier movement. This means the significance of the traveling nature of the seismic wave on the structural response in connection with the vibration modes. For the succeeding response analysis the reasonable modal truncation is made by adopting the first thirty modes whose natural frequencies fall in the essential frequency range in view of the frequency contents of the earthquake motion. The damping factors for these modes are assumed 2 percent of their critical values.

The foundation impedance functions which are computed from those at the pile head are shown in Figure 8. The modified impedance functions which approximate the three dimensional characteristic in the low frequency range are indicated by the dashed lines. The spring effect in this range is substituted by the peak value since the rigorous solution is rather horizontal with some droppings at the natural frequencies of the soil layers which disappear as the internal damping inceases. The damping effect below the fundamental natural frequency of the soil layers on the other hand, is computed by subtracting the imaginary part of the impedance functions with no internal damping (Baranov solution) from the corresponding ones with a specified internal damping since there is no radiating wave in this frequency range.

Frequency response function

Figure 9 shows the frequency response amplification of the integrated soil-foundation-superstructure system in the situation of a uniform input at the base rock level. Only the representative nordal points responses are depicted together with the free field site response and the fixed base superstructural response in order to observe the soil-structure interaction effect. The footing top, denoted by F2, is noted to have been affected by the soil layers vibration rather than by the superstructure, indicating a predominant peak at the fundamental frequency of the former. The girder response, denoted by G13, shows two comparable peaks that relate the soil layers and the superstructural characteristics. It is understood that the inertial interaction which causes shifting of the natural frequencies of the fixed base superstructure occurs in the high frequency range where the pier modes are predominant,while the soil layers vibration amplifies the superstructural response in the low frequency range where the girder modes are predominant.

The frequency response with use of the Baranov solution or the plane strain solution for all the frequency range gives almost the same result except a minor discrepancy near the fundamental natural frequency of the soil layers, see Figure 10.

The engineering practice for the input motion is to take the surface response of the free field even though the soil impedance is considered. This accounts for only incorporating the soil amplification for the earthquake motion but discarding the kinematic interaction between soil and foundation. Figure 11 compares this approximation for the present system, which overestimates response beyond the fundamental frequency of the soil layers.

Earthquake response

Figure 12 shows the maximum acceleration response and Figure 13 the maximum displacement response of the integrated system for the Kaihoku Tr. accelerogram (modified as to give the maximum acceleration of 100 gal at the base rock level), Miyagi-ken-oki earthquake, June 1978. In order to compute the displacement time history from the accelerogram record, the base line correction is excuted from the least square fitting of acceleration and velocity such that no drift of the base line will occur for displacement. Furthermore, to take out the recording error, the high-pass filter (roll-off frequency f_{rL}=0.1 Hz and cut-off frequency f_{cL}=0.15 Hz) and the low-pass filter (roll-off frequency f_{rH}=10 Hz and cut-off frequency f_{cH}=12 Hz) are used. In the out-of-phase input situation of incident angle $\alpha=\pi/2$ in which the seismic wave travels from left to right, twice of the acceleration response is attained at the right-end pier and the girder on it when compared with the uniform input situation. This fact means that the higher modes are amplified in the former case. The approximate response with use of the surface soil response as an input overestimates response at all sections. The sensitivity of these input assumption is rather small for the displacement response.

From the structural design point of view, the internal forc are of interest,which are drawn in Figure 14 for the bending moment and in Figure 15 for the shearing force. The observation concerning the acceleration response holds for these internal forces. This fact emphasizes the importance of the traveling nature of the seismic wave in designing structures reported here

REFERENCES

1. Gutierrez,J.A. and Chopra,A. (1978) A substructure Method For Earthquake Analysis of Structures Including Structure Soil Interaction, Earthquake Engineering and Structural Dynamics, 6,: 51-69
2. Kausel,E., Whitman,R.V., Morray,J.P. and Elsabee,F. (1978) The Spring Method For Embedded Foundations, Nuclear Engineeri and Design, 48,: 377-392

3. Kukreti,A.R. and Feng,C.C. (1978) Dynamic Substructuring For Alternating Subsystems, J. Eng. Mech. Div., ASCE, 104, EM5: 1113-1129
4. Lysmer,J., Udaka,T., Tsai,C.F. and Seed,H.B. (1975) FLUSH: A Computer Program For Approximate 3-D Analysis of Soil-Structure Interaction Problems, EERC Report 75-30, Univ. of Calif., Berkeley
5. Novak,M., Nogami,T. and Aboul-Ella,F. (1978) Dynamic Soil Reactions for Plane Strain Case, J. Eng. Mech. Div., ASCE, 104, EM4: 953-959
6. Schnable,P.B., Lysmer,J. and Seed,H.B. (1972) SHAKE: A Computer Program For Earthquake Response Analysis of Horizontally Layered Sites, EERC Report 72-12, Univ. of Calif., Berkeley
7. Takemiya,H. (1981) Embedded Effect on Soil Structure Interaction, State-of-the-art in Earthquake Engineering, 1981, 7th WCEE, Istanbul, Turkey
8. Takemiya,H. and Yamada,Y. (1981) Layered Soil-Pile-Structure Dynamic Interaction, Earthquake Engineering and Structural Dynamics, 9: 437-457
9. Werner,S.D., Lee,L.C., Wong,H.L. and Trifunac,M.D. (1977) Structural Response to Traveling Seismic Waves, J. Struc. Div., ASCE, 105, ST12: 2547-2564

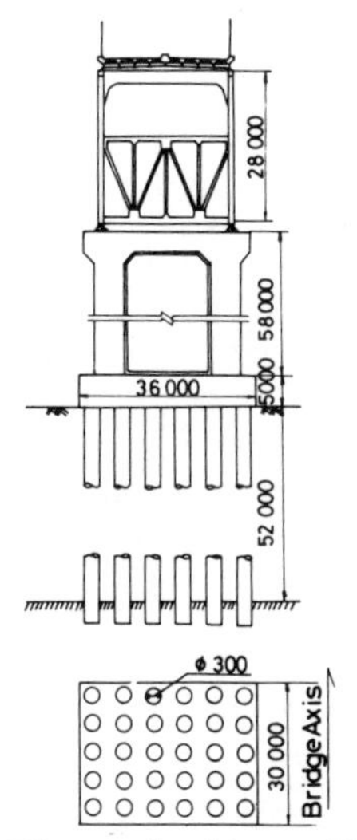

Figure 5. Pier-pile foundation

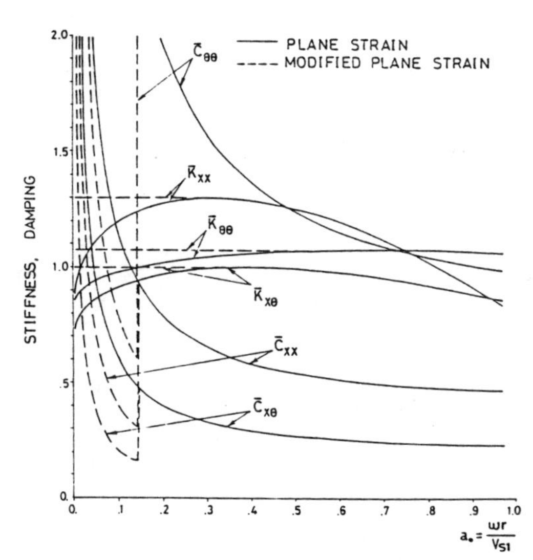

Figure 8. Subgrade impedance functions

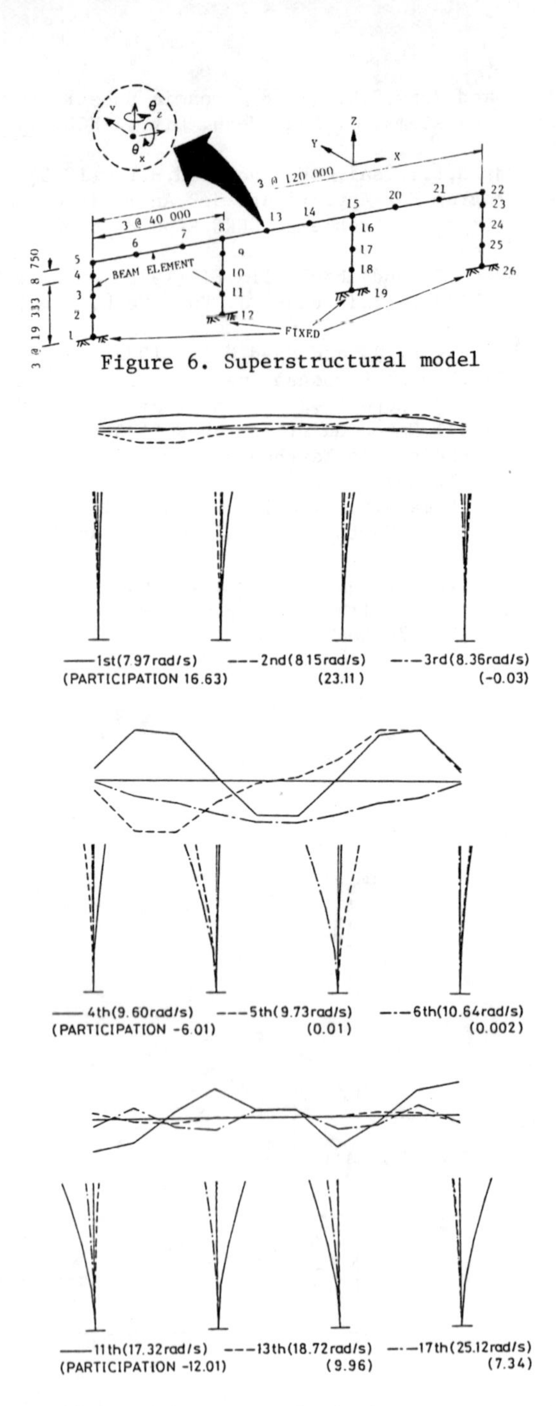

Figure 6. Superstructural model

Figure 7. Normal modes of superstructure

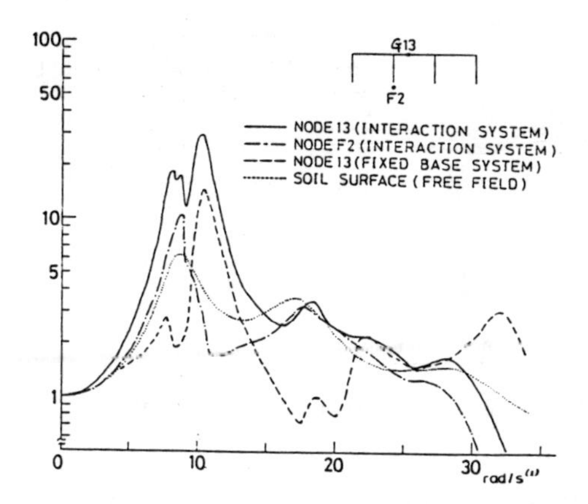

Figure 9. Frequency response function

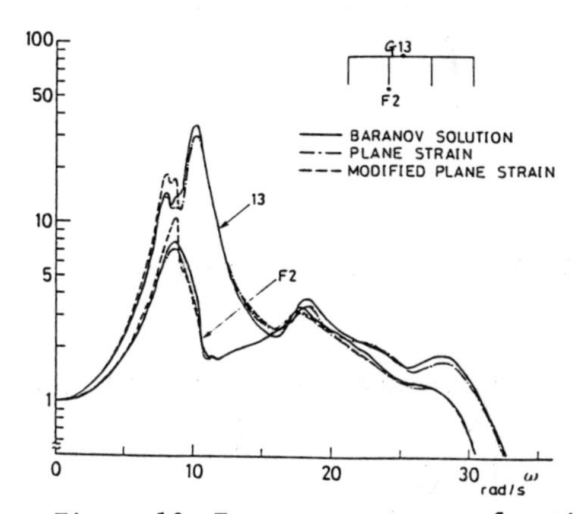

Figure 10. Frequency response function

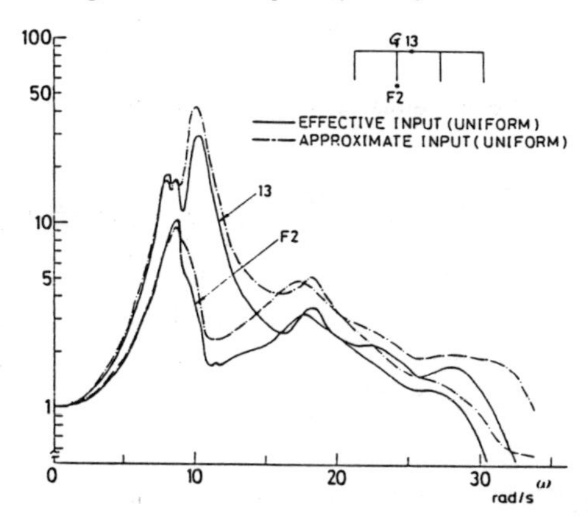

Figure 11. Frequency response function

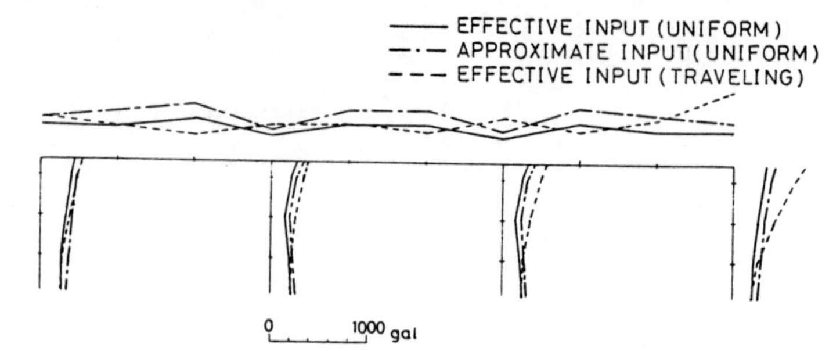

Figure 12. Maximum acceleration response

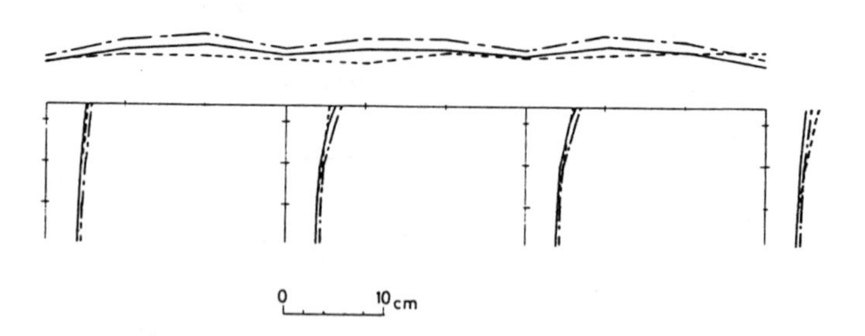

Figure 13. Maximum displacement response

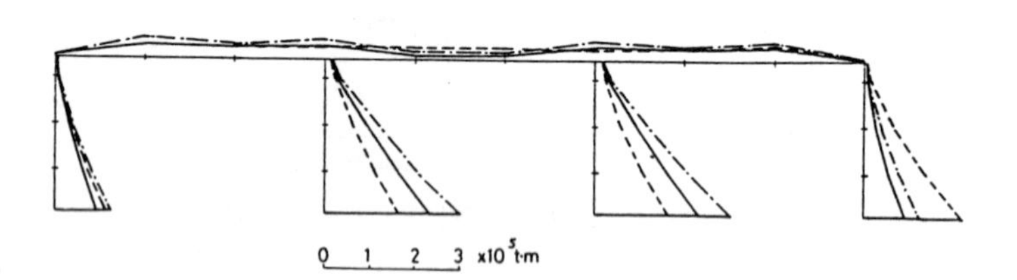

Figure 14. Maximum bending moment response

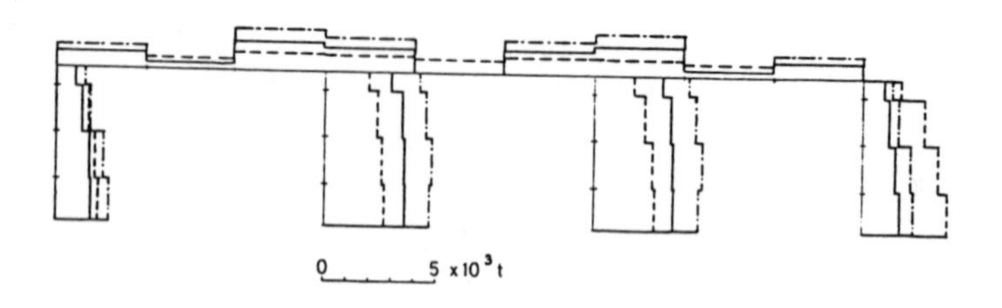

Figure 15. Maximum shearing force response

Table 1. Structural Dimensions

DIMENSIONS \ STRUCTURAL PARTS	Superstructure Girder	Piers 1 through 4
Span length ℓ(m)	120.0	58.0
Young's Modulus E(t/m^2)	2.1×10^7	2.69×10^6
Shear Rigidity G(t/m^2)	8.1×10^6	1.15×10^6
Inertia Moment (for bending) I(m^4)	76.9	4720.0
Inertia Moment (for torsion) I(m^4)	15.1	46.0
Weight w(t/m)	42.1	116.5
Cross Sectional Area A(m^2)	0.513	52.0

See Figure 5 for other relevant data

Table 2. Soil Layers Properties

Surface Layers	Thickness H(m)	Shear Velocity V_s(m/s)	Unit Weight w(t/m^3)	Damping D= 2ξ
1	2.5	91.24	2.05	0.270
2	5.7	76.42	2.05	0.528
3	4.0	178.03	1.85	0.136
4	2.6	219.51	2.05	0.170
5	3.3	261.64	1.90	0.112
6	6.3	196.12	1.90	0.140
7	7.0	296.80	2.10	0.146
8	9.2	341.41	2.20	0.142
9	6.0	339.88	1.90	0.112
10	7.0	498.19	1.90	0.088
11	Half-space	600.0		

Soil Dynamics & Earthquake Engineering Conference / Southampton / 1982.07.13-15

The Influence of Gravity Forces in the Response of Framed Structures to Earthquake Excitations

G.OLIVETO
Università di Catania, Italy

INTRODUCTION

It is well known that the effects of gravity forces may be separated from the dynamical response of structures if the reference configuration is taken coincident with that of static equilibrium, Clough (1975). It is also known that the instabilizing effects of axial forces may change considerably the frequencies of vibration of structures and that these usually vanish at the critical loads, Clough (1975), Warburton (1976).

Recent studies on arches, Johnson (1980), have shown that not only may the frequencies be changed but also the mode shapes by the presence of static loads.

Therefore the dynamical response of structures evaluated through the mode superposition method may be incorrect where these instabilizing effects are not included in the evaluation of the natural frequencies and related mode shapes.

This paper provides an algorithm for the evaluation of the natural frequencies and mode shapes of framed structures including the instabilizing effects of axial forces.

These mode shapes are then used in the mode superposition method to evaluate the dynamical response to earthquake excitations thus providing a comparison with the response when these effects are not included.

2. EQUATIONS OF MOTION

The differential equations of motion, for a prismatic frame member, allowing for second order effects but neglecting rotatory inertia and shear deformation may be written as follows

$$(EIy,_{xx}),_{xx} - (Ny,_x),_x + my,_{tt} = p(x) \qquad (1)$$

$$(EA(\chi,_x + (y,_x)^2/2),_x + m\chi,_{xx} = 0 \qquad (2)$$

where
- $EI(x)$ denotes the bending stiffness of the member
- $EA(x)$ the axial stiffness
- $N(x)$ the axial force
- $m(x)$ the mass per unit of length
- $y(x,t)$ the component of displacement normal to the centre line
- $\chi(x,t)$ the axial displacement
- $p(x)$ a static load distribution normal to the centre line.

Furthermore a comma denotes differentiation with respect to the following variables which in the present case are either the position coordinates x or the time variable t.

Axial load distributions would give rise to an unknown distribution of axial forces $N(x)$ along the member which would then further complicate the integration of the equations of motion.

As this variation is small compared to the average axial force in the member any distribution of axial loads will be ignored. The global effects of these forces may however be taken into account by applying their static equivalents to the nodal points.

In order to separate the static from the dynamic

response it is expedient to make the following positions

$$y(x,t) = z(x) + v(x,t)$$

$$\chi(x,t) = w(x) + u(x,t)$$

where z and w satisfy the static equilibrium equations

$$(EIz,_{xx}),_{xx} - (N(z)z,_{x}),_{x} = p(x) \quad (3)$$

$$(EA(w,_{x} + z^2,_{x}/2)),_{x} - N(z),_{x} = 0 \quad (4)$$

The equations of motion (1) and (2), taking into account (3) and (4) become

$$(EIv,_{xx}),_{xx} - (N(z)v,_{x}),_{x} + mv,_{tt} - (\Delta N(v)z,_{x}),_{x}$$

$$- (\Delta N(v)v,_{x}),_{x} = 0 \quad (5)$$

$$(EAu,_{x}),_{x} + mu,_{tt} + (EAz,_{x}v,_{x}),_{x} + (EAv^2,_{x}/2),_{x} = 0 \quad (6)$$

where $\Delta N(v) = N(z+v) - N(z)$.

For small motions the nonlinear terms in equations (5) and (6) may be neglected obtaining

$$(EIv,_{xx}),_{xx} - (N(z)v,_{x}),_{x} + mv,_{tt} = 0 \quad (7)$$

$$(EAu,_{x}),_{x} + mu,_{tt} = 0 \quad (8)$$

Any dynamical analysis through equations (7) and (8) requires beforehand the evaluation of the axial force N(z). This in turn requires the solution of the nonlinear static equilibrium equations (3) and (4).

3. STATIC ANALYSIS

In what follows the frame members will be considered prismatic and uniform and the static load p(x) independent of x. Equations (3) and (4) therefore become

$$EIz,_{xxxx} - (N(z)z,_{x}),_{x} = p \quad (9)$$

$$EA(w,_{xx} + z,_{x}z,_{xx}) = 0 \quad (10)$$

The general solution to equation (9) has been given elsewhere, Oliveto (1980), for the case when p=0.

For $p \neq 0$ it may be written

$$z = z_c + z_p$$

where

$$z_c = C_o + C_1\xi + C_2\frac{2(1-\cos\sigma\xi)}{\sigma^2} + C_3\frac{6(\sigma\xi - \sin\sigma\xi)}{\sigma^3} \quad (11)$$

$$z_p = \frac{qL}{6}\left[3\frac{\beta-\alpha+2}{\sigma^2}\xi^2 - \frac{6(\sigma\xi-\sin\sigma\xi)}{\sigma^3} + \frac{\alpha-\beta}{4} \cdot \right.$$

$$\left. \cdot \frac{12(\sigma^2\xi^2 - 2(1-\cos\xi))}{\sigma^4}\right] \quad (12)$$

and $\xi=\frac{x}{L}$ is a dimensionless position coordinate

$\sigma^2=-\frac{N(z)L^2}{EI}$ a dimensionless parameter depending on the axial force $N(z)$

α,β functions of σ which will be specified later

$q=\frac{pL^3}{2EI}$ is a loading dimensionless parater.

The constants of integration C_o, C_1, C_2, C_3 may be found by applying the kinematic boundary conditions as depending linearly on the nodal displacements z_i, ϕ_i, z_j, ϕ_j, (fig.1), and nonlinearly on the parameter σ, Oliveto (1981). The rather unusual form of equations (11) and (12) is advantageous in avoiding singularities in the behaviour as $\sigma \to 0$.

The application of the static boundary conditions provides the following relationship between the nodal forces and displacements,

$$\begin{bmatrix} T_i L \\ M_i \\ T_j L \\ M_j \end{bmatrix} = \frac{EI}{L}\left\{\begin{bmatrix} \delta & \theta & -\delta & \theta \\ \theta & \alpha & -\theta & \beta \\ -\delta & -\theta & \delta & -\theta \\ \theta & \beta & -\theta & \alpha \end{bmatrix}\begin{bmatrix} z_i/L \\ \phi_i \\ z_j/L \\ \phi_j \end{bmatrix} + q\begin{bmatrix} -1 \\ -\theta^{-1} \\ -1 \\ \theta^{-1} \end{bmatrix}\right\} \quad (13)$$

where the Blaszkowiak's stiffness functions $\alpha,\beta,\theta,\delta$, are defined as follows, Lightfoot (1979).

$\alpha = \sigma(s - \sigma c)/d$; $\beta = \sigma(\sigma - s)/d$

$\theta = \alpha+\beta$; $\delta = 2\theta - \sigma^2$;

$c = \cos\sigma$; $s = \sin\sigma$; $d = 2(1-c) - \sigma s$

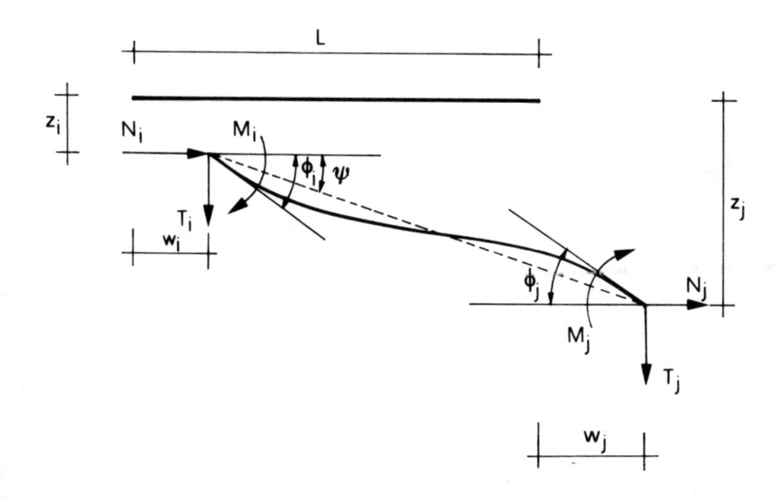

Figure 1.

Expressions (11) to (14) apply for $N(z)<0$ (compressive axial force). For $N(z)>0$ (tensile axial force) it suffices to set $\sigma^2=N(z)L^2/EI$ and substitute in the above expressions $i\sigma$ for σ. After solving equation (4) the axial force $N(z)$ may be given as follows, Oliveto (1981),

$$N(z) = \frac{EA}{L}\,(w_j - w_i + \eta L) \tag{14}$$

with

$$\eta = \frac{1}{2L^2}\int_0^1 z_{,\xi}^2 \, d_\xi \tag{15}$$

The relationship between nodal axial force and displacements therefore becomes :

$$\begin{bmatrix} N_i L \\ N_j L \end{bmatrix} = \frac{EI}{L}\left\{\lambda^2 \begin{bmatrix} 1 & -1 \\ -1 & 1 \end{bmatrix} \begin{bmatrix} w_i/L \\ w_j/L \end{bmatrix} + \eta \begin{bmatrix} -1 \\ 1 \end{bmatrix}\right\} \tag{16}$$

with $\lambda^2 = L^2 \dfrac{A}{I}$.

The dimensionless parameter η may be given also as

$$\eta = \eta_p + \eta_c + \eta_{pc} \tag{17}$$

where

$$\eta_p = \frac{1}{2L^2}\int_0^1 z_{p,\xi}^2 \, d\xi$$

$$\eta_c = \frac{1}{2L^2} \int_0^1 z_{c,\xi}^2 \, d\xi$$

$$\eta_{pc} = \frac{1}{L^2} \int_0^1 z_{p,\xi} \, z_{c,\xi} \, d\xi$$

After some algebraic manipulations it is found that η_p, η_c and η_{pc} may be written also as

$$\eta_p = \frac{1}{2} q^2 \underline{a}^T G \underline{a} \tag{18}$$

$$\eta_c = \frac{1}{2} \underline{d}^T H^T Y H \underline{d} \tag{19}$$

$$\eta_{pc} = q \underline{a}^T Z H \underline{d} \tag{20}$$

where the vectors and matrices in the above expressions are defined as follows

$$\underline{d}^T = \left[\phi_i, \phi_j, \psi \right]$$

$$\underline{a}^T = \left[\frac{\beta-\alpha+2}{\sigma^2}, \; -\frac{3}{2}, \; \frac{\alpha-\beta}{6} \right]$$

$$H = \begin{bmatrix} 1 & 0 & 0 \\ \frac{\beta-\theta}{2} & -\frac{\beta}{2} & \frac{\theta}{2} \\ \frac{\delta-\theta}{2} & \frac{\theta}{6} & -\frac{\delta}{6} \end{bmatrix} \; ; \quad Y = \begin{bmatrix} y_{11} & y_{12} & y_{13} \\ y_{12} & y_{22} & y_{23} \\ y_{13} & y_{23} & y_{33} \end{bmatrix}$$

$$G = \begin{bmatrix} g_{11} & g_{12} & g_{13} \\ g_{12} & g_{22} & g_{23} \\ g_{13} & g_{23} & g_{33} \end{bmatrix} \; ; \quad Z = \begin{bmatrix} z_{11} & z_{12} & z_{13} \\ z_{21} & z_{22} & z_{23} \\ z_{31} & z_{32} & z_{33} \end{bmatrix}$$

with

$y_{11}=1$; $y_{12}=2(1-c)/\sigma^2$; $y_{13}=6(\sigma-s)/\sigma^3$

$y_{22}=2(\sigma-sc)/\sigma^3$; $y_{23}=3(y_{12})^2/2$; $y_{33}=18(3\sigma-4s+sc)/\sigma^5$

$g_{11}=1/3$; $g_{12}=(\sigma^2+2(1-c-\sigma s))/\sigma^4$; $g_{13}=2(\sigma^3+3\sigma c-3s)/\sigma^5$

$g_{22}=y_{33}/9$; $g_{23}=y_{13}^2/6$; $g_{33}=6(2\sigma^3+3(\sigma-sc)-12(s-\sigma c))/\sigma^7$

$z_{11}=1/2;\ z_{12}=2(s-\sigma c)/\sigma^3;\ z_{13}=3(\sigma^2+2(1-c-\sigma s))/\sigma^4$
$z_{21}=y_{13}/3;\ z_{22}=y_{23}/3;\ z_{23}=y_{33}/3\ ;$
$z_{31}=3(\sigma^2-2(1-c))/\sigma^4\ ;\ z_{32}=6(2(s-\sigma c)-\sigma+sc))/\sigma^5;$
$z_{33}=y_{13}^2/2$

As most of the expressions given above take the indeterminate form 0/0 for $\sigma=0$, the following expressions should be used for small values of σ

$$\alpha=6(10-\sigma^2)/(15-\sigma^2);\quad \beta=3(20-\sigma^2)/(2(15-\sigma^2))$$

$$\theta=15(12-\sigma^2)/(2(15-\sigma^2))\ ;\ \delta=15(12-\sigma^2)/(15-\sigma^2)-\sigma^2$$

$$(\beta-\alpha+2)/\sigma^2 = (20-\sigma^2)/(8(15-\sigma^2))$$

$y_{12}=(12-\sigma^2)/12;\ y_{13}=(20-\sigma^2)/20;\ y_{22}=(80-11\sigma^2)/60$
$y_{23}=(6-\sigma^2)/4;\ y_{33}=(504-53\sigma^2)/280$

$g_{12}=(9-\sigma^2)/36;\ g_{13}=48(21-\sigma^2)/7!;\ g_{22}=2(504-53\sigma^2)/7!$
$g_{23}=2(10-\sigma^2)/5!;\quad g_{33}=8(90-7\sigma^2)/7!$

$z_{12}=8(10-\sigma^2)/5!;\ z_{13}=(9-\sigma^2)/12;\ z_{31}=(30-\sigma^2)/5!$
$z_{32}=4!\ (84-13\sigma^2)/7!;\quad z_{33}=(10-\sigma^2)/20$

The case of tensile axial forces may be treated in the same way as for the stiffness functions.

Equations (13) and (16) may be assembled together to give

$$\begin{bmatrix} N_i L \\ T_i L \\ M_i \\ \hline N_j L \\ T_j L \\ M_j \end{bmatrix} = \frac{EI}{L}\left\{\left[\begin{array}{ccc|ccc} \lambda^2 & 0 & 0 & -\lambda^2 & 0 & 0 \\ 0 & \delta & \theta & 0 & -\delta & \theta \\ 0 & \theta & \alpha & 0 & -\theta & \beta \\ \hline -\lambda^2 & 0 & 0 & \lambda^2 & 0 & 0 \\ 0 & -\delta & -\theta & 0 & \delta & -\theta \\ 0 & \theta & \beta & 0 & -\theta & \alpha \end{array}\right] \begin{bmatrix} w_i/L \\ z_i/L \\ \phi_i \\ \hline w_j/L \\ z_j/L \\ \phi_j \end{bmatrix} + \begin{bmatrix} -\lambda^2\eta \\ -q \\ -q\theta^{-1} \\ \hline \eta\lambda^2 \\ -q \\ q\theta^{-1} \end{bmatrix}\right\} \qquad (21)$$

Equations (21) may be assembled according to standard rules of matrix structural analysis to provide the following set of nonlinear equations for the structure

$$K(\underline{d})\underline{d} + \underline{h}(\underline{d}) = \underline{f} \qquad (22)$$

$\underline{f}$ being a vector of prescribed nodal forces.
The most direct way of solving equations (22) is the method of successive approximations though the Newton-Raphson method could be applied as well by use of the tangent stiffness matrix which has been formulated elsewhere, Oliveto (1981), (1982).

4. THE DYNAMIC STIFFNESS MATRIX

The static analysis outlined above allows one to evaluate the axial force N(z) for each member of the framework. This in turn may be used to define a stress dependant dynamic stiffness matrix through equations (7) and (8).

For a uniform member, the procedure to be followed may be found in standard textbooks on Dynamics of Structures as Clough (1975) and Warburton (1976). Only the essential results will be reported here.
Any flexural mode of vibration may be written in the form

$$\phi(\xi)=D_1 \sin\gamma\xi+D_2\cos\gamma\xi+D_3\sinh\varepsilon\xi+D_4\cosh\varepsilon\xi \tag{23}$$

where

$$\gamma=\sqrt{\frac{\sigma^2}{2}+\sqrt{\left(\frac{\sigma^2}{2}\right)^2+a^4}}$$

$$\varepsilon=\sqrt{-\frac{\sigma^2}{2}+\sqrt{\left(\frac{\sigma^2}{2}\right)^2+a^4}}$$

$$a^4=\frac{mL^4}{EI}\,\omega^2$$

and ω^2 is the square of the frequency of vibration.
By setting

$$\sin\gamma=s;\quad \cos\gamma=c$$
$$\sinh\varepsilon=S;\quad \cosh\varepsilon=C$$

The relationship between the nodal displacements and the constants D_1, D_2, D_3, D_4 may be written as

$$\begin{bmatrix} v_i \\ L\phi_i \\ v_j \\ L\phi_j \end{bmatrix} = \begin{bmatrix} 0 & 1 & 0 & 1 \\ \gamma & 0 & \varepsilon & 0 \\ s & c & S & C \\ c\gamma & -s\gamma & \varepsilon C & \varepsilon S \end{bmatrix} \begin{bmatrix} D_1 \\ D_2 \\ D_3 \\ D_4 \end{bmatrix} \tag{24}$$

or $\underline{v}=W\underline{e}$.

A similar relationship may be found for the nodal forces. By setting

$$U = \frac{EI}{L^2}\begin{bmatrix} -\gamma(\gamma^2-\sigma^2) & 0 & \varepsilon(\varepsilon^2+\sigma^2) & 0 \\ 0 & \gamma^2 & 0 & -\varepsilon^2 \\ \gamma c(\gamma^2-\sigma^2) & -s(\gamma^2-\sigma^2) & -\varepsilon C(\varepsilon^2+\sigma^2) & -\varepsilon S(\varepsilon^2+\sigma^2) \\ -\gamma^2 s & -\gamma^2 c & \varepsilon^2 S & \varepsilon^2 C \end{bmatrix}$$

and

$$\underline{f}^T = \left[T_i L, M_i, T_j L, M_j\right]$$

this may be written as

$$\underline{f} = U\underline{e} \tag{25}$$

By solving equations (24) and substituting into equations (25) the following relationship is obtained

$$\underline{f} = UW^{-1}\underline{v} = K\underline{v} \tag{26}$$

where the matrix K is as follows

$$K = \frac{EI}{L^2}\begin{bmatrix} \delta & \theta & -\bar{\delta} & \bar{\theta} \\ \theta & \alpha & -\bar{\theta} & \beta \\ -\bar{\delta} & -\bar{\theta} & \delta & -\theta \\ \bar{\theta} & \beta & -\theta & \alpha \end{bmatrix} \tag{27}$$

and

$$\begin{aligned} &\delta=\varepsilon\gamma(\varepsilon^2+\gamma^2)(\varepsilon c S+\gamma s C)/d \\ &\bar{\delta}=\varepsilon\gamma(\varepsilon^2+\gamma^2)(\gamma s+\varepsilon S)/d \\ &\theta=\varepsilon\gamma((\gamma^2-\varepsilon^2)(1-cC)+2\varepsilon\gamma sS)/d \\ &\bar{\theta}=\varepsilon\gamma((\varepsilon^2+\gamma^2)(C-c))/d \\ &\alpha=(\varepsilon^2+\gamma^2)(\varepsilon sC-\gamma cS)/d \\ &\beta=(\varepsilon^2+\gamma^2)(\gamma S-\varepsilon s)/d \\ &d=2\varepsilon\gamma(1-cC)+(\varepsilon^2-\gamma^2)sS \end{aligned} \tag{28}$$

It should be noticed that in the limit as $\sigma\to 0$ equations (28) provide the well known dynamic stiffness functions; also as $\omega\to 0$ it may be easily checked that

$$\varepsilon\to 0; \quad \gamma\to\sigma; \quad \bar{\delta}\to\delta; \quad \bar{\theta}\to\theta$$

and the Blaszkowiak's stiffness functions are obtained.

The dynamic axial-deformation stiffness matrix which may be obtained from equation (8) relates nodal axial forces and displacements as follows

$$\begin{bmatrix} N_i L \\ N_j L \end{bmatrix} = \frac{EI}{L^2}\lambda^2 \begin{bmatrix} \mu & -\upsilon \\ -\upsilon & \mu \end{bmatrix}\begin{bmatrix} \mu_i \\ \mu_j \end{bmatrix} \tag{29}$$

where the dynamic stiffness functions μ and υ are given as follows

$$\mu = bc/s; \qquad \upsilon = b/s$$
$$s = \sin b; \qquad c = \cos b \tag{30}$$
$$b^2 = \frac{mL^2}{EA}\,\omega^2$$

By assembling flexural and axial nodal forces and displacements the following member relationship is found

$$\begin{bmatrix} N_i L \\ T_i L \\ M_i \\ \hline N_j L \\ T_j L \\ M_j \end{bmatrix} = \frac{EI}{L^2} \left[\begin{array}{ccc|ccc} \lambda^2\mu & 0 & 0 & -\lambda^2\upsilon & 0 & 0 \\ 0 & \delta & \theta & 0 & -\bar{\delta} & \bar{\theta} \\ 0 & \theta & \alpha & 0 & -\bar{\theta} & \beta \\ \hline -\lambda^2\upsilon & 0 & 0 & \lambda^2\mu & 0 & 0 \\ 0 & -\bar{\delta} & -\bar{\theta} & 0 & \delta & -\theta \\ 0 & \bar{\theta} & \beta & 0 & -\theta & \alpha \end{array}\right] \begin{bmatrix} u_i \\ v_i \\ L\phi_i \\ \hline u_j \\ v_j \\ L\phi_j \end{bmatrix} \tag{31}$$

The equations of motion for the undamped free vibrations of the structure obtained by assembling the member stiffness matrices and allowing also for nodal lumped masses may be written as

$$(K(\omega^2) - \omega^2 M)\underline{d} = \underline{0} \tag{32}$$

The nonlinear eigenproblem stated by equation (32) may be solved for as many frequencies and mode shapes as required by means of the algorithm developed by Wittrick and Williams (1971).

Slight modifications are required in the organization of the computational details with respect to those given in an earlier paper, Williams and Wittrick (1970). The total number of natural frequencies exceeded by a specified frequency, which is the key of the algorithm, is given by

$$J = J_K + J_o$$

where J_K is the number of sign changes between consecutive principal minors of the dynamic matrix $(K(\omega^2)-\omega^2 M)$ and J_o is the total number of natural frequencies of the structure obtained by setting $\underline{d}=\underline{0}$. J_K may be evaluated as suggested by Wittrick and Williams (1971) or even better by a method reported by Martin and Wilkinson (1971) which, allowing for row interchanges in the decomposition of the dynamic matrix, provides a better performance against numerical instabilities and ill-conditioning. The same paper may be consulted for the calculation of the mode shapes.

As far as J_o is concerned it was given as $J_o = \Sigma J_m$ where $J_m = J_a + J_b$ should be calculated for each member of the structure.

If the structure is stable under the static loads, then J_a is the highest integer such that

$$J_a < \frac{\omega L}{\pi}\sqrt{\frac{m}{EA}}$$

and

$$J_b = i - \frac{1}{2}\,(1-(-1)^i sg(d))$$

where d is the denominator of the dynamic stiffness functions in equations (28) and i is the highest integer $<\gamma/\pi$.

It should be noticed that while the expression for J_a is identical, that for J_b contains the one provided by Williams and Wittrick (1970) as a special case.

5. DYNAMIC ANALYSIS

Any dynamic analysis accounting for the instabilizing effects of axial forces should be performed according to the equations (5) and (6) and be consequently nonlinear in nature.

However, assuming that the displacements in the configuration of static equilibrium are small and allowing for only small motions, the nonlinear terms in the above quoted equations may be neglected.

After solving the nonlinear eigenvalue problem (32) for a convenient number of frequencies and mode shapes, the standard mode superposition method for distributed-parameter systems may be used in the calculation of the dynamic response.

The equations of motion in principal coordinates may be written as

$$\ddot{y}_i+2\xi_i\omega_i\dot{y}_i+\omega_i^2 y_i = \frac{L_i}{\hat{M}_i}\ddot{v}_g(t) \qquad i=1,2.... \qquad (33)$$

where ξ_i is a modal damping factor and $\ddot{v}_g(t)$ is the acceleration of the ground motion. The generalized masses $\hat{M}_i$ and the modal earthquake-excitation factors L_i may be written as

$$\begin{aligned} M_i &= \sum_j \int_0^L \phi_i^2(x)m(x)dx + \underline{\phi}_i^T M\underline{\phi}_i \\ L_i &= \sum_j \int_0^L \phi_i(x)r(x)m(x)dx + \underline{\phi}_i^T M\underline{r}_i \end{aligned} \qquad (34)$$

Besides the modal shapes function $\phi_i(x)$ and the member distributed masses m(x),the function r(x) is the static displacement influence function representing the displacements resulting from a unit displacement of the ground $v_g=1$.

The second terms on the right-hand side of equations (34) allow for nodal lumped masses and the vectors $\underline{\phi}_i$ and $\underline{r}_i$ are constructed from the nodal values of $\phi_i(x)$ and r(x). The actual calculation of the integrals in equations (34) may be carried out in closed form for members with uniformly distributed parameters. By using equation (23) and the companion for axial vibrations

$$\psi(x)=D_5\sin b\xi+D_6\cos b\xi \qquad (35)$$

these integrals may be written as

$$\begin{aligned} &\int_0^L m\phi^2(x)dx=mL(\underline{e}^T B\underline{e} + \underline{c}^T A\underline{c}) \\ &\int_0^L m\phi(x)r(x)ds=mL(\alpha_x\underline{e}^T\underline{b} + \alpha_y\underline{c}^T\underline{a}) \end{aligned} \qquad (36)$$

where α_x and α_y are direction cosines depending on the orientation of the member and the vectors and matrices are given as follows for flexural vibrations

$$\underline{e}^T= [D_1,D_2,D_3,D_4]=\underline{v}^T W^{-T}$$

$$\underline{v}^T= [v_i,L\phi_i,v_j,L\phi_j]$$

$$\underline{b}^T= [(1-c)/\gamma,s/\gamma,-(1-C)/\varepsilon,S/\varepsilon]$$

$$W^{-1} = \frac{1}{\varepsilon^2+\gamma^2} \begin{bmatrix} -\delta/\gamma & (\gamma^2-\theta^2)/\gamma & \bar{\delta}/\gamma & -\bar{\theta}/\gamma \\ (\theta+\varepsilon^2) & \alpha & -\bar{\theta} & \beta \\ \delta/\gamma & (\theta+\varepsilon^2)/\varepsilon & -\bar{\delta}/\varepsilon & \bar{\theta}/\varepsilon \\ (\gamma^2-\theta) & -\alpha & \bar{\theta} & -\beta \end{bmatrix}$$

$$B = \begin{bmatrix} \frac{\gamma-sc}{2\gamma} & \frac{s^2}{2\gamma} & \frac{\varepsilon sC-\gamma cS}{\varepsilon^2+\gamma^2} & \frac{\varepsilon sS+\gamma(1-cC)}{\varepsilon^2+\gamma^2} \\ & \frac{\gamma+sc}{2\gamma} & \frac{\gamma sS-\varepsilon(1-cC)}{\varepsilon^2+\gamma^2} & \frac{\gamma sC+\varepsilon cS}{\varepsilon^2+\gamma^2} \\ & \text{Symm} & \frac{SC-\varepsilon}{2\varepsilon} & \frac{S^2}{2\varepsilon} \\ & & & \frac{SC+\varepsilon}{2\varepsilon} \end{bmatrix}$$

$s=\sin\gamma$, $c=\cos\gamma$, $S=\sinh\varepsilon$, $C=\cosh\varepsilon$

and for axial vibrations as follows

$$\underline{c}^T=[D_5,D_6]=\underline{u}^T\bar{W}^{-T};\quad \underline{u}^T=[u_i,u_j];\quad \underline{a}^T=\left[\frac{1-s}{b},\ \frac{s}{b}\right]$$

$$A = \begin{bmatrix} \frac{b-sc}{2b} & \frac{s^2}{2b} \\ \frac{s^2}{2b} & \frac{b+sc}{2b} \end{bmatrix};\qquad \bar{W}^{-1} = \begin{bmatrix} -\frac{c}{s} & \frac{1}{s} \\ 1 & 0 \end{bmatrix}$$

$s=\sin b$; $c=\cos b$.

The generalized masses $\hat{M}_i$ and modal earthquake-excitation factors L_i may be used for both response-spectrum analyses and time-history analyses.

6. NUMERICAL APPLICATIONS

Several numerical applications have been performed in order to investigate the effects of gravity on the frequencies of vibration and related mode shapes. The change in the frequencies has always been the most significant while the mode shapes have changed considerably only for particular distributions of static loads.

A noticeable change in the mode shapes has been observed for tall frames. A 20 storey,one bay frame

GRAVITY FORCES NEGLECTED

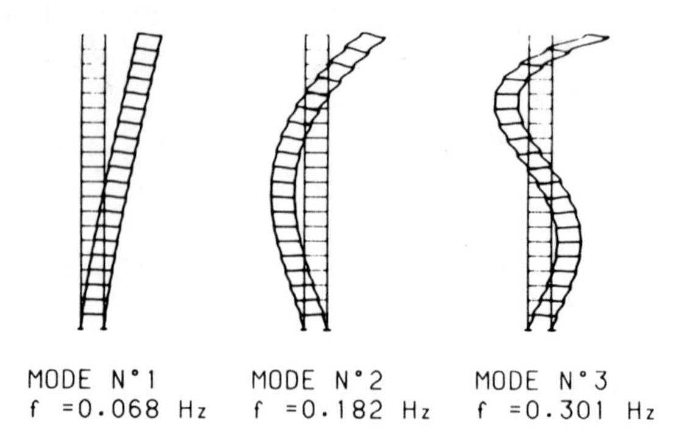

GRAVITY FORCES INCLUDED

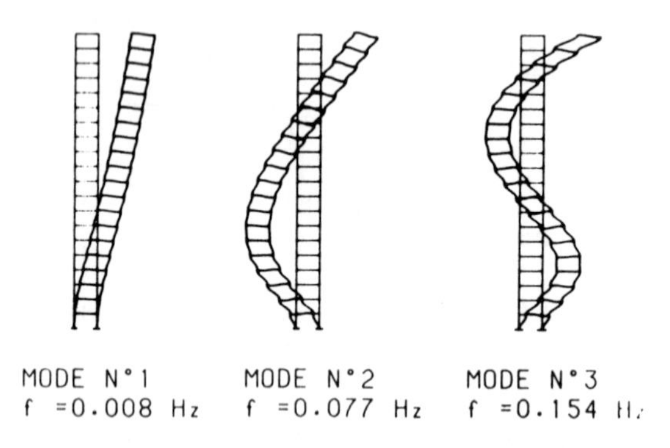

Figure 2. Modes of vibration.

has been designed in a way so that the ratio between the Euler load and the axial forces in the columns is 3.9. The first three modes of vibration and the corresponding frequencies have been reported in Fig.2 for the two cases of gravity forces included and neglected for comparison. An artificially generated ground motion, Vanmarke (1976), has then been used to simulate an earthquake excitation and the dynamic response of the framework has been calculated

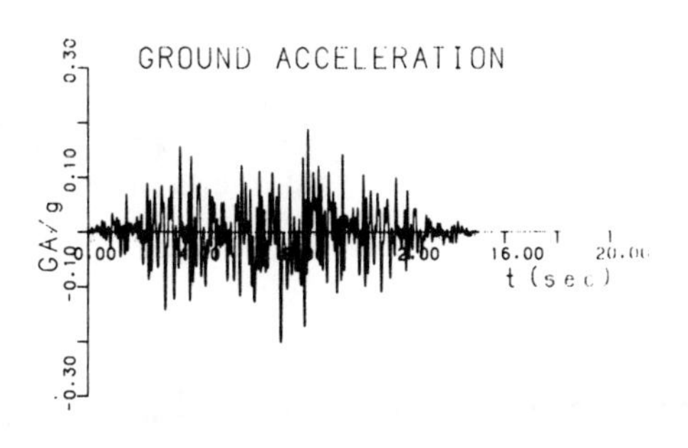

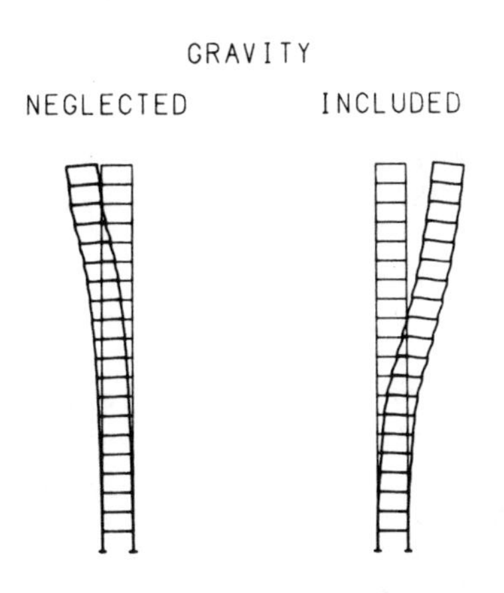

Figure 3. Dynamic response at the time of the maximum displacement of the top storey calculated from the contribution of the first 12 modes.Damping is neglected.

from the superposition of the first 12 modes.
The response at the times of the maximum displacement of the top storey is reported in Fig.3 together with the ratio of the ground acceleration GA to the gravity acceleration g. It may be noticed that by neglecting the effect of the gravity forces this displacement would be underestimated by a factor greater than 2 .

ACKNOWLEDGEMENTS

The author would like to thank his student M.Granata for his substantial help in the computations and the National Research Council of Italy (CNR) for financial support.

REFERENCES

Blaszkowiak,S. and Kaczkowski,Z. (1966) Iterative Methods in Structural Analysis.Pergamon Press,Oxford.

Clough,R.W. and Penzien,J. (1975) Dynamics of Structures. Mc Graw-Hill,Inc. New York.

Johnson,E.R.,Plaut,R.H. and Deel II,C.C. (1980) Load-Frequency Relationships for Shallow Elastic Structures. Recent Advances in Structural Dynamics,Institute of Sound & Vibration Research,University of Southampton.

Lightfoot,E.,Mc Pharlin,R.D.,Le Messurier,A.P.(1979) Framework Instability Analysis Using Blaszkowiak's Stiffness Functions.Int.J.Mech.Sci.Vol.21,pp.547-555.

Oliveto,G.(1981) Second Order Analysis of Elastic Frames. Meccanica,Vol.16,No.2 June.

Oliveto,G.(1982) The Response of Space Frames to Earthquake Excitations Including Gravity Forces. VI Congresso Nazionale AIMETA. Genova.

Martin,R.S. and Wilkinson,J.H. (1971) Solution of Symmetric and Unsymmetric Band Equations and the Calculations of Eigenvectors of Band Matrices. Linear Algebra,Wilkinson and Reinsch Eds,Springer-Verlag.

Vanmarke,E.H. (1976) Structural Response to earthquakes. In Seismic Risk and Engineering Decisions, C.Lomnitz and E.Rosenblueth (Editors),Elsevier.

Warburton,G.(1976) The Dynamical Behaviour of Structures. Pergamon Press,Oxford.

Williams,F.W. and Wittrick,W.H.(1970) An Automatic Computational Procedure for Calculating Natural Frequencies of Skeletal Structures. Int.J.Mech.Sci. Pergamon Press,Vol.12,pp.781-791.

Wittrick,W.H. and Williams,F.W. (1971) A General Algorithm for Computing Natural Frequencies of Elastic Structures. Quart.Journ.Mech. and Applied Math.,Vol.XXIV,Pt.3. Clarendon Press,Oxford.

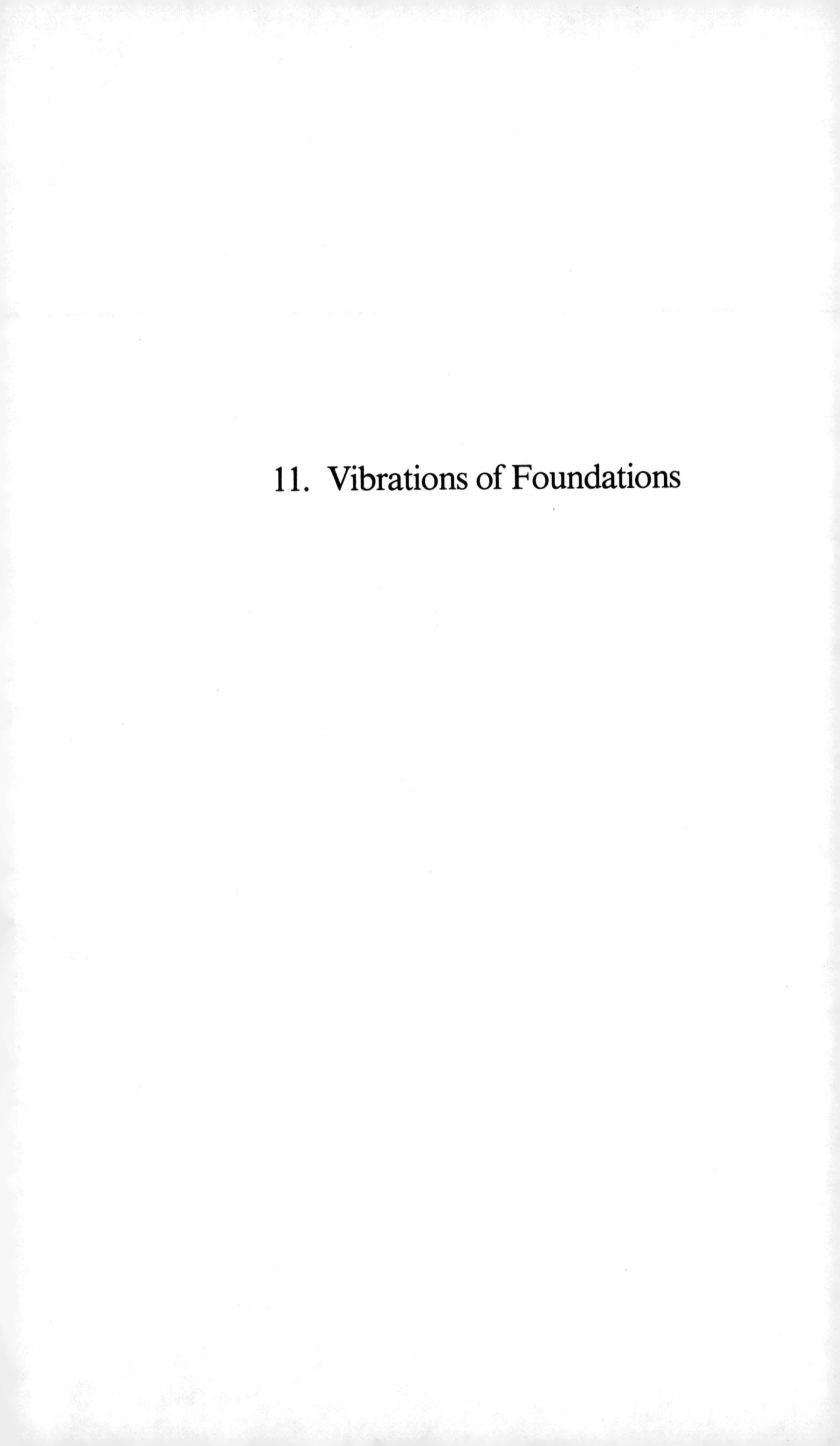

11. Vibrations of Foundations

Soil Dynamics & Earthquake Engineering Conference / Southampton / 1982.07.13-15

Response of Hammer Foundations

M.NOVAK
University of Western Ontario, Canada

INTRODUCTION

Many types of machines produce transient dynamic forces that are quite short in duration and can be characterized as pulses or shocks. Typical machines producing this type of load are forging hammers, presses, crushers and mills. The forces generated by the operation of these machines are often very powerful and can result in many undesirable effects such as large settlement of the foundation, cracking of the foundation and heavy vibration. To recognize these hazards and secure optimum operation of the facility vibration analysis is conducted at the design stage.

The methods of analysis of the foundations of shock producing machines are well developed but suffer from inconsistent description of foundation properties and particularly from the omission or arbitrary definition of damping. This paper attempts to alleviate these difficulties with the additional objective of keeping the analysis as simple as possible. Hammers are most typical of the shock-producing machines and therefore this presentation is limited to them.

TYPES OF HAMMER FOUNDATIONS

There are many types of hammers. According to their function, they can be divided into forging hammers and hammers for die stamping. The basic elements of the hammer-foundation system are the frame, head (tup), anvil and foundation block (Fig. 1a). The frame of forging hammers is separated from the anvil. In die stamping hammers, the frame is usually connected to the anvil.

The forging action of hammers is generated by the impact of the falling head against the anvil. The head is allowed to fall freely or its velocity is enhanced using steam or compressed air. To reduce the stress in the concrete and shock transmission into the frame, viscoelastic mounting of the anvil is usually provided (Fig. 1). This may have the form of a pad of hard industrial felt, a layer of hardwood or, with very powerful hammers, a set of special absorbers such as coil springs and dampers.

The foundation block is most often cast directly on soil as indicated in Fig. 1a. When the bearing capacity of soil is not sufficient or undesirable settlement is anticipated, the block may be installed on piles. When the transmission of vibration and shock forces in the vicinity and adjoining facilities is of concern, a softer mounting for the foundation may be desirable. This can be achieved by supporting the block on a pad of viscoelastic material such as cork or rubber (Fig. 1b) or on vibration isolating elements. A trough is needed to protect these elements from the environment. For reasons of easy access and convenience, the springs are sometimes positioned higher up and the footing is suspended on hangers or cantilevers (Fig. 1c).

STIFFNESS AND DAMPING CONSTANTS OF THE SYSTEM

The prediction of the response of the hammer foundation requires the description of the stiffness and damping of the foundation and the pad under the anvil.

Foundation on soil

Stiffness and damping of foundations supported on soil can be evaluated using elastic half-space theory. The principal advantages of this theory are that it accounts for energy dissipation through elastic waves (geometric damping), provides for systematic analysis and describes soil properties by basic constants such as shear modulus or shear wave velocity which can be established by independent experiments. However, some assumptions of the theory differ from real conditions and therefore, some adjustment of the theoretical results is necessary in order to account for the shape of the base, embedment, soil nonhomogeneity and limited thickness of the soil stratum.

The correction for the shape of the base can be made by the introduction of the equivalent radius of a circular base. For a rectangular base and vertical vibration, the equivalent radius is $r_o = \sqrt{ab/\pi}$ in which a and b are the width and length of the base

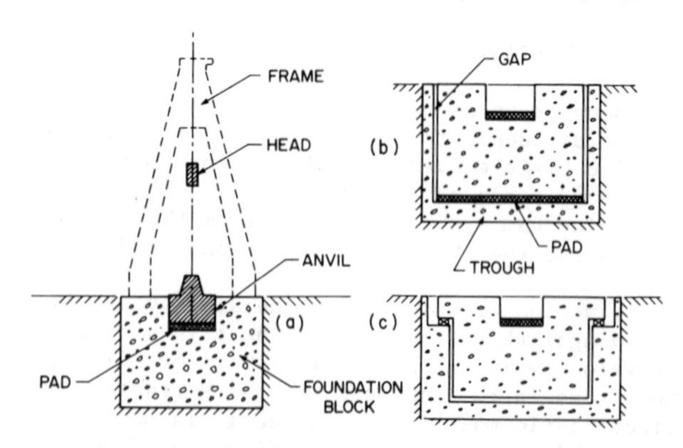

Figure 1 Schematic of hammer and foundations

respectively.

The vertical stiffness, k, and damping, c, of a foundation are defined as forces associated with a unit amplitude and unit vibration velocity, respectively. A number of approaches are available for the determination of foundation stiffness and damping. They were recently reviewed and compared by Roesset (1980). For an embedded foundation, stiffness and damping constants can be evaluated approximately as (Novak and Beredugo, 1972; Novak, 1974b)

$$k = Gr_o \left(C_1 + \frac{G_s}{G} \frac{\ell}{r_o} S_1\right) \tag{1a}$$

$$c = r_o^2 \sqrt{\rho G} \left(\bar{C}_2 + \bar{S}_2 \frac{\ell}{r_o} \sqrt{\frac{\rho_s}{\rho} \frac{G_s}{G}}\right) \tag{1b}$$

Here, C_1, $\bar{C}_2$ are dimensionless parameters related to the stiffness and damping, respectively, derived from the medium under the base (the elastic half-space) or a stratum; S_1, $\bar{S}_2$ are constants related respectively to the stiffness and damping derived from the reactions of the layer lying above the level of the base (Fig. 2) and acting on the vertical sides of the footing; G = soil shear modulus and ρ = mass density of the half-space while G_s, ρ_s are the shear modulus and mass density of the side layer (backfill), respectively. Finally, ℓ = effective embedment depth.

The parameters C_1 and $\bar{C}_2$ depend primarily on the depth of the stratum, h, Poisson's ratio, ν and the dimensionless frequency $a_o = r_o\omega/V_s$ in which ω = circular frequency and $V_s = \sqrt{G/\rho}$ = shear wave velocity. The parameters S depend on the dimensionless frequency. The parameters having the subscript 1 refer to stiffness while those with subscript 2 indicate damping. In the absence of material damping, the parameters $\bar{C}_2$, $\bar{S}_2$ correspond to equivalent viscous damping and stem from radiation (geometric) damping. The parameters S and C are given in Novak and Beredugo (1972). For layered media, the parameters C can be obtained using the approach due to Luco (1974) and others. The analysis can be simplified if the frequency dependent parameters are replaced by suitably chosen frequency independent constants.

Equations 1 give results that agree quite well with the finite element solution (Roesset, 1980) but some adjustment is desirable to bring the numerical results of the formulae closer to experimental observations. First, experiments indicate (Novak, 1970) that for deep deposits, the half-space theory tends to overestimate the geometric damping of surface foundations in the vertical direction by about 100 per cent. The reason for this discrepancy seems to be that soil usually features some layering which reflects elastic waves back to the foundation and reduces geometric damping. To be on the safe side, it appears advisable to divide the theoretical values of $\bar{C}_2$ by a factor of about two.

The second correction involves embedment effects. The theory indicates that embedment provides a significant source of geometric damping and contributes also to stiffness. These theoretical suggestions were, in general, confirmed by experiments. However, it was also observed that with the heavy vibration typical of hammers, the soil may separate from the footing sides and a gap may occur as indicated in Fig. 2. This gap is likely to develop close to the surface where the confining pressure is not sufficient to maintain the bond between the soil and the foundation. The separation may be accounted for by considering an effective embedment depth, ℓ, smaller than the actual embedment depth, L. Another way of accounting for footing separation is to assume a slippage zone around the footing (Novak and Sheta, 1980).

Soil material damping Foundation stiffness and damping are also affected by soil material damping. The material damping of soil is hysteretic and independent of frequency. It is conveniently described using the complex shear modulus

$$G^* = G + iG' = G(1 + i\tan\delta) \tag{2}$$

in which $i = \sqrt{-1}$, $\tan\delta = G'/G$ with δ = the loss angle and G' = the imaginary part of the complex soil modulus, G^*. Another measure of material damping is the damping ratio $\beta = 1/2 \tan\delta$. Material damping can be incorporated using the correspondence principle of viscoelasticity. In the sense of this principle, the shear modulus, G, in Equations 1 has to be replaced by the complex shear modulus defined by Equation 2. If only the explicitly appearing G is replaced by G^*, the stiffness and damping constants including material damping become approximately

$$k_h = k - \tan\delta c\omega \tag{3}$$

$$c_h = c + \tan\delta k/\omega \tag{4}$$

in which k and c are evaluated without regard to material damping. With shallow layers, the incorporation of material damping is important because the geometric damping is quite small or may not materialize at all if the first natural frequency of the hammer foundation is lower than the first natural frequency of the soil layer.

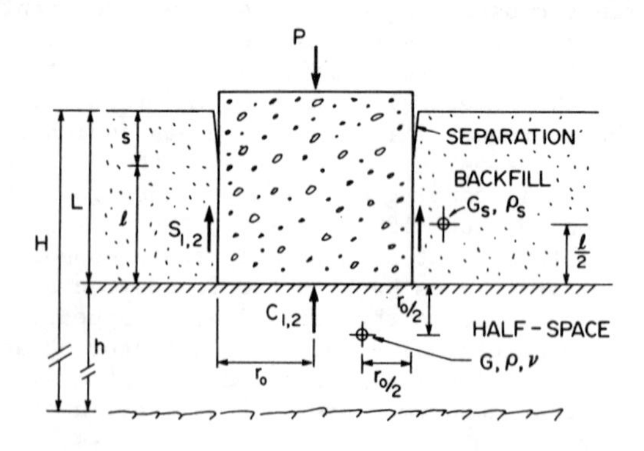

Figure 2 Embedded foundation and notation

Pile foundations

Vertical stiffness and damping of foundations supported by piles are readily available for endbearing piles in homogeneous soil (Novak, 1974), floating piles in homogeneous soil (Novak, 1977) and floating piles in soil with a parabolic variation of shear modulus with depth (Novak, Aboul-Ella, 1978b). For an arbitrary soil profile they can be calculated using the technique described by Novak and Aboul-Ella (1978a,b). A correction for pile separation can be made as described by Novak and Sheta (1980). When the piles are closely spaced, an allowance should be made for pile-soil-pile interaction effects (group effects). This factor is treated by Poulos and Davis (1980), Wolf and von Arx (1978), Waas and Hartmann (1981), Sheta and Novak (1982) and others.

Pads and absorbers

When the foundation block or the anvil rest on a pad of viscoelastic material, the vertical stiffness constant of the pad is

$$k_p = E_p A_p / d \qquad (5)$$

in which E_p = Young's modulus of the pad, A_p = area of the pad and d = its thickness. The damping constant can be calculated in terms of the complex Young's modulus and is

$$c_p = \tan\delta_p \; k_p / \omega_o \qquad (6)$$

where δ_p = the loss angle of the pad material and ω_o = the natural frequency of the block or anvil calculated with k_p.

IMPACT FORCES

The energy of the impact is determined by the weight of the head and its impact velocity. However, the foundation response to the impact also depends on the time history of the force resulting from the impact and transferred to the anvil in the form of a pulse. This pulse is a transient force, P(t), of short duration, T_p. The time history of the pulse and its duration depend on the conditions of forging and are to a high degree random; little is known about them. However, because the duration of the pulse is very short, in the order of 0.01 or 0.02 s, it may be possible for the preliminary considerations to replace the real pulse by a rectangular pulse having duration $t_p < T_p$ but the same power, i.e.

$$P_o t_p = \int_o^{T_p} P(t)\,dt \qquad (7)$$

For a rectangular pulse, the response of a one mass system can readily be obtained from the Duhamel integral and with damping neglected is

$$v(t) = \frac{P_o}{k}(1 - \cos\omega_o t) \qquad \text{for } t \leq t_p \qquad (8a)$$

and

$$v(t) = \frac{P_o}{k}\,[\cos\omega_o(t-t_p)-\cos\omega_o t] \qquad \text{for } t>t_p \tag{8b}$$

Examples of this response are plotted in Fig. 3. The response histories shown were calculated for rectangular pulses having different durations, t_p, but the same power $P_o t_p$ taken as unity (a unit pulse). The duration of the pulse is expressed as a fraction of the natural period of the system $T=2\pi/\omega_o$ also taken as unity. It can be seen that the peak response decreases as the unit pulse duration increases and that for the ratios t_p/T lower than about 0.1, the peak response is practically independent of pulse duration and equal to that obtained with an infinitely short pulse. Thus, it appears possible and conservative to predict the response using the assumption of an infinitely short pulse.

ANALYSIS OF HAMMER FOUNDATIONS

For the analysis, hammer foundations are represented by various lumped mass models whose number of degrees of freedom (independent displacements) depends on the foundation type and the impact eccentricity. In most foundations, the anvil rests on an elastic pad and the impact is centric. Then, a two-mass model shown in Fig. 4a is adequate. This model has two degrees of freedom and is most often used in hammer foundation design. Therefore, further discussion is based on this model.

Undamped vibration

Consider a two mass model (Fig. 4a) in which m_1 is the mass of the anvil, m_2 the mass of the footing and k_1 is the stiffness of the pad under the anvil (Equation 5); k_2 is the stiffness of

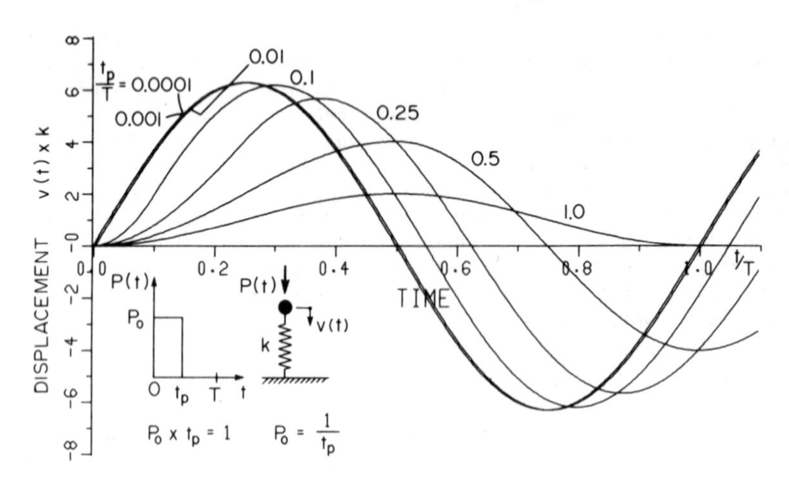

Figure 3 Response of one mass system to rectangular unit pulse of varying duration

the soil (Equation 1a), piles or any other support of the footing block. With the notation

$$k_{11} = k_1 \ , \ k_{22} = k_1+k_2 \ , \ k_{12} = k_{21} = -k_1 \tag{9}$$

the governing equations of the motion of the anvil, $v_1(t)$ and the foundation, $v_2(t)$, are

$$m_1 \frac{d^2 v_1(t)}{dt^2} + k_{11} v_1(t) + k_{12} v_2(t) = 0$$

$$m_2 \frac{d^2 v_2(t)}{dt^2} + k_{21} v_1(t) + k_{22} v_2(t) = 0 \tag{10}$$

The complete solution for the unknown displacements is well known and can be written as a sum of two independent particular solutions (modes), i.e.

$$\begin{Bmatrix} v_1(t) \\ v_2(t) \end{Bmatrix} = \begin{Bmatrix} v_{11} \\ v_{21} \end{Bmatrix} \sin\omega_1 t + \begin{Bmatrix} v_{12} \\ v_{22} \end{Bmatrix} \sin\omega_2 t \tag{11}$$

in which the two natural frequencies are

$$\omega_{1,2}^2 = \frac{1}{2} \left(\frac{k_{11}}{m_1} + \frac{k_{22}}{m_2}\right) \mp \sqrt{\frac{1}{4}\left(\frac{k_{11}}{m_1} - \frac{k_{22}}{m_2}\right)^2 + \frac{k_{12}^2}{m_1 m_2}} \tag{12}$$

The two corresponding ratios of displacements representing the undamped vibration modes are

$$a_j = \frac{v_{1j}}{v_{2j}} = \frac{-k_{12}}{k_{11}-m_1\omega_j^2} = \frac{k_{22}-m_2\omega_j^2}{-k_{21}} \quad \text{for } j=1 \text{ or } 2 \tag{13}$$

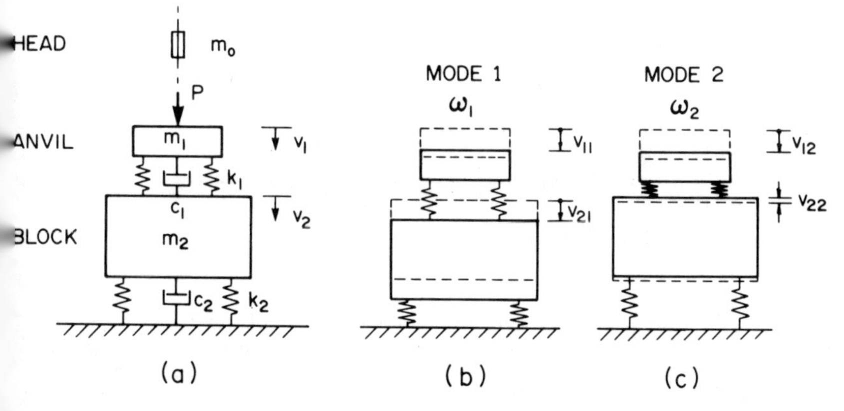

Figure 4 Hammer foundation as a system of two masses and its vibration modes

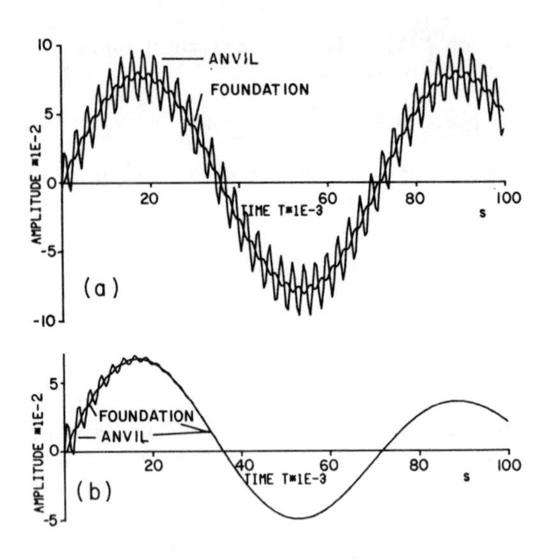

Figure 5 Response of two mass hammer foundation with stiff anvil pad: (a) - undamped, (b) - damped (D_1 = 10%, D_2 = 5%); (amplitude in in., 1 in. = 2.54 cm)

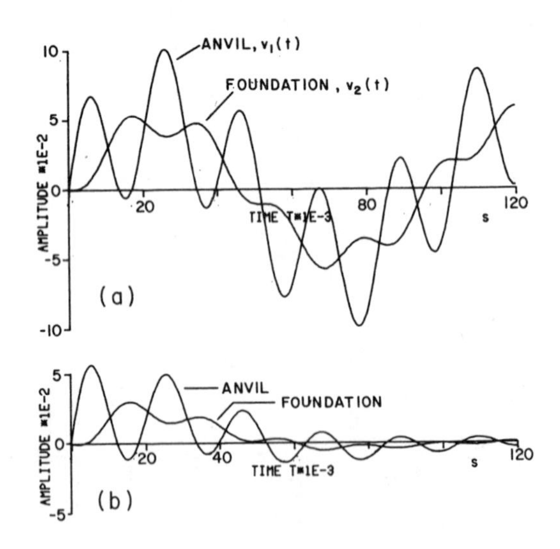

Figure 6 Response of two mass hammer foundation with elastic anvil pad: (a) - undamped, (b) - damped (D_1 = 51%, D_2 = 7.4%); (amplitude in in., 1 in. = 2.54 cm)

In the displacement v_{ij}, the first subscript identifies the amplitudes of mass m_1 or m_2 while the second subscript indicates the frequency and mode with which the amplitude v_{ij} is associated. The two vibration modes are shown in Fig. 4.

The magnitude of the amplitudes v_{ij} is determined by initial conditions. If both masses are at rest at time t=0 of the collision between the hammer head and the anvil, the initial conditions are

$$v_1(0)=0 \ , \ v_2(0)=0 \ , \ \dot{v}_1(0)=\hat{c} \ , \ \dot{v}_2(0)=0 \tag{14}$$

Here $\dot{v}_1 = dv_i(t)/dt$ and the initial anvil velocity $\hat{c}$ follows from the basic formula for collision,

$$\hat{c} = (1 + k_r) \frac{m_o}{m_o+m_1} c_o \tag{15}$$

where m_o, c_o are the mass and impact velocity of the hammer head respectively and k_r is the coefficient of restitution. The initial conditions, Equations 14, and Equation 13 written for two different frequencies ω_1 and ω_2 yield the amplitudes v_{ij}. Denoting the natural frequency of the anvil in case of a rigidly supported foundation block

$$\omega_a = \sqrt{k_1/m_1} \tag{16}$$

the vibration amplitudes appearing in Equation 11 are

$$v_{11} = \frac{\hat{c}}{\omega_1} \frac{\omega_2^2-\omega_a^2}{\omega_2^2-\omega_1^2} \tag{17a}$$

$$v_{12} = \frac{\hat{c}}{\omega_2} \frac{\omega_a^2-\omega_1^2}{\omega_2^2-\omega_1^2} \tag{17b}$$

for the anvil and

$$v_{21} = \frac{\hat{c}}{\omega_1\omega_a^2} \frac{(\omega_a^2-\omega_1^2)(\omega_2^2-\omega_a^2)}{\omega_2^2 - \omega_1^2} = v_{11}(1 - \frac{\omega_1^2}{\omega_a^2}) \tag{17c}$$

$$v_{22} = - \frac{\hat{c}}{\omega_2\omega_a^2} \frac{(\omega_a^2-\omega_1^2)(\omega_2^2-\omega_a^2)}{\omega_2^2 - \omega_1^2} = -v_{21} \frac{\omega_1}{\omega_2} \tag{17d}$$

for the foundation. With these amplitudes, the undamped response of both the anvil and the foundation can be calculated from Equation 11. Examples of the undamped response are shown in Figs. 5a and 6a. Figure 5 shows the response of a foundation having a rather stiff anvil pad while Figure 6 corresponds to a foundation with a pad of medium elasticity. The contribution of the second vibration mode depends on the elasticity of the pad but the magnitude of the amplitudes and the overall character of the response are greatly affected by damping.

Damped vibration

In the two mass model, damping is defined by the constants c_1 and c_2 (Fig. 4a) which can be calculated using Equations 1b and 6. These constants could be introduced into the governing equations of the motion but it is more convenient to predict the damped response approximately using the notions of modal analysis and modal damping.

The foundation response comprises the two vibration modes shown in Fig. 4. The damping ratio associated with vibration in each of these modes can be evaluated by means of an energy consideration if it is assumed that the damped mode is approximately the same as the undamped mode.

The work done during a period of vibration $T = 2\pi/\omega_j$ by the damping force $P(\dot{\delta})$ is, in general,

$$W = \int_0^T P(\dot{\delta})d\delta(t) \tag{18}$$

in which δ = the relative displacement between the bodies to which the dashpot is attached. For the hammer foundation shown in Fig. 4, vibrating in mode j with natural frequency ω_j,

$$\delta_{1j}(t) = \delta_{1j}\sin\omega_j t \ , \ \delta_{2j}(t) = \delta_{2j}\sin\omega_j t \tag{19}$$

in which the relative amplitudes are

$$\delta_{1j} = v_{1j}-v_{2j} \ , \ \delta_{2j} = v_{2j} \tag{20}$$

The damping forces generated by the dashpots are

$$\begin{aligned} P(\dot{\delta}_1) &= c_1\dot{\delta}_{1j} = c_1\delta_{1j}\omega_j\cos\omega_j t \\ P(\dot{\delta}_2) &= c_2\dot{\delta}_{2j} = c_2\delta_{2j}\omega_j \cos\omega_j t \end{aligned} \tag{21}$$

The total work done by these forces follows from Equation 18 as

$$W = \pi\ \omega_j[c_1(v_{1j}-v_{2j})^2 + c_2v_{2j}^2] \tag{22}$$

In free vibration, the maximum potential energy of the whole system is the same as the maximum kinetic energy and is

$$L = \frac{1}{2}\ (m_1\ v_{1j}^2 + m_2\ v_{2j}^2)\omega_j^2 = \frac{1}{2}\ M_j\omega_j^2 \tag{23}$$

where the generalized mass of mode j

$$M_j = m_1v_{1j}^2 + m_2v_{2j}^2 \tag{24}$$

Then, the damping ratio is defined as $D_j = W/4\pi L$. This yields the damping ratio associated with the foundation vibration in the jth mode,

$$D_j = \frac{1}{2\omega_jM_j}\ [c_1(v_{1j}-v_{2j})^2 + c_2v_{2j}^2] \tag{25}$$

where $j = 1,2$. In Equation 25, c_1 is the damping constant of the anvil and c_2 is the damping constant of the foundation. The frequencies appearing in these equations are the natural frequencies ω_1 and ω_2 calculated from Equation 12. Thus, for each natural mode, one set of c_1, c_2 may be necessary due to converting a constant hysteretic damping to equivalent viscous damping. The amplitudes v_{1j} and v_{2j} are the undamped amplitudes given by Equation 17. Alternatively, arbitrary modal amplitudes complying with Equation 13 may be used in Equation 25; in this case, one amplitude can be chosen for each mode, e.g. $v_{1j}=1$, and the other calculated using the ratio a_j.

If the frequency $\omega_2 >> \omega_1$, which can be the case with a stiff anvil pad, then also $v_{11} \cong v_{21}$, $v_{22} << v_{12}$ and

$$\omega_1 \cong \sqrt{k_2/(m_1+m_2)} \ , \ \omega_2 \cong \sqrt{k_1/m_1}$$

Consequently, Equation 25 simplifies to approximate expressions for modal damping ratios,

$$D_1 = \frac{c_2}{2\sqrt{k_2(m_1+m_2)}} \ , \ D_2 = \frac{c_1}{2\sqrt{k_1 m_1}} = \frac{1}{2} \tan\delta_p \tag{26}$$

Because Equations 11 represent the superposition of vibration modes, the damped vibration of the anvil and the foundation block can be written as

$$\begin{Bmatrix} v_1(t) \\ v_2(t) \end{Bmatrix} = \begin{Bmatrix} v_{11} \\ v_{21} \end{Bmatrix} e^{-D_1\omega_1 t} \sin\omega_1' t + \begin{Bmatrix} v_{12} \\ v_{22} \end{Bmatrix} e^{-D_2\omega_2 t} \sin\omega_2' t \tag{27}$$

in which the damped natural frequencies

$$\omega_j' = \omega_j \sqrt{1-D_j^2} \tag{28}$$

and v_{1j}, v_{2j} are the undamped amplitudes established from Equations 17; the damping ratios D_j are given by Equations 25 or 26. In Figs. 5 and 6, examples of the damped response are plotted to complement the undamped response. The only difference is the inclusion of damping. Even the modest damping incorporated smooths the response, quickly eliminating the second harmonic component in the case of a stiff anvil pad (Fig. 5b). With an elastic anvil pad (Fig. 6b), the contribution of the second harmonic component remains significant. In both cases, damping can reduce the amplitudes quite markedly, particularly for large, embedded foundations. Such a foundation was considered when calculating the response shown in Fig. 6.

The peak displacements can be obtained from the time histories of the motion, Equation 27, or evaluated approximately. Because ω_2 is always much larger than ω_1, the peak displacements occur approximately at time

$$t_m = 1/\omega_1' \arctan\left(\sqrt{1-D_1^2}/D_1\right)$$

and are

$$\begin{Bmatrix} \hat{v}_1 \\ \hat{v}_2 \end{Bmatrix} = \begin{Bmatrix} v_{11} \\ v_{21} \end{Bmatrix} e^{-D_1\omega_1 t_m} \sin\omega_1' t_m + \begin{Bmatrix} v_{12} \\ |v_{22}| \end{Bmatrix} e^{-D_2\omega_2 t_m} \tag{29}$$

The peak relative displacement of the anvil, determining the stress in the anvil pad, is approximately

$$\hat{v}_a = (v_{12} + |v_{22}|)e^{-\frac{\pi}{2}D_2} \tag{30}$$

The total peak force transmitted into the ground (or piles) comprises the restoring force and the damping force and, with respect to the 90° phase shift between them is approximately

$$F = \sum_j \hat{v}_{2j} \sqrt{k_2^2+(c_2\omega_j')^2} \tag{31}$$

in which $\hat{v}_{2j}$ are the modal contributions to $\hat{v}_2$ obtained from Equation 29; thus, $\hat{v}_2 = \hat{v}_{21} + \hat{v}_{22}$.

The force acting on the pad is

$$F_a = \hat{v}_a \sqrt{k_1^2 + (c_1\omega_2')^2} \tag{32}$$

More complicated systems The above approach can readily be extended to include more complicated systems. When the foundation block is supported by absorbers and protected by a trough, a three mass model shown in Fig. 7 can be used to analyze the response.

First, the eigenvalue problem of the system is solved to yield three vibration modes with amplitudes v_{ij} and frequencies ω_j. The modal damping is produced by the three dashpots. The modal damping associated with the three vibration modes is obtained by extension of the energy argument and Equation 25, becoming

$$D_j = \frac{1}{2\omega_j M_j}\left[c_1(v_{1j}-v_{2j})^2+c_2(v_{2j}-v_{3j})^2+c_3 v_{3j}^2\right] \tag{33}$$

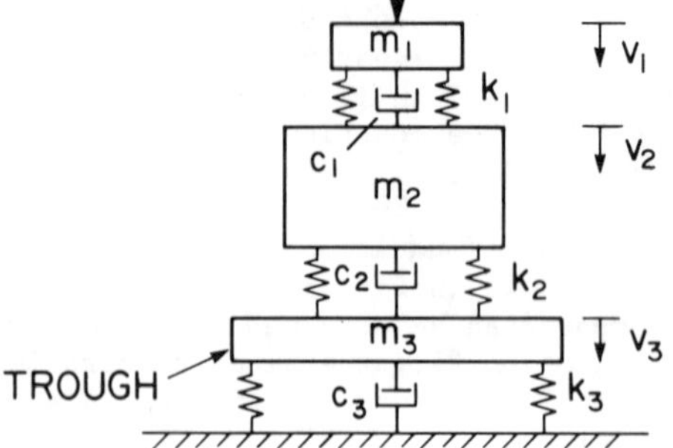

Figure 7 Three mass model of hammer foundation with vibration absorbers and trough

in which

$$M_j = \sum_{i=1}^{3} m_i v_{ij}^2 \qquad (34)$$

and j = 1, 2 and 3. The damped response of mass m_i can be written as

$$v_i(t) = \sum_{j=1}^{3} v_{ij} e^{-D_j \omega_j t} \sin\omega_j' t \qquad (35)$$

in which v_{ij} are the undamped amplitudes obtained from the initial conditions and the modal ratios. (The undamped amplitudes can also be used in Equation 33 in lieu of modal displacements.)

Complex eigenvalue approach The damping ratios, D_j, obtained from the energy consideration can be verified by introducing the damping constants, c_i, into the governing equations of the motion, Equations 10, and solving the complex eigenvalue problem. The governing equation of damped motion is

$$\lceil m \rfloor \{\ddot{v}\} + [c]\{\dot{v}\} + [k]\{v\} = 0 \qquad (36)$$

in which the damping matrix, [c], is symmetrical and has elements $c_{11} = c_1$, $c_{22} = c_1 + c_2$, $c_{12} = c_{21} = -c_1$. All the other matrices are determined by Equations 10. The complex natural frequencies obtained from Equation 36 are

$$\omega_j^* = -D_j\omega_j \pm i\ \omega_j\sqrt{1-D_j^2} = -D_j\omega_j \pm i\ \omega_j' \qquad (37)$$

in which $i = \sqrt{-1}$ and the damping ratio

$$D_j = -\mathrm{Re}(\omega_j^*)/\omega_j \qquad (38)$$

The damping ratios, calculated in this way for the two mass foundation pertinent to Fig. 6, are plotted for different values of foundation damping, c_2, in Fig. 8. This damping is described by the ratio $A=c_2/c_2(max)$ in which $c_2(max)$ is the full amount of damping calculated from Equation 1b and leading to the damping ratios D_1 = 7.4% and D_2 = 51%. The damping ratios obtained from Equation 25 based on energy consideration are practically identical with those calculated from Equation 38 and shown in Fig. 8.

Shallow vs. deep foundations The progress in the evaluation of stiffness and damping of both shallow and pile foundations makes it possible to compare the anticipated performance of such foundations. For two simple foundations differing only by the type of their support, such a comparison is shown in Fig. 9. The shallow foundation features larger amplitudes and larger damping; the pile foundation exhibits higher frequencies, smaller amplitudes but larger forces transmitted into the piles.

Other approaches The results presented indicate that the approximate approach based on initial velocity and modal damping from an energy consideration is adequate for practical purposes. More complicated solutions, such as those using a presumed time history of the pulse or the complex eigenvalue approach, do not appear necessary because there are other physical limitations to the accuracy of the theory. The uplift of the anvil, whose incorporation would require a nonlinear solution, is one of them.

CONCLUSIONS

An approach is presented that makes it possible to predict damped response of hammer foundations with any number of degrees of freedom. The inclusion of damping makes the analysis more realistic.

Modal damping evaluated on the basis of an energy consideration agrees very well with that obtained from the complex eigenvalues.

ACKNOWLEDGEMENTS

The research into foundation dynamics has been supported by a grant in aid of research from the Natural Sciences and Engineering Research Council of Canada. The assistance of L. El Hifnawy is gratefully acknowledged.

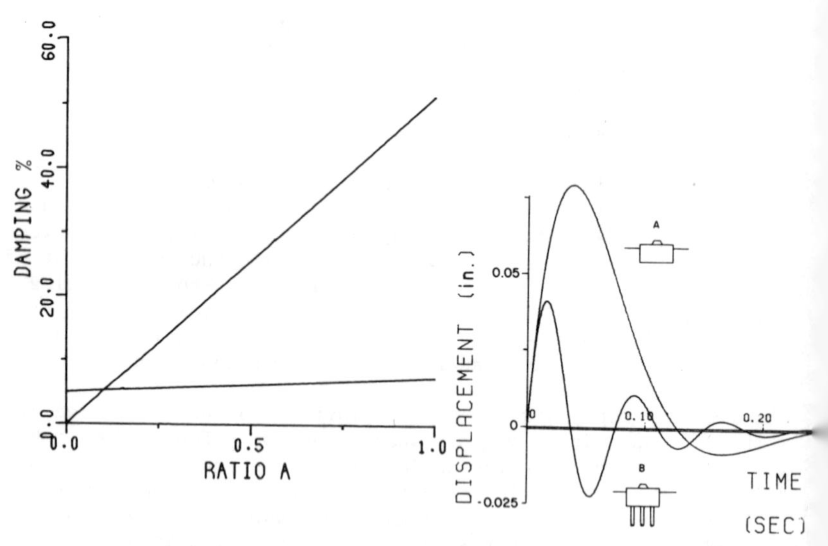

Figure 8 Damping ratios of two mass foundation vs. damping intensity $A=c_2/c_2(max)$

Figure 9 Response for two hammer foundations A - shallow foundation; B - pile foundation

REFERENCES

Luco, J.E. (1974) Impedance Functions for a Rigid Foundation On a Layered Medium. Nuclear Engineering and Design 31, 204-217.

Novak, M. (1970) Prediction of Footing Vibrations. J. of Soil Mechanics and Foundations Division, ASCE, Vol. 96, SM3, 837-861.

Novak, M. (1974a) Dynamic Stiffness and Damping of Piles. Canadian Geotechnical Journal, Vol. II, 574-598.

Novak, M. (1974b) Effect of Soil on Structural Response to Wind and Earthquake. J. of Earthquake Engineering and Structural Dynamics, Vol. 3, No. 1, 79-96.

Novak, M. (1977) Vertical Vibration of Floating Piles. J. of Engineering Mechanics Division, ASCE, Vol. 103, EM1, 153-168.

Novak, M. and Beredugo, Y.O. (1972) Vertical Vibration of Embedded Footings. J. of Soil Mechanics and Foundations Division, ASCE, SM12, 1291-1310.

Novak, M. and Aboul-Ella, F. (1978a) Impedance Functions of Piles in Layered Media. J. of Engineering Mechanics Division, ASCE, Vol. 104, EM3, 643-661.

Novak, M. and Aboul-Ella, F. (1978b) Stiffness and Damping of Piles in Layered Media. Proc. of the Earthquake Engineering and Soil Dynamics, ASCE Specialty Conference, Pasadena, California, 704-719.

Novak, M. and Sheta, M. (1980) Approximate Approach to Contact Effects of Piles. Proc. of Geotechnical Engineering Division ASCE National Convention "Dynamic Response of Pile Foundations: Analytical Aspects", Florida, 53-79.

Poulos, H.G. and Davis, E.H. (1980) Pile Foundation Analysis and Design. John Wiley and Sons, 397.

Roesset, J.M. (1980) Stiffness and Damping Coefficients of Foundations. Dynamic Response of Pile Foundations: Analytical Aspects, Proc. of a Specialty Session, ASCE National Convention, Florida, 1-30.

Sheta, M. and Novak, M. (1982) Vertical Vibration of Pile Groups. J. of the Geotechnical Engineering Division, ASCE, April (to appear).

Waas, G. and Hartmann, H.G. (1981) Analysis of Pile Foundations Under Dynamic Loads. SMIRT, Paris, 10.

Wolf, J.P. and von Arx, G.A. (1978) Impedance Functions of a Group of Vertical Piles. Proc. ASCE Specialty Conference on Earthquake Engineering and Soil Dynamics, Pasadena, 1024-1041.

Soil Dynamics & Earthquake Engineering Conference / Southampton / 1982.07.13-15

Frequency Dependent Dynamic Response of Footings

M.OTTENSTREUER
Ruhr-Universität, Bochum, Germany

INTRODUCTION

The dynamic loading of a building causes stress and displacement fields in the structure as well as in the supporting or surrounding soil.

Stresses and displacements of both soil and structure however influence each other through the contact area of the foundation and the soil. This phenomenon is understood as soil-structure interaction.

The loading can stem from various excitations e.g. seismic loading or out-of-balance forces of machines which transmit dynamic forces on the foundation.

There is a variety of methods to calculate soil-structure interaction problems, e.g. the half space method (Wong and Luco, 1976, Gaul, 1980) and the Finite element method (Waas, 1972). Special advantages and disadvantages are discussed there.

As another tool the Boundary Element Method proved to be well suited to handle soil dynamics problems. It is able to calculate embedded structures (Dominguez, 1978) as well as viscoelastic properties of soil and interaction of several foundations (Ottenstreuer, 1981) but is equally well applicable to arbitrarily layered media.

BOUNDARY ELEMENTS FOR SOIL DYNAMICS

Starting with the system of integral equations given e.g. by Cruse (1969) and written here in the notation of Antes, Ottenstreuer, Schmid (1982)

$$c u_{i}^{\alpha} = t_{j}^{\beta} \int_{\Gamma} u_{ij}^{\alpha\beta}\, d\Gamma - u_{j}^{\beta} \int_{\Gamma} t_{ij}^{\alpha\beta}\, d\Gamma \qquad (1)$$

the fundamental solutions $u_{ij}^{\alpha\beta}$ and $t_{ij}^{\alpha\beta}$ give a relation between displacements u_i^{α} at the nodal point α in the direction of coordinate x_i and tractions t_j^{β} and surface displacements u_j^{β} at the nodal point β in the direction of coordinate x_j.

C depends on the position of α; it becomes 1 inside the domain and $\frac{1}{2}$ on the boundary. The fundamental solution $u_{ij}^{\alpha\beta}$ itself is the displacement at β due to a harmonic single force in α in an infinite domain (Cruse and Rizzo, 1968):

$$u_{ij}^{\alpha\beta} = \frac{1}{4\pi\rho \ c_s^2}\left(\psi \ \delta_{ij} - \chi r,_i \ r,_j\right) \quad ,$$

$$\psi = \left(1 - \frac{c_s^{\ 2}}{\omega^2 r^2} + \frac{c_s}{i\omega r}\right)\frac{e^{\frac{-i\omega r}{c_s}}}{r}$$

$$- \frac{c_s^{\ 2}}{c_p^{\ 2}}\left(- \frac{c_p^{\ 2}}{\omega^2 r^2} + \frac{c_p}{i\omega r}\right)\frac{e^{\frac{-i\omega r}{c_p}}}{r} \quad ,$$

$$\chi = \left(- \frac{3c_s^{\ 2}}{\omega^2 r^2} + \frac{3c_s}{i\omega r} + 1\right)\frac{e^{\frac{-i\omega r}{c_s}}}{r} - \frac{c_s^{\ 2}}{c_p^{\ 2}}\left(- \frac{3c_p^{\ 2}}{\omega^2 r^2}\right.$$

$$\left. + \frac{3c_p}{i\omega r} + 1\right)\frac{e^{\frac{-i\omega r}{c_p}}}{r} \ . \qquad (2)$$

The tractions are derived as

$$t_{ij}^{\alpha\beta} = \frac{1}{4\pi}\left\{\left(\frac{d\psi}{dr} - \frac{1}{r}\chi\right)\left(\delta_{ij}\frac{\partial r}{\partial n} + r,_j \ n_i\right)\right.$$

$$- \frac{2}{r}\chi\left(n_j r,_i - 2r,_i \ r,_j \frac{\partial r}{\partial n}\right) - 2\frac{d\chi}{dr} r,_i \ r,_j \frac{\partial r}{\partial n}$$

$$\left. + \left(\frac{c_p^{\ 2}}{c_s^{\ 2}} - 2\right)\left(\frac{d\psi}{dr} - \frac{d\chi}{dr} - \frac{2}{r}\chi\right) r,_i \ n_j\right\} \ . \qquad (3)$$

r is the distance between the singular point α and a generic point on the surface, n is the outward unit normal, and ω the frequency of the vibrating force. The propagation velocities of dilatational and distortional waves are given as

$$c_p^{\ 2} = \frac{\lambda + 2\mu}{\rho} \qquad (4)$$

and

$$c_s^{\ 2} = \frac{\mu}{\rho} \qquad (5)$$

where λ and μ are Lamé constants.

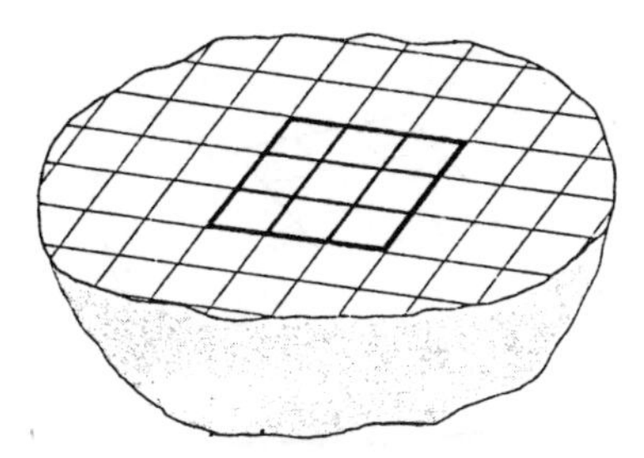

Figure 1 Discretization of the soil surface

For the numerical solution of the problem the soil surface including the contact area between soil and foundation is divided into elements (Figure 1). In each element tractions and displacements are assumed to be constant.

As the influence of elements far away from the foundation diminishes due to increasing distance r (Equation 2, 3) only part of the surface near the foundation has to be discretized.

Formulating Equation 1 in matrix notation gives (Brebbia, 1978)

$$\frac{1}{2}\,\underline{I}\,\underline{u} = \underline{U}\,\underline{t} - \underline{T}\,\underline{u}\,, \tag{6}$$

where $\underline{u}$ and $\underline{t}$ are the vectors of nodal displacements and tractions respectively. $\underline{U}$ and $\underline{T}$ are influence matrices and $\underline{I}$ represents the identity matrix.

The weak coupling between horizontal and vertical vibrations leads to a coupling of horizontal and rocking vibrations for rigid foundations. This mutual influence is usually small and will completely be neglected here. This assumption results in zero elements in the $\underline{T}$ matrix as given in Equation 6 (Ottenstreuer and Schmid, 1981) which then reduces to

$$\frac{1}{2}\,\underline{I}\,\underline{u} = \underline{U}\,\underline{t} \tag{7}$$

The tractions outside the foundation area are zero. This leads to a decoupling of the equations of the foundation area and those of the unloaded soil surface in Equation 7.Then there is no coupling between the elements of the foundation area and those of the unloaded soil surface. Thus the application of so called relaxed boundary conditions only requires the discretization of the foundation area itself (Figure 2).

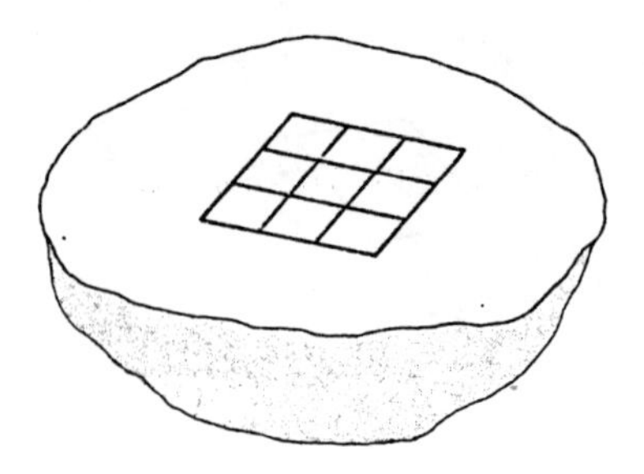

Figure 2 Discretization of the foundation area

If the halfspace is assumed to be homogeneous all its dynamic properties including radiation damping of infinite domains are already taken care of in the boundary integral equation method.

Displacement fields of deformable foundations

With prescribed loading Equation 7 immediately gives the displacement field of a limp foundation as a function of the dimensionless frequency

$$a_o = \frac{\omega \cdot b}{c_s} \tag{8}$$

with b as characteristic length of the foundation (b is half the side length of a square foundation).

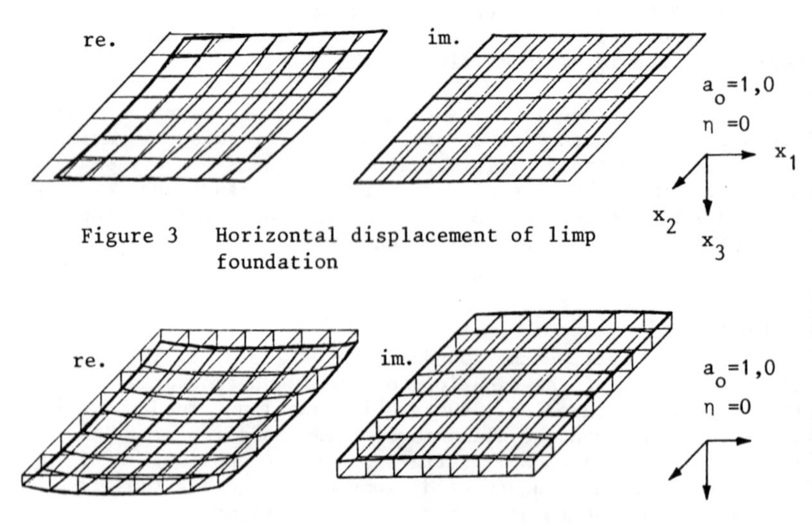

Figure 3 Horizontal displacement of limp foundation

Figure 4 Vertical displacement of limp foundation

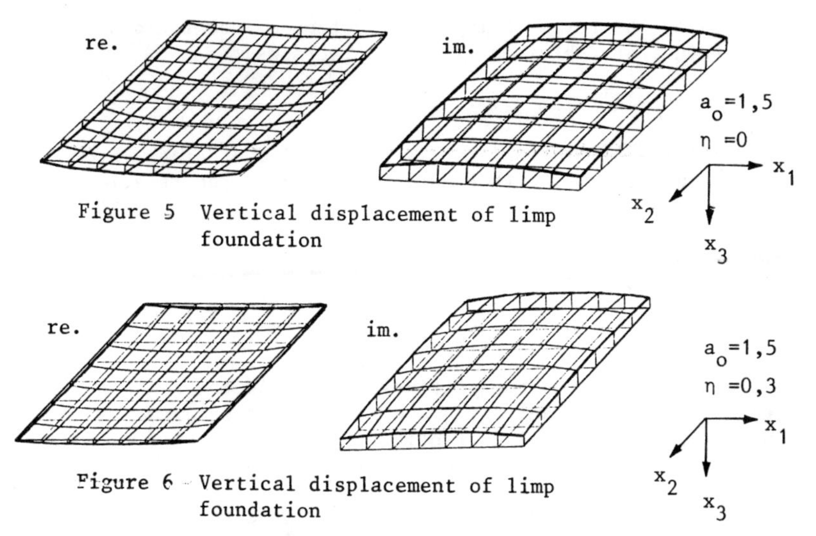

Figure 5 Vertical displacement of limp foundation

Figure 6 Vertical displacement of limp foundation

Figures 3 ÷ 6 show the displacements of a square foundation under uniform stress in x_1- and x_3-direction for special values of a_o. The displacement fields are separated into real and imaginary parts. The real part corresponds to the elastic displacement which is in phase with the load. The influence of damping can be estimated by the imaginary part being $\pi/2$ out of phase. Whereas the imaginary part in Figures 3 through 5 is the result of radiation damping only, Figure 6 shows the behaviour of the vibrating foundation on soil with added material damping. Material damping is taken into consideration by introducing a complex modulus of elasticity (Achenbach, 1975)

$$\hat{E} = E(1+i\eta) \tag{9}$$

where

$$\frac{\eta}{\omega} = \theta = \text{constant} \tag{10}$$

for a "Voigt" damping model and

$$\eta = \omega \cdot \theta = \text{constant} \tag{11}$$

for a constant hysteretic damping model.

Displacements of rigid foundations

Massless foundations Compact block foundations on soft soil allow the assumption of rigid base plates. This leads to a reduction of the degrees of freedom to six rigid body displacements. The matrix $\underline{a}$ maps the rigid body motions $\underline{\tilde{u}}$ into element displacements $\underline{u}$:

$$\underline{u} = \underline{a}\,\underline{\tilde{u}} \,. \tag{12}$$

Similarly the forces $\underline{P}$ in the direction of the rigid body displacements are obtained from the tractions $\underline{t}$ as

$$\underline{P} = \underline{a}^T \cdot \underline{A}\ \underline{t} \tag{13}$$

where $\underline{A}$ is the diagonal matrix of the element areas. From Equation 7 follows with Equations 12 and 13 the relation

$$\underline{K}\ \underline{u} = \underline{P} \tag{14}$$

where the complex matrix

$$\underline{K} = \frac{1}{2}\ \underline{a}^T\ \underline{A}\ \underline{U}^{-1}\ \underline{a} \tag{15}$$

is generally called the stiffness matrix of the soil. A clear relation between excitation and resulting displacements can now be given for a rigid footing. The components of $\underline{K}$ are presented for one and two foundations by Ottenstreuer (1981).

In the Argand diagram (Figure 7) the resulting amplitude $\hat{u}$ is given by addition of the real and imaginary part of the rotating vectors.

For a single harmonic force this yields in

$$\hat{u}_j = \frac{1}{\sqrt{(K_j^R)^2 + (K_j^I)^2}} \tag{16}$$

where

$$K_j = K_j^R + iK_j^I \qquad i = \sqrt{-1}\ . \tag{17}$$

In Equation 16 the fact has been used that $\underline{K}$ is only a diagonal matrix with elements K_j. In dimensionless form the components are

$$k_j = \frac{K_j}{\mu b} \tag{18.a}$$

for the displacements and

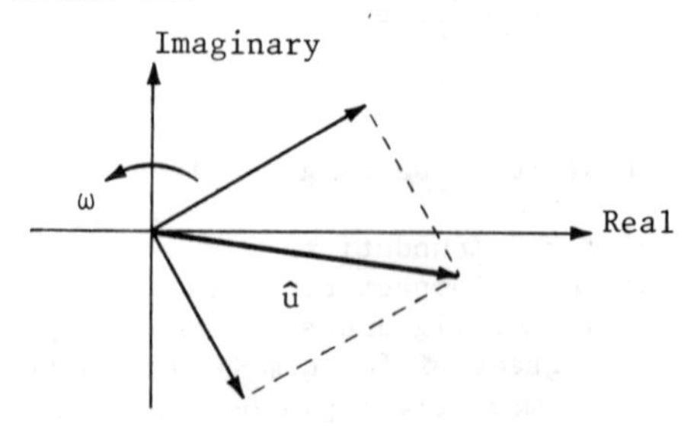

Figure 7 Rotating vector representation

$$k = \frac{K}{\mu b^3} \qquad (18\ b)$$

for the rotations.

A dimensionless notation for the displacement amplitude runs as follows

$$\hat{u}_j\ \mu b = \frac{1}{\sqrt{(k_j^R)^2 + (k_j^I)^2}} \quad . \qquad (19)$$

The displacement response related to the dimensionless frequency a_o is portrayed in Figures 8 and 9, where the loading is a single force in x_1- and x_3-direction. It can be noticed that the maximum displacement decreases with increasing frequency until it finally goes to zero at sufficiently high frequency. The same behaviour is to be seen at a single spring-dashpot-system.

Dynamic magnification factor

Magnification of displacements has to be calculated considering the mass of the foundation. This requires the coupling with the massless foundation to one entire system. If the center of gravity of the mass and the point of application of the external force coincide the equation of motion is

$$(-\omega^2\ \underline{M} + \underline{K})\ \underline{\tilde{u}} = \underline{P} \quad . \qquad (20)$$

In this case $\underline{M}$ and $\underline{K}$ are diagonal matrices and the equations are decoupled. Looking at the dimensionless notation leads for the translational degrees of freedom to equations of the type

$$(-a_o{}^2\ m_j + k_j)\ \mu \cdot b \cdot \tilde{u}_j = P_j \qquad (21)$$

where

$$m_j = \frac{M_j}{\rho \cdot b^3} \qquad (22)$$

represents a dimensionless mass

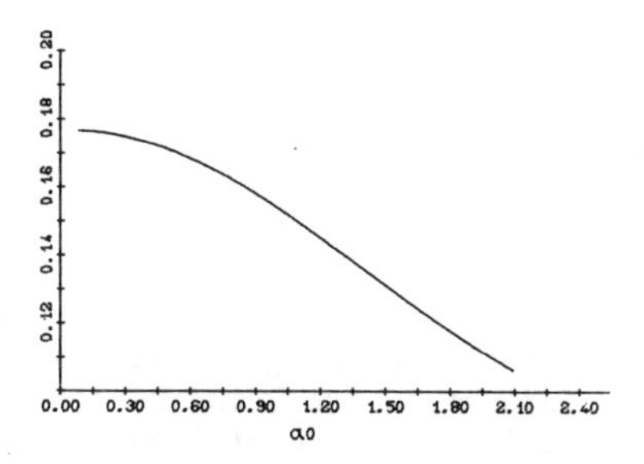

Figure 8 Amplitude of horizontal vibration

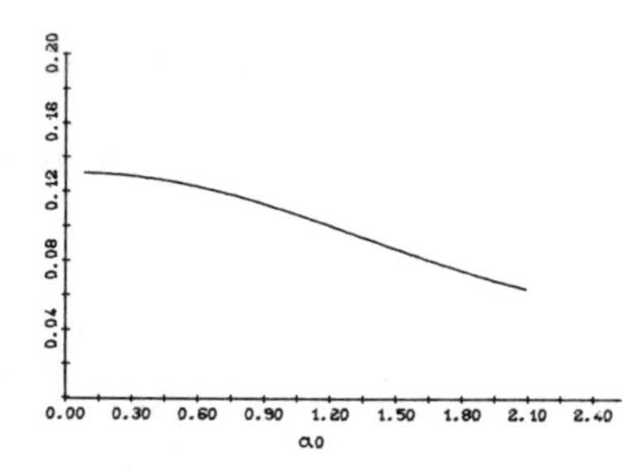

Figure 9 Amplitude of vertical vibration

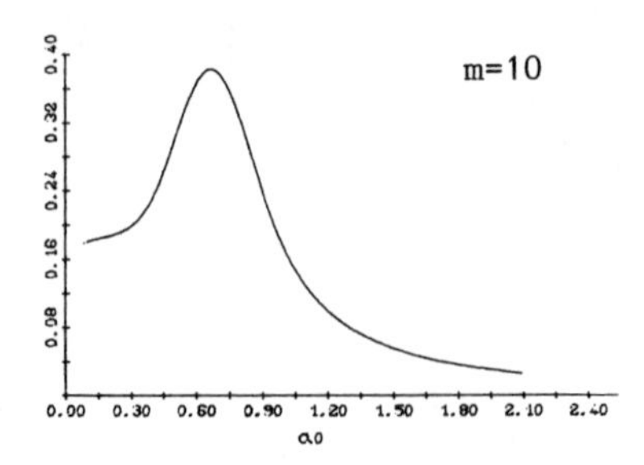

Figure 10 Amplitude of horizontal vibration

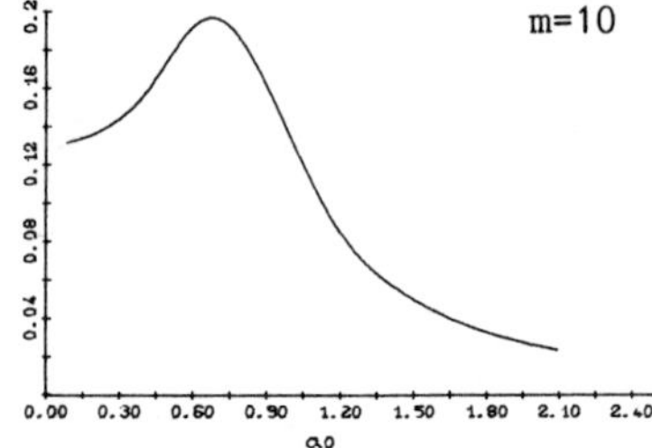

Figure 11 Amplitude of vertical vibration

The displacement response for a single harmonic force is then

$$\hat{u}\ \mu\ b = \frac{1}{\sqrt{(-a_o{}^2\ m + k^R)^2 + (k^I)^2}} \ . \tag{23}$$

The response functions shown in Figures 10 and 11 illustrate the influence of the foundation mass on the maximum amplitudes of the vibrating system.

A strong magnification is found in the neighbourhood of $a_o \approx 0,7$ which characterizes the first natural frequency of the system. The finite amplitude at resonance frequency is due to radiation damping. As it should be the frequency yields the same value for the foundation with and without mass in the static case (see Figures 8 and 9). If the dead load of the mass is taken into consideration the displacements have to be superposed.

Centrifugal force

The displacements of a foundation excited by a rotating mass are discussed in the following. In contrast to a single force a rotating mass produces forces

$$P = \omega^2\ e\ m_e \tag{24}$$

in x_1- or x_2-direction and forces

$$P = -\ i\ \omega^2\ e\ m_e \tag{25}$$

in x_3-direction, where e is the distance between the point of application and the center of the rotating mass m_e. Under the assumption that the center of the foundation area, the center of the mass and the point of application of the centrifugal force coincide it follows that the three degrees of freedom for the displacements are decoupled.

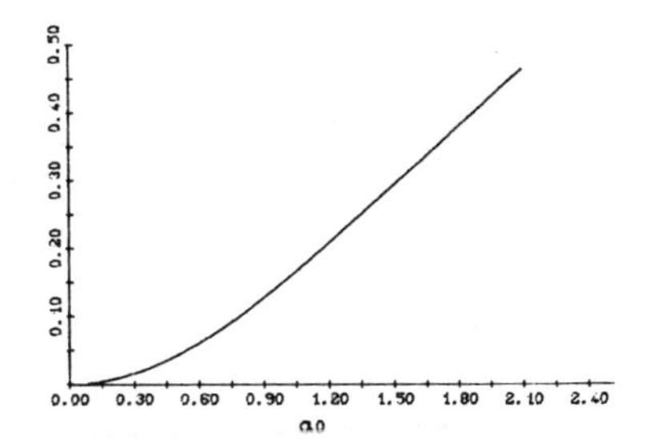

Figure 12 Amplitude of horizontal vibration

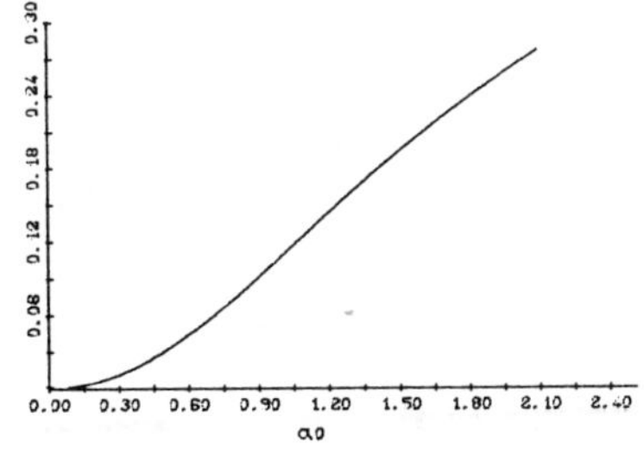

Figure 13 Amplitude of vertical vibration

With the definitions for P in Equations 24 and 25 one obtains again

$$\underline{K}\ \underline{u} = \underline{P} \tag{26}$$

which leads to the dimensionless maximum displacement response

$$\frac{\hat{u} \cdot \rho \cdot b^3}{e\ m_e} = \frac{a_o^{\,2}}{\sqrt{(k^R)^2 + (k^I)^2}} \tag{27}$$

The amplitudes are shown in Figures 12 and 13. They differ from those in Figures 8 and 9 due to the frequency dependent growing force.

If the mass of the foundation is taken into account as shown in Equation 21 the dimensionless amplitudes may be written as

$$\frac{\hat{u}\ M}{e\ m_e} = \frac{a_o^{\,2}\ m}{\sqrt{(-a_o^{\,2}\ m + k^R)^2 + (k^I)^2}} \tag{28}$$

Figures 14 and 15 illustrate the magnification factor for the footing with mass. The natural frequency of the mass-foundation system is clearly indicated.

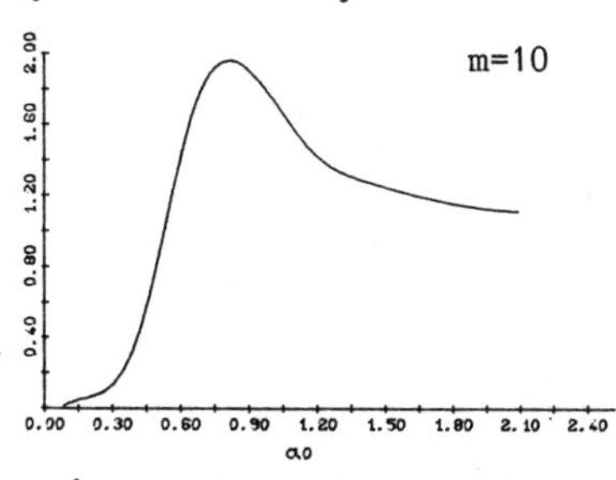

Figure 14 Amplitude of horizontal vibration

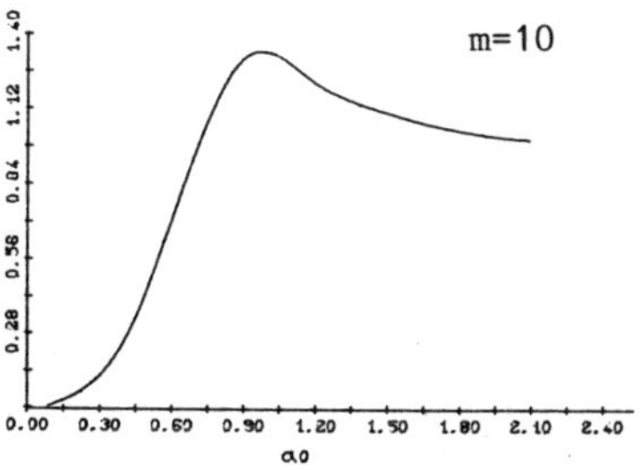

Figure 15 Amplitude of vertical vibration

CONCLUSION

The results shown here demonstrate the ability of the boundary element method to handle soil dynamics problems. Without any difficulties its application given above can be extended to several foundations. The method is not limited to surface foundations but is equally well applicable to embedded foundations and layered soil.

ACKNOWLEDGEMENT

The programming work was done by Margret Gibhardt and typing by Christa Hoogterp.

REFERENCES

Achenbach, J.D. (1975) "Wave Propagation in Elastic Solids" Amsterdam: North Publishing Company

Antes, H.; Ottenstreuer, M.; Schmid, G. (1982) "Randelemente" Techn.-Wiss. Mitteilungen Nr. 82-2 des Inst. für Konstr. Ingenieurbau der Ruhr-Universität Bochum

Brebbia, C.A. (1978) "The Boundary Element Methods for Engineers" Pentech Press, London

Cruse, T.A. (1969) "Numerical Solutions in Three Dimensional Elastostatics", Int. Journal of Solids and Structures, 5, 1259-1274

Cruse, T.A. and Rizzo, F.J. (1968) "A Direct Formulation and Numerical Solution of the General Transient Elastodynamic Problem I" Journal of Mathem. Anal. and Appl., 22, 244-259

Dominguez, J. (1978) "Dynamic Stiffness of Rectangular Foundations" Publication No. R 78-20, MIT, Department of Civil Engineering

Gaul, L. (1980) "Zur Dynamik der Wechselwirkung von Strukturen mit dem Baugrund" Habilitationsschrift, Universität Hannover

Ottenstreuer, M. (1981) "Ein Beitrag zur Darstellung der Wechselwirkung zwischen Bauwerk und Baugrund unter Verwendung des Verfahrens der Randelemente" Techn.-Wiss. Mitteilungen Nr. 81-6 des Inst. für Konstr. Ingenieurbau der Ruhr-Universität Bochum

Ottenstreuer, M. and Schmid, G. (1981) "Boundary Elements Applied to Soil-Foundation Interaction" in Boundary Element Methods (ed. C.A. Brebbia) Springer-Verlag, Berlin Heidelberg New York

Waas, G. (1972) "Linear Two Dimensional Analysis of Soil Dynamics Problems in Semi-Infinite Layered Media" Ph. D. thesis, University of California, Berkeley

Wong, H.L. and Luco, J.E. (1976) "Dynamic Response of Rigid Footings of Arbitrary Shape" Earthquake Eng. and Structural Dynamics, 4, 579-587

Soil Dynamics & Earthquake Engineering Conference / Southampton / 1982.07.13-15

Foundations for Auto Shredders

FRANK E.RICHART, Jr. & RICHARD D.WOODS
University of Michigan, Ann Arbor, USA

SYNOPSIS

Unbalanced forces developed by hammer wear and impact must be resisted by auto shredder foundations. Methods for estimating the impact forces are described. Because of different soil conditions, a concrete mat, a concrete block, and a pile-supported foundation system were adopted at three different construction sites. The design procedures involved in determining the dynamic response for each type of foundation are illustrated by examples. Vibration measurements made on the pile-supported foundation after construction permitted comparisons of prototype motions with design predictions.

INTRODUCTION

Automobile shredders consist of rows of rotating hammers which pass between a slotted anvil as shown in Fig. 1. The automobile body which is fed into this system is reduced to scrap metal by the impact and shearing forces developed when the body interferes with the hammer motion. Thus, during the process large impact forces are developed by the machine. Also, because of uneven hammer wear, large steady-state unbalanced rotating forces are produced. These dynamic forces are transmitted through the foundation system into the underlying soil.

This paper describes briefly the design of three types of auto shredder foundations, (a) a rigid concrete mat, (b) a deeply embedded rigid concrete block, and (c) a pile-supported concrete mat. Each foundation system reduces the machine vibration to tolerable levels.

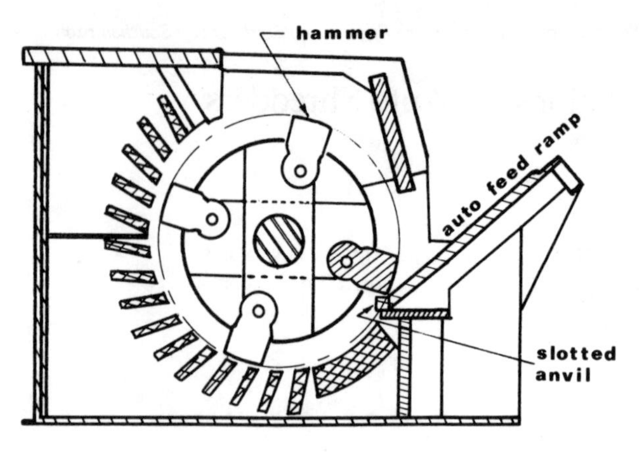

AUTO SHREDDER

Fig. 1 Cross-Section of Auto Shredder

DYNAMIC FORCES

Uneven wear of the hammers produces an unbalanced force vector rotating about the shaft. The limits for this type of unbalanced force are established by the machine manufacturer and the control of the magnitude of this force depends upon the owner's maintenance procedure. Hammers must be replaced periodically because of wear, and the machine can be nearly balanced after each hammer replacement but will become more unbalanced with time of operation.

Each type of shredder has a different allowable unbalance, depending on the number and size of the hammers and the operating speed. For a machine carrying 34 hammers, two rows of 9 each weighing 240 lb (1067N) and two rows of 8 at 144 lb (640N) each, and operating at 700 RPM, the vertical and horizontal (i.e. centrifugal) force amounts to 56,400 lb (2.51×10^5N). Another machine with 34 hammers each weighing 450 lb (2002N) and operating at 600 RPM has a limiting force of 112,500 lb (5.0×10^5N).

Impact forces are developed when each row of hammers hits the auto body, thus the frequency of impacts is <u>four times</u> the operating frequency. The limiting value of this impact force depends upon the impulse required to stop the hammer at the point of impact. This occurs occasionally when hard chunks of metal cannot be shredded by one impact. Figure 2(a) shows a single hammer and the dynamic forces which act on it. The point of impact is taken as 12 in. (0. 30m) from the centerline of the 4 in. (0. 10m) diameter hammer bolt. The moment of the impact force ($M = Q_I r_q$) tends to rotate the hammer about the hammer bolt. This rotation is resisted by the frictional moment $fF\, r_b$

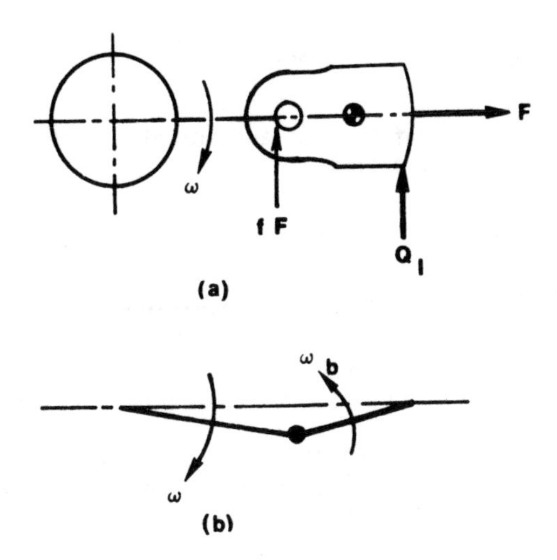

Fig. 2 Hammer Action (a) Forces on Hammer
(b) Motion of Hammer and Arm.

at the hammerbolt plus the inertial resistance of the hammer to being rotated about the hammerbolt. At a rotating speed of 600 RPM the centrifugal force developed by each hammer is

$$F = \frac{W}{g}r_1\omega^2 = \frac{450}{386}(28.5 + 8.5)(2\pi 10)^2 = 1.70\text{x}10^5\text{lb} \quad (7.56\text{x}10^5\text{N})$$

Then the impact force which can be developed just to overcome frictional resistance is

$$Q_f = fF\left[r_b/r_q\right] = 0.15\text{x}1.70\text{x}10^5\text{x}1.5/12 = 3200 \text{ lb } (1.42\text{x}10^4\text{N})$$

for each hammer.

The dynamic force needed to rotate the hammer about the hammerbolt depends on the time interval involved. At the contact point the linear velocity is

$$v_c = r_c\omega = (28.5 + 12) \times 2\pi\text{x}10 = 2545 \text{ in/sec } (64.6 \text{ m/sec})$$

Then if the contact point is brought to rest in a distance of 20 in (0.51m) which is assumed as the compressed height of a mashed auto body, then the time interval needed for this contact point to be stopped is 0.008 sec. During this 0.008 sec. the hammerbolt has travelled through an angle of $\theta_b = 0.008 \times 2\pi 10 = 0.50$ radians. Then if the hammer must swing through an angle of 60° before it can pass by the hard material (See Fig. 2b), the average angular rotation of the hammer about the hammerbolt is

$$\omega_b = \frac{\Theta n}{\Delta t} = \frac{\pi/3}{0.008} = 131 \text{ rad/sec}$$

From considerations of impulse and momentum, the average impact force can be estimated from

$$I_b \omega_b = Q_I \, r_q \, \Delta t$$

$$\text{or } Q_I = \frac{90 \times 131}{12 \times 0.008} = 1.23 \times 10^5 \text{lb} \quad (5.47 \times 10^5 \text{N})$$

This discussion of the impact forces illustrates that the magnitude of the force transmitted to the anvil can be significant and must be considered.

Additional dynamic forces are transmitted to the foundation by vibrating conveyors. Typical values are 13,750 lb (6.12×10^4N) for horizontal dynamic force and 11,000 lb (4.89×10^4N) for vertical dynamic force, both at 720 RPM. However, different conveyors will have different force and frequency outputs.

MAT FOUNDATION FOR SHREDDER

The soil at the site was loose fine sand and silt and the water table was near the surface. Thus clean fill was required and both the natural soil and the fill were compacted with surface vibratory compaction equipment.

For this installation a rigid concrete foundation mat 3.5 ft (1.067m) thick was chosen. The mat provided the large surface contact area which was the important criterion for resisting the overturning moments. The mass of the mat was of secondary importance. General plan dimensions of the mat are shown in Fig. 3 and significant data are listed below

Weight of mat and machinery	= 1.70×10^6lb (7.56×10^6N)
Plan area	= 1842 ft^2 ($171m^2$)
Radius of circle with same area, r_o	= 24.2 ft (7.38m)
Mass moment of inertia for rotation of foundation about line on base parallel to axis of shredder I_o	= 1.050×10^7ft lb sec^2 (1.424×10^7mNsec^2)
Radius of circle having same moment of inertia as plan r_o	= 27.0 ft (8.23m) for rocking

The soil properties were influenced by the confining pressures developed by the weight of the installation plus the fill. The soil properties were

Shear modulus	G	= 13,500 lb/in^2 (9.3×10^7 N/m^2)
Poisson's Ratio	ν	= 1/3
Saturated unit weight	γ_{sat}	= 125 lb/ft^3 (1.96×10^4N/m^3)

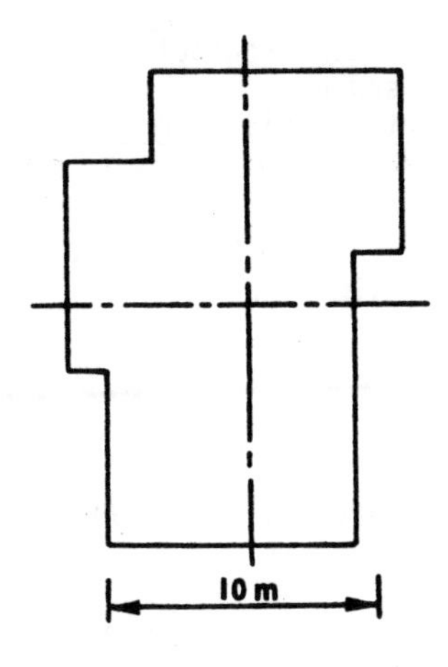

Fig. 3 Plan of Mat Foundation

The shredder for this installation had an allowable unbalanced force of 56,400 lb (2.51×10^5N) rotating at 720 RPM from wear of 34 hammers. Two rows of 9 hammers each weighing 240 lb (1067N) and two rows of 8 each weighing 144 lb (640N) each constituted the hammer system. The average impulsive forces which could act over a time interval of 0.085 sec. were 7.8×10^4 lb (3.46×10^5N) for the 240 lb (1067N) and 4.7×10^4 lb (4.70×10^4N) for the 144 lb (640N) hammers.

Four vibratory conveyors each operating at 360 RPM were aligned perpendicular to the shredder axis. Thus these vibratory forces could be superposed on the shredder forces, with the final resultant force depending on the phase relationship of each vibratory motion. The vertical and horizontal components of dynamic conveyor forces were

Conveyor No.	Dead Weight (lb*)	Horizontal Force (lb*)	Vertical Force (lb*)
1	13,000	± 18,000	± 10,500
2	3,300	± 4,850	± 2,800
3	3,200	± 4,850	± 2,800
4	4,800	± 5,500	± 3,200

*Note: No. of Newtons = No. lb x 4.448.

Vertical Motion

The procedures for evaluation of dynamic motions of the foundation are described in detail in Ref. (1), thus only the basic elements are treated here. The mat foundation was treated as an effective rigid circular foundation resting at the surface of an elastic half-space.

For response to vertical vibrations, the mass ratio, B_Z, and damping ratio, D_Z, are

$$B_Z = \frac{(1-\nu)}{4}\frac{W}{\gamma_{sat}(r_o)^3} = \frac{0.67x1.7x10^6}{4x125(24.2)^3} = 0.16;$$

$$D_Z = \frac{0.425}{\sqrt{B_Z}} = 1.06$$

The static vertical displacement caused by the 56,400 lb (2.51 x10^5N) vertical unbalanced force was

$$z_s = \frac{(1-\nu)Q_z}{4GR_o} = \frac{0.67x56400}{4x13,500(24.2x12)} = 0.0024 \text{ in. } (6.1x10^{-6}m)$$

and the dynamic magnifications factor, M_Z, amounted to 1.0. The low value of dynamic magnification factor follows from the high value of damping ratio in vertical motion and indicates that the dynamic motion is essentially the same as the static displacement. For the unbalanced force at 720 RPM, the dynamic motion amounted to $A_Z = z_s M_Z$ = 0.0024 in (6.1x10^{-5}m). If all the vertical components of the conveyor forces were in phase they would produce a sinusoidal motion of 0.0009 in (2.3x10^{-6}m). This sinusoidal motion would be superposed on that developed by the shredder to give the total motion. However, it is possible to arrange the phases of the conveyor motions to minimize the resultant force output.

Rigid-Body Rocking of the Foundation

Because of the high water table at the site, it was necessary to mount the shredder on a pedestal to provide space for conveyors beneath. The centerline of the shredder was located 11.33 ft (3.45m) above the top of the concrete mat. Thus large overturning moments were introduced by the unbalanced forces and the mat dimensions were selected to provide resistance to these forces.

The mass-ratio B_ψ for rigid body rocking, and associated damping ratio, D_ψ, are

$$B_\psi = \frac{3(1-\nu)I_g g}{8\gamma_{sat}(r_o)^5} = \frac{3(0.67)1.05x10^7x32.2}{8x125(27.0)^5} = 0.047; \quad D_\psi = \frac{0.15}{(1B_\psi)\sqrt{B_\psi}} = 0.66$$

$$M \approx 1.0; \quad \psi s = \frac{3(1-\nu)M_\psi}{8Gr_o^3} = \frac{3x0.67x56,400x14.83}{8x13500x144(27)^3} = 5.5x10^{-6} \text{ rad}$$

Then

$$A_\psi = M_\psi \ \psi_s = 5.5\text{x}10^{-6} \text{ rad, or}$$

$$A_x = 14.83(12)A_\psi = 0.00098 \text{ in } (2.5\text{x}10^{-6}\text{m}).$$

Thus, the amplitude of horizontal motion at the shredder centerline amounts to about 0.001 in ($2.5\text{x}10^{-6}$m) at the operating speed of 720 RPM.

These two calculations show that the mat develops high values of geometrical damping in vertical and rocking motions. The impulsive loads produced smaller motions than those calculated above, and the solution by the phase-plane method (see ref. 1) will not be treated here.

BLOCK FOUNDATION FOR SHREDDER

The site for this shredder installation had competent stiff clay (G=21,000 lb/in^2)($1.45\text{x}10^8$N/m^2), ν = 0.4, γ_{sat} = 125 lb/ft^3 ($1.96\text{x}10^4$N/m^3) at a depth of 21.5 ft (6.55m) below grade. Between the thin surface crust and the stiff clay was a soft alluvium with an effective value of G of 1/20 of that for the stiff clay.

One economically feasible solution for this foundation was to use a mass concrete block foundation as shown in Fig. 4.

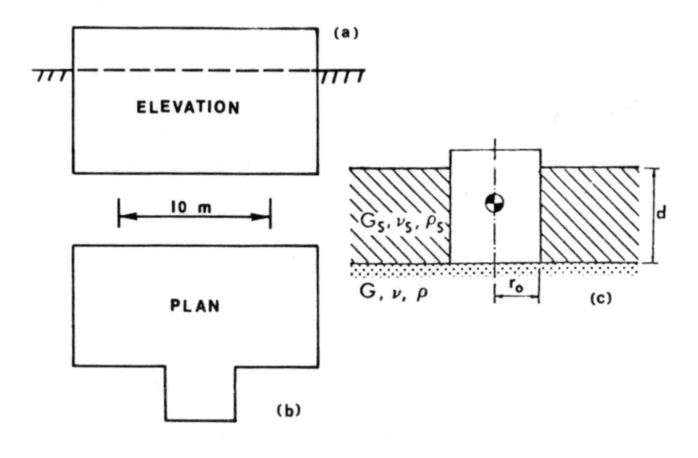

Fig. 4 Block Foundation, (a) Elevation, (b) Plan, (c) Equivalent Cylindrical Foundation.

The response of this embedded block to the imposed dynamic loads was studied using Novak's analysis for embedded foundation (ref. 2).

This shredder had two rows of 8 and two rows of 9 hammers, each weighing 450 lb (2002N). The unbalanced radial force had an amplitude of 112,500 lb ($5.0x10^5$N) @ 600 RPM. Impact forces developed as each hammer was temporarily stopped were as noted in the previous section on DYNAMIC FORCES.

Figure 4(c) shows the equivalent cylindrical foundation of radius, r_o, which is embedded a depth, d, into soil having a shear modulus, G_s, and Poisson's ratio, ν_s. Beneath the base of the foundation the soil had the properties G, and ν. For this equivalent foundation, the values of r_o, d, and h_o were

d = 21.5 ft (6.55m); r_o = 22 ft (6.71m); h_o = 37 ft (11.28m)

and the dead weight, W, and the mass moment of inertia, I_g, about the C.G. were

$W = 6.40x10^6$ lb ($2.85x10^7$N) and,

$I_g = 6.26x10^8$ in lb sec^2 ($7.07x10^7$mNsec^2)

Vertical Vibrations

Novak's expressions for the spring and damping factors in vertical vibration include the influence of the soil acting along the sides of the foundation as well as that below the base. Thus,

$$k_{zz} = 4Gr_o/(1-\nu) + 2.7G_s d$$

$$c_{zz} = 3.4r_o^2 \sqrt{\rho G/(1-\nu)}$$

Following through the calculations for damping ratio,

$D = c_{zz}/2\sqrt{K_{zz}m}$. $M = 1/(2D\sqrt{1-D^2})$, $z_s = Q_z/k_{zz}$,

leads to

$A_z = Mz_s$ = 0.0031 in ($7.9x10^{-6}$ in).

Rocking and Horizontal Vibrations

The horizontal force Q_x was applied above the center of gravity of the foundation and above the center of soil resistance. Consequently, coupled rocking and horizontal motions were developed. The motion depends upon the resonant frequencies and damping of the soil-foundation system and the frequency of the exciting force and moment. The calculations follow the procedure clearly described in ref. (2) and are not included here. The resulting vibrations can be described by a horizontal translation of the center of gravity of amplitude, u_g, and a rotation about the center of gravity, ψ_g. At the frequency of 600 RPM,

$$u_g = 0.003 \text{ in } (7.6\text{x}10^{-5}\text{m}), \text{ and}$$

$$\psi_g = 5.6\text{x}10^{-6} \text{ radians.}$$

Thus this particular block foundation was considered to perform satisfactorily.

PILE-SUPPORTED FOUNDATION FOR SHREDDER

For this installation the soil at the site consisted of an upper crust of competent material then a zone of soft cohesive soil overlying a bed of firm sand. Thus piles were required to bypass the soft zone and to transmit the static and dynamic loads to the sand.

The shredder used at this site had the same dynamic force outputs as those described for the mat foundation. This shredder was mounted on a pile cap and pedestal system as shown in Fig. 5, with the centerline of the shredder at a distance of 13.81 ft. (4.21m) above the top of the pile cap. Thus, operation of the shredder developed steady state forces at 720 RPM which caused vertical, horizontal, and rocking motions of the foundation, and developed impact forces at a frequency of four times the operating speed.

Figure 5(a) shows the final pile pattern selected following several cycles of analysis and modification of the geometry. The key resisting members of this system were the 52 concrete-filled 12 in (0.30m) OD pipe piles, and secondary restraint was provided by clean, cohensionless soil compacted to a dense condition against the vertical faces of the pile cap. Pile loading tests were run in the field on a pile which had been driven to capacity then redriven to minimize the effects of subsequent dynamic loads. Repeated cyclic loading about the static load provided values for the vertical stiffness of the individual piles. From these cyclic loading tests, the vertical static stiffness of a single pile 50 ft (15.24m) long was found to be $k_p = 2.5\text{x}10^6$ lb/in ($4.38\text{x}10^8$ N/m). This value was used for each pile when calculating the combined effects of all piles for the vertical and rocking modes of vibration.

Vertical Vibrations

The dead weight of the components of the foundation system was $1.0\text{x}10^6$ lb ($4.5\text{x}10^6$N) for the pile cap, $3.5\text{x}10^6$ lb ($15.57 \text{ x } 10^6$ N) for the pedestals, and $2.25\text{x}10^5$ lb ($10.0\text{x}10^5$N) for the shredder and motor. The total mass to participate in translational vibrations was m = 4080 lbsec^2/in ($7.14\text{x}10^5$Nsec^2/m). Then, using the vertical stiffness of the 52 piles, and the steady-state unbalanced force, the static vertical deflection was

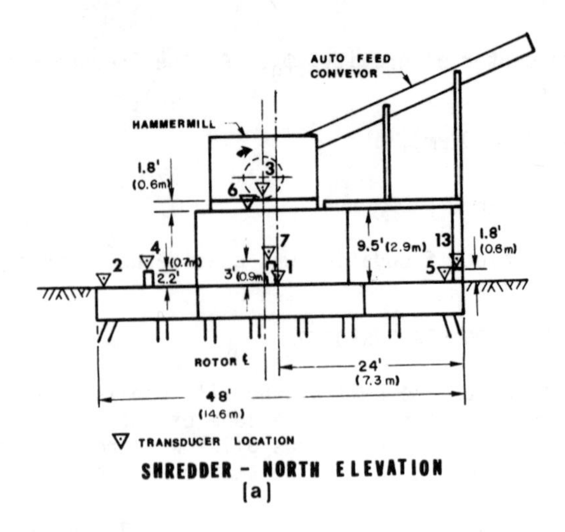

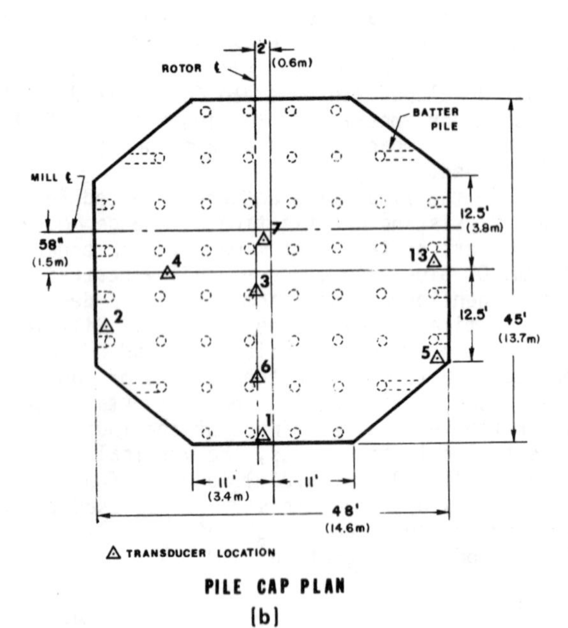

Fig. 5 Pile-Supported Shredder, (a) North Elevation, (b) Pile Cap Plan.

$z_s = Q_z/k_z = 56400/52x2.5x10^6x0.65 = 0.00067$ in ($1.7x10^{-5}$m)

with an undamped natural frequency of f_n=28.4 Hz. Thus the magnification factor, M, was about 1.2, and the dynamic motion amounted to 0.0008 in ($2.0x10^{-5}$m). In the equation above for Z_s, the number 0.65 represents the pile group effect.

For average values of impact loads of $Q_I = 1.0x10^{-5}$ lb $4.45x10^5$N acting over a time interval of 0.008 sec., the phase-plane procedure (see ref. 1) gave an estimated vertical motion of 0.001 in ($2.54x10^{-5}$m) at the frequency of 2800 RPM. This force-time pattern corresponded to the action of one hammer. For the phase-plane solution, a damping factor of c = 6000 lb-sec/in ($1.05x10^6$Nsec/m) per pile was established from the PILAY program (ref. 3).

The soil adjacent to the vertical face of the foundation was not considered to add to the stiffness or damping of the system because of the underlying soft soil.

Rocking Vibration

Because of the horizontal unbalanced force applied at a distance of 17.3 ft (5.27m) above the base of the pile cap, the overturning moment was $T_\psi = 9.76x10^5$ ft lb ($1.32x10^6$mN).

The mass moment of inertia of the foundation system in rocking about the centerline of the base was I_ψ=$8.5x10^6$ftlbsec^2 (11.52mNsec^2) and the resisting spring constant provided by vertical deformation of the 52 piles as the foundation rotated was $k_{\psi p} = 2.88x10^{12}$inlb/rad ($3.25x10^{11}$mN/rad). Thus, the static rotation was $\psi_s = T_\psi/k_{\psi p} = 4.0x10^{-6}$rad. This rotation contributes a horizontal motion of x_s = 0.001 in ($2.54x10^{-5}$m) at the centerline of the shredder. The natural frequency in rocking about this base centerline was f_n = 26.7 Hz. Then the dynamic magnification factor was 1.25, even for the undamped case, and the horizontal dynamic motion at the shredder centerline would be A_x = 0.00125 in ($3.2x10^{-5}$m).

MEASUREMENTS ON PILE-SUPPORTED SHREDDER

Opportunities for comparing performance of constructed facilities with predicted performance are rare. However, the real test of any analytical technique lies in how well measurements match prediction. The pile-supported shredder described above was the subject of such a comparison. Vibration measurements were made while the shredder was idling and while it shredded cars.

Instrumentation

Velocity transducers and a strip chart recorder were used to make the vibration measurements. The velocity transducers were Electro-Tech, 4.5 Hz units, two of which detected vertical motions and one detected horizontal motions. The strip chart recorder was a Hewlett-Packard Model 320, dual channel, hot-pen writing, amplifier recorder.

Measurements

The locations at which measurements were made are shown in Fig 5 (a) and (b). By recording two transducers simultaneously, it wa possible to compare phases and determine the mode of motion as wel as amplitude. Figure 6 (a) shows the vertical motion at opposite ends of the pile cap while no cars were being shredded (idling con dition) and shows that these two points were 180° out-of-phase. The lower trace in Fig. 6 (a) at location 2 shows a different signature because that location was near a support for a vibrating conveyor which was operating at all times.

Figure 6 (b) shows vertical (upper) and horizontal (lower) motion at location 3 near the axis of the shredder when cars were be ing shredded. This record shows that the two directions of motio

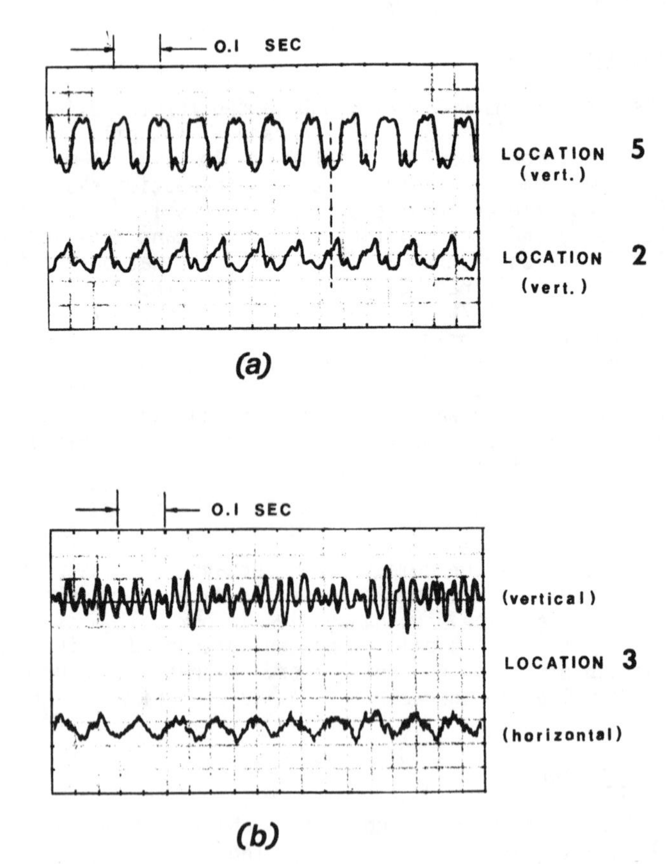

Fig. 6 Velocity-Time Traces, (a) Vertical Motions from Locations 2 and 5, (b) Vertical and Horizontal Motions from Location 3.

occurred at different frequencies. The horizontal motion was at the rotational speed of the machine (12 Hz) while the vertical motion was at 4 times that speed.

Discussion

From these vibration measurements, it was concluded that when the shredder was running at idle, the predominant motion was rocking at a frequency of about 12 Hz. The maximum vertical displacement was about 0.00165 in ($4.2x10^{-5}$m) peak and the maximum horizontal rocking displacment was 0.0019 in ($4.8x10^{-5}$m) peak. The axis for rocking was estimated to be about 15 ft (4.57m) below the pile cap.

When cars were being shredded, the mode of motion at maximum amplitude was vertical translation. This vibration occurred at about 48 Hz or 4 times the primary frequency of the shredder. This frequency represents the rate at which hammers shear through metal at the anvil. The maximum vertical displacement in this mode was 0.0026 in ($6.6x10^{-5}$m) peak.

CONCLUSION

Analytical results for a mat, block, and pile-supported foundation system have been described by examples. The most important factor is identifying the maximum loads and associated frequencies which act as excitation. Then translational, and coupled rocking and horizontal modes of vibration must be studied. The analytical procedure to be used depends on the geometry of the system, and elastic solutions are acceptable because of the small strains developed in the soil. A critical parameter in the analysis is the shear modulus of the soil, which should be established by in situ measurements if possible.

Measurements were made on the pile-supported foundation during idling and during shredding. During idling the unbalanced force at about 12 Hz produced vertical, and rocking and horizontal motions. At location 2 at the edge of the pile cap, the maximum vertical displacement was 0.0017 in ($4.3x10^{-5}$m) during idling, and for the same condition, the maximum horizontal displacement at location 3 was 0.0019 in ($4.83x10^{-5}$m). The predicted vertical motion was .0022 in ($5.58x10^{-5}$m) from the combined effects of vertical and rocking motions developed by the steady-state unbalanced force (idling).

The most significant finding from the field measurements was the vertical motion of 0.0026 in ($6.6x10^{-5}$m) at a frequency of about 48 Hz, or four times times the operating speed. Based on data supplied by the shredder manufacturer, the expected mode should have been rocking at about 12 Hz. Because the vertical motion caused by the impact loads was about 2.6 times greater than that estimated in the example, methods of measuring or estimating the impact forces need to be improved.

REFERENCES

1. Richart, F.E., Jr., Hall, J.R., Jr., and Woods, R.D. (1970), "Vibrations of Soils and Foundations", Prentice-Hall, Inc., Englewood Cliffs, N.J.

2. Novak, M. (1974), "Effect of Soil on Structural Response to Wind and Earthquake", Earthquake Eng. and Structural Dynamics, V. 3, No. 1, pp. 79-96.

3. PILAY, Computer Code for Pile Vibrations in Layered Soils, By M. Novak and F. Aboul-Ella, SACDA, U of Western Ontario, London, Ontario, N6A 5B9, Canada.

4. Kuhlemeyer, R. L. (1976), "Static and Dynamic Laterally Loaded Piles", Res. Rep. CE76-9, U of Calgary, Canada, March, 65 pp.

Soil Dynamics & Earthquake Engineering Conference / Southampton / 1982.07.13-15

Dynamic Soil Properties for Machine Foundation Design

SHAMSHER PRAKASH
University of Roorkee, India

VIJAY K.PURI
University of Missouri-Rolla, USA

INTRODUCTION

Several problems in engineering practice require a knowledge of dynamic soil properties. In general, problems of dynamic loading of soils are divided into either small strain amplitude or large strain amplitude response. In machine foundations the amplitude of dynamic motion and consequently the strain in the soil is usually small, while structures subjected to earthquake or blast must be able to tolerate large strain levels.

A large number of field and laboratory methods have been developed for evaluating dynamic soil properties. The major dynamic soil properties that need to be determined are: (1) Shear strength evaluated in terms of strain rates and stress-strain characteristics (2) Liquefaction parameters: cyclic shearing stress ratio, cyclic deformation, and pore-pressure response (3) Dynamic moduli, Young's modulus, shear modulus, bulk modulus, and constrained modulus and corresponding spring constants. (4) Damping (5) Poisson's ratio.

In machine foundations, dynamic soil moduli (and corresponding elastic spring constants) and damping are frequently needed. Knowledge of Poisson's ratio is also needed although it is not as frequently determined. Excellent reviews on the subject have been prepared by Richart (1977),Woods(1978),Silver (1981) and Prakash (1981) and Prakash and Puri (1984).

Shear strength evaluated in terms of strain rates has not found much application in practice and the dynamic properties are analysed more on the basis of deformations. Pertinent literature has been critically reviewed by Prakash (1981). Ishihara (1981) has reviewed strength of cohesive soils under earthquake type loading. Stress-strain characteristics are defined more by the dynamic moduli and its dependence on strain, confining pressure and other parameters.

In this paper developments in the determination of dynamic soil properties, in-situ and by laboratory methods since 1978 have been described. The importance of in-situ methods has been highlighted. Liquefaction parameters need a separate presentation by virtue of its importance and nature of investigations.

ELASTIC CONSTANTS OF SOIL

The soils behaviour is non-linear from the beginning of stress application. For practical purposes, the actual nonlinear stress-strain curve of soil is "linearized", i.e., replaced by straight lines. Therefore, modulus and Poisson's ratio are not constants for a soil, but rather are quantities which approximately describe the behaviour of a soil for a particular set of stresses, loading conditions and geometry. Different values of modulus and Poisson's ratio will apply for any other set of stresses and geometry. Figures 1 and 2 illustrate the basic concepts of different types of soil moduli.

FACTORS AFFECTING ELASTIC CONSTANTS

The factors affecting dynamic elastic constants are: (1) Type of soil and its properties e.g. water content, dry density and state of disturbance. (2) Effective confining pressure. (3) Strain level. (4) Time effects. (5) Degree of saturation. (6) Magnitude of dynamic stress. (7) Number of repetitions of dynamic stress. (8) Others. The effects of these factors on dynamic shear modulus has been discussed by different investigators, Hardin (1978), Hardin and Black (1969), Ishihara (1971), Iwasaki and Tatsuoka (1977), Prakash (1975, 1981), Prakash and Puri (1977, 1981, 1984), Richart (1977), Whitman and Lawrence (1963) and Woods (1978).

Soil type and confining pressure. The effect of soil type and confining pressure may be expressed by equation 1, Hardin and Black (1969)

$$G_{max} = 1230\,(OCR)^k \cdot \frac{(2.973-e)^2}{1+e} \cdot (\bar{\sigma}_o)^{0.5} \qquad (1)$$

in which,

OCR = Over consolidation ratio
$\bar{\sigma}_o$ = mean effective confining pressure
e = void ratio
k = a factor depending on plasticity index
and G_{max} = value by dynamic shear modulus 'G' for shear strains of 10^{-4} or less (usually 10^{-6}).

The values of dynamic shear moduli G_1 and G_2 corresponding to confining pressures $\bar{\sigma}_{01}$ and $\bar{\sigma}_{02}$ may be expressed by equation 2.

$$\frac{G_1}{G_2} = \left(\frac{\bar{\sigma}_{01}}{\bar{\sigma}_{02}}\right)^n \qquad (2)$$

The value of index 'n' varies from 0.3 to 0.7 and a value of $n = 0.5$ has been recommended by most investigators.

Effect of strain level. Shear modulus decreases with increase in shear strain level. Typical variation of 'G' with shear strain from in-situ tests is shown in Fig. 3. It is customary to plot a graph between normalized modulus (defined as G - value at a particular strain divided by G_{max}) and shear strain as shown in Fig. 4.

Effects of time. Time effects are important in some cases Anderson and Stokoe (1977), Anderson and Woods (1975). For most soils time dependent behaviour at low strain levels can be characterized by an initial phase when modulus changes rapidly with time. This is followed by a second phase when modulus increases almost linearly with the logarithm of the time, Fig. 5, Anderson and Stokoe, (1977). For the most part, the initial phase results from void ratio changes during primary consolidation, and is, therefore, referred to as 'primary consolidation'. The second phase, in which modulus increases almost linearly with the logarithm of time, is believed to result largely from a strengthening of physico-chemical bonds in the case of cohesive soils and an increase in particle contact for cohesionless soils and is referred to as the 'long-term time effect'. The long-term effect represents the increase in modulus with time which occurs after primary consolidation is completed.

Two methods are used to describe the long-term effect. The long-term time effect is expressed in an absolute sense as a coefficient of shear modulus increase with time, 1_G (Fig. 5).

$$1_G = \Delta G/\log_{10}(t_2/t_1) \tag{3}$$

in which,

t_1, t_2 = times after primary consolidation, and

ΔG = change in low-amplitude shear modulus from t_1 to t_2

The long-term time effect is also expressed in relative terms by the normalized shear modulus increase with time, N_G, (Fig. 5).

$$N_G = \left(\frac{\Delta G}{\log_{10}(t_2/t_1)}\right)\left(\frac{1}{G_{1000}}\right) 100\% = \frac{1_G}{G_{1000}} 100\% \tag{4}$$

in which,

G_{1000} = Shear modulus at 1000 minutes after completion of primary consolidation.

Effect of different factors on values of I_G and N_G has been discussed by Anderson and Stokoe (1977), Anderson and Woods (1976) and Prakash and Puri (1984). Time effects are significant at practically all strain levels. Because of time effects, the values of soil modulus in the field are higher than those determined in the laboratory, Richart (1977).

Degree of saturation. A study of the effect of degree of saturation on shear wave velocity for a sample of Ottawa sand shows that much of the difference between the values for the dry and saturated conditions can be accounted for by the effect of the weight of water. Therefore, it is sufficient for an evaluation of V_s or G for cohesionless soils to consider the in-situ unit weight and the effective pressure.

Magnitude of dynamic stress and number of stress cycles. In case of machine foundations oscillatory stresses are small compared to static stresses, Prakash and Puri (1969). The strain under each application of oscillatory stress goes on decreasing compared to the strain under previous stress application and after first few applications of dynamic load, the stress-strain loop stabilizes even though the stress-strain behaviour is hysteritic.

EQUIVALENT SPRING CONSTANTS

The two approaches commonly used for analysis and design of machine foundations are: (1) Linear elastic weightless spring approach, Barkan (1962) and (2) Elastic half space approach, Richart, Hall and Woods (1970). In the linear elastic weightless spring approach, the soil below the footing is replaced by equivalent weightless elastic springs depending upon the mode of vibration. Coefficients have been defined to represent the elastic resistance of soil for different modes of vibration from which the equivalent springs can be computed, Barkan (1962), Prakash (1981) and Prakash and Puri (1984). The coefficients used to define elastic soil resistance depend upon soil properties and geometery of the problem, for example, the coefficient of elastic uniform compression C_u may be expressed by equation 5 (Barkan, 1962).

$$C_u = \frac{1.13E}{(1-\nu^2)} \cdot \frac{1}{\sqrt{A}} = \frac{1.13(2G)}{(1-\nu)} \cdot \frac{1}{\sqrt{A}} \qquad (5)$$

in which,

E = Youngs' modulus
ν = Poissons' ratio
A = Area of the foundation.

In the elastic half space approach, the vibrating footing is analysed as an oscillator resting on a semi-infinite, homogeneous, isotropic, elastic medium defined by the shear modulus 'G', poisson's ratio ν and mass density 'ρ'. These two solutions may apparently appear to be widely different, though the relevant soil properties are related to each other by simple relationships if the geometery of the problem can be exactly defined. Based upon this concept, analog solutions to the elastic half space approach have been obtained, Richart, Hall and Woods (1970). Further details on computation of equivalent soil springs are discussed elsewhere, Barkan (1962), Richart, Hall and Woods (1970), Prakash (1975, 1981), Prakash and Puri (1984).

METHODS FOR DETERMINATION OF DYNAMIC SOIL CONSTANTS

Methods for determination of dynamic soil properties may be discussed under two main categories (1) Laboratory methods and (2) Field methods.

Labatory methods

Several laboratory tests have been devised to measure dynamic elastic constants and damping of soils. It must be understood that effort is directed to simulate the loading conditions from the field to the laboratory, in relation to stress level rates of loading, confining pressure and other factors described previously. It is almost impossible to simulate completely the field loading conditions. Still laboratory testing has a considerable merit to study the effect of various factors affecting dynamic properties and to identify significant variable. The following laboratory methods are employed for determination of dynamic elastic constants and damping values of soils. (1) Resonant column tests. (2) Ultrasonic pulse tests. (3) Cyclic triaxial compression tests. (4) Cyclic simple shear. (5) Cyclic torsional simple shear. (6) Shake table tests.

Resonant column tests. The resonant column test for determining modulus and damping characteristics of soils in the low strain range is based on the theory of wave propagation in prismatic rods, Richart, Hall and Woods (1970). Either compression waves or shear waves can be propagated through the soil specimen from which either Young's modulus or shear modulus can be determined. Woods (1978) has discussed different devices in use around the world. Hollow samples may be used with advantage, Drnevich (1967, 1972, 1977).

Ultrasonic pulse tests. Using peizolectric crystals, it is possible to generate and receive ultrasonic waves in soils. Crystals can be obtained which generate either compression or shear waves. Stephenson (1977) described a set up which includes a pulse generator, an oscilloscope and two ultrasonic probes (transmitter and receiver). One of the drawbacks of this technique is the identification and interpretation of exact wave arrival times. More importantly, however, the strain amplitudes which can be achieved with the pulse technique are only in the very low region. The primary advantage of the pulse technique is that tests can be performed on very soft seafloor sediments still retained in a core liner, Woods(1978). This technique is not currently used routinely for measurement of soil properties.

Cyclic triaxial compression tests. Cyclic triaxial tests have been extensively used for study of values of E, G and damping for strain levels higher than 5×10^{-4} (liquefaction of sands) Seed, (1976, 1979). Also, Young's modulus, E, and damping ratio, have often been measured in the cyclic triaxial test by performing strain-controlled tests. This test has little value from point of view of machine foundations.

Cyclic simple shear tests. Cyclic simple shear tests may be used for determination of dynamic soil properties for machine foundations or for earthquake loading conditions. The simple shear device consists essentially of (1) a sample box, (2) an arrangement for applying a cyclic load to the soil, and (3) an electronic recording system. A detailed discussion of various devices in use has been presented by Woods (1978) and Prakash (1981).

Cyclic torsional simple shear test. Details of torsional simple shear devices have been described by Woods (1978) and Prakash and Puri (1984). Use of this device eliminates some of the drawbacks of simple shear device.

Shake table tests. In order to avoid difficulties associated with other simple shear tests, because of the potential stress concentrations, non-uniform stress conditions, and void ratio variations which occur in small-scale simple shear tests Castro (1969), many researchers suggested that a large scale simple shear test might provide a superior mode of testing soils for liquefaction behavior. In machine foundation work such devices have limited application.

Silver (1981) has prepared Table 1 indicating the relative quality of each test techniques for measuring dynamic soil properties.

Table 1
RELATIVE QUALITY LABORATORY TECHNIQUES FOR MEASURING DYNAMIC SOIL PROPERTIES (Silver, 1981)

	Relative Qulaity of Test Results				
	Shear Modulus	Young's Modulus	Material Damping	Effect of No.of Cycles	Attenuation
Resonant Column	Good	Good	Good	Good	--
with Adaptation	--	--	--	--	Fair
Ultrasonic Pulse	Fair	Fair	--	--	Poor
Cyclic Triaxial	--	Good	Good	Good	--
Cyclic Simple Shear	Good	--	--	--	--

Table 1 (cont)

	Shear Modulus	Young's Modulus	Material Damping	Effect of No.of Cycles	Attenuation
Cyclic Torsional Shear	Good	--	Good	Good	--
Shake Table	Fair	--	--	Good	--

FIELD METHODS

The following field methods are in use in different parts of the world for determination of dynamic soil properties. (1) Cross-bore hole wave propagation test. (2) Up-hole or down hole wave progagation test. (3) Surface wave propagation test. (4) Vertical footing resonance test. (5) Horizontal footing resonance test. (6) Free vibration test on footings. (7) Cyclic plate load test. (8) Standard penetration test. A brief description of these tests will be presented.

Cross bore hole wave propagation test. In the cross-bore hole method (Stokoe and Woods, 1972), the velocity of wave propagation is measured from one subsurface boring to a second subsurface boring. A minimum of two bore holes are required one for generating an impulse and the other for sensors. For more details reference is invited to Woods (1978). Hoar and Stokoe (1981) have described in-expensive in-hole source for wave velocity measurements.

Up-hole or down-hole wave propagation tests. Up-hole and down-hole tests are performed by using only one bore hole. In the up-hole method, the sensor is placed at the surface and shear waves are generated at different depths within the bore hole. In the down-hole method, the excitation is applied at the surface and one or more sensors are placed at different depths within the borehole. Both the up-hole and the down-hole methods give average values of wave velocities for the soil between the excitation and the sensor if one sensor is used, or between the sensors, if more than one is used in the bore hole (Richart, 1977).

Surface-wave propagation test. Rayleigh waves travel in a zone close to the surface and their velocity of propagation is approximately the same as that of shear waves. Rayleigh waves can generated by a source of steady state vibrations Ballard (1964), Fry (1963), Richart, Hall and Woods (1970).

Vertical footing resonance test. Suitably designed concrete footing are excited into vertical vibrations using mechanical oscillator and from the observed amplitude frequency response, the soil constants and damping are computed IS 5249-1978, Prakash (1981), Prakash and Puri (1981, 1984).

Horizontal footing resonance tests. These are similar to vertical resonance tests except for the difference that the footing is excited using horizontal unbalanced forces.

Free vibration tests. The footing is excited into vertical or horizontal vibration by impact and its response is monitored.

Cyclic plate load test. The equipment for a cyclic-plate-load test is similar to that used in a static-plate-load test. It is assembled according to details given in ASTM D-1194-72 (1977) or IS 1888-1971 and Barkan (1962). A cyclic plate load test is performed as described by Prakash 1981. A typical cyclic load and settlement graph is shown in Fig. 6. The load intensity versus the elastic rebound is plotted as shown in Fig. 7. The value of C_u is calculated from the slope of the curve.

Standard penetration test. From observed uncorrected SPT values or N - values, shear wave velocity V_s may be computed from equation 6 Imai (1977).

$$V_s = 91(N)^{0.337} \text{ m/s} \tag{6}$$

Advantages and limitations of SPT have been adequately discussed by several investigators Fletcher (1965), Kovac (1975) Mohr (1966), De Mello (1971), Schmertmann (1975, 1977) and Seed (1981)

EVALUATION OF TEST DATA

Values of soil constants as obtained from different field and laboratory tests generally are quite different. This is to be expected in view of the different factors affecting their value. The design values need be chosen consistent with expected confining pressures and strain amplitudes likely to occur in the field problem. Detailed procedure for this purpose have been suggested by Prakash (1981), Prakash and Puri (1977, 1980, 1981, 1984).

DAMPING IN SOILS

Damping to the motion of the footing may be caused by two specific energy losses (1) the internal damping or absorption of energy within the soil mass and (2) dissipation of energy associated with the geometery of the foundation soil system. The former is known as the material or internal damping and the later dispersion or geometrical damping, Richart, Hall and Woods (1970), Prakash (1981). For a vibrating footing both geometrical and material damping come into play simultaneously. However, their relative magnitudes depend upon foundation geometery, soil type and mode of vibration.

Factors affecting damping.
There are several factors which affect damping. The more important of these are: (1) Strain level (2) Confining pressure (3) Void ratio (4) Number of cycles of oscillatory stress (5) Frequency of motion (6) Time effects.

The damping factor ζ increases, in general, with (1) increasing strain amplitude (2) decreasing confining pressure $\bar{\sigma}_o$ (3) decreasing void ratio (4) and increasing number of cycles of loading. Time effects for damping have also been studied, Marcusson and Wahls, (1977). Hardin and Drnevich (1972) have proposed charts and expressions for determination of damping in a particular problem. Considerable judgement is still needed to fix the value of damping to be taken in a design problem. Fortunately, the choice is not that bad and varied in machine foundation design.

Determination of damping. Value of material damping may be determined from free vibration decay curve in a resonance column test Richart, Hall and Woods (1970) or footing vibration test Prakash and Puri (1984). Geometerical damping may be estimated for any particular mode of vibration using charts or equations after Richart, Hall and Woods (1970). Over all value of damping (combined material plus geometerical damping) may be obtained from steady state field vibration tests, Prakash and Puri (1984).

FINAL COMMENTS

Methods of determining of dynamic soil constants and damping have been briefly reviewed. Factors affecting their values have been discussed. These factors must be considered in evaluating the test data. Mention must be made here of the comparison of modulus values obtained from laboratory and field tests Cunny and Fry (1973) reported values of G_{max} from 14 sites, determined both by laboratory and field tests. The stesdy state surface vibration method was used for evaluating G_{max} in the field while the resonant column test was used in the laboratory. The laboratory-determined shear and compression moduli were found to range within $\pm$ 50% of the in-situ moduli. It was observed that the cross-hole method should give better values of shear wave velocity (v_s) at depths from which undisturbed samples were taken, and that inclusion of the secondary time effect would bring the laboratory values nearer to **the** field falues (cohesive soils). The secondary time effect is negligible for sands. Stokoe and Richart (1973) and Iwasaki and Tatsuoka (1977) found agreement between the resonant column and the cross-hole field test values.

No serious efforts have been directed to determination Poisson's ratio of soils. Also better correlations need be established between damping and different factors which affect it. Also further studies need be directed toward evaluations of the geometrical damping related to vibrations of footings supported by layered media as well as of footings supported by soils which vary in stiffness with depth or confining pressure.

REFERENCES

Anderson, D.G. and K.H. Stokoe II (1977), "Shear Modulus: A time-Dependent Soil Property", Dynamic Geotechnical Testing, ASTM Spec. Tech. Pub. 654, Denver, CO, June, pp. 66-89.

Anderson, D.G. and R.D. Woods, (1976), "Time Dependent Increase in Shear Modulus of Clay", J. of the Geo.Eng.Div., A.S.C.E., vol.102, No. GT5, May 1976, pp. 525-537.

ASTM-D1194-(72), Reapproved 1977, "Standard Test Method for Bearing Capacity of Soil for Static Load on Spread Footings".

Ballard, R.F., Jr. (1964), "Determination of Soil Shear Moduli at Depth by In-Situ Vibratory Techniques", W.E.S., Misc. Paper No. 4-691, December.

Barkan, D.D. (1962), "Dynamics of Bases and Foundations", McGraw-Hill Book Co., New York.

Castro, G. (1969), "Liquefaction of Sands", Harvard Soil Mechanics Series No.81, Cambridge, MA, Jan.

Cunny, R.W. and Z.B. Fry, (1973), "Vibratory In-Situ and Laboratory Moduli Compared", JSMFD, ASCE, Vol.99, SM12, Dec.,pp 1055-1076.

DeMello, V. (1971), "The Standard Penetration Test--A State-of-the-Art Report", Fourth Pan American Conference of Soil Mechanics and Foundation Engineering, Puerto Rico, vol. 1, pp. 1-86.

Drnevich, V.P. (1967), "Effect of Strain History on the Dynamic Properties of Sand", Ph.D. Dissertation, Univ. of Michigan,151 pp

Drnevich, V.P. (1972), "Undrained Cyclic Shear of Saturated Sand" Journal of the Soil Mechanics and Foundations Division, ASCE, Vol. 98, No. SM8, Aug., pp. 807-825.

Drnevich, V.P. (1977), "Resonant Column Testing - Problems and Solutions", ASTM Symposium on Dynamic Soil and Rock Testing in the Field and Laboratory for Seismic Studies, Denver,June, pp. 384-398.

Fletcher, G. (1965), "Standard Penetration Test: Its Uses and Abuses", Journal of the Soil Mechanics and Foundations Division, ASCE, vol. 91, No. SM4, July, pp. 67-75.

Fry, Z.B. (1963), "Development and Evaluation of Soil Bearing Capacity, Foundations of Structures", W.E.S., Technical Report No. 3-622, rept. no. 1, July.

Hardin, B.O. (1978), "The Nature of Stress-Strain Behavior of Soils", State of the Art Report, Proc. ASCE Spec. Conf. on Earthquake Engrg. and Soil Dynamics, Pasadena, pp. 3-90, June.

Hardin, B.O. and W.L. Black (1969), "Closure to Vibration Modulus of Normally Consolidated Clays", J. Soil Mech. Found. Div., ASCE, vol. 95, no. SM6, pp. 1531-1537, November.

Hardin, B.O. and V.P. Drnevich (1972), "Shear Modulus and Damping in Soils, Design Equations and Curves", JSMFD, ASCE, Vol.98, No. SM7, pp. 667-692, July.

Hoar, R.J. and K.H. Stokoe (1981), "Crosshole Measurement and Analysis of Shear Waves", Proc. Tenth International Conf. SMFE, Stockholm, June 15-19, Vol. 3, pp. 223-226.

Imai, T. (1977), "Velocities of p- and s- Waves in Subsurface Layers of Ground in Japan", Proc. 9th International Conference of Soil Mech and Found., Tokyo, vol. 2, pp. 257-260.

Indian Standard Method of Load Test on Soils, IS 1888-1971, 1st rev, Indian Standards Institution, New Delhi.

Indian Standard Method of Test for Determination of Dynamic Properties of Soil, IS 5249-1978, 1st rev., Indian Standards Institution, New Delhi.

Ishihara, K. (1971),"Factors Affecting Dynamic Properties of Soils",Proc. Fourth Asian Regional Conference on Soil Mechanics and Foundation Engineering, Bangkok, vol. 2, August.

Ishihera, K (1981), "Strength of Cohesive Soils Under Transient Cyclic Loading Condition", State of the Art in Earthquake Engrg. edited by O. Erguney and M. Erdik, Turkish National Committee on Earthquake Engineering, Istanbul, Turkey, pp. 155-169.

Iwasaki, T, and F. Tatsuoka, (1977), "Dynamic Soil Properties with Emphasis on Comparison on Laboratory Tests with Field Measurements, Proc. Sixth World Conf. on Earthquake Engineering, New Delhi, vol.1, pp. 153-158, January.

Knovacs, W.D. (1975), Discussion of "On Dynamic Shear Moduli and Poisson's Ratios of Soil Deposits", Soils and Foundations, Vol. 15, No. 1, March.

Prakash, S (1975), "Analysis and Design of Vibrating Footings", Soil Mechanics Recent Developments , Proceeding of the General Session of the Symposium, University of New South Wales, Sydney, pp. 295-326, July 14-18.

Prakash, S. (1981), "Soil Dynamics", McGraw Hill Book Co., New York, NY.

Prakash, S and V.K. Puri (1969), "Design of a Typical Machine Foundation by Different Methods", Bulletin ISET, Vol.VI, No.3, pp. 109-136, September.

Prakash, S. and V.K. Puri (1977), "Critical Evaluation of IS: 5249-1969", Indian Geotechnical Journal, Vol. VII, No.1, pp. 43-56.

Prakash, S. and V.K. Puri (1980), "Dynamic Properties of Soils from In-situ Tests", unpublished report, University of Missouri-Rolla, MO., July.

Prakash, S. and V.K. Puri, (1981), "Dynamic Properties of Soils from In-situ Tests", J., Geotechnical Conf. Div. ASCE, vol. 107, no. GT7, July, pp. 943-963.

Prakash, S. and V.K. Puri (1984), "Analysis and Design of Machine Foundations", McGraw Hill Book Co., New York, NY.

Richart, F.E., Jr. (1977), "Dynamic Stress-Strain Relations for Soils", State of the Art Report, Proc. Ninth International Conf. on Soil Mechanics and Foundation Engineering, Tokyo, vol. 2, pp. 605-612.

Richart, F.E., Jr., D.G. Anderson and K.H. Stokeo II, (1977), "Predicting In-Situ Strain-Dependent Shear Moduli of Soil", Proc. Sixth World Conf. on Earthquake Engrg., New Delhi, Jan. pp. 6-159 to 6-164.

Richart, F.E. Jr., J.R. Hall and R.D. Woods, (1970), "Vibrations of Soils and Foundations", Prentice-Hall Inc, Englewood Cliffs, NJ.

Schmertmann, J.H. (1975), "Measurement of In-situ Shear Strength" Proceedings of the Conference on In-situ Measurement of Soil Properties, ASCE Spec. Conf., Raleigh, N.C., Vol.2, pp. 56-138.

Schmertmann, J.H. (1977), "Use the SPT to Measure Dynamic Soil Properties? - Yes, But...!" Dynamic Geotechnical Testing, ASTM, Spec. Tech. Pub. 654, pp. 341-355, June, Denver.

Seed, H.B. and I.M. Idriss (1981), "Evaluation of Liquefaction potential of Sand Deposits Based on Observation of Performance in Previous Earthquakes", In-Situ Testing to Evaluate Liquefaction Suceptibility, ASCE Annual Convention, St. Louis, Preprint #81-544.

Silver, M.L. (1981), "Load Deformation and Strength Behavior of Soils Under Dynamic Loading", State -of-the-Art Paper, Intern. Conf. on Recent Advances in Geotechnical Earthquake Engineering and Soil Dynamics, St. Louis, MO vol.3, pp. 873-897, April-May 1981.

Stephenson, R.W. (1977), "Ultrasonic Testing for Determining Dynamic Soil Moduli", Dynamic Geotechnical Testing, ASTM, Special Tech. Pub. No. 654, pp. 179-195, June, Denver.

Stokoe, K.H. and R.D. Woods (1972), "In-situ Shear Wave Velocity by Cross-Hole Method", J. Soil Mech. Found. Div., ASCE, vol.98, no. SM5, pp. 443-460.

Wahls, H.E. and W.F. Marcusson III (1977), "Effects of Time on Damping Ratio of Clays", Dynamic Geotechnical Testing, ASTM, Special Tech. Pub. No. 654, Denver, June.

Whitman, R.V. and F.V. Lawrence (1963), "Discussion", J. Soil Mech. Found. Div., ASCE, vol 89, no. SM5, pp. 112-115.

Woods, R.D. (1978), "Measurement of Dynamic Soil Properties - State of the Art", Proc. ASCE Spec. Conf. on Earthquake Engrg. and Soil Dynamics, Pasadena, vol.I, pp. 91-178, June.

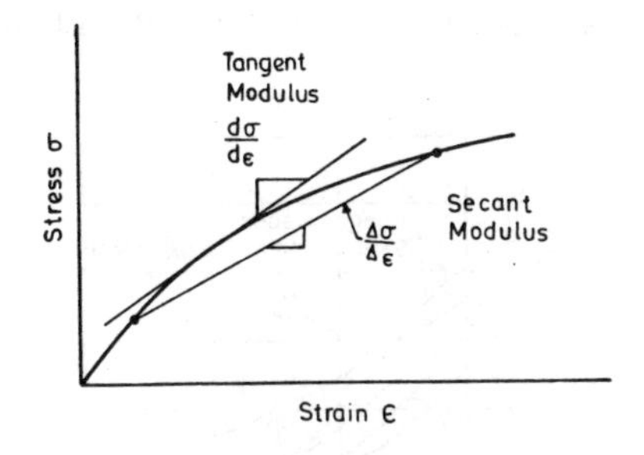

Fig. 1 -Definition of secant and tangent modulus.

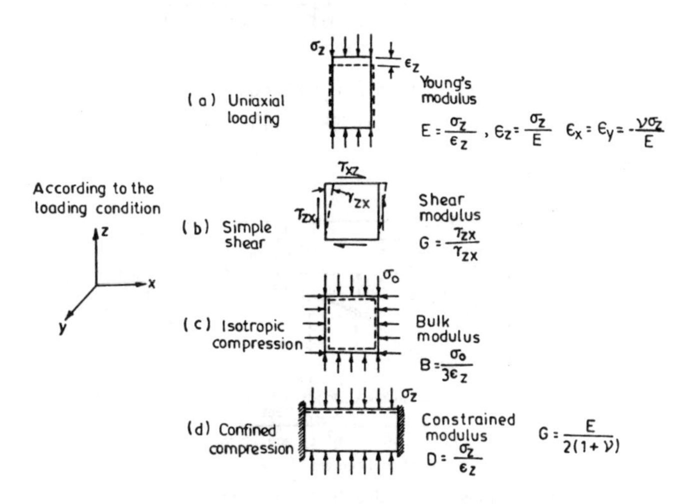

Fig. 2 - Various types of modulus

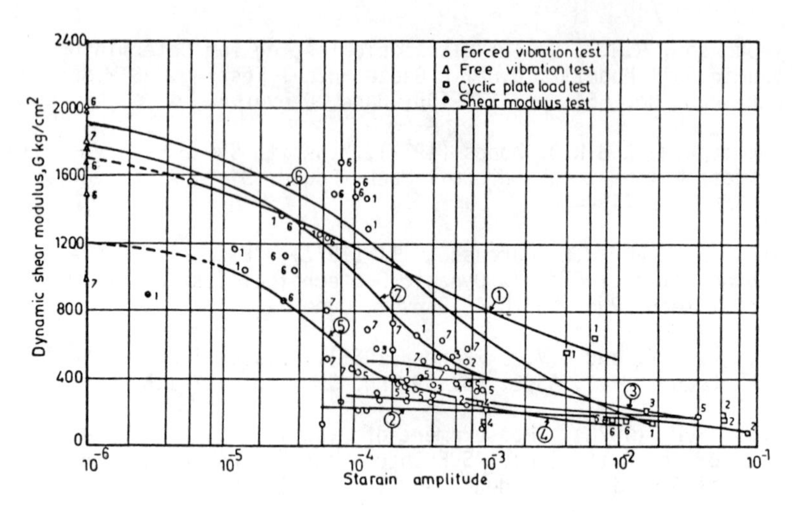

Fig. 3 - Dynamic shear modulus vs strain.(After Prakash and Puri, 1980)

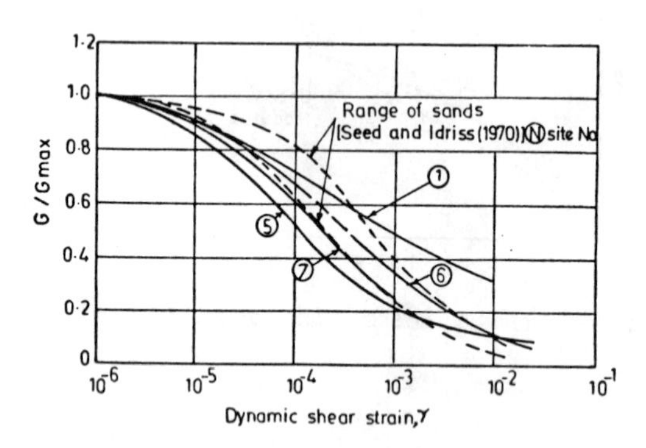

Fig.4 - Normalized shear modulus (G/Gmax) vs. shear strain
After Prakash and Puri, 1980, 1981)

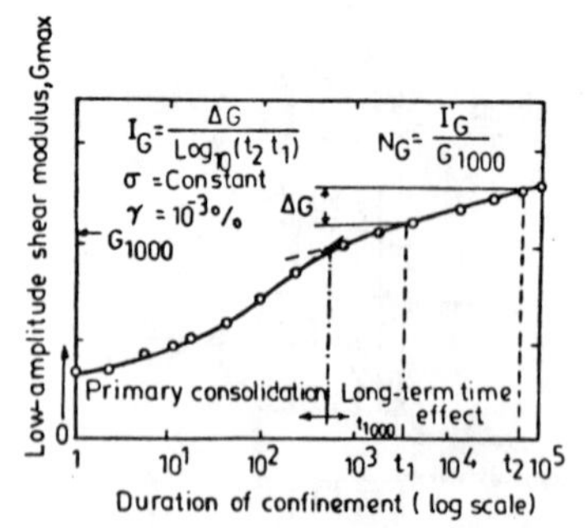

Fig.5 - Phases of modulus time response.
(After Stokoe and Anderson, 1977)

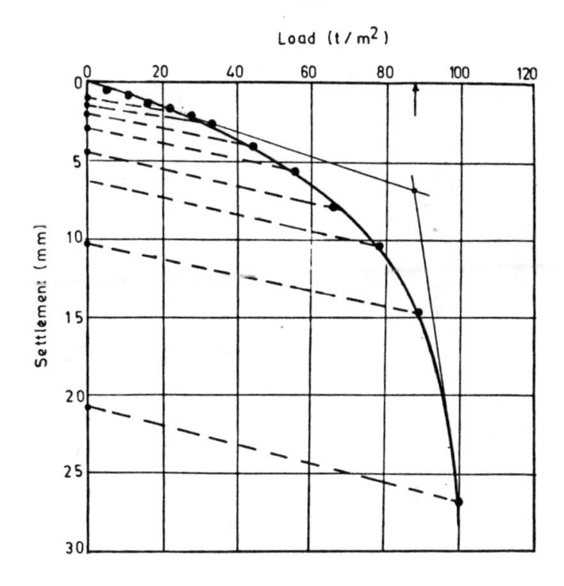

Fig. 6 - Load settlement of a vertical plate at Blending silo site of Akaltara cement factory.

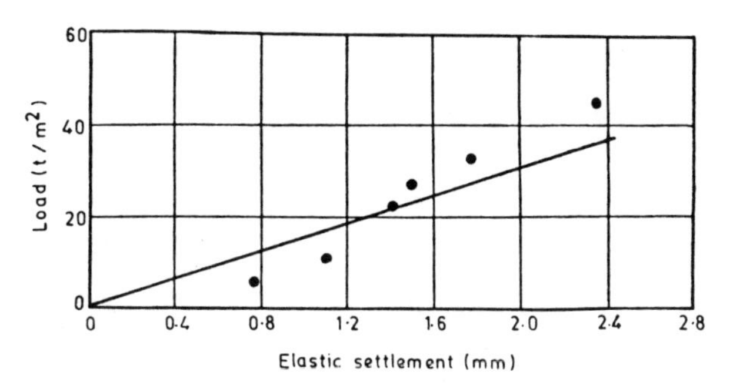

Fig. 7 - Load - elastic settlement plot.

12. Soil Liquefaction

Soil Dynamics & Earthquake Engineering Conference / Southampton / 1982.07.13-15

Large Scale Shaking Table Tests on the Effectiveness of Gravel Drains for Liquefiable Sand Deposits

YASUSHI SASAKI
Public Works Research Institute, Japan

EIICHI TANIGUCHI
Massachusetts Institute of Technology, USA

INTRODUCTION

Countermeasures to liquefaction of sand deposits during earthquakes are not completely understood at the present time, although there have been notable recent investigations in this area.

The most reliable countermeasure to liquefaction of sand deposits is to increase their density; this method has been widely used. In order to increase the density of such deposits the sand is usually compacted. However, this procedure is accompanied by noise and vibration problems, and so in some cases, cannot be used in urban areas.

On the other hand, the gravel drain system can be constructed without causing unpleasant noise or vibration. In general, in this method, vertical gravel piles or walls or beds are installed in the sand deposit to reduce the buildup of pore water pressure during earthquakes. This procedure can be carried out by using machines with lower noise and vibration, since it is not necessary to compact the soil.

In the research field, Seed and Booker (1977) investigated gravel drain systems and presented a simplified theory with resulting calculations for evaluating the possible effectiveness of such a system. Tokimatsu and Yoshimi (1980) performed some model tests on the stabilizing effects of gravel drains on potentially liquefiable sand deposits under buildings. These investigations offer a basis to evaluate the drainage effects of gravel drains; nevertheless, large scale model tests are needed to confirm the effectiveness of this system for preventing liquefaction of sand deposits; we need also to develop a reasonable design procedure for this system. For this purpose, large scale shaking table tests were performed in the Public Works Research Institute, Japan.

SHAKING TABLE TESTS

Purpose

The purposes of shaking table tests are:

i) to know the generation and dissipation characteristics of pore water pressure in the sand deposit by cyclic loading;

ii) to clarify the effective area of the gravel drain from the viewpoint of preventing liquefaction;

iii) to know whether the gravel drain method is effective in preventing the liquefaction of subsoils under a road that is partially buried.

Shaking Table

The size of the shaking table is 12m x 12m and the horizontal and vertical excitation are available.

Models

Six large scale models shown in Figure 1 were used in the tests. The model ground was set up in a steel box, the size of which is 12m in length, 3.5m in height and 2m in width. The height of the model ground was 3m. Gravel (size No. 6, see Table 1) was laid at the bottom of the box, and sand (properties of which are shown in Table 1) was put into the box in an air dried condition and compacted by human feet at every thickness of 50cm. Water was slowly infiltrated from the bottom to saturate the sand deposit after setting it up. The water table level is about 10cm below the model ground surface. Table 1 shows physical properties of the sand and gravel used in the test.

The gravel drains in these tests are wall instead of pile type, to make analysis in the two-dimensional domain possible. No. 5 gravel (see Table 1) 20cm in width had two filter zones of 10cm thick, No. 7 gravel on both sides.

Two models out of six contain the half buried type road model which is made of reinforced concrete.

Test Procedure

The horizontal sinusoidally fluctuating load was applied to the models longitudinally. The acceleration of loading was 200 gal, the duration time was 1 minute, and the frequency was 5Hz.

Seventeen accelerometers and 35 pore water pressure meters were installed in the model ground, and horizontal acceleration and pore water pressure were measured.

Results and Discussions

Figure 2 shows an example of change of acceleration, pore water pressure, and displacement with time. It is conceivable that the vibratory part of pore water pressure is caused by the rigid side walls and the mean value of the record signifies the accumulated pore water pressure by cyclic loading. Figure 3 shows the pore water pressure distribution in model 2 containing one gravel drain, 20 seconds after the beginning of the excitation. Figure 3 indicates that pore water pressure within 50cm from the edge of the gravel drain is smaller than that far away from the gravel drain. It can be noticed that the effective area of the gravel drain in this case is about 50cm from its edge. Sand and water spouting was observed in the sand area during and after excitation. This spouting continued about 5 minutes after the excitation ended.

Figure 4 compares the generation and dissipation of pore water pressure in models 1, 3 and 4. In this case the generation characteristics of pore water pressure are almost the same for the three models; here the gravel drain did not effectively prevent the generation of pore water pressure. The reason is thought to be that the frequency of loading was high and the generation of pore water pressure overcame the effect of water drainage.

However, Figure 4 indicates that the dissipation of pore water pressure in model 3 is larger than that of the model ground without a gravel drain.

Figure 5 shows the variation of pore water pressure with time in model 3. From this figure it can be noted that less pore water pressure is generated in an area surrounded by gravel drains than in the outer area. It is thought that the gravel drain becomes a kind of barrier to prevent water flow from outside the gravel drain.

Final settlement of the model ground surface was measured by using a scale after cyclic loading; settlement of the model ground was 5-15cm for models 1, 2 and 3.

Remarkable effects of the gravel drain on the half buried type road model were observed in comparing test results for models 5 and 6. Figure 6 shows the final displacement of the half buried type road model caused by cyclic sinusoidal loading. This model rose up for about 24cm and ran against the horizontal beam in the absence of a gravel drain, whereas it went up for only 5.0-6.2cm in the case where the gravel drain was set up below and beside the model. The apparent density of the half buried type road model is 1.07 and the density of the liquefied sand is about 1.9; therefore, the half buried type road model rose up. Thus, the gravel drain was very effective in preventing the half buried type road model from rising.

Figures 7 and 8 illustrate the distribution of pore water pressure in models 5 and 6. These figures indicate that the pore water pressure at the points in the gravel drain just below the half buried type road model with the gravel drain is very low compared with that at the same points in the case where there is no gravel drain. The uplift to the half buried type road model can be calculated by integrating the pore water pressure just below the base of the model. The uplift for the case with gravel drains is 1.89kN and that for the case without gravel drains is 36.2kN at 20 seconds after the beginning of excitation. This means that the rapid decrease of pore water pressure just below the half buried type road model resulted only in a small uplift to the base of the half buried type road model. Therefore, it is very effective to set up the gravel drain layer below the base for preventing the rise of the half buried type road model due to liquefaction of the surrounding sand deposit.

The gravel drain standing at both sides of the model no doubt functioned well to prevent water flow into the part underneath. In Figure 7 the pore water pressure is concentrated near the corner of the half buried type road model. It is conceivable that this high pressure is generated by the dynamic stress concentration at this part of the sand deposit.

Figure 9 shows the comparison of pore water pressure in cases with and without gravel drains. It can be noted from this figure that pore water pressure below the half buried type road model is decreased by the gravel drain both during and after excitation. It is especially interesting that the dissipation of pore water pressure after loading is very rapid.

LABORATORY SOIL TESTS

Types of Tests

Cyclic triaxial tests, cyclic torsional tests, resonant column tests and static triaxial tests were performed in the laboratory for the Sengen-yama sand used in the shaking table tests, to determine the dynamic soil constants which are necessary for the finite element analyses.

Test Procedure

The cyclic torsional tests were performed by using disturbed sand samples. It can be judged that the model ground was overconsolidated, because it was compacted before infiltrating water into the model. Considering this condition, the air dried soil sample was compacted and isotropically consolidated by a confining pressure of 33.9kPa which corresponds to the overburden pressure at the depth of 2m in the model ground before infiltrating water. Then water was filtered into the specimen until the degree of saturation of the specimen became 60%, which was estimated both from measurement of the density and water content before the liquefaction tests, and from a compaction test conducted by using a small cylindrical box (36.8cm in inner diameter and 80cm in height). Subsequently the confining pressure was reduced to 19.6kPa, which corresponds to the effective overburden pressure at a depth of 2m in the model ground after water infiltration. The average relative density of the specimens was 60.5%. After the specimen was prepared, a sinusoidally fluctuating cyclic torsional load was applied under stress controlled and undrained conditions.

The coefficient of volume compressibility was measured by using the cyclic triaxial testing apparatus.

Test Results

Figure 10 shows the relationship between the cyclic stress ratio τ/σ_o' and the number of cyclic loadings for the specimen with an overconsolidation ratio of 1.73 and degree of saturation of 60%. It can be noted from Figure 10, for example, that the cyclic stress ratio τ/σ_o' , which generates the double amplitude of axial strain of 5% by 20 cyclic loading, is 0.34. This value is about 50% higher than that for a normally consolidated and perfectly saturated specimen.

The buildup of the pore water pressure by cyclic loadings was represented by a curve offered by Seed _et al_. (1976), expressed by

$$\frac{u}{\sigma_o'} = \frac{1}{2} + \frac{1}{\pi} \arcsin \{ 2 (\frac{N}{N_\ell})^{1/\alpha} - 1 \} \qquad (1)$$

where u=pore water pressure, σ_o'=effective confining pressure, N=number of cyclic loadings, N_ℓ=number of cyclic loadings for liquefaction, and α=constant.

The shear modulus of Sengen-yama sand at the shear strain of 10^{-6} was 34.4 MPa, 39.9 MPa and 53.9 MPa for the confining pressure of 9.8kPa, 19.6kPa and 29.4kPa.

The static shearing strength was determined from the static triaxial tests; the cohesion c'=0 and the angle of shearing resistance ϕ'=36.6°.

FINITE ELEMENT ANALYSES

Purpose

The purpose of the finite element analyses performed here is to confirm the validity of the finite element technique developed, taking into account the generation and dissipation of pore water pressure.

Program

The two-dimensional finite element computer program designated as "SADAP" (Koga _et al_. [1981]) was modified to contain the generation and dissipation of pore water pressure during dynamic loading. The computer program "SADAP" employs the direct integration method in taking into account the nonlinearity of the stress-strain relationship of soils.

Models

The Hardin-Drnevich model was used in this series of analyses to express the stress-strain relationship of soils.

The methods presented by Ishihara and Towhata (1981) and Seed _et al_. (1976) were used in this calculation to express the generation of pore water pressure in the sand layer during cyclic loading. In this paper the former is designated the Ishihara model and the latter, the Seed model. The Ishihara model is based on the effective stress path during liquefaction. The stress path for the virgin loading can be assumed to be a parabola given by

$$\sigma_v' = m - \frac{B_p'}{m}\tau^2 \qquad (2)$$

where σ_v' = effective vertical stress, B_p' = soil constant representing the characteristics of the pore water pressure buildup, and m = parameter for locating a current parabolic stress path at each time step of computation. The increase of pore water pressure occurs during unloading and reloading, and it results in the decrease of effective stress given by

$$\Delta\sigma_v' = -B_u' \left(\frac{\tau}{\sigma_{v_o}'} - \frac{\tau_m}{\sigma_{v_o}'} \right) \left(\frac{\sigma_v'}{\sigma_{v_o}'} - \kappa \right) \Delta\tau \qquad \text{for } \sigma_v' \geq \kappa\sigma_{v_o}'$$

$$\Delta\sigma_v' = 0 \qquad \text{for } \sigma_v' < \kappa\sigma_{v_o}' \tag{3}$$

where τ_m = maximum shearing stress applied to the soil in the most recent cycle, B_u' = soil constant representing pore water pressure buildup during unloading and reloading, σ_{v_o}' = initial effective vertical stress, κ = parameter representing the point at which the pore water pressure ceases to build up when the vertical effective stress decreases to a certain value. After the stress path crosses the phase transformation line, the stress path during the increase in shearing stress is assumed to trace a hyperbolic curve given by

$$\left(\frac{\sigma_v'}{a} \right)^2 - \left(\frac{\tau}{a \tan\phi_\ell'} \right)^2 = 1 \tag{4}$$

where ϕ_ℓ' = the angle of shearing resistance in the low confining pressure, and a = parameter for locating a current hyperbola. For the unloading phase, the stress path is assumed to follow a straight line which is tangent to the hyperbolic curve.

The Seed model is based on undrained cyclic laboratory tests on liquefaction. As the first step, the number of cyclic loadings required to cause liquefaction should be determined from the shearing stress calculated by the response analysis. It was assumed that the relationship between the logarithm of the cyclic stress ratio τ/σ_v' and the logarithm of the number of cyclic loadings required to cause liquefaction N_ℓ can be approximated by a straight line given by

$$\log_{10} \frac{\tau}{\sigma_v'} = A \log_{10} \frac{N}{N_\ell} + B \tag{5}$$

where A and B are constants determined from the experiments.

In the Seed model, the relationship between pore water pressure buildup and the number of cyclic loadings is expressed by Equation (1). By using Equations (1) and (5) the increment of the pore water pressure can be calculated for the time increment.

The dissipation of pore water pressure in the ground can be expressed by the following equation, if it is assumed that the flow of water is governed by Darcy's law.

$$\frac{\partial}{\partial x} \left(\frac{k_x}{\rho_w g} \frac{\partial u}{\partial x} \right) + \frac{\partial}{\partial y} \left(\frac{k_y}{\rho_w g} \frac{\partial u}{\partial y} \right) = m_v \left\{ \frac{\partial u}{\partial t} - \frac{\partial u_g}{\partial t} \right\} \tag{6}$$

where ρ_w = density of water, g = acceleration of gravity, x,y = coordinates, m_v = coefficient of volume compressibility, k_x, k_y = coefficient of permeability in the x and y direction, respectively, and u_g = pore water pressure generated by cyclic shearing stress. Figure 11 shows the finite element modeling used in the calculation.

Soil Constants

The soil constants required for the Ishihara model were determined from the results of the cyclic torsional tests and the static triaxial tests; $B_p' = 0.211$ $B_u' = 0.072$ and the angle of transformation line $\theta_s' = 33°$.

The soil constants required for the Seed model are α in Equation (1) and A, B in Equation (5). These constants were taken as $\alpha = 1.0$, $A = -0.123$ and $B = -0.529$, which were determined from the cyclic torsional test data.

The soil constants used in the dynamic response analyses are shown in Table 2.

Results and Discussions

Figure 12 shows the comparison of the calculated amplification of acceleration in the model ground with measured values at a distance of 1m from the center of the model at 6 seconds after the beginning of excitation. Acceleration on the shaking table was 214.7 gal at that time. Figure 12 indicates that the calculated acceleration increases monotonically with the decreasing depth and the calculated values are larger than the measured values in the shallower part.

Figure 13 illustrates the buildup of pore water pressure during excitation. This figure shows pore water pressure calculated by the Ishihara and Seed models, and measured values of the shaking table tests at a distance of 2m from the gravel drain in model 2.

Figure 13 denotes that the pore water pressure calculated by the Ishihara model agrees relatively well with the measured values, although the calculated buildup of the pore water pressure is a little more rapid than that of the measured one at a depth of 1.5m. The accumulation of pore water pressure by the Seed model is more rapid than the measured value.

The pore water pressure becomes equal to the effective overburden pressure at 1 second after the beginning of excitation in the calculation, while it occurs at 5-6 seconds after the beginning of excitation in the shaking table tests. One possible reason for this difference could be that the model ground is not perfectly saturated with water and the effects of degree of saturation on the generation of pore water pressure cannot be appropriately evaluated in the calculation.

Figure 14 shows the comparison of the distribution of the pore water pressure between the calculation and experiment at 6 seconds after the beginning of excitation. The distribution of pore water pressure by the Ishihara and Seed models is similar to that obtained from this experiment.

CONCLUSIONS

(1) The dissipation of pore water pressure after the end of excitation was accelerated by the gravel drain.
(2) There was large uplift to the half buried type road model as a result of the pore water pressure generated during excitation in the case where no countermeasure to liquefaction was installed in the model ground nearby. In the other

case, uplift was remarkably reduced by installing gravel drains below the half buried type road model. Vertical gravel drains installed at both sides under the half buried type road model functioned effectively to stabilize it by preventing water and sand flow from the area outside the gravel drains.

(3) The generation of the pore water pressure was calculated by using the two-dimensional finite element computer program "SADAP" which considers the dissipation of the pore water pressure. The calculation offered relatively reasonable results on pore water pressure buildup during cyclic loading.

REFERENCES

Ishihara, K. and Towhata, I. (1981) "One-Dimensional Soil Response Analyses During Earthquakes Based on Effective Stress Method," Journal of the Faculty of Engineering, the University of Tokyo, Ser.(B), 36.

Koga, Y., Tateyama, S. and Karasawa, Y. (1981) "SADAP-1, A Computer Program for the Static and Dynamic Analyses of Earth Structures Considering Non linear Stress-Strain Relationship of Soils," Memorandom of the Public Works Research Institute No.1688 (in Japanese).

Seed, H.B. and Booker, J.R. (1977) "Stabilization of Potentially Liquefiable Sand Deposits Using Gravel Drains," Journal of the Geotechnical Engineering Division, ASCE, 103, GT7:757-768.

Seed, H.B., Martin, P.P. and Lysmer, J. (1976) "Pore Water Pressure Change During Soil Liquefaction," Journal of the Geotechnical Engineering Division, ASCE, 102, GT4:323-346.

Tokimatsu, K. and Yoshimi, Y. (1980) "Effects of Vertical Drains on the Bearing Capacity of Saturated Sand During Earthquakes," Proceedings of International Conference of Engineering for Protection from Natural Disasters, Bangkok, 643-655.

Table 1. Physical properties of sand and gravel

Model	Dry Density ρ_d (g/cm^3)	Wet Density ρ_t (g/cm^3)	Void Ratio e	Relative Density Dr(%)	Vertical Permeability Coefficient k_V(cm/sec)	Horizontal Permeability Coefficient k_H(cm/sec)
1	1.46	1.92	0.873	46.9	-	-
2	1.46	1.91	0.893	41.9	1.57×10^{-2}	9.02×10^{-3}
3	1.45	1.92	0.882	44.7	-	-
4	1.49	1.88	0.827	58.3	7.90×10^{-3}	-
5	1.47	1.93	0.855	51.3	-	-
6	1.48	1.94	0.846	53.6	-	-

	Sand	Gravel		
		NO.5	NO.6	NO.7
Specific Gravity	2.73	2.69	2.52	2.69
Maximum Grain Size(mm)	5	39	20	9.5
D_{50}(mm)	0.28	22	9.5	3.3
Uniformity Coefficient	2.91	-	-	-
Permeability Coefficient(cm/sec)	1.18×10^{-2}	6.1 x 10	1.31 x 10	5.79

Table 2. Soil constants used in the calculation

soil name	density ρ (g/cm^3)	cohesion c (kPa)	angle of shearing resistance ϕ' (degree)	coefficient of earth pressure at rest K_0	static Poisson's ratio ν_s	dynamic Poisson's ratio ν_d
sand	1.92	0	36.6	0.5	0.3	0.49
No.5 gravel	1.43	0	45.0	0.5	0.3	0.49

soil name	β in the Hardin Drnevich model	horizontal coefficient of permeability k_x (cm/sec)	vertical coefficient of permeability k_y (cm/sec)	coefficient of volume compressibility (m^2/MN)
sand	1	0.011	0.011	1.53
No. 5 gravel	1	61.0	61.0	0.765

soil name	$G_0(\text{MPa})=K\{(\sigma_0'(\text{MPa})\}^n$	
	K	n
sand	92.2	0.262
No. 5 gravel	45.8	0

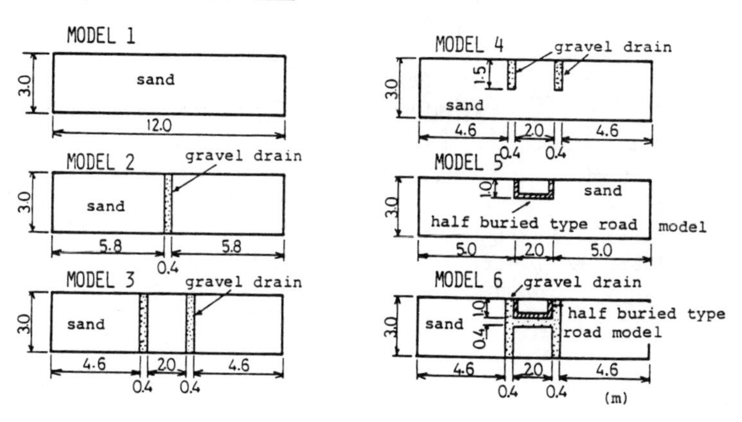

Figure 1. Models used in the shaking table tests

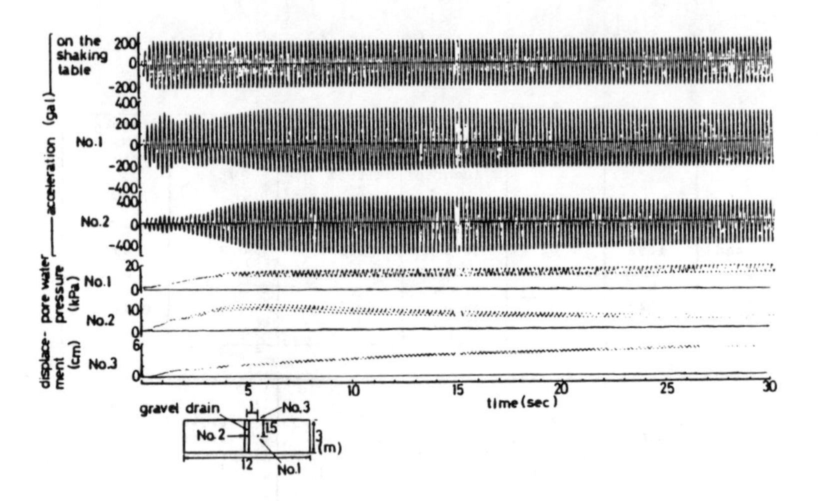

Figure 2. Records of acceleration, pore water pressure, and displacement

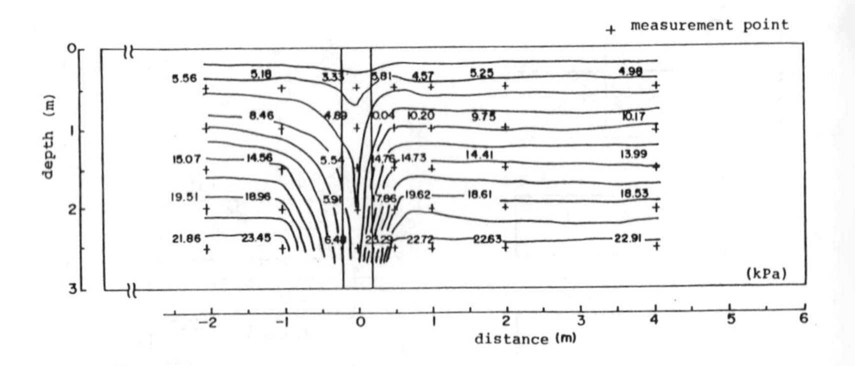

Figure 3. Distribution of pore water pressure (model 2, at 20 seconds after the beginning of excitation)

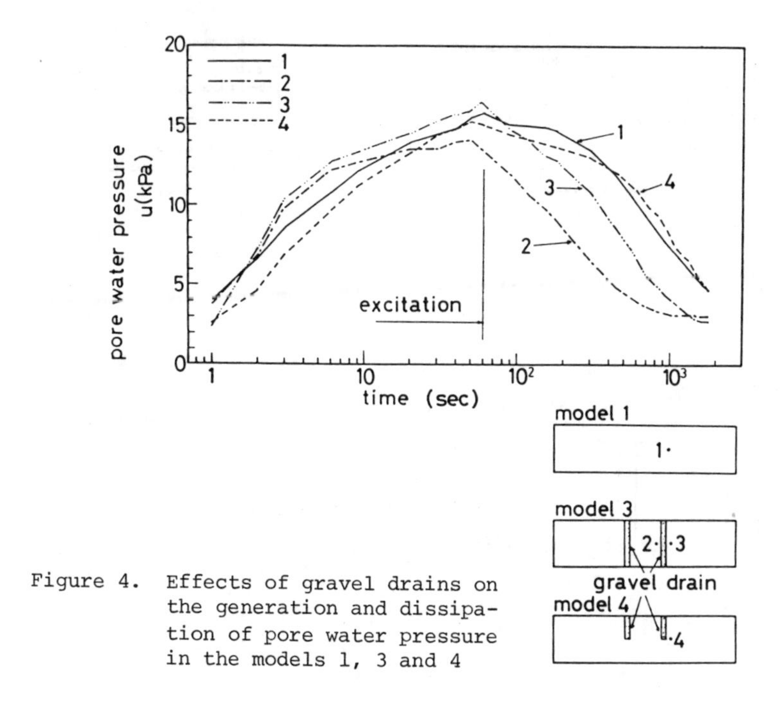

Figure 4. Effects of gravel drains on the generation and dissipation of pore water pressure in the models 1, 3 and 4

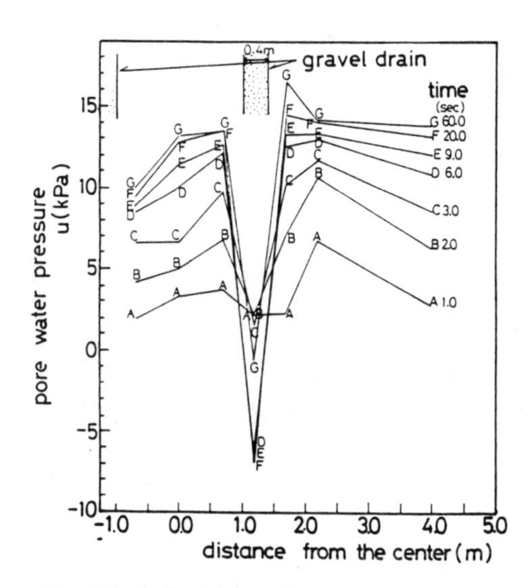

Figure 5. Distribution of pore water pressure in the horizontal direction during excitation (model 3)

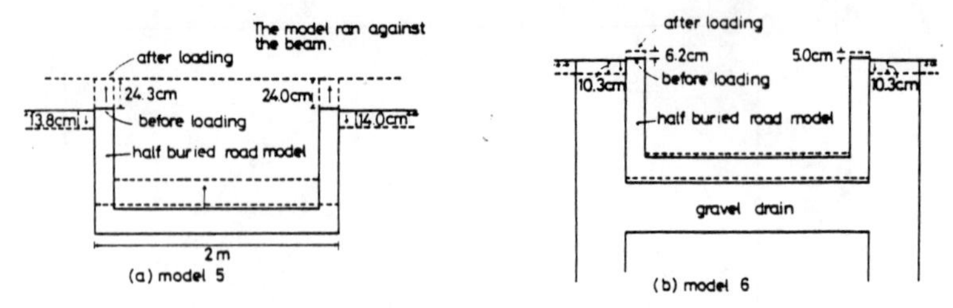

Figure 6. Final displacement of the half buried type road model with and without gravel drains caused by liquefaction

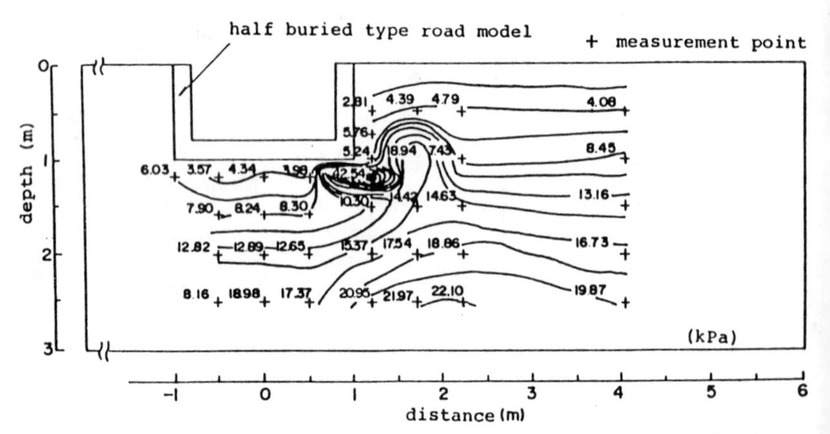

Figure 7. Distribution of pore water pressure (model 5, at 20 seconds after the beginning of excitation)

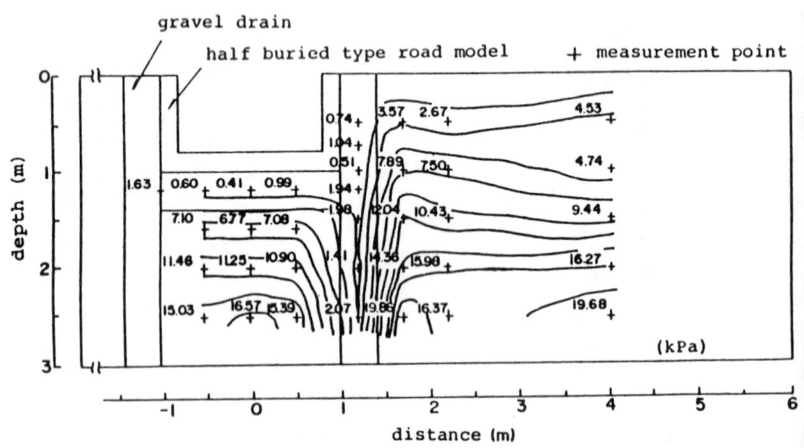

Figure 8. Distribution of pore water pressure (model 6, at 20 seconds after the beginning of excitation)

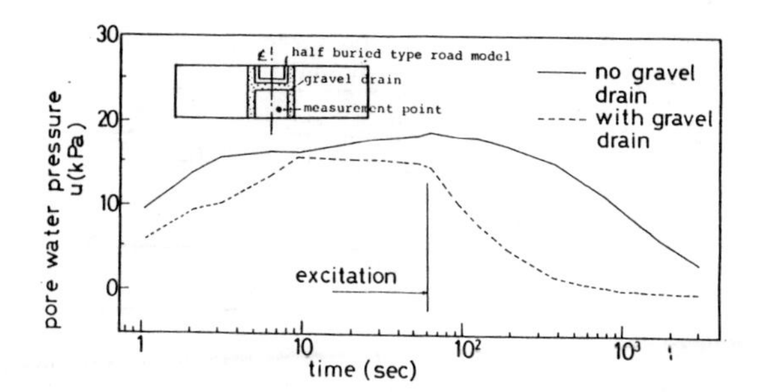

Figure 9. Effects of gravel drains on the generation and dissipation of pore water pressure below the half buried type road model (model 5 and 6)

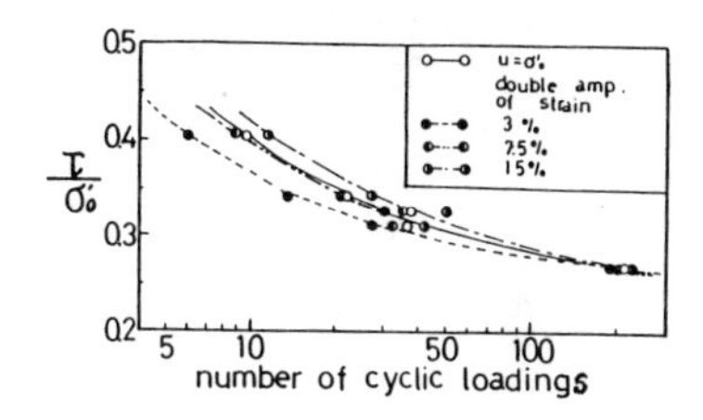

Figure 10. Cyclic stress ratio and number of cyclic loadings (Sengen-yama sand)

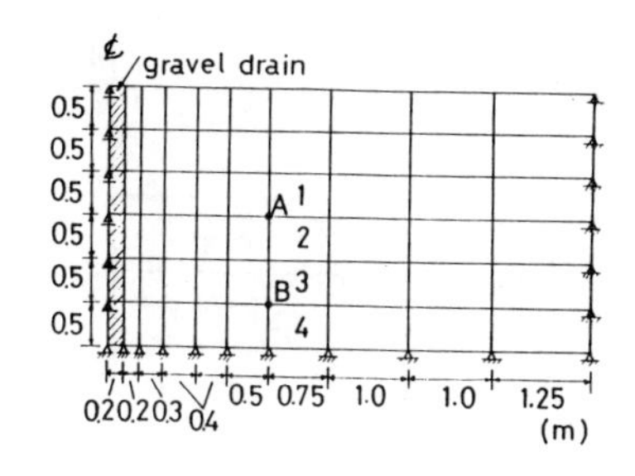

Figure 11. Model used in the finite element analyses

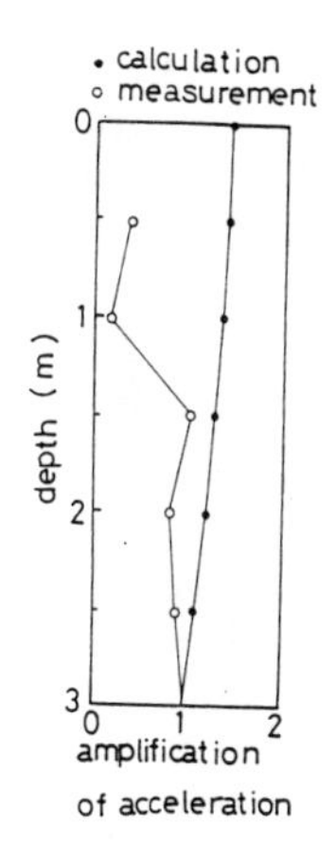

Figure 12. Calculated and measured acceleration (model 2, 1 m from gravel drain)

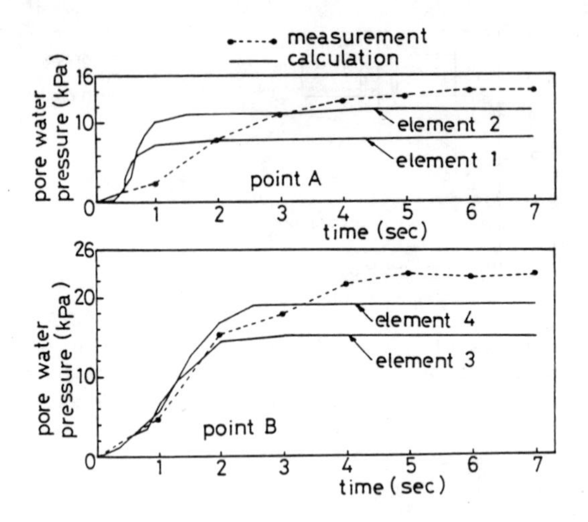

Figure 13(a). Comparison of pore water pressure between measurement and calculation by the Ishihara model (The location of the point and element is shown in Figure 11.)

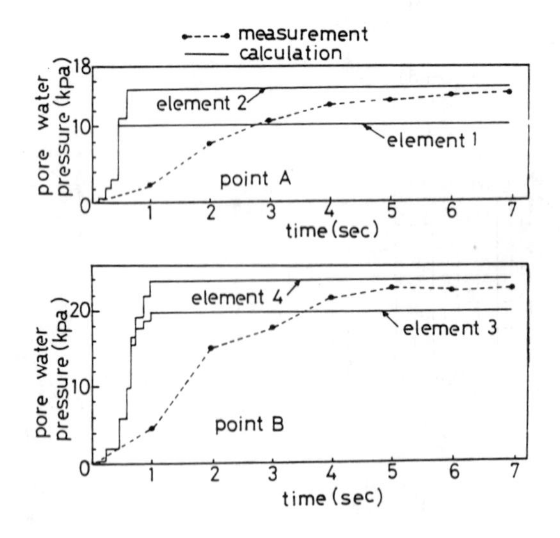

Figure 13(b). Comparison of pore water pressure between measurement and calculation by the Seed model (The location of the point and element is shown in Figure 11.)

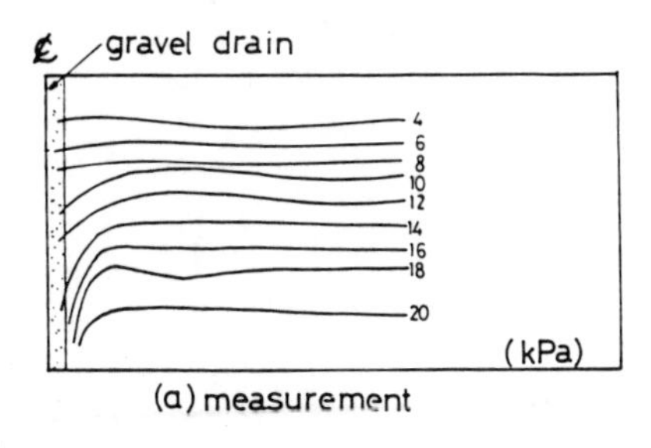

(a) measurement

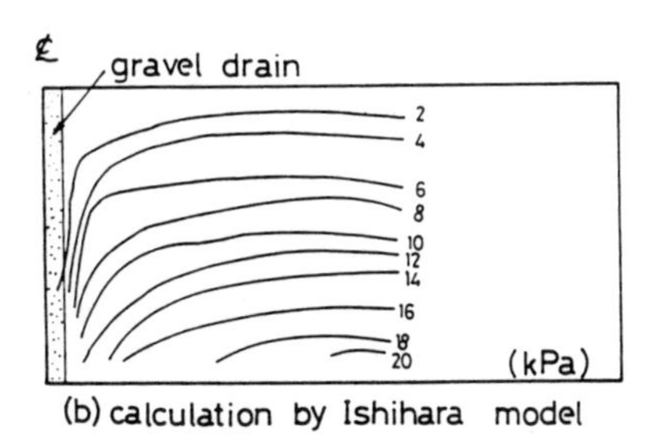

(b) calculation by Ishihara model

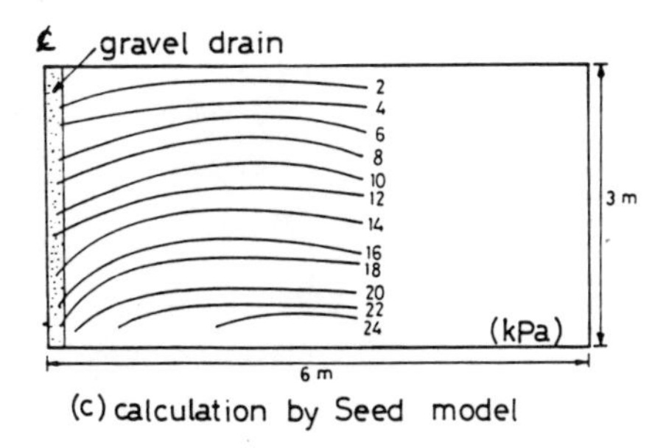

(c) calculation by Seed model

Figure 14. Distribution of pore water pressure of the measurement in model 2 and the calculation by the Ishihara model and the Seed model (at 6 seconds after the beginning of excitation)

Soil Dynamics & Earthquake Engineering Conference / Southampton / 1982.07.13-15

Analyses of Earthquake-Induced Flow Slide Movements

JEY K.JEYAPALAN
Texas A&M University, College Station, USA

SYNOPSIS

Substantial ground failure can occur due to an earthquake as a result of liquefaction. If the sand deposit is loose, the shear strength will not be recovered during deformation and a flow slide will result. In this phenomenon, the shear strength of the soil mass will be from viscous shearing and the debris will flow over substantial distances, ranging from hundreds of feet to several thousands of feet with devastating effects on people and structures in the path of the flowing deposit. In order to be able to assess the potential for damage in case of such an earthquake-induced ground failure, it is necessary to be able to predict the characteristics of the deformation and movement of these catastrophic events. In this paper, a procedure that has some potential for such analyses is described. The short term solutions to the dynamic wave theory are matched with the long term pseudo-steady state solutions, using singular perturbation techniques. The liquefied soil mass is represented by a Newotonian viscous model in the short term solutions and by an effective stress based visco-frictional model in the long term solutions. The increase in shearing resistance due to the dissipation of excess pore water pressures is also taken into consideration in the analyses.

INTRODUCTION

Certain soil deposits are vulnerable to earthquake-induced ground failures due to excessive pore pressure build-up and the associated loss of shearing resistance. The soil mass generally liquefies and flows over substantial distances, ranging from hundreds of feet to several thousands of feet with devastating effects on people and structures in its path.

The New Madrid earthquake sequence was a series of shocks that occurred during 1811 and 1812 in the Central Mississippi Valley. Marked ground disturbances characterized by fissures, sand boils and large landslides, occurred in an area more than 100 miles long and 50 miles wide along the Mississippi Valley. Seed (1967) estimated that major slides occurred within 30 miles of the epicentral region. Fuller (1912) described several geologic phenomenon that were the results of soil liquefaction. Quick conditions and lateral spreading were recorded in the flat alluvial valleys, whereas major landslides and some flow slides were observed on the steep bluffs. The Charleston, South Carolina earthquake (1886) was estimated to be of magnitude of 7+. Eye-witness accounts provide evidence of ground failures due to earthquake-induced liquefaction. Lateral spreading of alluvial deposits is indicated by the movement of railroad tracks and by the convergence of river banks. Several flow slides occurred in the vicinity of Half Moon Bay about 3 miles southwest of the San Andreas fault due to the 1906 San Francisco earthquake. In one of these liquefaction flow slides 2,000 yd^3 of material was involved. Another slide involved more than 10,000 yd^3. The largest of these flow failures moved 440 yds down a canyon and filled up the lower reach of this canyon to a depth of 10 to 15 ft. These examples are by no means unique. There were several other major liquefaction ground failures at other locations due to this 1906 San Francisco earthquake. Also there were many catastrophic liquefaction ground failures due to major seismic shocks in Mino-Owasi, Japan (1891), Tonanki, Japan (1944), Fukui, Japan (1948), Chile (1960), Alaska (1960), Nigata, Japan (1964), Takachioki, Japan (1968), and San Fernando, California (1971). More details of these serious incidents are given in Youd (1971, 1973).

Furthermore, during the 1971 Chilean earthquake, liquefaction of tailings dams was reported at several locations, including the famous El Cobre flow slide in which the flow of over 2 million yd^3 of silty sand over a distance of about 12 km took the lives of more than 200 people living downstream of the dam. One of the more recent of such failures was the catastrophic failure of the Mochikoshi Tailings dam in Japan as a result of earthquake shaking in March 1978. Liquefaction of the soil deposits after the earthquake has led to a major flow slide which carried about 100,000 tons of material about 2,000 ft to the Mochikoshi river and thence about 25 to 30 km to the Gulf of Suraga. The liquefaction failure which in this case contained a high proportion of sodium cyanide, posed a major threat to the health and safety of nearby residents and to acquatic life.

In all these cases, a significant characteristic of these failures and the damage has been the extensive flow movements of the materials involved in the slides as a result of the phenomenon of liquefaction. It becomes of extreme importance, therefore, to determine the potential extent of the movements which may result when earthquake-induced ground failure occurs. In this paper, the details of an analytical procedure that has some form of flow slide movement predictions are presented. The liquefied soil deposit is represented as a viscous fluid in this study and solutions are developed for velocity, displacement, and free surface profile of the flowing mass.

ANALYSIS OF FLOW SLIDE MOVEMENT

In order to analyze the flow phenomena associated with events of this type, it is necessary to use a suitable rheological model to represent the behavior of liquefied soil mass during flow. The subject of behavior of loose sand deposits under undrained loading has been studied thoroughly and is well understood. Extensive results on this topic are reported in Casagrande (1936), Castro (1969), Seed and Lee (1966, 1967), Seed and Peacock (1971), Seed et al. (1975), and Seed (1979). Because, the volume of a saturated sample of loose sand can not change during an undrained loading, the contractive volume change tendency causes a large positive change in pore pressure. The peak strength of the sample is reached at a small value of axial strain and the shearing resistance then decreases rapidly to a small residual value. Therefore, during a liquefaction flow slide loose sands will not lose all static strength. Furthermore, due to the shearing that takes place during flow, the sand will generate appreciable viscous resistance. However, in this work the residual shear strength will be ignored during the initial stages of flow and a purely viscous model will be chosen in order to keep the analyses simple. For the long term solution, a visco-frictional material model is chosen.

The analysis of the flow problem can be done in two parts:

(a) Short Term Solutions, and
(b) Long Term Solutions.

These two solutions will be developed separately using different mechanisms of flow and will be matched using a simple perturbation approach to develop solutions which are uniformly valid.

SHORT TERM SOLUTIONS

The short term solutions can be developed by treating the flow slide as a dam-break problem with some allowance for viscous resistance. The sand deposit susceptible to liquefaction is assumed to be of height, H_o is located on a slope, β as shown in Figure 1. Plane strain conditions are assumed and in order to simplify the analyses a vertical unsupported face in static equilibrium is assumed before the earthquake. Complete liquefaction of the sand deposit and its subsequent deformation can be represented in the analyses as an instantaneous complete removal of a vertical barrier holding a highly viscous fluid of viscosity, η. The equations governing the flow depth and flow velocity of the liquefied sand

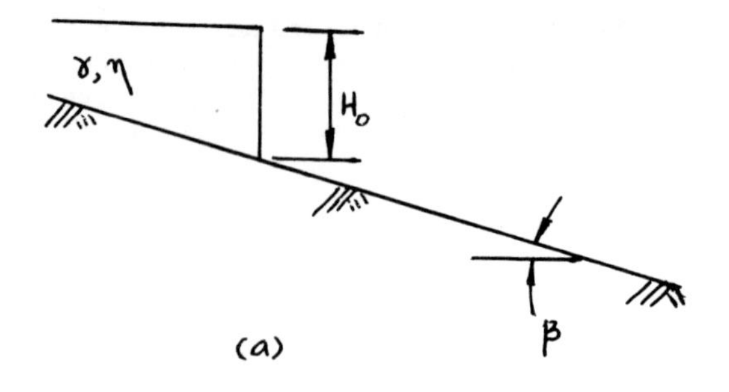

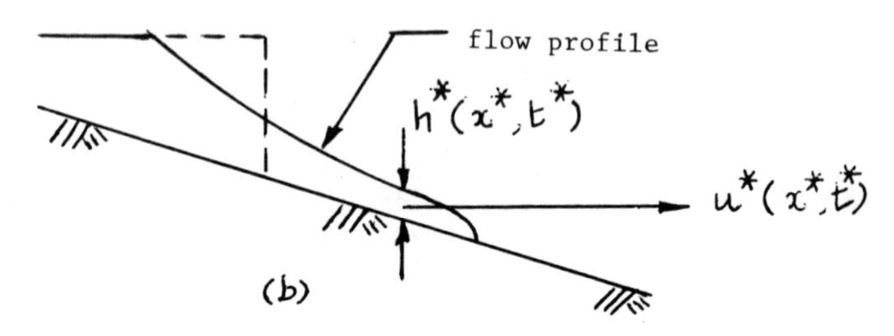

Figure 1. Flow Slide Model for Short Term Solutions

deposit are one-dimensional momentum equation and continuity equation given by:

$$\frac{\partial u^*}{\partial t^*} + u^* \frac{\partial u^*}{\partial x^*} + 2c^* \frac{\partial c^*}{\partial x^*} + 2g \frac{\eta c^*}{\gamma_{h_*}{}^2} - g \sin \beta = 0 \tag{1}$$

and

$$c^* \frac{\partial u^*}{\partial x^*} + 2 \frac{\partial c^*}{\partial t^*} + 2u^* \frac{\partial c^*}{\partial x^*} = 0 \tag{2}$$

where $c^* = \sqrt{gh^*}$, the celerity,
u^* = flow velocity,
h^* = flow depth,
x^* = horizontal distance, (see Fig. 1),
t^* = time,

and g = acceleration due to gravity.
Using a group of dimensionless variables,

$$x = x^*/H_o$$
$$h = h^*/H_o$$
$$c = c^*/\sqrt{gH_o}$$
$$u = u^*/\sqrt{gH_o}$$
$$t = t^*\sqrt{g/H_o}$$
and
$$R = 2\eta/\gamma H_o\sqrt{g/H_o} \qquad (3)$$

the governing equations (1) and (2) can be written in dimensionless form,

$$\frac{\partial u}{\partial t} + u\,\frac{\partial u}{\partial x} + 2c\,\frac{\partial c}{\partial x} + R\,\frac{u}{c^4} - \sin\beta = 0 \qquad (4)$$

$$c\,\frac{\partial u}{\partial x} + 2\,\frac{\partial c}{\partial t} + 2u\,\frac{\partial c}{\partial x} = 0 \qquad (5)$$

The above dimensionless equations govern the flow slide of a liquefied sand deposit. The initial conditions are

$$u\,(x,\,o) = 0 \qquad (6)$$

$$c\,(x,\,o) = 1 \qquad x < 0 \qquad (7)$$

$$c\,(x,\,o) = 0 \qquad x > 0$$

and the boundary conditions are,

$$u\,(x,\,t) = 0 \text{ for } t > 0 \text{ and } x = -\,t \qquad (8)$$

$$c\,(x,\,t) = 1 \text{ for } t > 0 \text{ and } x = -\,t \qquad (9)$$

Asymptotic series solutions can be developed for this initial value problem as given in Jeyapalan (1982), and the results are:

$$u(x,t,R) \doteq \frac{2}{3}\,(1+m) - \frac{216\,(1+m)\,Rt}{11\,(2-m)^4} + \frac{30\,Rt}{22\,(2-m)^3} + \frac{10\,Rt}{198}\left(\frac{2}{3} - \frac{m}{3}\right)^{3/2} + \frac{2t\,\sin\beta}{3} - \frac{10t\,\sin\beta}{15}\left(\frac{2}{3} - \frac{m}{3}\right)^{3/2} \qquad (10)$$

$$\text{and} \quad c(x,t,R) = \frac{1}{3}(2-m) + \frac{27(1+m)Rt}{11(2-m)^4}$$

$$+ \frac{21\,Rt}{22(2-m)^3} + \frac{7\,Rt}{198}\left(\frac{2}{3} - \frac{m}{3}\right)^{3/2}$$

$$- \frac{t \sin\beta}{30} - \frac{7t \sin\beta}{15}\left(\frac{2}{3} - \frac{m}{3}\right)^{3/2}, \qquad (11)$$

where $m = x/t$. (12)

The dimensionless flow depth at a desired location x is determined by the relation

$$h(x,t) = c^2(x,t,R), \qquad (13)$$

and the flow velocity at this section is given by equation (10). These perturbed solutions given approximate values of the flow velocity and the depth at all sections of the disturbed free mass of liquefied soil as functions of time for small values of R.

The velocity in expression (10) gives the inviscid behavior at the time of flow slide initiation, t=0, as shown in Figure 2. After the loss of significant shearing resistance, for later instants of time t_1, t_2, etc., the slope

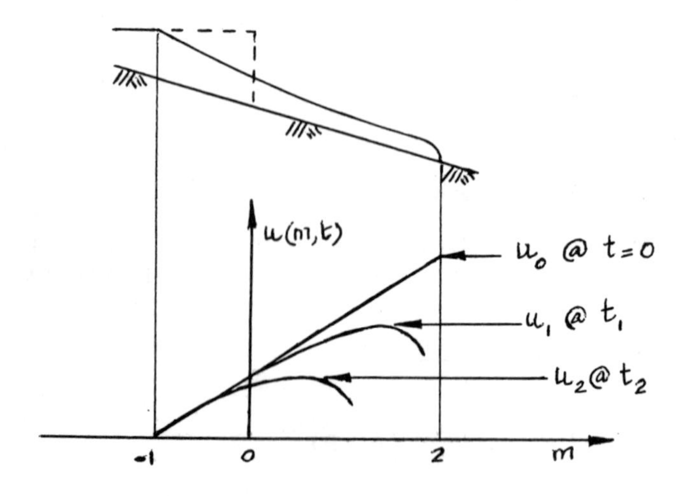

Figure 2. Flow Slide Velocity Distribution

will deform as shown in Figure 2 and the flow depth is increased by the presence of viscous boundary resistance.

Consequently, the velocity is decreased as shown in Figure 2. In the tip region, the viscous forces and the pressure gradients are of the same order of magnitude. Therefore, the velocity of flow does not change appreciably in this region. Thus, the maximum value of the velocity function is a good approximation of the tip velocity as shown in Figure 3. These

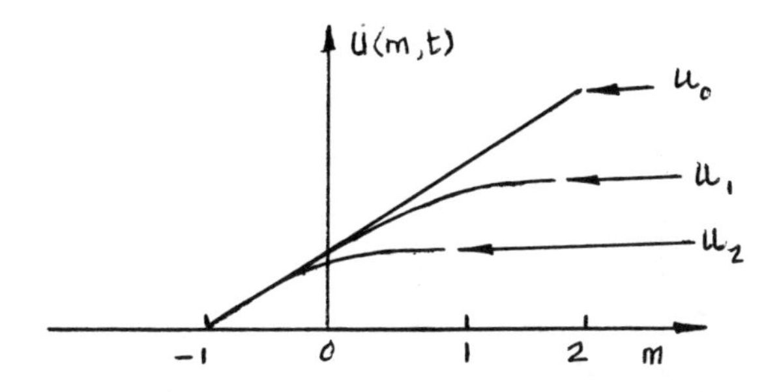

Figure 3. Flow Slide Tip Velocity

maximum velocities at different instants of time after the initiation of the flow slide have been plotted against time as shown in Figure 4. The corresponding tip displacement history can be evaluated using a numerical procedure and plotted as shown in Figure 4. However, the approximate location of the maximum velocity can be found by setting

$$\frac{\partial u}{\partial m}(x,t,R) = 0 \tag{14}$$

which yields from equation (10),

$$\frac{2}{3} - \frac{9\,(134 + 77m)\,Rt}{11\,(2-m)^5} + \left(\frac{5R}{198} + \frac{\sin\beta}{3}\right)\left(\frac{2}{3} - \frac{m}{3}\right)^{3/2} t = 0. \tag{15}$$

For small values of Rt << 1.0, the location of maximum flow velocity will occur near m = 2. Therefore, by an order of magnitude analysis, the equation (15) reduces to

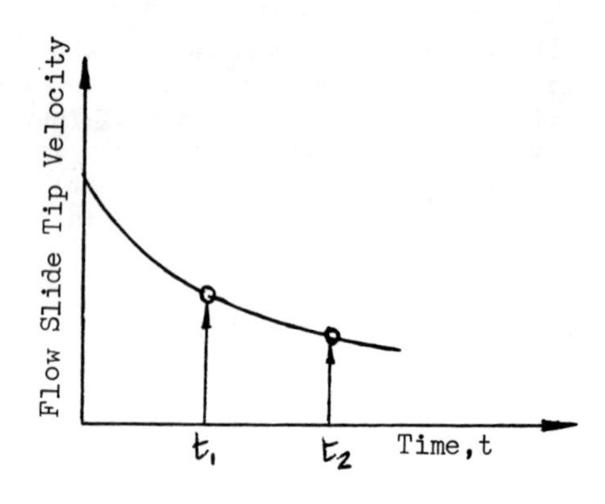

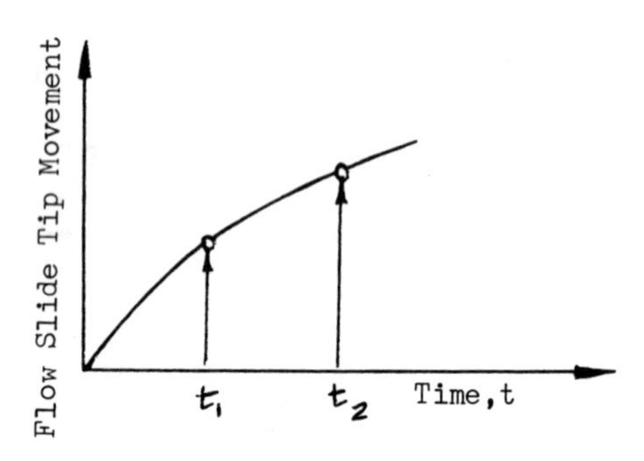

Figure 4. Flow Slide Tip Velocity and Displacement

$$p^5 - 353.45\ Rt = 0 \tag{16}$$

where

$$p = 2-m. \tag{17}$$

From equation (16)

$$p = 3.2334\ (Rt)^{1/5} \tag{18}$$

and the location of maximum velocity is given by,

$$m = 2 - 3.2334\ (Rt)^{1/5}. \tag{19}$$

The wave front velocity can be found by substituting equation (19) in (10), and this gives

$$u_{wave} = 2\left[1 - 1.3473\ (Rt)^{1/5}\right] + \frac{2t\ \sin\beta}{3} - 0.7460\ \sin\beta\ (Rt)^{3/10} + \ldots \quad (20)$$

Therefore, for $0(Rt)^{1/5}$ accuracy, the flood wave velocity can be written as

$$u_{wave} = \frac{2t\ \sin\beta}{3} + 2\left[1 - 1.3473\ (Rt)^{1/5}\right] \quad (21)$$

and the flood displacement is given by

$$x_{wave} = \frac{t^2}{3}\sin\beta + 2t\left[1 - 1.1228\ (Rt)^{1/5}\right]. \quad (22)$$

The short term solutions given by equation (21) and (22) are applicable only during the initial stages of flow.

LONG TERM SOLUTIONS

The above procedure for calculating the flow slide tip velocity is reasonable only until the location of the maximum velocity of flow reaches the origin of flow slide, at time t_c. After t_c, a simple pseudo-steady state model shown in Figure 5 can be used for evaluating the tip velocity. Using the free surface profile shown in Figure 5, at time t_c, the tip velocity is calculated with the pressure gradient and the gravitational force downstream of the slope. An estimate of t_c can be found readily by setting m in Equation (19) to zero as follows:

$$2 - 3.2334\ (Rt_c)^{1/5} = 0 \quad (23)$$

$$\text{Therefore } t_c = \frac{0.0905}{R}. \quad (24)$$

During flow, the excess pore water pressures will dissipate with time and a visco-frictional material model of the form (Fig. 6) given by the apparent viscosity equation,

$$\eta_a = \eta + \frac{\tau_R}{\dot{\gamma}} \quad (25)$$

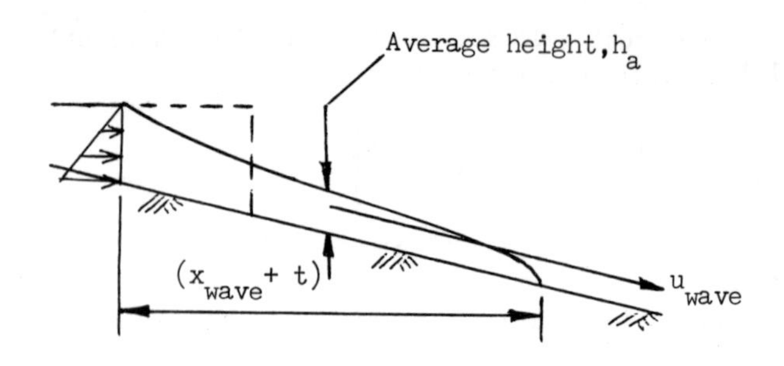

Figure 5. - Pseudo-steady State Flow Model

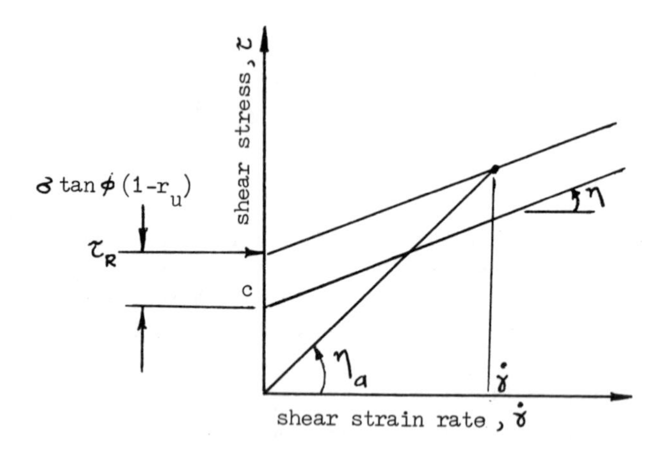

Figure 6. - Visco-frictional Model

where $\tau_R = c + \sigma(1-r_u) \tan \phi$, (26)

will better represent the behavior of flowing soil mass in the analyses. In Equation (26), the pore pressure ratio, r_u can be represented empirically by the equation

$$r_u = e^{-kt} \tag{27}$$

where k will be a function of coefficient of consolidation of the soil mass and the length of the longest drainage path. The dimensionless wave velocity can be calculated using the pseudo-steady state model as

$$u_{wave} = \frac{h_a^2 \; H_o}{2\eta_a} \sqrt{\frac{H_o}{g}} \left[\frac{(1-t \tan \beta)}{2(x_{wave} + t)} + \sin \beta\right] \quad (28)$$

$$\text{where } h_a = \frac{t \; (2 - t \tan \beta)}{2(x_{wave} + t) \cos \beta} , \quad (29)$$

which is the average height of the soil mass in motion.

Hence Equation (28) becomes

$$u_{wave} = \frac{\gamma H_o t^2}{8\eta_a} \frac{(2 - t \tan \beta)^2}{(x_{wave} + t)^2 \cos^2 \beta} \left[\frac{(1-t \tan \beta)}{2 \; (x_{wave} + t)} + \sin \beta\right] \sqrt{\frac{H_o}{g}} \quad (30)$$

Therefore, integrating Equation (30) gives

$$(x_{wave} - \underset{@\, t_c}{x_{wave}}) = \frac{\gamma H_o}{8\eta_a \cos^2 \beta} \sqrt{\frac{H_o}{g}} \int_{t_c}^{t} \frac{(2-t \tan \beta)^2}{(x_{wave} + t)^2} \left[\frac{(1-t \tan \beta)}{2 \; (t+x_{wave})} + \sin \beta\right] dt \quad (31)$$

By matching the short term solution for x_{wave} given in Equation (22)

$$x_{wave} = \frac{t_c^2}{3} \sin \beta + 2t_c \left[1 - 1.1228 \; (Rt)^{1/5}\right] + \frac{\gamma H_o}{8\eta_a \cos^2 \beta} \sqrt{\frac{H_o}{g}} \int_{t_c}^{t} \frac{(2-t \tan\beta)^2}{(x_{wave} + t)^2} \left[\frac{(1-t \tan \beta)}{2(t + x_{wave})} + \sin \beta\right] dt \quad (32)$$

where t_c was given in Equation (24) and η_a in Equation (25) in which the shear strain rate,

$$\dot{\gamma} = \frac{2 \; u_{wave}}{h_a} . \quad (33)$$

The Equations (32), (33), and (25) can be solved numerically.

RESULTS

In order to check the validity of these analytical solutions, a series of flume experiments were performed. Impoundments of viscous oil of different viscosities were released instantaneously simulating flow slides. A 16 mm movie camera was used to record the movement of the oil as a function of time as shown in Figure 7. Typical results from the flume experiments are compared with analytical predictions in Figure 8. The results from the analyses for x_{wave} were obtained by an iterative numerical scheme. The movement of the tip of the flowing oil mass from the analyses are consistently higher than that from the flume experiment in Figure 8. This is due to the presence of the side wall

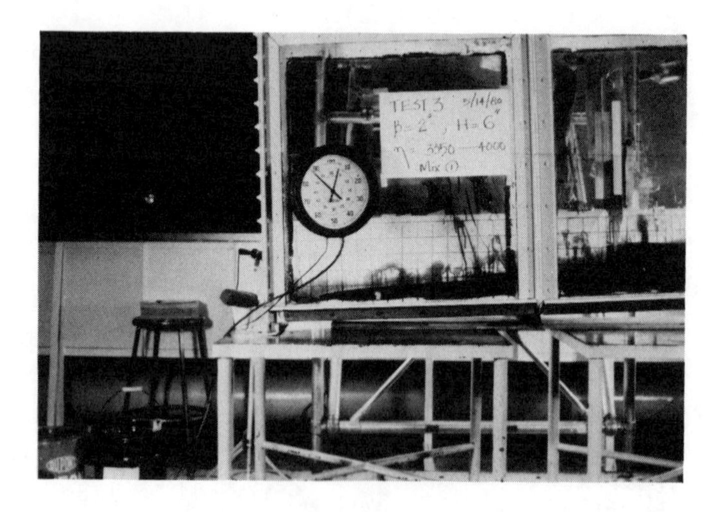

Figure 7. - Flume Experiment

boundary resistance in the flume experiment. Plane strain flow conditions are assumed and therefore, there is no account of the viscous resistance effects of the side wall boundary layers in the analytical procedure presented in this paper. Consideration for three-dimensional effects were made as part of this research and the details will be presented in a paper published elsewhere.

CONCLUSIONS

Cohesionless soil deposits are vulnerable to earthquake-induced

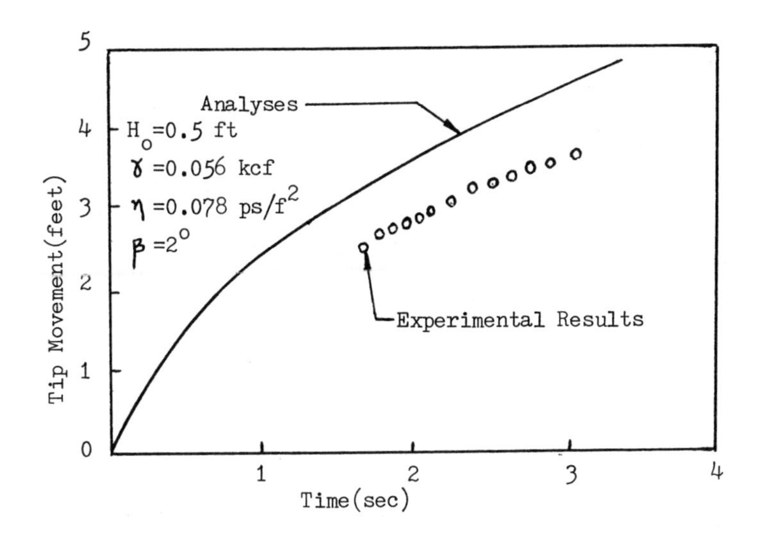

Figure 8. - Comparison of Analyses with Experimental Results

ground failures due to excessive pore pressure build-up and the associated loss of shearing resistance. The soil mass liquefies and flows over substantial distances with devastating effects on people and structures in its path. In this paper, the details of an analytical procedure that has some potential for flow slide movement predictions are discussed. The solutions for the initial stages and the long term conditions of flow are developed independently assuming a viscous model and a visco-frictional model respectively. The results from these analyses are compared with that from a laboratory flume experiment, using a highly viscous oil as the material involved in a flow slide. The theoretical results compare well with the observations from the flume experiment

ACKNOWLEDGMENTS

A portion of this work was completed while the author was at the University of California at Berkeley. The author gratefully acknowledges Professors J. M. Duncan and H. Bolton Seed for a number of stimulating discussions. Mrs. Joy Taylor typed the manuscript.

REFERENCES

Casagrande, A. (1936), "Characteristics of Cohesionless Soils Affecting the Stability of Slopes and Earth Fills," Journal of the Boston Society of Civil Engineers, January.

Castro, G. (1969), "Liquefaction of Sands," Ph.D. dissertation, Harvard University, January.

Fuller, M. L. (1912), "The New Madrid Earthquake," U.S. Geological Survey, Bulletin No. 494.

Jeyapalan, J. K. (1982), "A Viscous Model for Flow Slides of Tailings Dams," Proceedings of the Fourth International Conference on Numerical Methods in Geomechanics, Edmonton, Alberta, April.

Seed, H. B., and Lee, K. H. (1966), "Liquefaction of Sands During Cyclic Loading," Journal of the Soil Mechanics and Foundations Division, ASCE, SM6, November, pp. 105-134.

Seed, H. B., and Lee, K. H. (1967), "Undrained Strength Characteristics of Cohesionless Soils," Journal of Soil Mechanics and Foundations Division, SM6, November, pp. 333-360.

Seed, H. B., and Peacock, W. H. (1971), "Test Procedures for Measuring Soil Liquefaction Characteristics," Journal of the Soil Mechanics and Foundations Division, ASCE, Vol. 97, SM8.

Seed, H. B., Pyke, R., and Martin, G. R. (1975), "Effects of Multidirectional Shaking on Liquefaction of Sands," University of California at Berkeley, EERC Report 75-41.

Seed, H. B. (1979), "Considerations in the Earthquake-Resistant Design of Earth and Rockfill Dams," 19th Rankine Lecture presented in England and published in Geotechnique, Vol. XXIX, No. 3, September.

Youd, T. L. (1971), "Landsliding in the Vicinity of the Van Norman Lake in the San Fernando, California Earthquake of February 9, 1971," U. S. Geological Survey, Paper No. 733.

Youd, T. L. (1973a), "Liquefaction, Flow and Associated Ground Failure," U. S. Geological Survey, Circular No. 688.

Youd, T. L. (1973b) "Ground Movements in the Van Norman Lake Vicinity in the San Fernando, California Earthquake of February 9, 1971," U.S. Department of Commerce, NOAA.

Soil Dynamics & Earthquake Engineering Conference / Southampton / 1982.07.13-15

Laboratory Modelling of Blast-Induced Liquefaction

RICHARD J.FRAGASZY & MICHAEL E.VOSS
San Diego State University, CA, USA

INTRODUCTION

The phenomenon of earthquake-induced liquefaction has been observed and intensely studied for over two decades. Considerable progress has been made toward an understanding of the mechanism causing it and toward developing methods of predicting the probability of its occurrence at a specific site. In contrast, the related phenomenon of blast-induced liquefaction has received very little attention. Most of the literature on blast-induced liquefaction has been concerned with its possible effects on the size and shape of nuclear and high energy explosives (Melzer, 1978; Blouin, 1979) or with its association with compaction of loose sands by blasting (Kok, 1977 and 1981). To date, there is no generally accepted theory to explain the blast-induced liquefaction mechanism, or to predict the occurrence and effects of blast-induced liquefaction on a site-specific basis.

The objective of the research described here is to verify a proposed mechanism for blast-induced liquefaction. It is only after this has been accomplished that researchers can begin to understand the factors which influence the susceptibility of a specific soil deposit to blast-induced liquefaction.

DESCRIPTION OF PROPOSED MECHANISM

Prater (1977) and Rischbieter et al (1977) have proposed a mechanism to explain blast-induced liquefaction based on the behavior of saturated sands under drained cycles of compressive loading. It is well known that granular soils deform inelastically during drained compressive loading. However, the normal assumption by Geotechnical engineers is that under undrained conditions no plastic volume change will occur in the soil skeleton due to compressive loading because the effective stress does not change. This is only a good assumption if the increase in total stress is small.

The actual rise in effective stress can be determined if the relative compressibilities of the soil skeleton and the pore water are known and Skempton's pore pressure parameter, B, is calculated. The change in effective stress $\Delta\sigma'_3$ is given by the following equation:

$$\Delta\sigma'_3 = (1 - B)\Delta\sigma_3 \quad \ldots\ldots\ldots\ldots \quad (1)$$

where $\Delta\sigma_3$ is the increase in total stress. The value of B is calculated as follows (Skempton, 1954):

$$B = \frac{1}{1 + n(c_w/m_v)} \quad \ldots\ldots\ldots\ldots \quad (2)$$

where c_w and m_v are the bulk compressibilities of water and the soil skeleton, respectively, and n is the porosity. Typical values of m_v range between 2×10^{-3} m^2/KN for a very soft clay to $5 \times 10^{-6} m^2/KN$ for a moderately dense sand (Lade and Hernandez, 1977). For these two values and a porosity of 50%, the values of B are 0.9999 and 0.955, respectively ($c_w = 4.67 \times 10^{-7} m^2/KN$). If the soil behaves inelastically during a cycle of compressive loading, the compressibility of soil will be different during loading and unloading. This could lead to liquefaction as illustrated in Fig. 1. Assume an element of soil exists with a total stress of 5 MPa acting on it, an effective stress of 3MPa, and a pore water pressure of 2MPa. The soil is subjected to an increase in total stress from 5 to 35MPa, as shown by the path AB in Fig. 1. If the soil is a dense sand and has an average B value during loading of 0.95, the pore water pressure will increase to 30.5MPa, and the effective stress will rise to 4.5MPa. These increases are shown by paths DE and GH, respectively. Plastic volume change in the soil skeleton requires a smaller compressibility and therefore a smaller B value during unloading. If the average B value is reduced to 0.75, the pore pressure will drop more slowly than it rose and will equal the total stress when they both reach 17MPa. At this point, F, the change in pore pressure is -13.5MPa, which is 75% of the change in total stress (-18MPa), and liquefaction occurs.

Two major assumptions are made in this theory. The first is that the dynamic strain is the same in the soil skeleton and the pore water. This is possible only if there is no separation of the wave fronts in the soil skeleton and the water. This assumption has been verified for blast-induced waves both mathematically by Ishihara (1977) and experimentally by Risch-

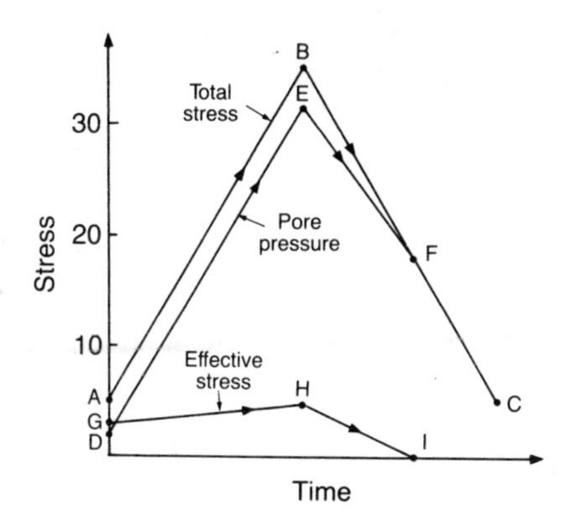

Figure 1. Model for Blast-Induced Liquefaction

bieter et al (1977) and Lyakhov and Polyakova (1967). The second assumption is that the hysteresis observed in the stress-strain curve during drained loading is also present during undrained conditions. This assumption has not been proven, and in fact, Cristescu (1967) states that a saturated soil behaves elastically and therefore undergoes no permanent volume change. If this is true, the proposed mechanism is not possible.

PREVIOUS LABORATORY AND FIELD INVESTIGATIONS

The laboratory and field investigations performed to date have supported the proposed liquefaction mechanism but have neither conclusively proven its validity nor provided the data needed to predict large scale behavior. Kok (1977) has caused liquefaction to occur in a laboratory experiment in which a plexiglass permeameter was filled with sand and saturated with deaired water. The cylinder was then struck by a pendulum, and both change in void ratio and pore water pressure were recorded. Kok (1977) also conducted small scale field experiments with up to 100 Kg of TNT. These field tests indicate that the horizontal zone of liquefaction increases as the cube root of the charge weight. Both of these experiments, however, were concerned with the compaction of soil. No data on the magnitude of the stress waves were obtained.

Studer and Hunziker (1977) have conducted shock tube experiments in which liquefaction was observed. They were unable, however, to produce 100% saturation in their test apparatus. Rischbieter et al (1977) also were unable to obtain 100% saturation in their field tests. This is very important because the compressibility of water is greatly increased by even a small amount of undissolved air. For example, a change in the degree of saturation from 100% to 99.9% increases compressibility from $4.67 \times 10^{-7} m^2/KN$ to $7.44 \times 10^{-6} m^2/KN$ (Richart et al, 1970), resulting in a drop in B value from 0.95 to approximately 0.6 for a moderately dense sand.

As the compressibility of the pore water increases, it becomes more difficult to achieve liquefaction. Since natural soils below the permanent ground water table are saturated, it is important to conduct liquefaction experiments with completely saturated soils, otherwise, liquefaction potential will be underestimated. Rischbieter (1977) cites the difficulty in obtaining 100% saturation as one of the major problems in performing blast-induced liquefaction studies. In the experiments described below, particular attention was paid to the problem of sample saturation.

EXPERIMENTAL METHOD

As a first step toward verifying and quantifying the mechanism proposed above, a series of undrained high pressure isotropic compression tests were performed on samples of both Eniwetok Beach Sand and Ottawa Sand. Quasi-static tests were conducted to verify the central assumption of the theory that elastic volume change in the soil skeleton can occur in undrained loading. It was felt that only if these tests were successful would more expensive and complicated dynamic tests be justified. To achieve the desired stress levels (35MPa), a high pressure triaxial system with a capacity of 70MPa was used. Standard triaxial samples of loose sand, 35.6mm (1.4in) in diameter were formed and saturated using the CO_2 method of saturation described by Lade and Duncan (1973). Brass shim stock 0.05mm thick was placed between the sand and the rubber membrane to reduce the effects of membrane penetration. The saturated samples were brought to an initial effective stress of 1.03MPa and a pore water pressure of 0.69MPa. A typical test consisted of a loading stage where the confining pressure was increased from 1.72MPa to 34.5MPa, and an unloading stage where the confining pressure was returned to its original value. During this cycle of loading, the drainage line from the sample was closed. The confining pressure and the pore water pressure were monitored with pressure transducers located just outside the triaxial cell, and were recorded on an X-Y recorder. The entire process took about two minutes. Figure 2 shows the typical form of the output from the X-Y recorder. It should be noted that the 45° line in the figure represents

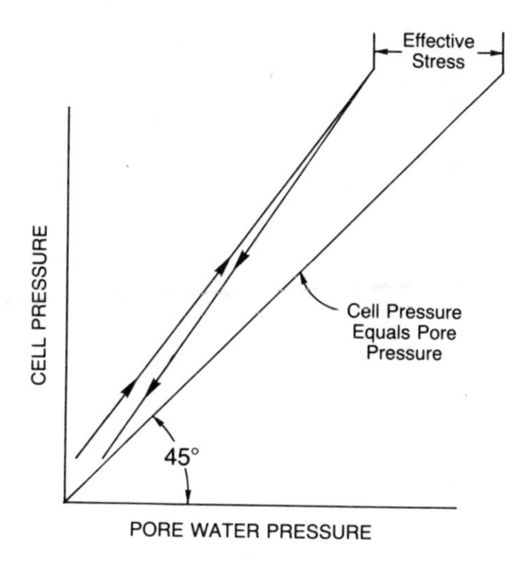

Figure 2. Typical Form of X-Y Recorder Output

liquefaction; i.e., cell pressure equals confining pressure and effective stress equals zero. Even if the liquefaction line is not reached during unloading, any excess pore pressure at the end of the test (above the initial 0.69MPa) represents permanent volume change of the soil skeleton. Detailed information on the experimental method is given in Fragaszy and Voss (1981).

TEST RESULTS

Nine tests were conducted using Eniwetok Beach sand, a coral sand with a specific gravity of 2.71, a mean grain size of 0.35 mm, and a minimum dry density of 1.30g/cc. These tests covered a range of dry densities between 1.30 and 1.45g/cc. The results of these tests are summarized in Table 1. In each of the tests, the increase in effective stress is approximately the same (0.4MPa), and liquefaction occurred at approximately the same cell pressure (25MPa). Figure 3 presents the results from a typical test.

Test E-7 was used to estimate the stress increase required to cause liquefaction. After the first cycle of loading from 1.69 to 34.5MPa and back, the drainage line was opened and the pore water pressure was allowed to return to 0.69MPa, thus reestablishing the initial stress conditions. A second test was then conducted, but with a cycle to only 6.9MPa. This resulted in a residual pore water pressure of 350KPa, only

Table 1. Results of Tests on Eniwetok Sand

		Initial		Peak		End		
Dry Density (g/cc)	Test	Confining Pressure (σ_3, MPa)	Water Pressure (u, MPa)	Confining Pressure (σ_3, MPa)	Water Pressure (u, MPa)	Confining Pressure (σ_3, MPa)	Water Pressure (u, MPa)	Residual Water Pressure (MPa)
1.35	E-6	1.72	0.69	34.5	33.1	1.72	1.72	1.03
1.31	E-7	1.72	0.69	34.5	33.1	1.72	1.72	1.03
	"	1.72	0.69	6.9	5.7	1.72	1.03	0.35
	"	1.72	0.69	13.8	11.7	1.72	1.38	0.70
	"	1.72	0.69	20.7	18.4	1.72	1.55	0.86
	"	1.72	0.69	27.6	25.5	1.72	1.72	1.03
	"	1.72	0.69	34.5	31.9	1.72	1.72	1.03
1.33	E-8	1.72	0.69	34.5	33.1	1.72	1.72	1.03
1.41	E-9	1.72	0.69	34.5	32.8	1.72	1.72	1.03
1.45	E-10	1.72	0.69	34.5	33.1	1.72	1.72	1.03
1.30	E-11	1.72	0.69	34.5	32.9	1.72	1.72	1.03
1.32	E-12	1.72	0.69	34.5	33.1	1.72	1.72	1.03
1.30	E-15	1.72	0.69	34.5	32.2	1.72	1.72	1.03

Table 2. Results of Tests on Ottawa Sand

			Initial		Peak		End		
Soil	Dry Density (g/cc)	Test	Confining Pressure (σ_3, MPa)	Water Pressure (u, MPa)	Confining Pressure (σ_3, MPa)	Water Pressure (u, MPa)	Confining Pressure (σ_3, MPa)	Water Pressure (u, MPa)	Residua Water Pressur (MPa)
Ottawa Flintshot	1.59	F-4	1.72	0.69	34.5	31.0	1.72	1.21	0.52
		"	1.72	0.69	6.9	5.3	1.72	0.69	0
		"	1.72	0.69	13.8	11.4	1.72	0.69	0
		"	1.72	0.69	20.7	17.6	1.72	0.86	0.17
		"	1.72	0.69	27.6	23.4	1.72	1.03	0.34
Ottawa Sawing	1.59	S-1	1.72	0.69	34.5	30.5	1.72	1.21	0.52
Ottawa Banding	1.46	B-1	1.72	0.69	34.5	31.4	1.72	1.55	0.86
		"	1.72	0.69	6.9	5.3	1.72	0.69	0
		"	1.72	0.69	13.8	11.5	1.72	0.86	0.17
		"	1.72	0.69	20.7	17.6	1.72	0.90	0.21
		"	1.72	0.69	27.6	24.1	1.72	0.90	0.21

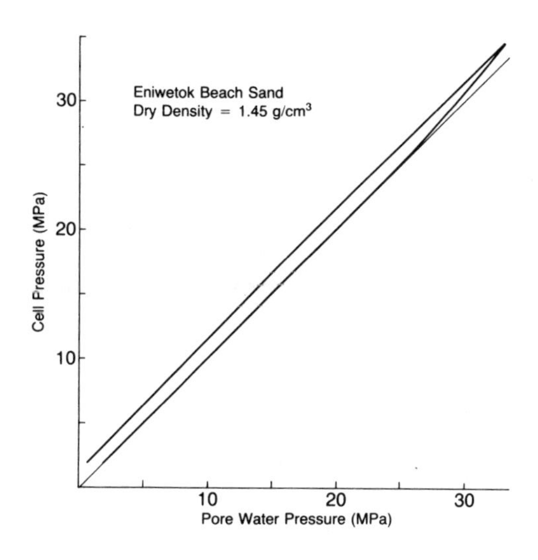

Figure 3. Results of Test E-10

one-third of the amount required to cause liquefaction. The initial stress conditions were then reimposed and a cycle of loading to 13.8MPa was applied, causing a residual pore pressure of 700KPa. In a like manner, additional cycles to 20.7, 27.6, and 34.5MPa were applied. The cycle to 20.7MPa produced a residual pore pressure of 860KPa. The last two cycles caused liquefaction.

Tests were also performed on three types of Ottawa sand: Flintshot, Sawing and Banding. The results of these tests are summarized in Table 2. The three sands differ only in their mean grain sizes. Flintshot is the coarsest, with a mean grain size of 0.60MM. The Sawing and Banding sands have mean grain sizes of 0.50 and 0.25mm, respectively. All the tests on Ottawa sands were conducted at or near minimum dry density.

The results of a test conducted on Ottawa Sawing sand are presented in Figure 4. The residual pore pressure observed at the end of this test was approximately 520KPa, only one-half the pressure required to cause liquefaction. The results of a test on a loose sample of Flintshot sand are very similar. The residual pore pressure observed was 520KPa, one-half the initial effective stress. This test is similar to test E-7 in that the same cycles of loading were performed, with reinstatement of the initial stress conditions after each cycle.

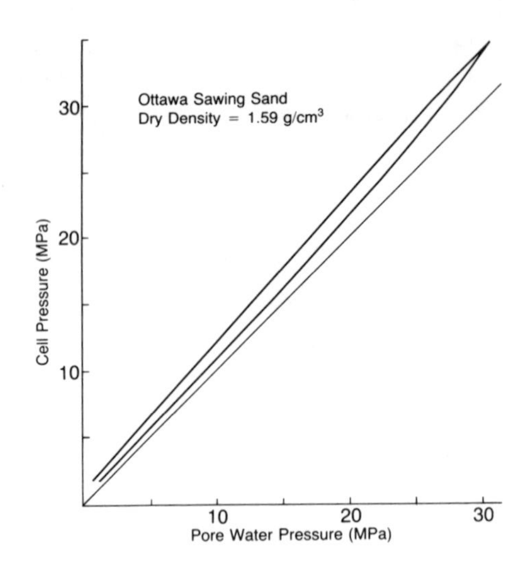

Figure 4. Results of Test S-1.

The cycles to 6.9 and 13.8MPa produced no observable residual pore pressure, but cycles to 20.7 and 27.6MPa produced residual pore pressures of 172 and 210KPa, respectively. The results of a third test, conducted on Banding sand, were somewhat different. The residual pore pressure observed was 860KPa, approximately 80% of that required to cause liquefaction. After the initial cycle the same series of cycles applied to the Flintshot sand was performed. The cycle to 6.9MPa produced no measurable pore pressure, but the cycles to 13.8, 20.7 and 27.6MPa produced residual pore pressures of 170, 210 and 210MPa, respectively.

DISCUSSION

To properly model the proposed liquefaction mechanism it is necessary to perform completely undrained tests on samples which are 100% saturated. In this section the deviations from these conditions are discussed so that a proper interpretation of the test results can be made. For reasons discussed below, it is felt that the samples were completely saturated. The deviations from "perfect" test conditions derive from drainage conditions. Truly undrained conditions were not present in the tests described above because of a) expansion of the steel tubing between the triaxial cell and the pore pressure transducer, b) deflection of the pressure transducer diaphragm, c) compression of the water in the tubing and valve between the sample and the transducer, and d) membrane penetration. The first three can be considered together as the compliance of the pore pressure measuring system. They have the effect of increasing the effective stress developed during the loading cycle and increasing the development of pore water pressure during unloading. Membrane penetration has the opposite effect. An analysis of the errors produced by these deviations from perfectly undrained conditions is presented below.

Sample saturation

As mentioned above, past liquefaction experiments have been less than completely successful because the soil being tested was not 100% saturated. The carbon dioxide method of sample saturation described by Lade and Duncan (1973) has been successfully used by earthquake liquefaction researchers for several years, and produces 100% saturation when done properly (Houston, personal communication). As a check to determine if the time allowed for sample saturation (a minimum of 18 hours) was sufficient, a sample was prepared and allowed to saturate for 72 hours under a backpressure of 0.6MPa. The results of a test on this sample were identical to a previous test in which only 18 hours were allowed for sample saturation. Because the effect of partial saturation is to lower the susceptibility of a soil to liquefaction, and because liquefaction actually occurred in the Eniwetok tests, it was felt that additional tests were not required to prove that 100% saturation was accomplished.

Effects of compliance of the pore pressure measuring system

The flexibility of the pore pressure transducer, the tubing and valves connecting it to the sample, and the compressibility of the water in the measuring system all combine to allow water to flow out of the sample during the loading portion of the test. The effect of this is to produce a smaller change in pore pressure in the sample during loading, and hence a larger effective stress compared to an inflexible system. Wissa (1969) has expanded Equation 2 to account for these effects:

$$B = \frac{1}{1 + \frac{nC_w}{m_v} + \frac{f_s}{v_0 m_v}} \quad \ldots\ldots\ldots (3)$$

in which V_0 is the total volume of the sample and f_s is the total flexibility of the pore pressure measuring system. The flexibility of the system is measured in units of cubic centimeters per unit increase in pore pressure. To calculate the flexibility for our test apparatus, the compression of water in the tubing and valves, the expansion of the tubing and the deflection of the transducer were determined for a rise in water pressure of 32.4MPa. The results of these calculations are a reduction in volume of water in the sample of 0.054, 0.0021, and 0.0002 cubic centimeters, respectively. This results in a calcualted flexibility of 1.71 x 10^{-6}cc/KPa. Using the same maximum and minimum values for the volume compressibility of the soil skeleton and the above value of flexibility, the range of effects of flexibility on the pore pressure generated during loading can be calculated. The calculated values of B are 0.9999 and 0.952 The difference in generated pore pressure for an increase in cell pressure of 32.4 MPa is no more than 97KPa, only 0.3% of the rise in cell pressure for a very stiff soil. There is virtually no difference for a very soft soil.

Effects of membrane penetration

Penetration of the membrane enclosing the triaxial specimen into the voids of the soil causes volume changes in tests where the effective confining pressure changes. The major difficulty in liquefaction tests arises when the effective stress in the sample drops. This causes the membrane to move out from the soil voids, thereby increasing the volume of the sample and reducing the pore pressure compared to the magnitude it would have reached without membrane penetration. This phenomenon causes an underestimate of the susceptibility of a soil to liquefaction. The magnitude of the errors caused by membrane penetration is a function of the grain size, the void ratio, the changes in effective stress during the test and the surface area to volume ratio of the triaxial specimen.

Frydman et al (1973) have conducted tests to determine the effects of membrane penetration. They found that volume change due to membrane penetration increases linearly with the logarithm of effective stress. For this reason membrane penetration is most important at low effective stresses, when the soil is near liquefaction. On the basis of these tests, they developed a chart to estimate the volume change per unit surface area due to membrane penetration as a function of soil grain size and changes in effective stress. Lade and Hernandez (1977) used brass shim stock plates to reduce the effects of membrane penetration by 70%. In the tests performed in our laboratory, brass shim stock was also used to reduce the effects of membrane penetration. The chart developed by Frydman et al (1973) was used to estimate the change in volume due to membrane penetration, with a reduction of two-thirds to account for the effects of the brass. For the loading portion of the test on Eniwetok sand, the estimated flexibility due to membrane penetration is 9.89 x $10^{-5}cm^3$ per KPa. This value is only correct for an increase in effective stress from 1.03MPa to 1.4MPa. As the effective stress drops below 1.03MPa (the starting value), membrane flexibility will rise rapidly.

The combined effect of measuring system flexibility and membrane penetration can be determined by the following equation developed by Lade and Hernandez (1977):

$$B = \frac{1 + \dfrac{f_m}{V_o m_v}}{1 + \dfrac{nc_w}{m_v} + \dfrac{f_s}{V_o m_v} + \dfrac{f_m}{V_o m_v}} \quad \ldots\ldots\ldots(4)$$

where f_m is the flexibility of the system due to membrane penetration.

Using the estimated flexibilities calculated above, a range in B values can be obtained for the combined effects of membrane penetration and measuring system compliance. For a soil skeleton compressibility of $2.04 \times 10^{-3} m^2/KN$, the calculated B value is 0.9999. For a compressibility of 5.0×10^{-6} m^2/KN, the calculated B value is 0.9598. In both cases, the difference between the theoretical B value for a perfectly undrained test and for the B values which would be obtained with the predicted system compliance is negligible.

When the cell pressure is reduced during the unloading portion of the test, equation (4) can still be used to determine the change in water pressure as a function of change in cell pressure. The flexibility of the measuring system is the same during unloading as it is during loading, and the flexibility due to membrane penetration will be the same if the effective stress drops back to its original value (1.03MPa). If the soil skeleton is elastic, then no residual pore pressure can be generated because the B value is the same as it was during loading. If, however, the soil skeleton becomes stiffer, the value of B will decrease and, for the same change in cell pressure, will cause less change in water pressure. Ignoring the effects of membrane penetration, this means that when the cell pressure returns to its initial value of 1.7MPa, the water pressure will be higher than 0.69MPa, its original value. Liquefaction will occur if the difference in loading and unloading moduli is large enough.

Since the effects of membrane penetration increase rapidly as the effective stress nears zero, the actual generation of residual pore pressure will be less than would occur under undrained conditions. The fact that the Eniwetok sand did actually liquefy can, therefore, be taken as proof that the proposed mechanism can explain blast-induced liquefaction.

SUMMARY AND CONCLUSIONS

A series of high pressure undrained isotropic compression tests has been performed on Eniwetok Beach sand and three types of Ottawa sand--Flintshot, Banding and Sawing. The objective was to verify a mechanism which has been proposed to explain blast-induced liquefaction. The central assumption of this theory is that the sand skeleton will undergo plastic volume change during a cycle of undrained loading. The tests began with saturation of cylindrical samples of sand in a high pressure triaxial cell with a cell pressure of 1.72MPa and a pore water pressure of 0.69MPa. After saturation, the cell pressure was increased to 34.5MPa, then reduced to 1.72MPa. During this cycle the pore water pressure was measured and plotted vs. cell pressure on an X-Y recorder. The pore water pressure was found to be larger at the end of the cycle than at the beginning. In the tests on Eniwetok sand this difference was sufficient to cause liquefaction. An analysis of the errors caused by deviations from true undrained loading was also performed. It was shown that for the purpose of verifying the blast-induced liquefaction mechanism, these errors were not significant.

On the basis of these findings, the following conclusions can be reached:

1. The blast-induced liquefaction mechanism proposed by Prater (1977) and Rischbieter et al (1977) has been verified for quasi-static, isotropic loading.

2. Eniwetok beach sand is considerably more susceptible to blast-induced liquefaction compared to Ottawa sand.

3. The stress increase required to cause liquefaction in Eniwetok sand is well within the range of compressive stresses produced by high energy and thermonuclear explosions.

ACKNOWLEDGEMENT

Research sponsored by the Air Force Office of Scientific Research, Air Force Systems Command, USAF, under Grant Number AFOSR 81-0085. The United States Government is authorized to reproduce and distribute reprints for Governmental purposes notwithstanding any copyright notation thereon.

REFERENCES

Blouin, S.E. (1979) Blast-Induced Liquefaction. Civil Systems, Inc. Report SCI IR 79-001 (draft). South Royalton, Vermont.

Cristescu, N. (1967) Dynamic Plasticity. North-Holland Publishing Co., Amsterdam, p. 515.

Fragaszy, R.J. and Voss, M.E. (1981) Laboratory Verification of Blast-Induced Liquefaction Mechanism. Final Report to the USAF Office of Scientific Research, Grant AFOSR-81-0085, also San Diego State University Civil Engineering Series No. 81145, San Diego, California.

Frydman, S., Zeitlen, J.G., and Alpan, I. (1973) The Membrane Effect in Triaxial Testing of Granular Soil. Journal of Testing and Evaluation, USA, 1, 1:37-41.

Ishihara, K. (1977) Propagation of Compressional Waves in a Saturated Soil. Proceedings, Intl. Symposium on Wave Propagation and Dynamic Properties of Soils, Albuquerque, New Mexico, pp. 451-467.

Kok, L. (1977) The Effect of Blasting in Water Saturated Sands. Proceedings, Fifth Intl. Symposium on Military Applications of Blast Simulation, Stockholm, pp. 7:6:1-7:6:10.

Kok, L. (1981) Settlements Due to Contained Explosions in Water-Saturated Sands. Proceedings, Seventh Intl. Symposium on Military Applications of Blast Simulation, Medicine Hat, Alberta, Canada, Vol. 2, pp. 4.3-1 - 4.3-8.

Lade, P.V. and Duncan, J.M. (1973) Cubical Triaxial Tests on Cohesionless Soil. Journal of the Soil Mechanics and Foundation Engineering Division, ASCE, 99, SM2:793-812.

Lade, P.V. and Hernandez, S.B. (1977) Membrane Penetration Effects in Undrained Tests. Journal of the Geotechnical Engineering Division, ASCE, 103, GT2:109-125.

Lyakhov, G.M. and Polyakova, N.I. (1972) Waves in Solid Media and Loads on Structures, FTD-MT-24-1137-71, Defense Documentation Center, Alexandria, Virginia. (translated from Volny v Plotnykh Sredakhi Nagruzki na Sooruzheniya, 1967.

Melzer, L.S. (1978) Blast-Induced Liquefaction of Materials. AFWL-TR-78-110, Air Force Weapons Laboratory, Kirtland Air Force Base, New Mexico.

Prater, E.G. (1977) Pressure Wave Propagation in a Saturated Soil Layer with Special Reference to Soil Liquefaction. Proceedings, Fifth Intl. Symposium on Military Applications of Blast Simulation, Stockholm, pp. 7:3:1 - 7:3:23.

Richart, F.E., Hall, J.R. and Woods, R.D. (1970) Vibrations of Soils and Foundations, Prentice-Hall, Inc. Englewood Cliffs, New Jersey.

Rischbieter, F. (1977) Soil Liquefaction--A Survey of Research. Proceedings, Fifth Intl. Symposium on Military Applications of Blast Simulation, Stockholm, Sweden.

Rischbieter, F., Cowen, P., Metz, K. and Schapermeier (1977) Studies of Soil Liquefaction by Shock Wave Loading. Proceedings, Fifth Intl. Symposium on Military Applications on Blast Simulation, Stockholm, pp. 7:5:1 - 7:5:6.

Skempton, A.W. (1954) The Pore Pressure Coefficients A and B. Geotechnique, Lond., 4, 4:143-147.

Studer, J. and Hunziker, E. (1977) Experimental Investigation on Liquefaction of Saturated Sand Under Shock Loading. Proceedings, Fifth Intl. Symposium on Military Applications of Blast Simulation, Stockholm, pp. 7:2:1 - 7:2:19.

Wissa, A.E.Z. (1969) Pore Pressure Measurement in Saturated Stiff Soils. Journal of the Soil Mechanics and Foundation Engineering Division, ASCE, 95, SM4:1063-1073.

Soil Dynamics & Earthquake Engineering Conference / Southampton / 1982.07.13-15

Liquefaction Potential of Soils with Plastic Fines

KIN Y.C.CHUNG
Gilbert Commonwealth, Jackson, MI, USA

I.H.WONG
Ebasco Services Inc., New York, NY, USA

INTRODUCTION

Methods, both empirical and analytical, for evaluating liquefaction potential of granular soils during earthquakes are well developed. Highly plastic clays are considered not susceptible to liquefaction. However, the behavior during earthquakes of soils with high contents of plastic fines is not well understood. After the catastrophic 1976 Tangshan, China earthquake, engineers found that, not only granular soils but also soils containing large amounts of plastic fines had experienced liquefaction. In the case of soils containing plastic fines, the usual manifestations of liquefaction such as sand boils and large settlements of structures were present (Zhong, 1980).

The paper summarizes the results of laboratory cyclic triaxial tests on alluvial soils with high plastic fines from various sites in the United States and Asia. A tentative relationship between cyclic shear strength and the plasticity index of the soils is presented herein. The type of soils studied in the paper include mainly clayey silt (ML, ML-CL), clayey sand (SC), and low plastic silty clay (CL) with plasticity index less than 15. A simplified procedure for analyzing the liquefaction potential of a normally consolidated soil containing a high percentage of fines is also suggested.

LIQUEFACTION POTENTIAL OF GRANULAR SOILS

In his classic papers Seed (1976 and 1979) describes the two analysis procedures for evaluating the liquefaction potential of granular soils. One of these procedures was empirical and based on Standard Penetration Testing blow counts. This procedure is widely used by the geotechnical engineering profession. Similar empirical procedures have been developed by geotechnical engineers in other countries (Chinese Building Code, 1974, and Iwasaki and Tokida 1979). These empirical procedures are similiar in philosophy to Seed's method and have worked well in their respective countries.

The alternate procedure for evaluating liquefaction potential of granular soils relies on laboratory cyclic triaxial and simple shear tests for the determination of the cyclic shear strength of the soils.

FACTORS AFFECTING THE CYCLIC STRENGTH OF SOILS WITH PLASTIC FINES

The liquefaction potential of a granular soil deposit during earthquake motions depends on the mean grain size, density and void ratio of the soil, the initial stresses, stress history, and the characteristics of the earthquake involved. For silty or clayey soils, the additional significant factors which affect their liquefaction potential are plasticity index and clay content.

REVIEW OF EXISTING FINDINGS FOR EVALUATING LIQUEFACTION POTENTIAL OF SOILS WITH FINES

Soil types such as low-plastic silty clay and clayey silt, containing high percentages of plastic fines, have long been considered not susceptible to liquefaction during earthquake. D'Appolonia (1968) defined liquefiable soils as follows:

- The percentage of silt and clay-size particles should be less than 10 percent;

- The particle diameter at 60 percent passing should be between 0.2 mm and 1.0 mm;

- The uniformity coefficient should be between 2 and 5;

In 1969, Lee and Fitton, based on cyclic triaxial tests on remolded samples, found that very fine sands and silty sands are the weakest soils with respect to pulsating loading. The strength increases gradually with increased grain size. However, when soils contained large amount of plastic fines, their cyclic strengths also increase. They concluded that clayey fines might tend to increase the strength considerably, whereas silty fines might tend to reduce the strength.

Recently, Donovan and Singh (1976) developed a liquefaction criterion for the fine soils encountered along the Alaska Oil Pipeline. Based on some cyclic triaxial tests on undisturbed samples they found that soils consisting of clays and some fine-grained silts with sufficient cohesive strength were able to resist grain movement which could result in the development of excess pore pressures. This condition is considered to be satisfied when the plasticity index is greater than 5. Pyke, et al (1978) when investigating the liquefaction potential of hydraulic fills in Long Beach, California found that based on cyclic triaxial tests on undisturbed samples at the same relative density, the silty sands and sandy silts in the hydraulic fills would be expected to have a lower standard penetration resistance than the generally cleaner sands. They suggested that the use of Seed's 1976 correlation figure between historical occurrence of liquefaction and penetration resistance may be rather conservative for the more silty materials.

Prager and Lee (1978) concluded that in areas of high seismicity the liquefaction potential of soils containing a significant clay fraction is not generally a problem. Furthermore, Gupta and Gangopadhyay (1973) have shown that increased clay content results in an increased number of cycles to achieve a given level of strain and a lower liquefaction potential.

Iwasaki and Tokida (1979) established a procedure for evaluating the liquefaction resistance of soils with the standard penetration resistance based on their fine content.

After the catastrophic 1976 Tangshan Earthquake, engineers found that not only granular soils but also soils containing large amounts of plastic fines had experienced liquefaction. Wang (1979) has shown that certain types of clayey materials may be vulnerable to severe strength loss as a result of earthquake shaking. In a more specific way, Zhong (1980) established a method to make correction for the standard penetration resistance according to percent of clay content (finer than 0.005 mm).

More recently, Seed and Idriss (1981) summarized the work of Tokimatsu and Yoshimi (1981) and Wang (1979) which were basd on observations of performance in previous earthquakes and concluded that:

TABLE 1 - SOIL PROPERTIES

SITE	SOIL TYPE	DEPTH M	WATER CONTENT %	LL	PI	VOID RATIO	PERCENT .074 mm	PERCENT .005m	D_{50} mm	LIMITING VOID RATIO e_{max}	e_{min}	EQUIVALENT RELATIVE DENSITY	SAMPLE CONDITIONS
Waterford Louisiana,USA	ML ML-CL	20 to 28	21 to 30	0 to 28	0 to 6	.53 to 78	75 to 95	8 to 17	.038 to .055	1.45	.45	67 to 90	Undisturbed NC Soil
Taiwan West-Coast	ML ML-CL	16 to 23	21 to 35	0 to 24	0 to 9	.48 to 75	25 to 90	13 to 18	.055 to .1	1.25 to 1.45	.38	50 to 90	Undisturbed NC Soils
Castro & Poulos (1976)	SC	N/A	N/A	N/A	N/A	.51	44	20	.1	1.25	.45	93 (88 Modified Proctor)	Compacted
Virgil Summer NPP S Carolina,USA	SC-CL	.8	17 to 23	26 to 52	9 to 26	.54 to .6	20 to 56	10	.02 to .4	1.45	.4	73 to 86	Undisturbed
Lee & Fitton (1959)	ML	-	-	26	3	.82	75	0	.018	1.25	.48	50	Compacted
	CL			38	17	.82	80	40	.0053	1.25	.48	50	
Tangshan (Zhong, 1980)	CL-ML	4	27	- 28	9	.78	-	9.42	.054	1.45	.45	67	Undisturbed

a. For silty soils plotting below the A-line and with

$D_{50} \angle 0.15$ mm, use

$$N_1 = (N_1)_{measured} + 7.5$$

and apply the correlation figure in their paper.

b. For clayey soils plotting above A-line and with the following characteristics:

Percent finer than 0.005 mm	$< 15\%$
Liquid limit,	< 35
Water content	> 0.9 LL

Use laboratory tests such as cyclic triaxial or simple shear tests.

c. The clayey soil is nonliquefiable if the clay content (determined by 0.005 mm) is larger than 20 percent or the water content is less than 0.9 LL.

It appears that a distinction should be made between soils with nonplastic fines, and those with plastic fines. According to the work by Lee and Fitton the cyclic strength of the soils does not necessarily increase as the amount of nonplastic fines increase.

SUMMARY OF THE TESULTS OBTAINED FROM LABORATORY CYCLIC TRIAXIAL TESTS

Data on the cyclic shear strength of silty and clayey soils based on cyclic triaxial tests from several different sites in North America and Asia are discussed in this section. Data obtained from literature survey are also included. These cyclic shear strengths are presented in Figure 1, which show the cyclic shear stress ratio versus the number of cycles to cause failure. The index properties of the samples tested are shown in Table 1.

The failure criterion defined herein was 10 percent double amplitude strain. Due to the clayey-silty nature of the critical in-situ strata, these materials did not liquefy as granular soils did. Typically, the test specimens underwent progressive strain with continuous load cycles and did not exhibit typical liquefaction. It is, therefore, appropriate to apply the strain criterion to these soils with plastic fines. For some cases not tested to 10 percent double amplitude strain, the use of their results is on the conservative side.

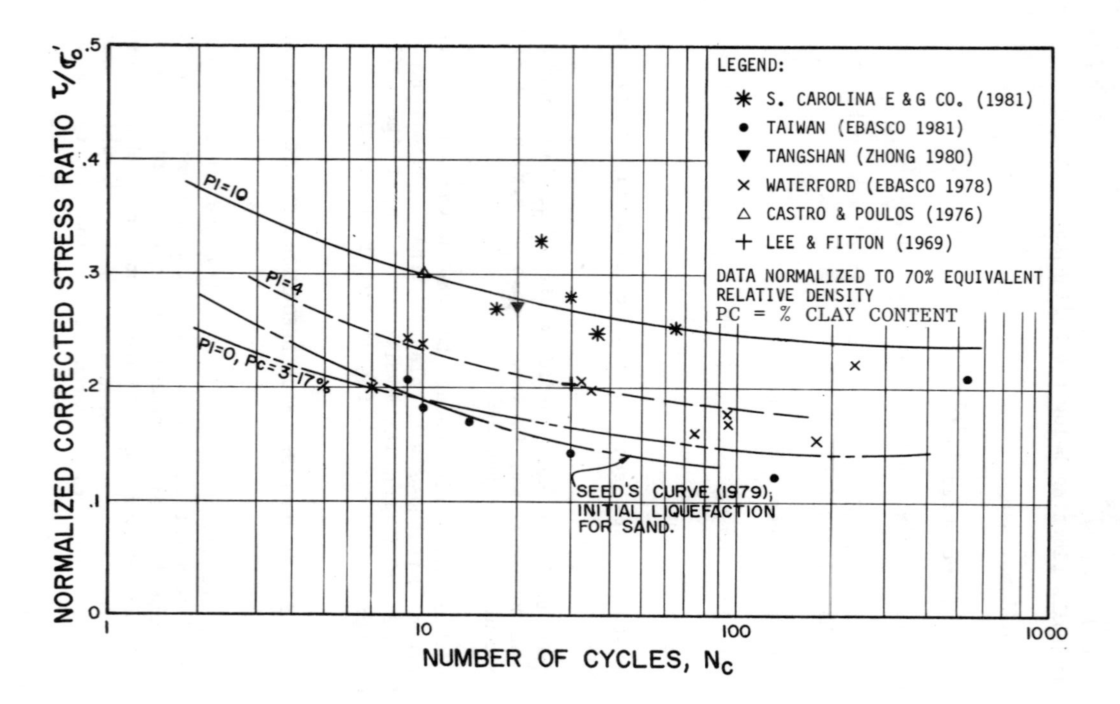

FIG. 1 CORRELATION BETWEEN NORMALIZED SHEAR STRESS RATIO AND NUMBER OF CYCLES

The test results shown in Figure 1 were based on samples with different densities. The data however have been normalized based on 70 percent relative density using the following procedure:

$$\left(\frac{\tau_h}{\sigma_0'}\right)_{normalized} = \left(\frac{\tau_h}{\sigma_0'}\right)_{field} \times \frac{70}{D_r}$$

in which $\left(\frac{\tau_h}{\sigma_0'}\right)$ field = corrected stress ratio representing field condition

D_r = Equivalent relative density for silty and clayey soils in percent

The equivalent relative density concept for fine-grained soils had been used by Lee and Fitton in 1969. For soils containing high amounts of fines, the maximum density used in the relative density determination was obtained from the Modified Proctor test, instead of the Shaking table tests required by ASTM D-2049. Also, Since there is no satisfactory method to determine a minimum density for soils with a large amount of plastic fines, a series of tests in accordance with ASTM on soils with nonplastic fines and with the mean grain size, D_{50}, ranging from 0.1 to 0.04 mm have been performed (Ebasco 1981). The maximum void ratio obtained varies from 1.25 to 1.45. These values were used for computing the equivalent relative density of the fine grained soils.

It is obvious that with the same equivalent relative density, the soils with higher plasticity index, PI, have higher cyclic shear strength. In order to study the limit of influence of the plasticity index on cyclic shear strength, the normalized shear strength versus PI is plotted on Figure 2. It seems that the cyclic shear strength of soils with PI larger than 15 needs not be considered.

APPLICATIONS OF RESULTS

Based on the above finding, a tentative procedure in evaluating liquefaction potential of soils with high plastic fines is proposed herein. It is described below:

a. In addition to the standard penetration test, the index properties of these soils should also be obtained. Index property tests should include Atterberg limits, natural water content, grain size distribution and clay content.

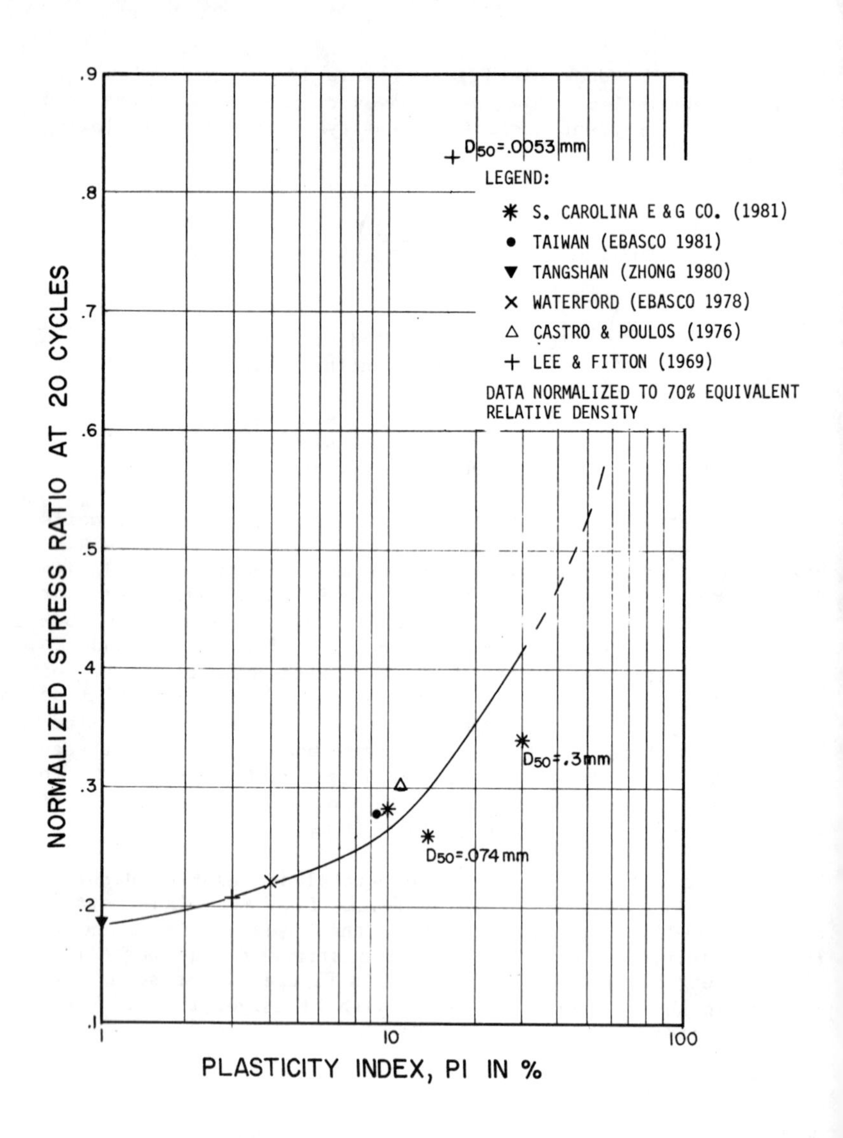

FIG.2 CORRELATION BETWEEN NORMALIZED SHEAR STRESS RATIO AND PLASTICITY INDEX.

b. Evaluate initial void ratio and determine equivalent relative density of the in-situ soils. Evaluate the minimum void ratio based on the modified Proctor density and select the maximum void ratio based on mean grain size (typical values are 1.25 for D_{50} = 0.1 mm and 1.45 for D_{50} = 0.04 mm) (Lee and Fitton, 1969, and Ebasco, 1981).

c. Use Figure 1 to obtain the normalized cyclic shear strength for soils with clay content higher than 3 percent.

d. For soils with clay content larger than 20 percent, check the PI value. It is highly possible that the soil is nonliquefiable.

CONCLUSIONS

This paper has described a simplified procedure for analyzing the liquefaction potential of a normally consolidated soil containing a high percentage of fines. This procedure is based on soil index properties and normalized cyclic shear strength. This procedure is a modification of the Seed and Idriss simplified procedure for granular soils. It correlates the earthquake resistance of the soils with plasticity and equivalent density.

This procedure has been utilized to evaluate the data from the Tangshan, China, Earthquake (Zhong, 1980) and seems to have good agreement. Since most of the published earthquake do not have all the necessary soil properties in order to utilize this method, further verification of this procedure is required.

The use of equivalent relative density as the normalizing parameter for the cyclic strength shown in Figure 1 is a temporary expediency for soils with a high percentage of plastic fines. Its applicability to other aspects of the behavior of cohesive soils is not well established. It is possible that as more data are generated, other parameters, such as undrained shear strength, may prove to be better suited as a normalizing factor.

REFERENCE

Castro, G., & Poulos, S. J., 1976, "Factors Affecting Liquefaction and Cyclic Mobility," ASCE Annual Convention and Expo., Preprint 2752, September, Philadelphia.

D'Appolonia, E., 1968, "Dynamic Loadings" ASCE Specialty Conference on Cambridge, Mass, August.

Donovan, N.C., & Singh, S., 1976, "Liquefaction Criteria For the Trans-Alaska Pipeline," ASCE Annual Convention and Expo., Preprint 2752, September, Philadelphia.

Ebasco Services Inc., 1978, "FSAR, Section 2.5, Waterford Stream Electric Station fo Louisiana P & L Co."

Ebasco Services Inc., 1980, "FSAR, Section 2.5, St. Lucie Nuclear Power Plant, Units 1 & 2 for Florida P&L Co."

Ebasco - CTCI Corp., 1981, "Site Study for Potential Nuclear Power Plant on Soil Foundation for Taiwan Power Company," 2 Vols.

Gupta, M.K., and Gangopadhygay, 1973," Liquefaction Characteristics of Sand Clay Mixtures," Symposium on Earth and Earth Structures Under Earthquakes and Dynamic Loads, University of Roorkee, Roorkee, India, March.

Iwasaki, T., & Tokida, K., 1979, "Studies on Soil Liquefaction Observed During the Miyagi - Ken Oki Earthquake of June 12, 1978," the 7th WCEE, Istanbul.

Lee, K. L., Fitton, J. A., 1969, "Factors Affecting the Cyclic Loading Strength of Soil," Vibration Effects of Earthquake on Soils and Foundations, STP 450, ASTM.

Prager, S. R. & Lee, K. L., 1978, "Post-Cyclic Strength of Marine Limestone Soils," ASCE Specialty Conference, June, Pasadena, CA.

Pyke, R. M., et al., 1978, "Liquefaction Potential of Hydraulic Fills," J. Geotechnical Eng. Div. ASCE, GT11, November.

Seed, H. B., 1976, "Evaluation of Soil Liquefaction Effects On Level Ground During Earthquakes," ASCE Annual Convention and Expo. Preprint 2752, September, Philadelphia.

Seed, H. B., 1979, "Soil Liquefaction and Cyclic Mobility Evaluation for Level Ground During Earthquakes," J. Geotechnical Eng. Div. ASCE, GT 2.

Seed, H. B., & Idriss, I. M., 1981, "Evaluation of Liquefaction Potential of Sand Deposits Based on Observations of Performance in Previous Earthquakes," ASCE, Preprint 81 - 544. on Insitu Testing To Evaluate Liquefaction Susceptibility, St. Louis, Missouri October.

South Carolina E & G Co., 1981, "FSAR, Virgil C. Summer Nuclear 3 Station, Section 2.5".

Tokimatsu, K., & Yoshimi, Y., 1981, "Field Correlation of Soil Liquefaction with SPT and Grain Size," Proc. International Conference on Recent Advances in Geotechnical Engineering and Soil Dynamics, St. Louis, Missouri, April.

Wang, Wenshaw, 1979, "Some Findings in Soil Liquefaction," Water Conservancy and Hydroelectric Power Scientific Research Institute, Beijing, China, August.

Zhong, L. H., 1980, "Analysis For Evaluating Liquefaction of Low Plasticity Clays (CL) During Earthquake," Chinese Journal of Geotechnical Engineering, Vol, 2, No. 3, September, in Chinese.

Soil Dynamics & Earthquake Engineering Conference / Southampton / 1982.07.13-15

Development of a Liquefaction Potential Map

LOREN R.ANDERSON
Utah State University, USA

JEFFREY R.KEATON
Dames & Moore Consulting Engineer

INTRODUCTION

General

The effects of earthquakes include loss of life and costly property damage; therefore, in areas of high seismic activity, earthquake hazard reduction must be an important consideration for intelligent land use planning. Damage during earthquakes can result from surface faulting, ground shaking, ground failure, generation of large waves (tsunamies and seiches) in bodies of water, and regional subsidence or downwarping (Nichols and Buchanan-Banks, 1974). All of these causes of damage need to be considered in reducing earthquake hazards.

This paper describes a method that was used to develop a liquefaction potential map for Davis County, Utah, U.S.A. (Anderson and others, 1982). The method is an enhancement of the technique presented by Youd and Perkins (1978). The liquefaction potential was evaluated on the basis of subsurface data that was obtained from private engineering consultants, state and local government agencies and from a supplementary subsurface investigation performed as one of the tasks in the study. The results of the study were summarized on four maps: (1) Selected Geologic Data Map, (2) Soils and Ground Water Data Map, (3) Ground Slope and Critical Acceleration Map and (4) Liquefaction Potential Map. The base maps were U. S. Geological Survey 7.5-minute topographic quadrangles reduced to a scale of 1:48,000.

Liquefaction-Induced Ground Failure

In this paper, the term "liquefaction" implies the occurence of liquefaction-induced ground failure. This is an important distinction because it is the ground failure which causes damage, not soil liquefaction per se.

In general, three types of ground failure are commonly associated with liquefaction: (1) flow landslides, (2) lateral spread landslides, and (3) bearing capacity failures. Youd and others (1975) relate these types of ground failure with the slope of the ground surface. The most common type of liquefaction-induced ground failure is probably lateral spread landsliding.

METHODOLOGY

General

Subsurface data were collected for selected sites from throughout the Davis County study area. Ground surface accelerations required to induce liquefaction at each site were computed. These acceleration values are referred to as "critical accelerations." The liquefaction potential for each site was then classified as high, moderate, low or very low depending on the probability of the computed "critical" ground surface acceleration being exceeded in 100 years.

Evaluation of Liquefaction Potential

An evaluation of liquefaction potential at a given site by current state of the art methods involves comparing the predicted cyclic stress ratio (τ/σ_o') that would be induced by a given design earthquake with the cyclic stress ratio required to induce liquefaction. The predicted cyclic stress ratio can be computed using response analysis techniques or by a simplified procedure based on rigid body theory modified to account for the flexibililty of the soil profile (Seed, 1979). The simplified theory for computing the cyclic stress ratio induced by an earthquake is given by Equation 1.

$$\frac{\tau_{av}}{\sigma_o'} = 0.65 \frac{a_{max}}{g} \left(\frac{\sigma_o}{\sigma_o'}\right) r_d \qquad (1)$$

where,

a_{max} = maximum acceleration at ground surface

σ_o = total overburden pressure on sand layer under consideration

σ_o' = initial effective overburden pressure on sand layer under consideration

r_d = a stress reduction factor varying from a value of 1 at the ground surface to a value of 0.9 at a depth of about 30 ft.

The cyclic stress ratio required to cause liquefaction can be evaluated either by laboratory tests on undisturbed samples or by an emperical relationship between some insitu property of the soil and the cyclic stress ratio required to cause liquefaction. Seed, Mori and Chan (1977) have developed an emperical relationship between the cyclic stress ratio required to cause liquefaction and the standard penetration resistance of the soil.

Seed (1979) points out that the factors that tend to influence liquefaction susceptibility, such as relative density, age of the deposit, seismic history and soil structure, also tend to influence the standard penetration resistance in a like manner. Although the standard penetration test has its shortcomings, if used properly and with judgement it provides a convenient and rapid method of evaluating the insitu characteristics of sand. The standard penetration test also provides a convenient method to utilize existing data in evaluating liquefaction potential because in the past the most common method to obtain samples of sand has been the standard penetration test.

Gibbs and Holtz (1957) correlated standard penetration resistance with relative density and effective overburden pressure. Their work showed that the standard penetration resistance for a constant relative density was a function of the overburden pressure. Therefore, Seed, Mori and Chan (1977) used a standard penetration resistance corrected to an overburden pressure of one ton per square foot in developing the relationship between standard penetration resistance and the cyclic stress ratio required to cause liquefaction.

Critical Acceleration

Qualitative descriptions of the liquefaction potential in the Davis County study area were assigned on the basis of the probability that the computed values of critical acceleration would be exceeded during a 100-year time period. The critical acceleration for a given location is defined as the lowest value of maximum ground surface acceleration required to induce liquefaction.

The standard penetration test data from soil borings in conjunction with Eq. 1 and the standard penetration - cyclic stress ratio relationship by Seed, Mori and Chan (1977, Fig. 11) were used to compute the critical acceleration at numerous locations throughout the study area. Equation 1 can be solved for the critical acceleration and stated as:

$$(a_{max})_c = \left(\frac{\tau_{av}}{\sigma_o'}\right) \left(\frac{\sigma_o'}{\sigma_o}\right) \left(\frac{1}{0.65\ r_d}\right) \qquad (2)$$

where, $(a_{max})_c$ = critical acceleration (ground surface acceleration required to induce liquefaction at a given site)

$\left(\frac{\tau_{av}}{\sigma_o'}\right)$ = cyclic stress ratio required to cause liquefaction at the given site and obtained from the standard penetration resistance and Seed, Mori and Chan (1977, Fig. 11).

σ_o = total overburden pressure at the point where the standard penetration resistance is measured

σ_o' = effective overburden at the point where the standard penetration resistance is measured

Judgement was required in assigning critical acceleration values. Generally more than one boring log was available for a given site and many standard penetration values were reported for each boring. Therefore, several critical acceleration values were computed for each boring at each site. A value considered to be representative of the critical acceleration was then assigned to the site. In assigning this representative value, consideration was given to consistancy within and between borings, to the soil type and to the limitations of the standard penetration test. A single low critical acceleration value at a site was not considered representative if it was not consistant with other critical acceleration values at the site and in the general area.

Critical acceleration values were computed only for sand and silty sand, and for sandy silt with less than 15 percent clay-size material and a plastic index less than 5. Since the penetration value is the number of blows required to drive a standard sampler one foot, layers being evaluated must be at least one foot thick. For this reason, borings containing sand layers thinner than one foot could not be assigned cricital acceleration values by this procedure.

Liquefaction Potential Based on Critical Acceleration

As stated above, the liquefaction potential was assigned on the basis of the probability that the critical acceleration would be exceeded in 100 years. The probabilities used in assigning liquefaction potential are presented in Table 1.

Table 1. Liquefaction potential related to exceedance probability

Probability of Exceeding Critical Acceleration in 100 years	Liquefaction Potential
> 50%	High
10 - 50%	Moderate
5 - 10%	Low
< 5%	Very Low

The probability values delineating liquefaction potential were selected partially on the basis of probability limits frequently used in selecting accelerations for structural design purposes. In structural engineering the concept of dual levels of design accelerations has become widely accepted in recent years. This concept first considers an earthquake with a moderate probability of occurrence during the projected lifetime of a structure; the structure should be designed to remain elastic (completely functional) during the earthquake event. The structure as designed for this first event should then be analyzed to estimate its probable response to a larger event which has a smaller probability of occurrence. The structure would be expected to develop ductility (be damaged) during its response to the second and larger motion but not expected to collapse.

The acceleration values usually chosen for these two levels of design are (1) that which has a 50 percent probability of being exceeded during the projected life of the structure (the elastic design motion) and (2)that which has only about a 10 percent probability of being exceeded during the life of the structure (the largest acceleration for which the structure may develop ductility). These probability values of 50 percent and 10 percent were selected as the limits delineating the high and moderate liquefaction potential categories. A probability value of 5 percent was then arbitrarily selected to separate low and very low liquefaction potential. For planning purposes a 100-year time period was considered appropriate.

Other probability limits could have been selected and this would have had some effect on the configuration of liquefaction potential zones on the map. However, regardless of the probability values used to define the high, moderate, low and very low classifications, those selected clearly allow a relative assessment of the liquefaction hazard within the study area.

Liquefaction Potential Map

Computed critical acceleration values for specific sites were plotted on a map of the study area. A two-step procedure was then used to develop the Liquefaction Potential Map. Contours were first drawn on the basis of the critical acceleration values and used to divide the study area into preliminary zones of high, moderate, low and very low liquefaction potential. The contours represented the critical accelerations that had exceedance probabilities of 50, 10 and 5 percent in 100 years.

After the liquefaction potential zones were initially identified from the critical acceleration contours, they were adjusted to reflect the geology of the area. This adjustment was particularly important because boring data and critical acceleration values were available only at selected locations and did not necessarily reflect specific geologic features such as the locations of stream beds and the distribution of sediments deposited in late Pleistocene Lake Bonneville.

Ground Failure Mode

The Liquefaction Potential Map delineates the various liquefaction potential zones. It can be used in conjunction with soil data and ground slope maps to predict the probable type of ground failure. Youd (1978, 1981) suggested that the type of ground failure induced by liquefaction is related to the ground surface slope and proposed the relationships between ground slope and failure mode shown in Table 2.

Table 2. Ground slope and expected failure mode

Ground Surface Slope	Failure Mode
<0.5%	Bearing capacity
0.5 - 5.0%	Lateral spread
>5.0%	Flow landslide

The thickness and setting of the sand deposit should also be cosidered in determining the probable mode of ground failure. For example, a 1-meter-thick, loose sand layer at a depth of 10 meters in an otherwise clay soil profile is not likely to cause a flow landslide or a significant bearing capacity failure regardless of the ground surface slope. However, this condition might induce a translational landslide in steep slopes or magnify ground surface movement due to lurching in flat areas.

A Ground Surface Slope Map and a Soil Properties Map were prepared for the study area. These maps can be used in conjunction with the Liquefaction Potential Map to evaluate the potential for various types of ground surface failure.

ASSESSING LIQUEFACTION POTENTIAL IN DAVIS COUNTY, UTAH

General Subsurface Conditions in Davis County

Virtually all of the urbanized area of Davis County was inundated by Pleistocene Lake Bonneville, of which the Great Salt Lake and Utah Lake are remnants. Consequently, most of the sediments in Davis County are probably late Pleistocene or younger in age.

The lake bed sediments of the region generally consist of deposits of sand, silt and clay. The liquefiable sand and silt deposits vary in thickness from several millimeters to several meters and occur throughout Davis County. Extensive gravel deposits are present along the east side of the study area on the upper shore lines of Lake Bonneville.

The ground water table in much of the study area is within a few feet of the ground surface and local areas of artesian conditions are present. In the bench areas along the higher shore lines of ancient Lake Bonneville, the water table is generally much deeper but cases of perched ground water are known to exist. Lawn sprinkling and other effects of additional development along the bench areas will probably contribute to the occurence of perched ground water.

Available Subsurface Data

Soils considered to be susceptible to liquefaction were found throughout the study area. Consequently, site specific analyses were required to delineate zones of differing susceptibility. The necessary soil boring data required to perform a liquefaction analysis included accurate descriptions of the soil profiles, standard penetration resistance data, ground water depth and the grain size characteristics of granular layers. Such information was sought from existing records and supplemented by field and laboratory testing programs.

Soil boring data were obtained from various consulting firms and government agencies which had performed subsurface investigations within the study area. Numerous techniques had been used to obtain subsurface information. The standard penetration test was not used by all investigators to measure field densities and to obtain samples. Therefore, it was necessary to convert various (non-standard) penetration values to standard penetration blow count values so that the data could be used with the Seed, Mori and Chan (1977) relationship between standard penetration resistance and the cyclic stress ratio required to cause liquefaction. The energy conversion technique presented by Lowe and Zaccheo (1975) was used to convert various penetration records to standard penetration blow counts.

Soils and Ground Water Data Map

A Soils and Ground Water Data Map was prepared to summarize the aerial extent and vertical depth of the liquefiable soil deposits and the ground water conditions in Davis County. The soils and ground water data included:

1. Depths at which liquefiable layers exist
2. Thickness of liquefiable layers
3. Depth to ground water
4. Minimum and average standard penetration values obtained in liquefiable deposits.

Ground water contour lines showing zones of various depths to first ground water were also drawn on the map.

Critical Acceleration Map

Critical accelerations were computed for specific sites within the study area and were plotted on a critical Acceleration Map. Each location where a critical acceleration was computed was assigned a liquefaction potential classification according to the probability that the critical acceleration would be exceeded during the next 100 years. Different map symbols were used in plotting the critical accelerations to indicate the liquefaction potential classification.

Seismic risk studies by Algermissen and Perkins (1972) and Dames & Moore (1978) indicate that the probability that a given ground surface acceleration will be exceeded is nearly the same throughout Davis County, Utah. The exceedance probability (Dames & Moore, 1978) summarized in Table 3 was used to assign liquefaction potential classifications in the study area.

Table 3. Liquefaction potential related to critical acceleration.

Liquefaction Potential	Critical Acceleration	Approximate 100-year Exceedance Probability
High	< 0.12 g	> 50%
Moderate	0.12 - 0.20 g	50 - 10%
Low	0.20 - 0.30 g	10 - 5%
Very Low	> 0.30 g	< 5%

A number of boring locations within the study area had granular soil with a very deep water table. Critical accelerations were not assigned to these locations. However, these locations were identified with a special map symbol because a possible rise in the ground water table or the future development of perched ground water made the areas potentially susceptible to liquefaction. Locations with sand layers less than one foot thick were also identified with a special map symbol even though a critical acceleration could not be assigned.

CONCLUSIONS AND RECOMMENDATIONS

Ground failure caused by liquefaction is a primary hazard associated with earthquakes. The first step in avoiding this hazard is to recognize where liquefaction might occur. A Liquefaction Potential Map was compiled for Davis County, Utah showing areas where conditions are favorable for liquefaction to occur.

Fine sand and silty sand are the soil types most conducive to liquefaction and they are found throughout Davis County. Soil type alone, however, does not determine the liquefaction potential of a given site. Several important factors influencing liquefaction potential were considered in this study. The standard penetration test provided a useful means for evaluating the influence of the soil structure, previous seismic history, and age of the deposit as well as the relative density of the soil. A decrease in the susceptibility to liquefaction from any of these factors is reflected by a corresponding increase in the standard penetration resistance.

The standard penetration resistance was used to compute the ground surface acceleration that would be required to induce liquefaction (critical acceleration). The liquefaction potential was then assigned on the basis of the probability that the critical acceleration would be exceeded in 100 years. Local geologic conditions were also considered in refining liquefaction potential boundaries.

The information generated by this study should prove to be valuable for those concerned with future land development. Planners and other concerned parties should realize that areas showing a high liquefaction potential need not be ruled out as possible sites for construction. However, we believe further analyses should be required for these sites including an economic analysis of preventive or protective measures that can be used to reduce the liquefaction potential. Haldar (1980) has developed a decision analysis framework which considers both the technical and economic aspects of limiting or eliminating damage associated with liquefaction.

One problem often encountered during this study was how to assess the susceptibility of thin sand layers and lenses. Since the standard penetration test was used as a basis for this study, reliable data could be obtained only for sand strata greater than one foot thick. Damages associated with the liquefaction of thin sand layers and lenses are not uncommon (Seed, 1968) but an accurate means for identifying the relative density of such strata has not been developed. A method that can be used and one that is gaining popularity in the United States is the cone penetrometer test. The cone penetrometer not only offers an economical means for continuous subsurface soil profiling (Baligh, Vivatrat and Ladd, 1980), but could also provide a direct means for continuously identifying the liquefaction potential in the soil profile (Schmertmann, 1978).

It is recommended that the liquefaction potential map be updated continually as more soil boring information becomes available and as new and improved techniques are developed for analyzing liquefaction.

ACKNOWLEDGMENTS

This project was sponsored by the U.S. Geological Survey as part of their Earthquake Hazards Reduction Program with Mr. Gordon Greene as the Contracting Officer's Representative. Dr. T. Leslie Youd of the U.S. Geological Survey reviewed the work during the project and made valuable suggestions. The financial support and the technical advice of the Survey is gratefully acknowledged.

Mr. Kevin Aubry and Mr. Stanley J. Ellis, while graduate students at Utah State University, collected all of the data, conducted the field work and compiled the maps for the Davis County study. Mr. Andrew C. Allen assisted in preparing the liquefaction potential map and the final report.

REFERENCES

Algermissen, S. T., and D. M. Perkins. (1972) A technique for seismic zoning: general considerations and parameters: Proc. Internat. Conference on Microzonation, Seattle, Wa. p. 865-878.

Anderson, L.R., J.R. Keaton, Kevin Aubry and S.J. Ellis. (1982). Liquefaction potential map for Davis County, Utah; Final Report to the U.S. Geological Survey by Utah State University, Contract No. 14-08-0001-19127.

Baligh, M. M., V. Vivatrat and C. C. Ladd. (1980) Cone penetration in soil profiling. Journal of the Geotechnical Engineering Division, ASCE, Vol. 106, No. GT4, Proc. Paper 15377, April. pp 447-461.

Dames & Moore. (1978) Final report, Development of criteria Seismic risk mapping in Utah, for State of Utah Department of Natural Resources Seismic Safety Advisory Council. Unpublished Consultants Report. July 20. 14 p.

Gibbs, H. J. and W. G. Holtz. (1957) Research on determining the density of sands by spoon penetration testing. Proceedings of the 4th International Conference on Soil Mechanics and Foundation Engineering. London.

Haldar, A. (1980) Liquefaction study - a decision analysis framework. Journal of the Geotechnical Engineering Division, ASCE, Vol. 106 No. GT12m Proc. paper 15925. December. pp. 1297-1312.

Lowe, John and Philip Zaccheo. (1975) Subsurface explorations and sampling. Chapter 1 of Foundation Engineering Handbook, Hans F. Winterkorn and Hsai-Yang, Eds. Van Nostrand Reinhold Co., New York, New York. pp. 1-66.

Nichols, D. R. and J. M. Buchanan-Banks. (1974) Seismic hazards and land-use planning: U.S. Geol. Surv., Circular 690.

Schmertmann, J. H. (1978) Study of feasibility of using Wissa-type piezometer probe to identify liquefaction potential of saturated fine sands. Technical Report S-78-2, U.S. Army Engineer Waterways Experiment Station. February. 73 p.

Seed, H. B. (1979) Soil liquefaction and cyclic mobility for level ground during earthquakes. Journal of the Geotechnical Engineering Division, ASCE, Vol. 105, No. GT2, Proc. Paper 14380. February. pp. 201-255.

Seed, H. B. (1968) Landslides during earthquakes due to soil liquefaction. Journal of the Soil Mechanics and Foundations Division, ASCE, Vol. 93, No. SM5, Proc. Paper 6110. September. pp. 1053-1122.

Seed, H. B., K. Mori and C. K. Chan. (1977) Influence of seismic history on liquefaction of sands. Journal of The Geotechnical Engineering Division, ASCE, Vol. 103, No. GT4, Proc. Paper 12841. April. pp. 257-270.

Youd, T. L. (1981) Engineer, U.S. Geological Survey, Menlo Park, California, personal communication.

Youd, T. L. (1978) Major cause of earthquake damage is ground failure. Civil Engineering, ASCE, Volume 48, No. 4. April. pp. 47-51.

Youd, T. L., D. R. Nichols, E. J. Helley and K. R. Lajoie. (1975) Liquefaction potental: in Borcherdt, R. D., ed, Studies for seismic zonation of the San Frncisco Bay Region, U. S Geol. Surv. Professional Paper 941-A, p. A68-A74.

Youd, T. L. and D. M. Perkins. (1978) Mapping liquefaction-induced ground failure potential: American Society of Civil Engineers, Geotechnical Engineering Division, Journal v. 104, no. GT4, p. 433-446.

Soil Dynamics & Earthquake Engineering Conference / Southampton / 1982.07.13-15

Dynamic Behaviour of Buried Structures in Liquefaction Process

TOSHIYUKI KATADA
Musashi Institute of Technology, Tokyo, Japan

MOTOHIKO HAKUNO
University of Tokyo, Japan

SYNOPSIS

Many of the underground conduits or tubes come up to the surface owing to the liquefaction of the surface ground, and the damage such as subsidence, inflection, or breakage is reported. Notwithstanding this fact that the damage is created by the liquefaction, it has not been so sufficiently explained how the soil influences the underground structures in the process of liquefaction and, as the result, how the underground structures will behave. In this study, from the liquefaction experiment using a sand box, the behavior of underground structures in the liquefaction process of the surface ground was intended to empirically explain.

INTRODUCTION

In Niigata Earthquake(June 6, 1964, M=7.5), the saturated sandy ground was liquefied owing to a seismic move, and as the result, many of the structures subsided or inclined and furthermore serious damage was caused everywhere in the city. The catastrophe has motivated many studies on liquefaction of the saturated sandy ground, and it can be said that the mechanism of the liquefaction have now been explained to a considerable extent. Figure 1 is a distribution map illustrating a probability of the liquefaction of the surface ground in the 23 wards of Tokyo made by Disaster Prevention Conference of Tokyo. In geographically low lying towns of the capital of this contry, a probability of the creation of a liquefaction phenomena is very high. To make the matters worse, there are too many types of underground structures or buried conduits or tubes such as subways, underground markets, city water and drainage, gas tubes, etc. Many of the underground conduits or tubes come up to the surface owing to the liquefaction of the surface ground, and the damage such as subsidence, inflection, or breakage is reported.

Notwithstanding the above fact that the damage is created by the liquefaction, it has not been so sufficiently explained how the soil influences the underground structures in the process of liquefaction and,as the result, how the underground structures will behave. Also in case that the seismic move is not so strong or in case that the moisture content is low and thereore the soil is hard to be liquefied, the soil does not develop to a liquefied state and will supposedly be just softened. As mentioned above, it has not yet been explained what behavior the underground structures will show even with reference to the case that the soil is softened. Some major underground structures such as subways have never experienced such soil destruction as the liquefaction during an earthquake. Therefore it is very important to explain how the underground structures will be affected by the soil destruction such as liquefaction or softening.

Thus we wish to attempt herewith a liquefaction experiment to ascertain what response characteristics the underground structures will show under the liquefaction process of the soil using a sand box. That is to say, we explain the acceleration response characteristics by conducting the liquefaction experiment making use of a model sandy ground in which a small-sized accelerometer is buried. In the above experiment, a sine wave(6 Hz) is used. Regardless of that, however, we will make a description of the response characteristics when a random wave with component wave of 0-20Hz which is close to actual seismic wave is used. Furthermore, we will carry out an observation on the response mechanism of the ground vs. buried structures system in the liquefaction process.

In addition, the established characteristics in the liquefaction process have not yet been obtained with regard to the horizontally long and narrow underground structures such as subways or buried constructions as well. Therefore, we will empirically explain the response characteristics of inflection strain(strain caused by inflection) by burying a vinylchrolide pipe in the model surface ground

THE ACCELERATION OF THE BURIED SMALL-SIZED ACCELEROMETER

Experimetal aparatus and method

The experimental apparatus consists of a shaking table, oscillator, sand box and model surface ground, measuring device (small-sized accelerometer and pore-water gauge), and recorder. For the experiment, two types of sand boxes were used. The box size of the type A is 360mm width by 500mm length with 250mm height, respectively. Since the material of the box is transparent, the changes of the sand can be observed very easily during the experiment. As the box of the type B, a box with longer length 1,000mm length with 250mm height and was of wood.

The model surface ground was made with a given quantity of water and sand. In other words, 13 kg of water and 50 kg of

sand were used for the sand box of the type A as the standard, whereas 26 kg of water and 100 kg of sand were used for the sand box of the type B as the standard. Into the saturated model ground. water was poured first and sand was put.
This procedure was repeatedly followed and the ground was made up. The thickness of the laminate of the model surface ground was about 20cm.
Into the unsaturated model ground, sand was first put and later water was sprinkled with a watering pot. This procedure was also repeatedly followed and the ground was made up. The relation of the moisture content vs. relative density is illustrated in Figure 2. The measuring apparatus(small-sized accelerometer and pore-water pressure gauge) was placed on the center of the surface ground whose height from the bottom is 10 cm. The small-sized accelerometer used was a measuring instrument of the plumb-bob type with a cantilever-type strain gauge conversion system. By regarding this small-sized accelerometer as an underground structure,its response acceleration can be measured and concurrently the state of the liquefaction of the surface ground can be judged. The shaking table is given vibration by a sine wave (6Hz) generated by an oscillator or a random wave. On th shaking table, a small-sized accelerometer which is quite the same type as the one buried in the model surface ground is also placed. With these unit, comparision and examination of the acceleration wave on the shaking table and the response acceleration wave of the underground structure (i.e., small-sized accelerometer) can be made. Also the reponse characteristics in the liquefaction process are to be observed by comparing the change of the por-water pressure.

A sine wave is used as an input

The small-sized accelerometer buried in the model surface points different values of the acceleration response characteristics following the difference of the moisture content of the model surface ground. In short, in case of the saturated surface ground, the accelerometer points just temporarily a large acceleartion response when the liquefaction is incomplete (Figure 3 (a)), whereas no response can be seen when the liquefaction is complete. In case of the unsaturated surface ground, a large acceleration response can be seen until the input is finished (Figure 3(b)).

Therefore it is understood from the above observation that the liquefaction is harder to be developed in case of the unsaturated surface ground than in case of the saturated surface ground. However it might be expected that the underground structures will be much more badly influenced than in case of the saturated surface ground when the unsaturated surface ground is once liquefied or softened owing to the seismic move.

In Figures 4 and 5, how the response characteristics of the small-sized accelerometer will change according to the moisture content is quantitatively illustrated. Figure 4 especially reveals that the acceleration response magnification has an inclination, with the peak of about 3.0 power, to be decreased in parallel with the increase of the moisture content.

As shown above, it is now apparent that a great response is created when the liquefaction is incomplete. Therefore in addition to that, it is problematic how long such a great response during the incomplete liquefaction will last. Illustrated in Figure 5 is the relation of moisture content of the model surface ground vs. the time for which the response characteristics during the incomplete liquefaction is shown. Here, a focal point arises how the response time during the incomplete liquefaction will be defined. It might be permissible to consider that the response during the incomplete liquefaction is started when the pore-water pressure rises and the response waveform begins to be driven into irregular states. In this study, the end of the response during the incomplete liquefaction is defined as shown below. When the liquefaction is completely developed, the response becomes small. Therefore it is at that time when the great response during the complete liquefaction becomes small with a progress of the liquefaction of the surface ground and the response magnification against the input will be 1.0.

A straight line depiected in Figure 5 is the one connecting a point representing the maximum value of the incomplete liquefaction time with another such point. As judged from the grade of the straight line, the incomplete liquefaction time becomes short contrarily to the increase of the moisture content. In Figure 5, the points as shown that the incomplete liquefaction time is infinite indicate that the peculiar response characteristics during the incomplete liquefaction time appeared during whole the continuous time of the inputted acceleration. Such characteristics are noticed with relation to the surface ground of the moisture content of 0.230-0.240.

As revealed above, an experimental result that when the lowly saturated sandy ground is liquefied or softened, the response magnification becomes great in comparison with the completely saturated surface ground and the incomplete liquefaction time also becomes long was obtained. In addition, with relation to the surface ground whose moisture content is lower than 0.230, the surface ground was not so softened or liquefied, when vibration was given. At that time, the magnification of the response acceleration against the inputted acceleration is 1.0-1.5. Incidentally, the response waveform is quite the same as of the inputted one and neither irregularity nor phase tradiness can be witnessed.

A random wave is inputted

Next, we would like to explain what response characteristics the small-sized acceleration will show when a liquefaction experiment is conducted using as an input a random wave which is nearly similar to an actual seismic wave.

In Figure 6(a), an acceleration response waveform of the small-sized accelerometer buried in the completely saturated sandy ground of a moisture content of 0.280, a random waveform as an input, and a pore-water pressure waveform are illustrated. As the pore-water pressure begins to rise and the surface ground is reduced to the state of the incomplete liquefaction, a great response acceleration compared with the input is noticed. The maximum response magification of the response acceleration is about 2.3. However, when the surface ground is completely liquefied, the response acceleration becomes small compared with the input.

Such acceleration response characteristics quite coincide with the experimental results obtained with use of a sine wave. It can therefore be said that if a sine wave is used as an input, the characteristics will be same from the viewpoint that a great acceleration response can be seen in the liquefaction process.

Next, illustrated in Figure 6(b) are the experimental results acquired from the unsaturated sandy ground of the moisture content of 0.260. Even after the pore-water pressure is raised and the surface ground is liquefied, a great acceleration response can be seen. This inclination is quite the same when a sine wave is used as an input.

A RESPONSE MECHANISM OF THE SURFACE-GROUND/BURIED-STRUCTURES SYSTEM IN THE LIQUEFACTION PROCESS

As stated in the previous section, it has been made clear from the liquefaction experiment using a sand box that the small-sized accelerometer shows the great response acceleration in the liquefaction process. Various causes have been pointed out for the fact that the buried model of the structures shows a great response in a transient state of the liquefaction of the sandy ground.

For example, when the model sandy ground is softened and liquefied by a vibrating force, the surface ground rigidity is lowered. In connection with this, it can be said that the natural frequency of the model surface ground coincides with that of the inputted wave in this process and a resonant phenomenon is presented.

However, an important matter with reference to the completely buried model might be whether the surface ground and the buried bodel structures will always move en bloc in the liquefaction process, or whether they can be considered to be in motion en bloc. Furthermore it is not yet known what is the cause of non-linear vibration of the surface-ground/buried-structures system as illustrated in Figure 3(b).

In the saturated surface ground, stress concentration might be created near the buried small-sized accelerometer and it is imagined that pore-water pressure is raised somewhat rapidly than the place at a distance. It may be concluded from the that the liquefaction is rapidly developed around the small-sized accelerometer and the instrument is easy to move, and it co-llides with the not yet liquefied surface ground.

However, in the unsaturated surface ground, no tendency that the pore-water pressure is raised especially rapidly near the small-sized accelerometer can be exhibited. In spite of that, the acceleration response waveform which comes out is such a one as showing repeatedly the fixed acceleration and the pulse-shaped great acceleration.

It is therefore rather difficult to conceive of the phenomenon of collision between the surface ground and the small-sized accelerometer as a cause of the small-sized accelerometer re-vealing a great response during the complete liquefaction process. To analize the response mechanism of the surface-ground/structures system in the liquefaction, the following ex-periment is conducted.

Illustrated in Figure 7 are the records of the acceleration res-ponse of the small-sized accelerometer which is placed on the surface of the unsaturated surface ground (moisture content:0.25 0) and the small-sized accelerometer buried in the same surface ground. To the small-sized accelerometer placed on the surface of the ground an aluminum plate(8X12cm) is attached, so that the instrument will not be overturned. Also a bar(0.3cmøX2cm) is attached to the aluminum plate to fix it to the ground, so that the plate will not slip on the surface (Figure 8).

The moisture content of the model surface ground used for the experiment illustrated in Figure 7 is 0.250, which is believed to be considerably unsaturated, while the specific gravity of the small-sized accelerometer is about 1.3 and the measuring equipment should be said considerably light. Therefore it is supposed that the small-sized accelerometer and the surface ground will be moving en bloc. As can be judged from a magnified illustration in Figure 7, the response waveform on the surface of the ground is very similar to the one under the surface ground, which might suggest that the small-sized accelerometer placed on the surface of the ground and the one buried under the surface ground are vibrating with quite the same response mechanism.

In the liquefaction process, no collision phenomenon between the small-sized accelerometer placed on the surface ground has been observed. Therefore the reason why the small-sized accelerometer placed on the surface of the ground or under the surface ground shows a great acceleration response during the incomplete liquefaction can be attributed to the vibration characteristics of the surface ground in the liquefaction process. In other words, even if the surface ground and the buried structures produce a "collision phenomenon" with each other, it might be said that no influence to the acceleration response characteristics can be seen. Therefore it could be judged that the response characteristics of the buried structures in the liquefaction process will be determined by the vibration characteristics of the surface ground.

THE STRAIN RESPONSE OF A VINYLCHROLIDE PIPE IN THE LIQUEFACTION PROCESS

Next, we would like to analyze the strain response characteristics in the liquefaction process by burying a vinylchrolide pipe in the model ground. Vinylchrolide pipes are actually used as underground conduits. An analogy with horizontally long and narrow underground structures (e.g., shield tunnels such as subways or underground steel pipes such as gas pipes or water mains) can be made with regard to the response characteristics by making clear the response characteristics of a vinylchrolide pipe in the liquefaction process.

The vinylchrolide pipe used for the experiment is 76cm long and of 10mm inner diameter. A strain gauge is attached to the place of 38cm long from an end, and after that inflection strain responses are measured by the 2-gauge method. The vinylchrolide pipe is to be buried in the surface of the ground whose height is 10cm from the bottom. Beside the vinylchrolide pipe, a pore-water pressure gauge is placed. From the changes of the pore-water pressure recorded by the pore-water pressure gauge, how the liquefaction is developed will be judged.

The model surface ground in which the vinylchrolide pipe is buried is liquefied by giving vibration to the direction perpendicular to the pipe and to the horizontal direction parallel to the ground. Figure 9 illustrates the state of the vinylchrolide buried. For comparison, the same experiment is to be performed on the dry surface ground and the influence of softening and liquefaction of the model surface ground on the response characteristics will be observed.

A sine wave as an input

Illustrated in Figure 10(a) is an inflection strain response of the vinylchrolide pipe under the unsaturated surface ground (moisture content: 0.240). As the pore-water pressure is raised, the surface ground is softened and liable to move. In such a state of incomplete liquefaction, a great strain response can be noted. After that, when the surface ground is liquefied and a shear wave is hard to propagate, the strain response of the vinylchrolide pipe becomes small. Immediately after that, however, the strain response again becomes large. This might be because that since the moisture content of the surface ground is low and sand particles are immediately repiled, the external force begins to act on the vinylchrolide pipe.

The same experiment was performed with reference to both the saturated surface ground (moisture content: 0.270) (Figure 10(b)) and the dry surface ground (Figure 10(c)) as well. If comparison with the maximum inflection strain is instituted based upon the experimental results in the 3 types of the surface grounds, values of the maximum inflection strain in the saturated and unsaturated grounds were found to be about 15 times as large as the value of the dry surface ground. Moreover, this large strain response is continuously noticed as long as the input lasts.

<u>A random wave is inputted</u>

Illustrated in Figure 11(a) is an inflection strain response of the vinylchrolide pipe in the saturated surface ground(moisture content: 0.280). As the pore-water pressure is raised and the surface ground begins to be liquefied, an inflection strain response is seen. However, after the surface ground is completely liquefied, the inflection strain response has become small. Such response characteristics are quite similar to the case of a sine wave(6Hz) being used as an input. In comparison, the experimental results in the dry surface ground are to be illustrated in Figure 11(b).

As far as the results of the sand box experiment shown above are concerned, it seems likely that even when a random wave is used as an input, a great inflection strain response will manifest itself with reference to the buried pipe.

CONCLUSIONS

In this study, the behavior of underground structures in the liquefaction process of the surface ground was intended to empirically explain. From the liquefaction experiment using a sand box, the following results are obtained.

(1) When a sandy ground with low moisture content is softened or liquefied, the acceleration or response magnification of the inflection strain response becomes great.
(2) Besides that, so long as the surface ground is softened or liquefied, the lower the moisture content of the surface ground is, the longer the great response continues.
(3) This might be because that the rigidity of the surface ground was lowered according to the progress of the liquefaction, and the surface ground and input made a nonlinear resonance.
(4) The reason why this great response lasts in the unsaturated

surface ground might be that the resonance time becomes long.

From the results as shown above, it can be said that although the liquefaction has been had in mind by us just with reference to the saturated sandy ground until now, enough consideration should be given to the liquefaction of the unsaturated sandy ground. Therefore, when the response characteristics of the underground structures are discussed, it will be conceivable that the matter whether the surface ground is softened or liquefied by a seismic move will be a big problem.

REFERENCES

Hakuno, M. and Katada, T. (1978) Dynamic Behavior of Underground Structures during Liquefaction of Sand, Proc. of the 5th Japan Earthquake Engineering Symposium-1978, Tokyo, Japan, pp.649-656.

Hakuno, M. and Shikamori, M., Seismic Force of Imcompletely Liquefied Sand on Underground Structures, Bulletin of the Earthquake Research Institute, Vol.53, pp.243-254.

Iwasaki, T., Tatsuoka, F. and Yoshida, S. (1976) Vibration Test of Model Pile Buried in Sand Layer, Materials of Public Works Research Institute, Ministry of Construction, Japan No.1152.

Iwasaki, T., Tatsuoka, F. and Sakaba, Y. (1976) Vibration Test of Model Pile Buried in Sand Layer, Proc. of the 14th Japan National Conference on Earthquake Engineering, Tokyo, Japan, pp.37-40.

Katada, T. and Hakuno, M. (1978) Response Characteristics of Underground Structures During a Liquefaction, Proc. of the 33th Annual Convention of Japan Civil Engineers, pp.348-349.

Katada, T. (1979) Dynamic Behavior of Structures Buried in Liquefied Surface Ground, Proc. of the 15th Japan National Conference on Earthquake Engineering, Tokyo, Japan, pp.1-4.

Shikamori, M., Sato, Y. and Hakuno, M. (1973) Influence of Imcompletely Liquefied Sand on the Underground Structures, Proc. of the 28th Annual Convention of Japan Civil Engineers.

Yoshida, T. and Uematsu, M. (1978) Dynamic Behavior of a Pile in Liquefaction Sand, Proc. of the 5th Japan Earthquake Engineering Symposium-1978, Tokyo, Japan, pp.659-664.

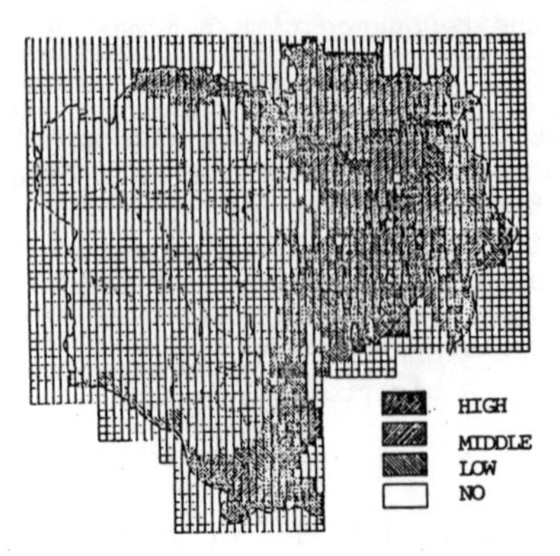

Fig. 1 PROBABILITY OF LIQUEFACTION IN TOKYO CITY

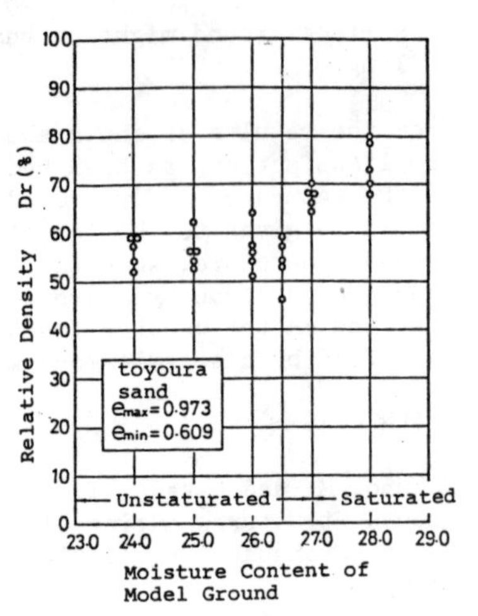

Fig. 2 RELATIVE DENSITY OF MODEL GROUND

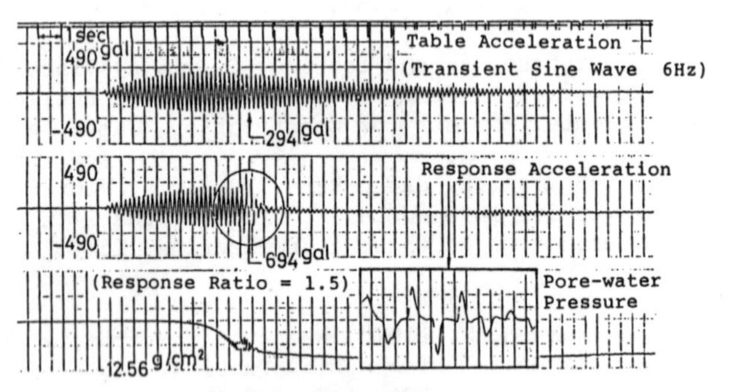

Fig. 3(a) ACCELERATION RESPONSE IN SATURATED GROUND -SINE INPUT-

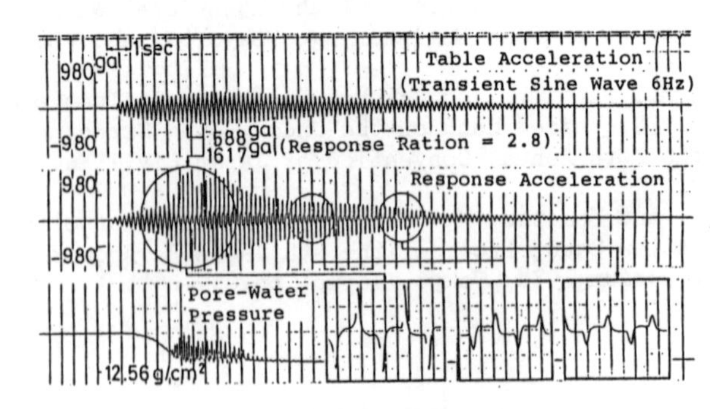

Fig. 3(b) ACCELERATION RESPONSE IN UNSATURATED GROUND -SINE WAVE-

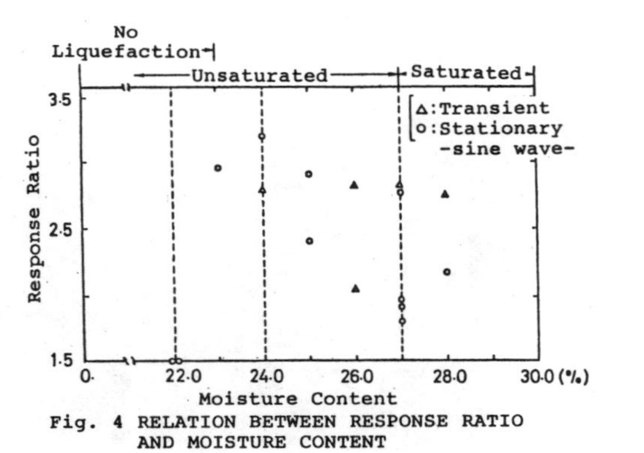

Fig. 4 RELATION BETWEEN RESPONSE RATIO AND MOISTURE CONTENT

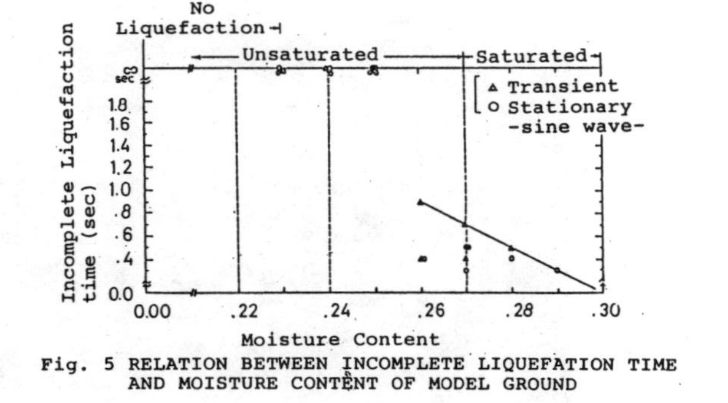

Fig. 5 RELATION BETWEEN INCOMPLETE LIQUEFATION TIME AND MOISTURE CONTENT OF MODEL GROUND

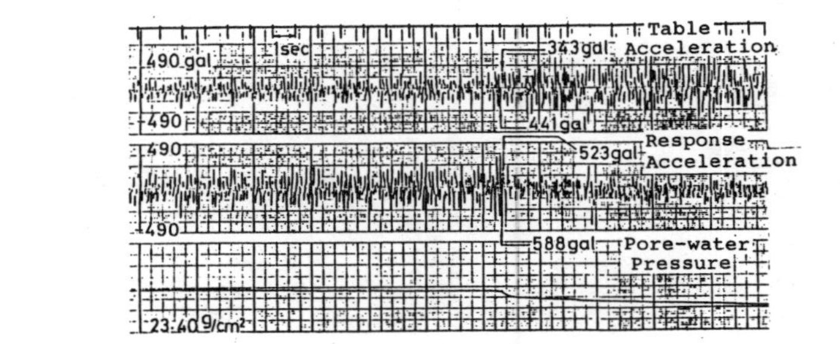

Fig. 6(a) ACCELERATION RESPONSE IN SATURATED GROUND -RANDOM WAVE-

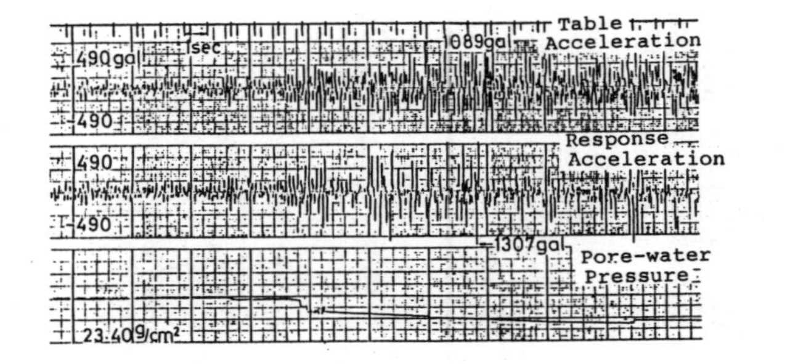

Fig. 6(b) ACCELERATION RESPONSE IN UNSATURATED GROUND -RANDOM WAVE-

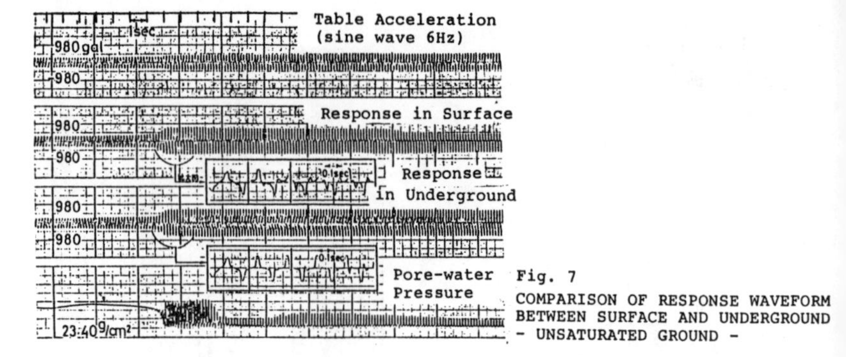

Fig. 7
COMPARISON OF RESPONSE WAVEFORM BETWEEN SURFACE AND UNDERGROUND
- UNSATURATED GROUND -

Small-sized accelerometer

plate

AM-2

ground

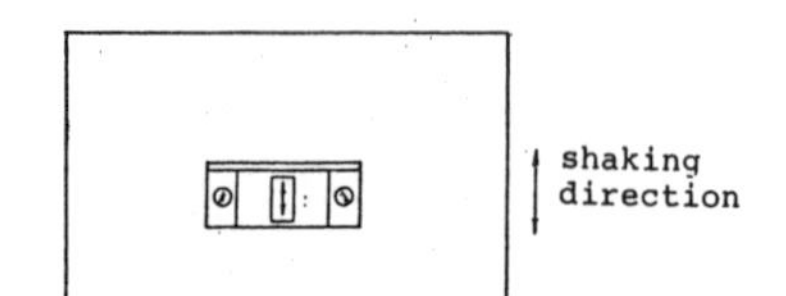

Fig. 8 SMALL-SIZED ACCELEROMETER ON SURFACE GROUND

Fig. 9 VINYLCHROLIDE PIPE BURIED IN SURFACE GROUND

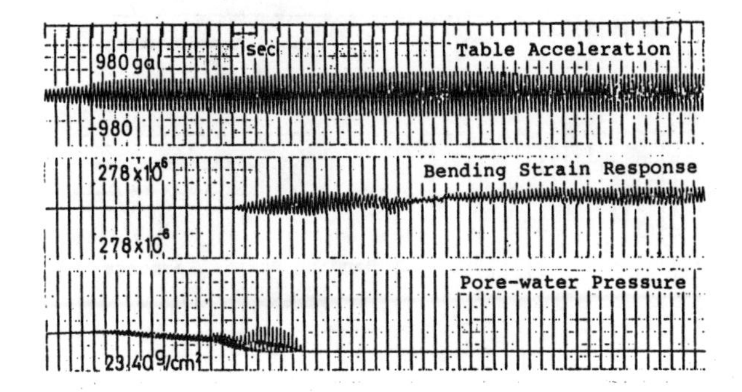

Fig. 10(a) BENDING STRAIN RESPONSE
IN UNSATURATED GROUND

Fig. 10(b) BENDING STRAIN RESPONSE
IN SATURATED GROUND

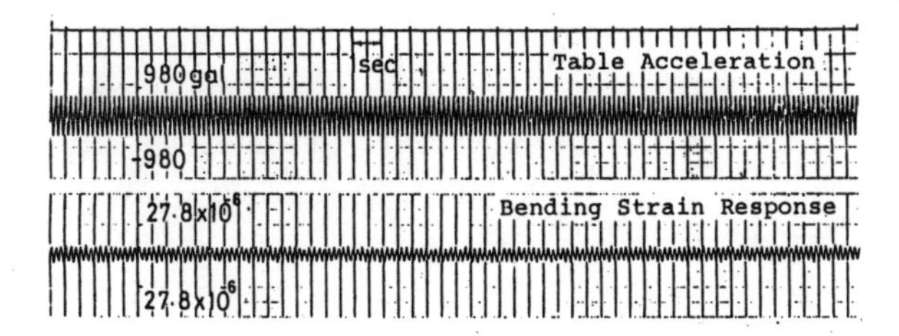

Fig. 10(c) BENDING STRAIN RESPONSE
IN DRY GROUND

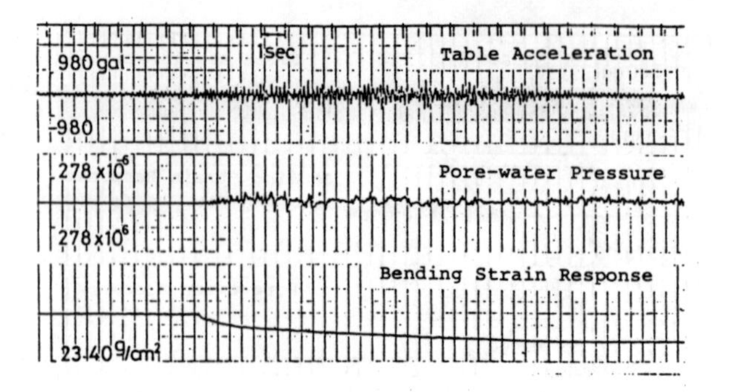

Fig. 11(a) BENDING STRAIN RESPONSE
IN SATURATED GROUND -RANDOM INPUT-

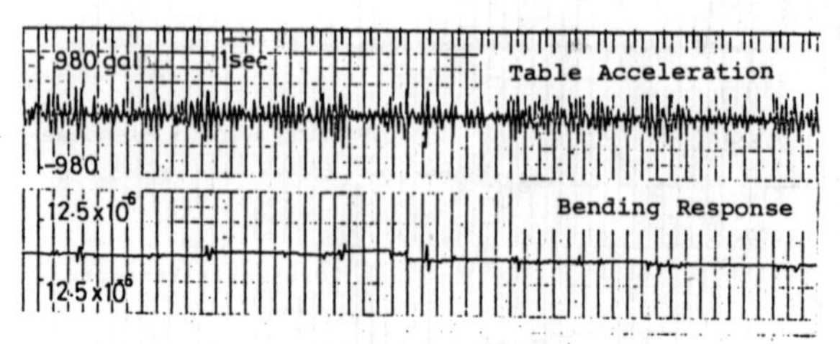

Fig. 11(b) BENDING STRAIN RESPONSE
IN DRY GROUND -RANDOM INPUT-

Soil Dynamics & Earthquake Engineering Conference / Southampton / 1982.07.13-15

Simplified Procedures for Assessing Soil Liquefaction During Earthquakes

TOSHIO IWASAKI, TADASHI ARAKAWA & KEN-ICHI TOKIDA
Ministry of Construction, Tsukuba, Japan

INTRODUCTION

Simplified methods to evaluate the effects of saturated sandy soil liquefaction are needed for the reasonable earthquake resistant design of structures considering the surrounding soil liquefaction. The authors, Iwasaki et al (1978), proposed two simplified methods with use of a liquefaction resistance factor F_L and a liquefaction potential index I_L to evaluate the liquefaction potential of saturated sandy soils. Basing on the proposed methods, the liquefaction potential can be estimated simply by using the fundamental properties of soils, i.e., N-values of the standard penetration test, unit weights, mean particle diameters, and the maximum acceleration at the ground surface. In this paper, the two simplified methods are firstly introduced, and to prove the effectiveness of the proposed methods the values of both F_L and I_L at 64 liquefied sites and 23 non-liquefied sites during past six earthquakes are calculated according to the simplified methods. Also shaking table tests on soil liquefaction are carried out for the saturated sandy model ground. Furthermore, several application methods using the factor F_L, the excessive pore water pressure induced in the saturated sandy soils and the effects of soil liquefaction on the resistance properties of grounds surrounding structures are described.

SIMPLIFIED METHODS

Liquefaction Resistance Factor (F_L)

An ability to resist the liquefaction of a soil element at an arbitrary depth may be expressed by the liquefaction resistance factor (F_L) indentified by Eq.(1).

$$F_L = \frac{R}{L} \quad \cdots\cdots\cdots\cdots\cdots\cdots \quad (1)$$

When the factor F_L at a certain soil is less than 1.0, the soil is judged to liquefy during earthquakes.
R in Eq.(1) is the in-situ resistance or undrain cyclic strength of a soil element to dynamic loads during earthquakes, and can be simply evaluated according to numerous undrained cyclic shear test results using undisturbed specimens, as follows,

for $0.04 \text{ mm} \leqq D_{50} \leqq 0.6 \text{ mm}$

$$R = 0.0882\sqrt{\frac{N}{\sigma_v'+0.7}} + 0.225 \log_{10} \frac{0.35}{D_{50}} \quad \cdots(2a)$$

for $0.6 \text{ mm} \leqq D_{50} \leqq 1.5 \text{ mm}$

$$R = 0.0882\sqrt{\frac{N}{\sigma_v'+0.7}} - 0.05 \quad \cdots\cdots\cdots\cdots\cdots\cdots(2b)$$

where N is the number of blows of the standard penetration test, σ_v' is the effective overburden pressure (in kgf/cm), and D_{50} is the mean particle diameter (in mm).
L in Eq.(1) is the dynamic load induced in the soil element by a seismic motion, and can be simply estimated by

$$L = \frac{\tau_{max}}{\sigma_v'} = \frac{\alpha_{s max}}{g} \cdot \frac{\sigma_v}{\sigma_v'} \cdot r_d \quad \cdots\cdots\cdots\cdots\cdots\cdots(3)$$

where τ_{max} is the maximum shear stress (in kgf/cm), $\alpha_{s\,max}$ is the maximum acceleration at the ground surface (in gals), g is the acceleration of gravity (=980 gals), σ_v is the total overburden pressure (in kgf/cm²), and r_d is the reduction factor of dynamic shear stress to account for the deformation of the ground. From a number of seismic response analyses for grounds, Iwasaki et al (1978) proposed the following relation for the factor r_d.

$$r_d = 1.0 - 0.015Z \quad \cdots\cdots\cdots\cdots\cdots\cdots\cdots\cdots\cdots(4)$$

where Z is the depth in meters.

Liquefaction Potential Index (I_L)

An ability to resist liquefaction at a given depth of grounds can be evaluated by the factor F_L. However it must be noticed that the damage to structures due to soil liquefaction is considerably affected by the severity of liquefaction degree. In view of this fact, Iwasaki et al (1978) also proposed the liquefaction potential index (I_L) defined by Eq.(5) to estimate the severity of liquefaction degree at a given site.

$$I_L = \int_0^{20} F \cdot W(Z) dZ \quad \cdots\cdots\cdots\cdots\cdots\cdots\cdots\cdots\cdots(5)$$

where $F=1-F_L$ for $F \leqq 1.0$ and $F=0$ for $F>1.0$, and $W(Z)=10-0.5Z$ (Z in meters)as shown in Fig.1. W(Z) accounts for the degree of soil liquefaction according to the depth, and the triangular shape of W(Z) and the depth of 20 meters are decided considering the liquefaction phenomena during the past earthquakes. For the case of $F_L=0$ for the entire depth, I_L becomes 100 being the highest, and for the case of $F \geqq 1.0$ for the entire depth, I_L becomes 0 being the lowest.

CASE STUDIES ON F_L AND I_L FOR PAST EARTHQUAKES

Both the liquefaction resistance factor F_L and the liquefaction potential index I_L were calculated using the proposed methods for 64 liquefied sites and 23 non-liquefied sites where geotechnical informations are available during the following six earthquakes: the Nobi Earthquake of 1891 (Magnitude=8.0), the Tonankai Earthquake of 1944 (M=8.0), the Fukui Earthquake of 1948 (M=7.8), the Niigata Earthquake of 1964 (M=7.5), the Tokachi-oki Earthquake of 1968 (M=7.9) and the Miyagi-ken-oki Earthquake of 1978(M=7.4). The liquefied sites and the non-liquefied sites for the case studies are summarized in Table 1 (A) and (B), respectively.

Characteristics of Factor F_L

Figs.2(A) and 2(B) show typical calculation results of F_L with depth at a liquefied site and a non-liquefied site, respectively. It can be seen that F_L is mostly less than 1.0 for the liquefied layers, and greater than 1.0 for the non-liquefied layers.

F_L-values with depth at all liquefied and non-liquefied sites in the Niigata Earthquake (see Table 1) are summarized in Fig.3. In this figure the liquefied layers are estimated basing on damages to structures (see Fig.2(A)). It can be also seen that F_L is mostly less than 1.0 for the liquefied layers and the liquefied layers are likely to locate at the depth of about 10 meters or shallower.

Fig.4 shows the frequency and the accumulative incidences of F_L-values calculated for both liquefied and non-liquefied layers at all sites in Table 1. According to Fig.4, it is found that the distribution of F_L at the liquefied layers is very different from that at the non-liquefied layers. At the liquefied layers most (about 87%) of F_L-values distribute in the range less than 1.0, while at the non-liquefied layers most (about 85%) of F_L-values distribute in the range more than 1.0. However it must be also noticed that about 13% of F_L-values exceed 1.0 at the liquefied layers and about 15% of F_L-values are less than 1.0 at the non-liquefied layers.

Application of the Simplified Method to In-Situ Site

In this paper one example of applications of the simplified method based on the facter F_L for an in-situ liquefied site during a past earthquake is introduced. Yuriage-kami Dyke along Natori River was damaged by the Miyagi-ken oki Earthquake of 1978 and sand boils were observed at numerous points near the Dyke shown in Fig.5. After the earthquake detail investigations on soil liquefaction were carried out at both the liquefied and the non-liquefied points, i.e., the liquefied points Y-1 and Y-2, and non-liquefied points Y-3 and Y-4, as shown in Fig.5.

Fig.6 illustrates the results of soil profieles, N-values, D_{50} and F_L-values at the liquefied points (Y-1, Y-2) and the non-liauefied point (Y-3). In this site both simplified analyses and detailed analyses are carried out to calculate the F_L-

values with depth. In simplified analyses, three levels of maximum ground surface accelerations i.e., 180, 240 and 300 (gals) are supposed, in view of the measured strong-motion records nearby. In detailed analyses, the ground acceleration recorded on the rocky layer during this earthquake is inputted at the estimated base. A maximum acceleration of 150 gals is taken. It is found from Fig.6 that soil liquefaction occurred at the points which had F_L-values almost less than 1.0. In this way it is clarified from the in-situ investigations that the liquefaction resitance factor F_L is effective to evaluate the liquefaction potential of saturated sandy soils.

Characteristics of Index I_L

Fig.7 summarizes both the relation between the number of cases and I_L, and the relation between the accumulative percentages and I_L, at all liquefied and non-liquefied sites in Table 1. It is found from this figure that I_L for liquefied sites seems to be higher than that at non-liquefied sites, i.e., for non-liquefied sites I_L is mostly less than 15 and the percentage that I_L is less than 5 is about 70%, and on the other hand for liquefied sites the percentage that I_L is less than 5 is only about 20% and at about 50% of the sites I_L is more than 15. From these results, the following simplified procedure for assessing soil liquefaction based on the index I_L may be proposed as a preliminary guideline.

$I_L = 0$: Liquefaction risk is very low,

$0 < I_L \leqq 5$: Liquefaction risk is low,

$5 < I_L \leqq 15$: Liquefaction risk is high,

$15 < I_L$: Liquefaction risk is very high.

As mentioned in the above, it is shown that the index I_L may be reasonably used to assess the liquefaction potential at a certain site.

SHAKING TABLE TESTS ON SOIL LIQUEFACTION

Shaking table tests were carried out to clarify the properties of soil liquefaction, the effects of soil liquefaction on structure foundations and the effectiveness of the proposed liquefaction resistance factor F_L. A loose saturated sand ground model with 0.95m deep, 6m long and 3m wide was prepard on a shaking table, and four kinds of pile foundation models were set up in the sand, as shown in Fig.8. In the tests, the table was shaked in the sinusoidal motion with a constant frequency of 7 Hz, and the input table accelerations ranged from 30 gals to 250 gals, as listed in Table 2. Accelerations and pore water pressures were measured in the sand. Accelerations, displacements and earth pressures of the pile models were also observed.
Fig.9 is an example of test results, i.e., the distribution of a degree of soil liquefaction and the time history for Test 2.

Hereupon the degree of soil liquefaction is defined by the factor L_u,

$$L_u = \Delta u/\sigma_v' \quad \cdots\cdots (6)$$

where Δu is the excessive pore water pressure. Sands with L_u of 1.0 are assumed to completely liquefy. From Fig.9 it seems that soil liquefaction spreads from the surface to the bottom of the ground gradually.

Fig.10 shows the relationship between the top displacement of the pile model (Model-2) and the degree of soil liquefaction with depth defined by H_L/H_0 (H_L:the area completely liquefied, H_0:the thickness of the ground model). It is seen that the displacement of the pile tends to increase as H_L/H_0 increases.

Figs.11 and 12 show the typical relationships between ground accelerations, pore water pressures and F_L-values for non-liquefied cases and liquefied cases, respectively. In these figures, F_L-values are estimated by Eqs. (2) and (3). It is found that F_L-values decrease according to the increase of pore water pressures, and that F_L-values are less than 1.0 for the liquefied layers and are higher than 1.0 for the non-liquefied layers.

Fig.13 summarizes the relation between F_L and L_u for the liquefied layers. From this figure, it is seen that F_L decreases as L_u increases and that F_L is less than 1.0 for L_u of 0.5 or higher and is more than 1.0 for L_u of 0.5 or lower. Furthermore, it is clarified that the sand layers are likely to completely liquefy when F_L decreases to less than 0.6.

From these shaking table tests, it is clarifed that the proposed factor F_L may be adequately used to estimate soil liquefaction potential of saturated sand layers.

METHODS FOR EVALUATING EFFECTS OF LIQUEFACTION

The effects of soil liquefaction on structures are required to be quantitatively clarified for establishing the concrete earthquake resistance design of structures. Several methods using the factor F_L are introduced to evaluate the effects of soil liquefaction.

Probability of Soil Liquefaction

It is clarified in Fig.4 that for a value of F_L both possibilities of liquefaction and non-liquefaction may be expected. Therefore it is needed to estimate the probability of soil liquefaction for a certain F_L-value. Fig.14 shows the relation between the probability of liquefaction or non-liquefaction and F_L which is estimated from the results in Fig.4. Hereupon 500 cases for both liquefied and non-liquefied layers in Fig.4 are collected. From this figure, the probability of liquefaction or non-liquefaction can be estimated. For example, the probability of liquefaction for F_L-value of 1.0 is about 50% and that for a F_L-value less than about 0.6 is almost 100%.

Pore Water Pressure

Excessive pore water pressures generated in sand layers is very important for soil liquefaction studies. In this paragraph the simplified procedures for evaluating excessive pore water pressure using a F_L-value are introduced.

Dynamic Soil Tests From dynamic triaxial tests on cyclic strength for soil liquefaction, a typical relation between the shear stress ratio τ/σ_v' (τ:shear stress) and the number of cycles N_l to generate liquefaction is shown in Fig.15, and the relation is approximately given by Eq.(7),

$$(\tau/\sigma_v') = a N_l^b \quad \cdots\cdots (7)$$

where constant values, a and b are decided basing on the dynamic triaxial tests as shown in Fig.15. If liquefaction assumes to occur for the cyclic strength R with the number of cycles N_R and for the dynamic load L with the number of cycles N_L, the relations on both R and L may be concluded in Figs.(8a) and (8b), respectively,

$$R = (\tau_R/\sigma_v') = a N_R^b \quad \cdots\cdots (8a)$$

$$L = (\tau_R/\sigma_v') = a N_L^b \quad \cdots\cdots (8b)$$

where τ_R and τ_L are the shear strength and the shear load, respectively. From Eqs.(1), (8a) and (8b), the following relation is obtained.

$$F_L = (N_R/N_L)^b \quad \cdots\cdots (9)$$

On the other hand, the relations between $\Delta u/\sigma_v'$ and N/N_l (N, N_l :number of cycles before liquefaction and that at complete liquefaction, respectively) are obtained, for example, as Fig.16. Hereupon because N_R and N_L are regarded as N and N_l, respectively, the following relation can be assumed.

$$(N_R/N_L) = (N/N_l) \quad \cdots\cdots (10)$$

From Eqs.(9) and (10), Eq.(11) can be concluded.

$$(N/N_l) = (F_L)^{1/b} \quad \cdots\cdots (11)$$

Therefore the pore water pressure can be estimated by the factor F_L as follows, according to the test results shown in Fig.16.

$$(\Delta u/\sigma_v') \sim (N/N_l) = (F_L)^{1/b} \quad \cdots\cdots (12)$$

Shaking Table Tests Pore water pressures can be estimated by the factor F_L basing on shaking table tests, i.e., by using the relation shown in Fig.3.

Relation between Pore Water Pressure and $\bar{F}_L$ Fig.17 summarizes a relation between pore water pressure and F_L according to the proposed methods, i.e., dynamic soil tests and shaking table tests. From this figure, pore water pressure can be simply evaluated by F_L-values.

Properties of Liquefied Sand Layer

For establishing a reasonable method of earthquake resistance design of structures considering soil liquefaction, it is important to clarify the properties of liquefied sand layers. For the purpose of estimating the properties quantitatively, simple static loading tests are carried out.

Fig.18 shows the outline of the test apparatus. The saturated sand specimens are consolidated by the air pressure through the rolling diaphragm seal. In the center of the rolling diaphragm seal, the loading plate (6cm in diameter) is connected. The specimens are consolidated by two kinds of confining pressures (σ_v'), i.e., 0.5 and 1.0 (kgf/cm²), and after three-houre consolidation the excessive pore water pressures (Δu) are supplied forcibly into the specimen. The ratio of the pore water pressure to the confining pressure is changed from 0 to 0.95 continuously. Under a certain forced ratio the static loading test is carried out, and the static load and the displacement of the loading plate are measured. Fig.19 shows a typical test result, i.e., the relation between the static load P (in kgf) and the displacement δ (in mm), according to the different ratios. From this figure it is found that the bearing calacity of the soil specimen is affected by the increase of pore water pressure.

Fig.20 summarizes the results of the overall tests. In this figure the bearing capacity of the soil is defined by

$$K = P/(A \cdot \delta) \quad \cdots\cdots(13)$$

where K is a bearing capacity (in Kgf/cm²/mm) and A is an area of the loading plate (in cm²). From this figure it is found that the bearing capacity of the soil decreases according to an increase of pore water pressure.

Fig.21 shows the relation between K/K_0 and L_u based on the results in Fig.20. K_0 is the bearing capacity when pore water pressures don't occur. From this figure the degree of decrease of the bearing capacity of the soil induced by liquefaction can be quantitatively evaluated from a degree of soil liquefaction. From the test results in Fig.13, the relation between F_L-value and L_u can be assumed as follows.

$$\left\{\begin{array}{ll} F_L \leqq 0.6 , & L_u = 1.0 \\ 0.6 < F_L \leqq 0.8 , & 0.9 \leqq L_u < 1.0 \\ 0.8 < F_L \leqq 1.0 , & 0.5 \leqq L_u < 0.9 \\ 1.0 < F_L , & L_u < 0.5 \end{array}\right\} \cdots\cdots(14)$$

Hereupon the average relation between K/K_0 and L_u may be proposed expediently as four stages shown in Fig.21 after considering the test results and the relation in Eq.(14). Therefore the relation between F_L and K/K_0 can be approximately estimated as

$$\left\{\begin{array}{ll} F_L \leqq 0.6 \;, & K/K_0 = 0 \\ 0.6 < F_L \leqq 0.8 \;, & K/K_0 = 1/3 \\ 0.8 < F_L \leqq 1.0 \;, & K/K_0 = 2/3 \\ 1.0 < F_L \;, & K/K_0 = 1 \end{array}\right\} \quad \cdots\cdots\cdots\cdots\cdots\cdots (15)$$

The results above-mentioned on the resistance propery of liquefied sand layers were already applied to the practical design of highway bridge (Japan Road Association) as shown in Table 3. In this table the reduction factor D_E is equivalent to K/K_0.

CONCLUSIONS

Two simplified methods based on the liquefaction resistance factor F_L and the liquefaction potential index I_L are proposed to assess the liquefaction potential. From the studies it is found that F_L-values are mostly less than 1.0 for liquefied layers and greater than 1.0 for non-liquefied layers, and that I_L-values at liquefied sites differ noticeably from those at non-liquefied sites.
From the experimental tests, it is also shown that the effects of liquefaction can be reasonably assessed by F_L-values.

ACKNOWLEDGEMENTS

The studies described above were greatly assisted by Professor Fumio Tatsuoka at the Institute of Industrial Science, the University of Tokyo, Dr. Susumu Yasuda at Kiso-jiban Consultants Co., Ltd., Mr. Toshio Kimata, Research Engineer at the Public Works Research Institute and Mr. Seiichi Yoshida, Former Assistant Researcher at the Public Works Research Institute. The authors wish to express sincere thanks to the four individuals.

REFERENCES

Iwasaki, T., Tatsuoka, F., Tokida, K. and Yasuda, S. (1978), "A Practical Method for Assessing Soil Liquefaction Potential Based on Case Studies at Various Sites in Japan," 2nd Inter-National Conference on Microzonation for Safer Construction Research and Application.

Iwasaki, T. and Tokida, K. (1980), "Studies on Soil Liquefaction Observed during the Miyagi-ken-oki Earthquake of June 12, 1978," Proc., 7th World Conference on Earthquake Engineering.

Iwasaki, T. and Tokida, K. (1981) "Soil Liquefaction Potential Evaluation with Use of the Simplified Procedure," Proc., International Conference on Recent Advances in Geotechnical Earthquake Engineering and Soil Dynamics.

Japan Road Association(1980), "Specifications for Highway Bridges, Part V. Earthquake Resistant Design (in Japanese).

Kuribayashi, E. and Tatsuoka, F. (1975), "Brief Review of

Liquefaction during Earthquakes in Japan," Soils and Foundations, Vol.15, No.4.
Tatsuoka, F., Iwasaki, T., Tokida, K., Yasuda, S., Hirose, M., Imai,T. and Kon-no, M. (1980), "Standard Penetration Tests and Soil Liquefaction Potential Evaluation," Soils and Foundations, Vol.20, No.4.

Table 1 Sites for Analysis of Soil Liquefaction Evaluation

(A) Liquefied Sites

No.	Site			Earthquake	Ref.
1	Shinano River	1	Niigata City	Niigata, 1964, M=7.5	
2	Railroad Bridge	2			
3	Higashi-Kosen Bridge	Br. 1			(3)
4		Be. 2			
5		Br. 4			
6	Bandai Bridge	Br. 6			
7	Yachiyo Bridge	Br. 1			
8		Br. 5			
9		Br. 7			
10	Shin-Matsuhama Bri.	Br. 2			
11	Taihei Bridge	Br. 1			
12		Br. 2			
13	Showa Bridge	Br. 1			
14		Br. 2			
15		Br. 3			
16		Br. 2			
17		Br. 3			
18	Niigata Airport				
19	Sekiya				(4)
20	Niigata Railroad Hospital	Br. 1			(1)
21		Br. 2			
22		Br. 1			
23		Br. 2			
24	Kawagishi-Cho	Br. 1			(5)
25		Br. 2			
26		Br. 3			
27		Br. 4			
28	Kawagishi-Cho	BC21-2			(1)
29		BC21-3			
30		BC104			
31		BC14			

No.	Site			Earthquake	Ref.
32	Nanae Beach	Br. 1	Hakodate City	Tokach-Oki, 1968, M=7.9	(6)
33		Br. 2			
34		Br. 3			
35	Hachinohe City				(7)
36	Gifu City		Gifu Pref.	Nobi, 1891, M=8.0	(2)
37	Unuma	Kagamigahara			
38	Ogaseike				
39	Mangoku, Ohgaki				
40	Meikodori		Nagoya City	Tonankai, 1944, M=8.0	
41	Kohmei				
42	Inaei				
43	Takaya 45		Fukui Pref.	Fukui, 1948, M=7.3	
44	Maruoka No. 2				
45	Takaya 2-168				
46	Abukuma Bridge	Br. 4	Miyagi Pref.	Miyagi-ken-Oki, 1978, M=7.4	(8)
47	Mouth of Abukuma River				
48	Yuriage-Kami	Y - 1			
49	"	Y - 2			
50	Yuriage Bridge	No. 1			
51	"	No. 2			
52	"	No. 3			
53	Yamazaki				
54	Oiri (1)				
55	"	No. 2			
56	Uomachi	B - 1			
57	"	B - 2			
58	Rifu	No. 12			
59	Shiomi	No. 1			
60	"	No. 2			
61	"	No. 3			
62	Nakamura	N - 4			
63	"	N - 5			
64	Wabuchi	W - 2			

(B) Non-Liquefied Sites

No.	Site			Earthquake	Ref.
1	Jindoji		Niigata City	Niigata, 1964, M=7.5	
2	Kogane-cho				
3	Higashi Kosen	Br. 5			(3)
4	Shin Matsuhama Bri.	Br. 1			
5	Omiya 266K 712M				
6	Showa Bridge	Br. 4			
7	Nishi Oh-Hata-Cho				
8	Gotanda Bridge	Br. 1			
9	"	Br. 2			
10	Maruoka		Fukui Pref.	Fukui, 1948, M=7.3	(2)
11	Nakamura	N - 1	Miyagi Pref.	Miyagi-ken-Oki, 1978, M=7.4	(8)
12	"	N - 2			
13	Yuiage-kami	Y - 3			
14	Kitakami River	No. 10			
15	Natori River, 3.2 Km				
16	Kinnou Bridge	P_8			
17	Abukuma Bridge	Br. 1			
18	"	Br. 2			
19	Eai Bridge	No. 1			
20	Minami Sendai	No. 2			
21	Uomachi	A - 1			
22	"	A - 2			
23	Wabuchi	W - 3			

Reference

(1) BRI (1965)
(2) BRI (1969)
(3) Japanese Society of Civil Engineers
(4) Ishihara (1976)
(5) J.S.S.M.F.E. (1976)
(6) Kishida (1970)
(7) Ohashi et al. (1977)
(8) Yasuda et al. (1980)

Table 2 Input Motion

Test No.	Input Motion	Input Acceleration (gals)
1	Sinusoidal 7 HZ	30
2		80
3		80
4		50
5		80
6		150
7		150
8		250

Table 3 F_L-D_E Relation

F_L	Depth, Z (m)	Reduction Factor, D_E
$F_L \leqq 0.6$	$Z \leqq 10$	0
	$10 < Z \leqq 20$	1/3
$0.6 < F_L \leqq 0.8$	$Z \leqq 10$	1/3
	$10 < Z \leqq 20$	2/3
$0.8 < F_L \leqq 1.0$	$Z \leqq 10$	2/3
	$10 < Z \leqq 20$	1

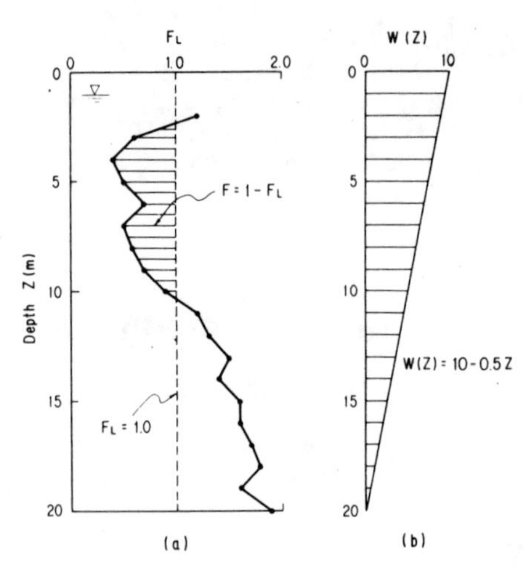

Fig. 1 Integration of F_L

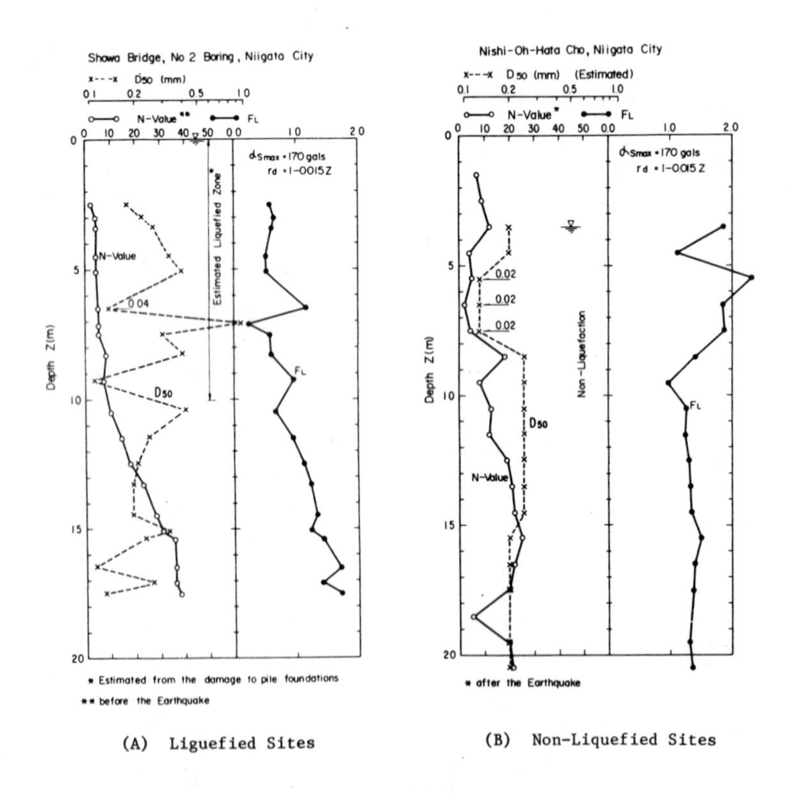

(A) Liguefied Sites

(B) Non-Liquefied Sites

Fig. 2 Relatinships between F_L and Depth

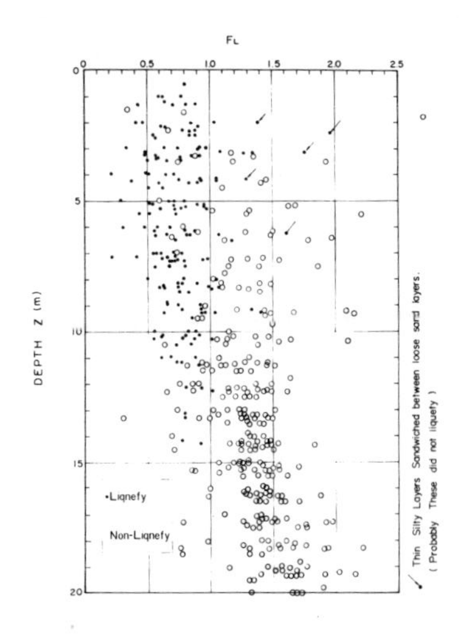

Fig. 3 Relationship between F_L and Depth at the Liquefied and Non-Liquefied Sites during the Niigata Earthquake

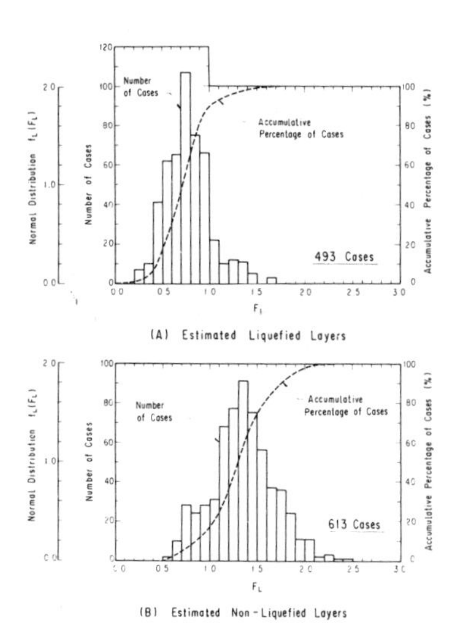

Fig. 4 Distribution of F_L Values and Their Accumulative Incidences, in Percentage, Comparing Liquefied Sites with Non-Liquefied Sites in Table 1

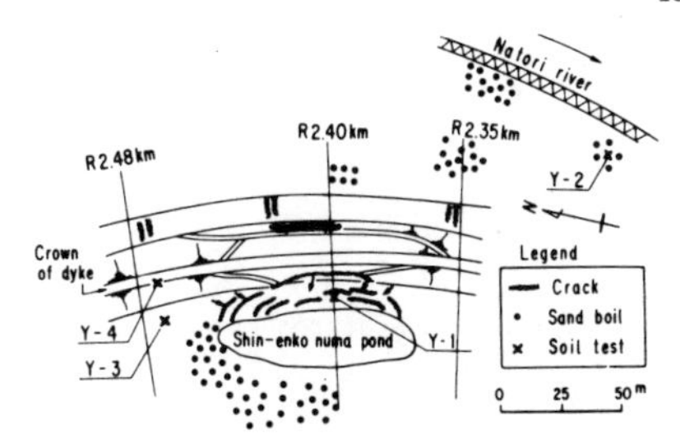

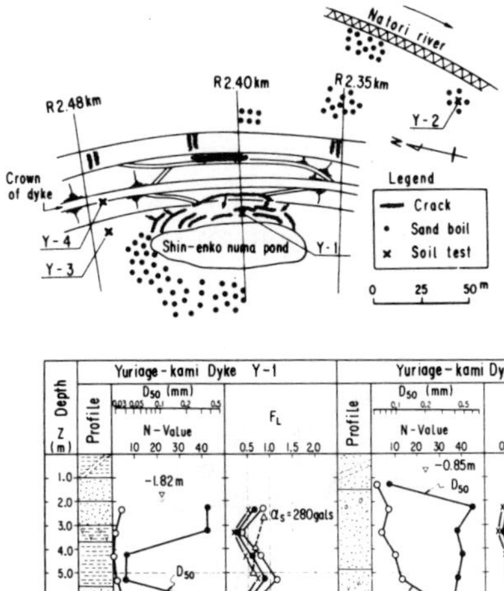

Fig. 5
Plan of Yuriage-kami Dyke and Points of Soil Tests

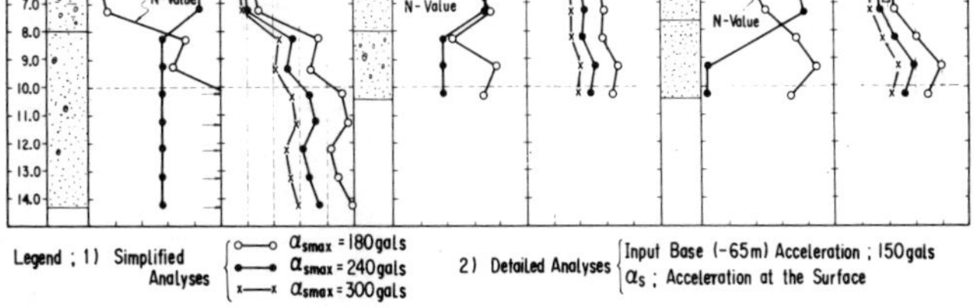

Fig. 6 F_L-values at Yuriage-kami Dyke
(Y-1, Y-2; liquefied, Y-3; non-liquefied)

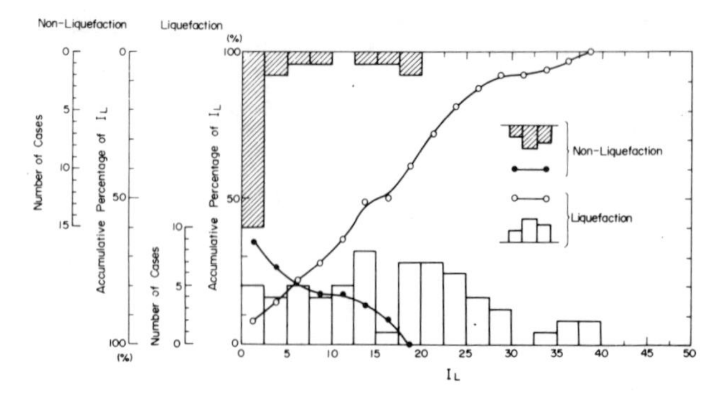

Fig. 7 Distribution of I_L Values and Their Accumulative Incidences, in Percentage, Comparing Liquefied Sites with Non-Liquefied Sites in Table 1

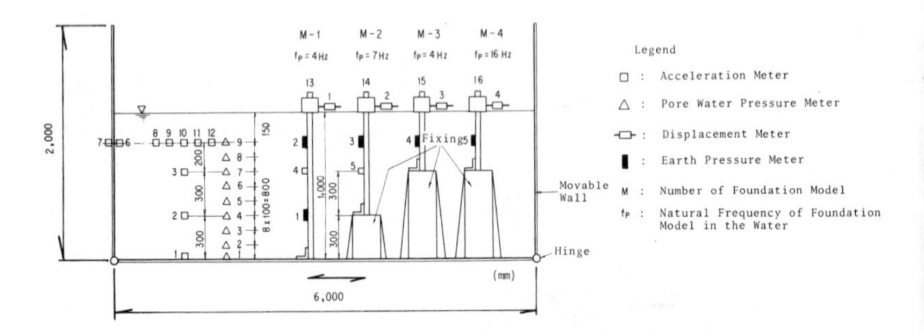

Fig. 8 Outline of Test Models and Measurement

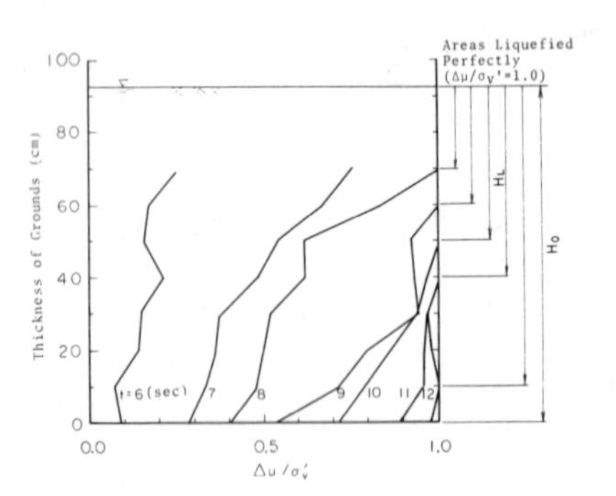

Fig. 9 Time History of the Rate of Ground Liquefaction $\Delta u/\sigma_v'$ (Test 2)

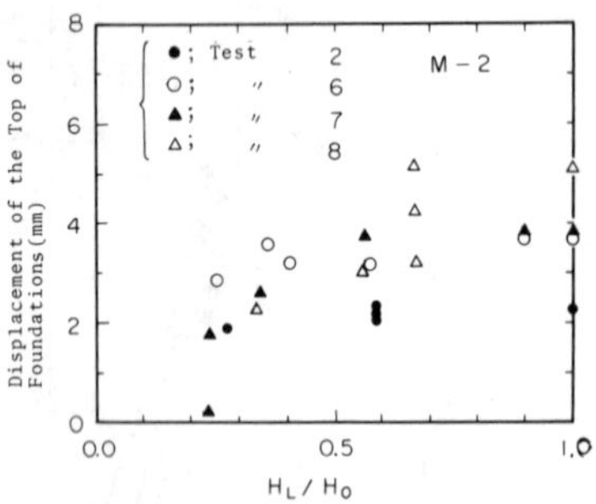

Fig. 10 Relationship between the Displacement of the Top of the Foundation Model and the Degree of Soil Liquefaction H_L/H_o

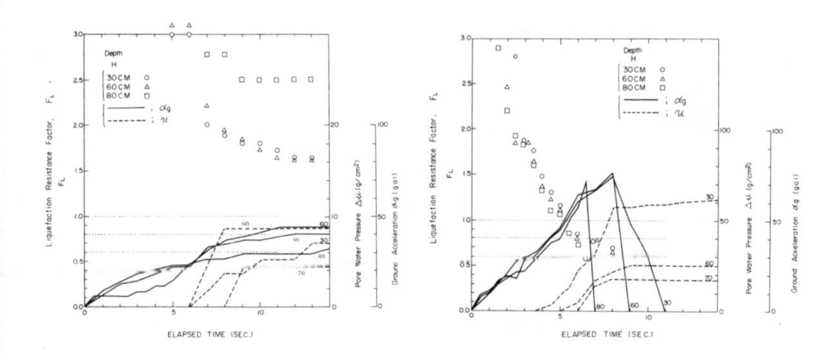

Fig. 11 Relationships between Pore Water Pressure and Acceleration of Non-Liquefied Sand Layers and F_L Values in Shaking Table Tests

Fig. 12 Relationships between Pore Water Pressure and Acceleration of Liquefied Sand Layers and F_L Values in Shaking Table Tests

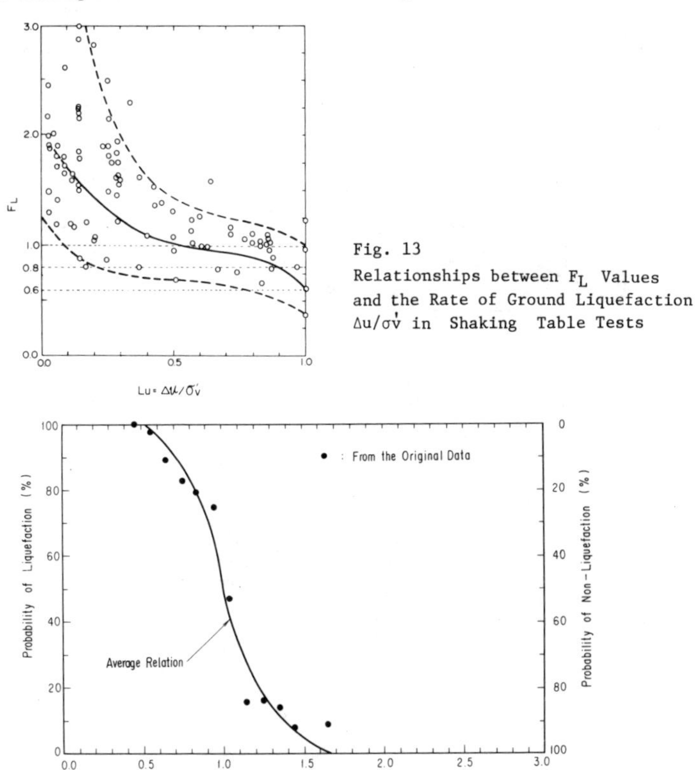

Fig. 13
Relationships between F_L Values and the Rate of Ground Liquefaction $\Delta u/\sigma_v'$ in Shaking Table Tests

Fig. 14 Relationship between Probability of Liquefaction or Non-Liquefaction and F_L

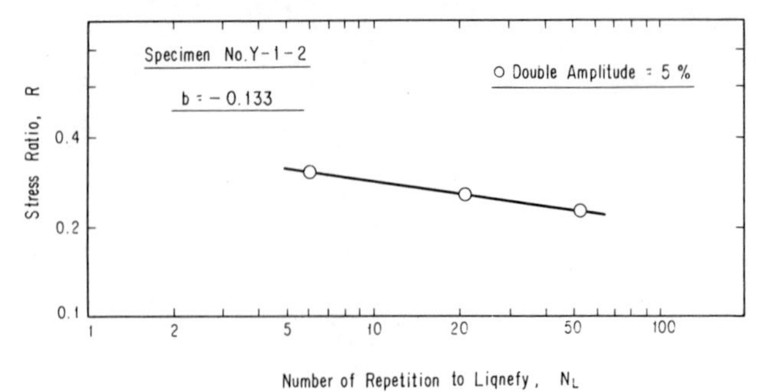

Fig. 15 An Example of Relationship between R and N_L (Dynamic Triaxial Test)

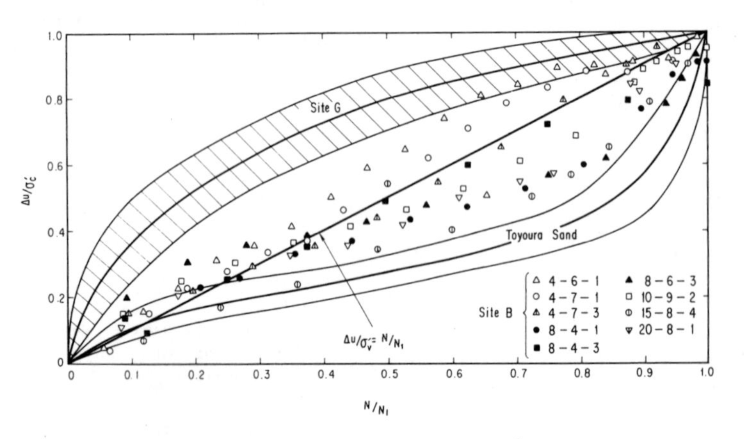

Fig. 16 Relationships between $\Delta u/\sigma_v'$ and N/N_L

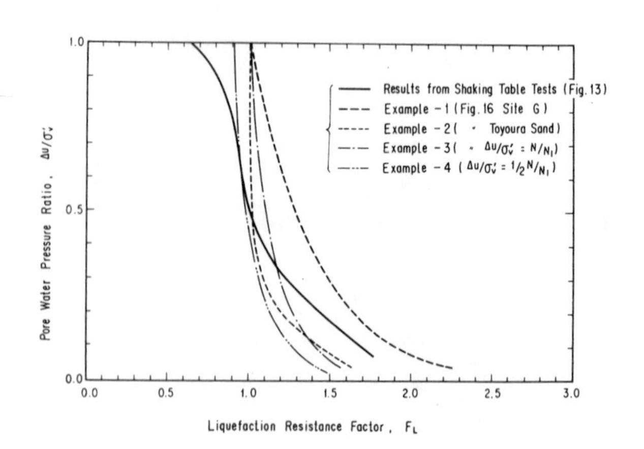

Fig. 17 Relationships between $\Delta u/\sigma_v'$ and F_L

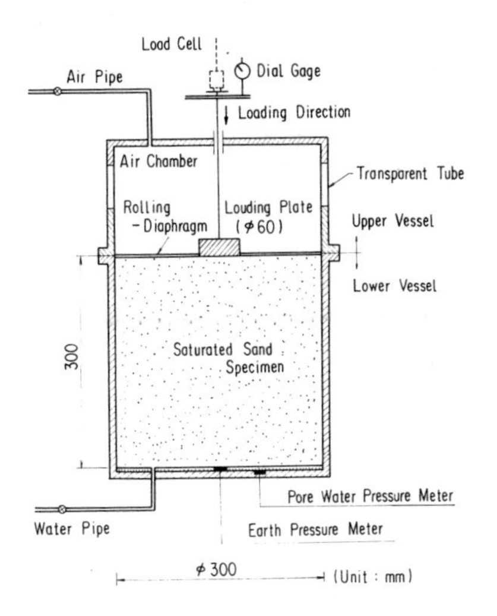

Fig. 18 Test Apparatus

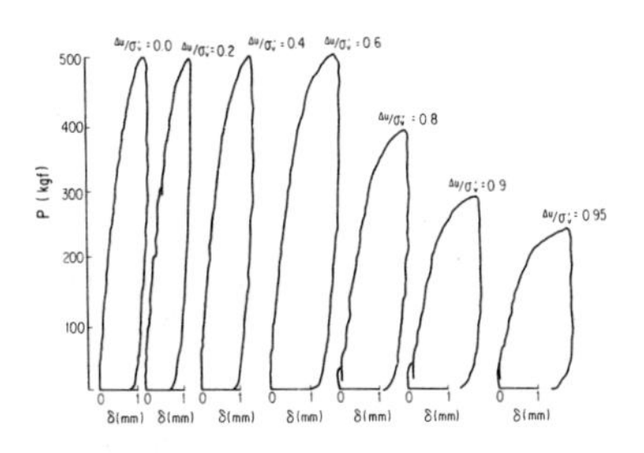

Fig. 19 An Example of Static Loading Test Results

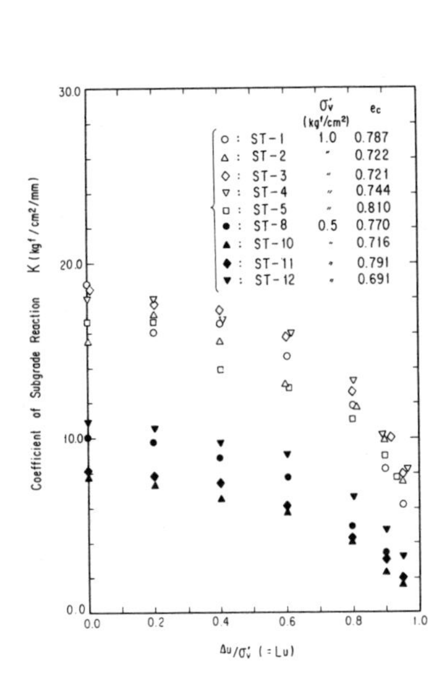

Fig. 20
Relationships between K and $\Delta u/\sigma_v'$

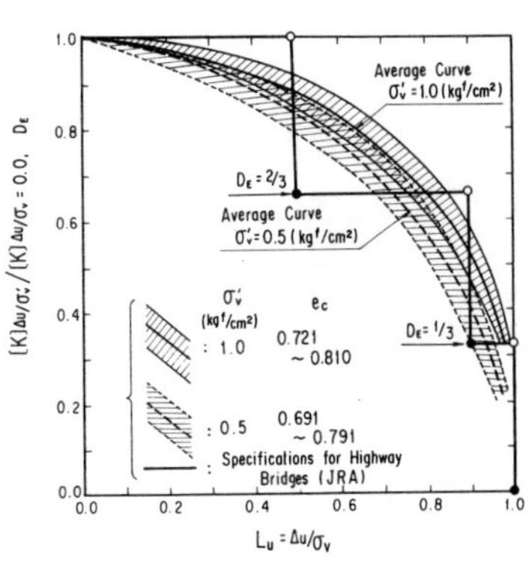

Fig. 21
Relationships between K/K_o, D_E and Lu $(=\Delta u/\sigma_v')$

Soil Dynamics & Earthquake Engineering Conference / Southampton / 1982.07.13-15

Liquefaction: Consequences for a Structure

ROBERT V.WHITMAN & PHILIP C.LAMBE
Massachusetts Institute of Technology, USA

INTRODUCTION

Several previous investigators have studied the settlement of structures founded upon a sandy soil in which large excess pore pressures are generated by earthquake ground shaking. De Alba et al. report the settlement of small footings placed on soil on a large shaking table (see Heller, 1977); large settlements begin when the pore pressures in the free field rise to about 90% of the initial vertical geostatic effective stress. The most detailed investigations of this matter have been conducted by Tokimatsu (1979) and Yoshima and Tokimatsu (1977). They employed shaking table tests with a narrow tank of soil, thus simulating the two-dimensional problem of a strip foundation. Tokimatsu and Yoshimi report that large settlements begin when the free-field pore pressures reach about 80% of the vertical effective stress, or when the pore pressures under the structures reach from 20% to 60% -- depending upon the bearing stress -- of the vertical effective stresses beneath the structure. This paper describes a brief series of tests carried out on a "shaking table" aboard a centrifuge. In this case the foundations were rigid circular cylinders. In general the results were quite similar to those described above. In addition, these new results have shed further light upon the phenomena which affect settlement.

TEST PROGRAM

These tests were carried out on the centrifuge at Cambridge University using the "bumpy road excitation" and a stack of rings to confine the soil. The equipment and instrumentation for such tests has been described in detail by Lambe and Whitman (1982) and other papers to Session IV of this Conference. The specific arrangement in this instance is shown in Figure 1. The centrifugal acceleration in all tests was 80g.

Pertinent dimensions for both model and corresponding prototype are as follows:

	Model-mm	Prototype-m
Diameter of soil mass	406	32.5
Height of soil mass	151	12.1
Diameter of foundation	113	9.0

For prototype conditions, the average bearing stress beneath the foundation was 130 kN/m^2. The ground shaking involved 10 more-or-less uniform pulses at a frequency of 83 Hz. For prototype conditions, this is a frequency of about 1 Hz, or a duration of shaking of 10 sec.

The soil was Leighton-Buzzard #120/200 sand, with a mixture of glycerin and water as the pore fluid (30% glycerin by weight). The soil was placed by pluviating the sand into water and then rodding to achieve the desired density. Several such layers were placed to construct the entire mass, and overall relative densities for the tests averaged 56% with a standard deviation of 4%. The consolidation characteristics of the soil are well known for these conditions (Whitman et al., 1982; Heidari and James, 1982). If the sand is only slightly disturbed by shaking, the consolidation time (non-dimensional time factor T=1) is about 6.5 sec. However, this time lengthens considerably after the sand has been shaken to the point of initial liquefaction.

There were basically four tests in the series. In one of the tests the soil was first shaken relatively weakly, with peak accelerations of about 0.09g prototype scale. Subsequent shakings in this test were progressively stronger. In the other three tests the initial shaking was relatively strong: 0.17 to 0.22g. The water table in these four tests initially was just below ground surface (0.5 meters deep prototype scale). For an additional test with a strong shaking, the water table was lowered to a depth of 1.3 meters. Prior to one of the tests, a number of short wires were inserted into the sand beneath the foundation; results indicated that these wires had no effect on either pore pressure response or settlement.

GENERAL NATURE OF RESULTS

A typical set of records for pore pressure and settlement is shown in Figure 2. The peak input acceleration in this case was 0.17g. The beginning of each pore pressure trace is at the location of the transducer. Both time and settlement have been scaled to prototype conditions; pore pressures are the same for both model and prototype. The most significant features of these results are:

* In the "free-field" to the sides of the mass, the pore pressure rises quickly, in about 3 cycles, to reach

the total vertical overburden stress. Thus a condition of "initial liquefaction" is reached early in the shaking.

* Directly under the structure, the pore pressure increases more slowly, and during shaking remains smaller than the pore pressure in the free-field at the same elevation.

* Immediately following shaking, the pore pressures beneath the footing increase to approach the free-field total vertical stress.

* Dissipation of excess pore pressures begins quickly following the end of shaking, beginning from the bottom of the soil. The free-field pore pressures nearest the surface remained in the full total overburden stress for about 15 sec. prototype time (about 0.2 sec. real time) following the end of shaking.

* Settlement of the structure begins about one cycle after the commencement of shaking, and increases continuously and almost linearly while shaking continues. There is no additional settlement once shaking stops.

* Settlement of the surface of the sand near the structure occurs mostly during shaking but continues after shaking ends.

During the first few cycles, horizontal accelerations measured on the foundation were less than input accelerations, but then increased to become about 25% larger than at the base. Similarly, the horizontal displacement of the foundation relative to the base initially was small but then increased.

Similar or generally consistent patterns were observed in the records from the other two tests with strong shaking. When the input accelerations were smaller, the settlements and the rates of pore pressure build-up were of course smaller. This tended to be true as well for shakings subsequent to the first.

DATA FOR PORE PRESSURE

Figure 3 displays composite results for pore pressure ratio from the three tests with strong initial shaking. The pore pressure ratio in this case is defined as the excess pore pressure at a point divided by the initial vertical effective stress at the point. Elastic theory was used to compute the contribution of the foundation to the initial effective stress. Where two values of the ratio appear, the upper value is that at the end of shaking while the lower value uses the maximum pore pressure recorded after shaking stops.

The shallower transducers located in the "free field" to the side of the foundation measured a pore pressure ratio of essentially unity, indicating that initial liquefaction occurred in this region with the stronger shaking. In the test with the smaller shaking, the maximum pore pressure ratio in the free-field was about 0.45.

Under the structure, the pore pressure ratios are much smaller than in the free-field, especially during shaking. The larger initial effective stresses in these locations clearly inhibit the build-up of excess pore pressures. Deep within the sand mass, partial dissipation may have reduced the peak excess pore pressures somewhat.

A different set of pore pressure ratios is displayed in Figure 4. Here the excess pore pressure has been divided only by that part of the initial vertical effective stress contributed by the weight of the soil. Well outside the foundation, this ratio is of course the same as that in Figure 3. The ratios under the foundation are all somewhat less than unity, indicating that the pore pressures there never did reach the values achieved in the free-field.

DATA REGARDING SETTLEMENT

Data for settlement of the foundation are summarized in Figure 5. As indicated by the legend on the figure, results from each of the several shakings applied to a mass of soil have been included. Data from one test in which the pore fluid was pure water have also been plotted, inasmuch as the settlements in this case were essentially the same as those for tests in which the glycerin-water mixture was used.

The dashed curve indicates the apparent trend to the settlement for an initial shaking. For shakings in which initial liquefaction does not occur in the free-field, the settlement remains small. Settlements increase rapidly once the condition of initial liquefaction is reached. While the surface of the sand itself settles, most of the settlement recorded in Figure 5 represents relative settlement between foundation and the surrounding soil.

In the one test with lowered water table, the settlements were at least 25% less than when the initial water table was near the surface.

DISCUSSION

As remarked in the introduction, the results from this brief series of tests were generally quite similar to those from model tests conducted in normal gravity. That is, large settlements begin to develop as the excess pore pressures in the free-field approach the initial vertical effective stress. These larger settlements are associated with an average pore pressure ratio beneath the foundation of about 0.2 during shaking.

The most interesting thing about these new results is the clear evidence that settlement occurs while shaking continues but ceases once the shaking stops -- even though the pore pressures under the foundation rise subsequent to the end of shaking to a level which should theoretically bring about a bearing capacity failure. No doubt the magnitude of the bearing stress is an important parameter. However, the mechanism of settlement is primarily one of the foundation working its way into the soil during each cycle and the settlement cannot be predicted from a model which considers conditions only after the shaking stops.

In these tests, equalization of pore pressures on a horizontal plane should occur rather rapidly. Thus, using values for the coefficient of consolidation which fit the observed rate of pore presure dissipation in the vertical direction, the time for essentially complete radial consolidation within a cylinder with radius equal to the foundation is only a few seconds. Such rapid adjustment is evident in the pore pressure records in Figure 2. Yet, while shaking continues the pore pressures beneath the foundation remain relatively low. Conditions in these tests were intermediate between "drained" and "undrained" conditions, and this fact no doubt contributed to the magnitude of the settlements which occurred.

While the settlements caused by the stronger initial shakings were relatively large, corresponding to about 300 mm at prototype scale, there was little or no tipping of the low, compact foundation. Such settlements can occur without necessarily damaging structures or causing them to be useless, although external connections to such structures might well be ruptured.

CONCLUSIONS

This paper has dealt with results from a preliminary series of tests, and generalizations from such limited data must be drawn with care. These tests have differed from previous model tests primarily by involving initial and dynamic stresses representative of those beneath actual prototype foundations. That the general conclusions are similar to those from the earlier tests lends credence to the results of model tests carried out at normal gravity. However, the complexity of the phenomena is evident in these new results and there is need to study the phenomenon of liquefaction-related settlement for a wider range of the pertinent variables.

ACKNOWLEDGEMENTS

These tests were sponsored by Toa Nenryo Kogyo of Tokyo, Japan, whose support is gratefully acknowledged. The work was carried out in cooperation with the faculty and staff of Cambridge University, especially Professor A.N. Schofield, Dr. R.G. James, Mr. Bruce Kutter and Mr. Chris Collison. Messrs. Massoud Heidari and Lawrence Lui, graduate student research assistants at MIT, assisted with preparations for the tests and in analysis of the results.

REFERENCES

Heidari, M. and R.G. James (1982) "Centrifuge Modelling of Earthquake Induced Pore Pressures in Saturated Sand," Proc., Int. Conf. Soil Dynamics and Earthquake Engineering, Southampton University, England.

Heller, L. (1977) Discussion of paper by De Alba, Seed and Chan, J. Geotechnical Eng. Div., ASCE, Vol. 103, No. GT7, pp. 834-836.

Lambe, P.C. and R.V. Whitman (1982) "Scaling for Earthquake Shaking Tests on a Centrifuge," Proc., Int. Conf. Soil Dynamics and Earthquake Engineering, Southampton University, England.

Tokimatsu, K. (1979) Generation and Dissipation of Pore Water Pressures in Sand Deposits During Earthquakes, Ph.D. thesis, Tokyo Institute of Technology.

Whitman, R.V., P.C. Lambe and J. Akiyama (1982) "Consolidation during Dynamic Tests on a Centrifuge," Paper presented to Las Vegas Convention, ASCE, April.

Yoshimi, Y. and K. Tokimatsu (1977) "Settlement of Buildings on Saturated Sand during Earthquakes," Soils and Foundations, Vol. 17, No. 1, pp. 23-38.

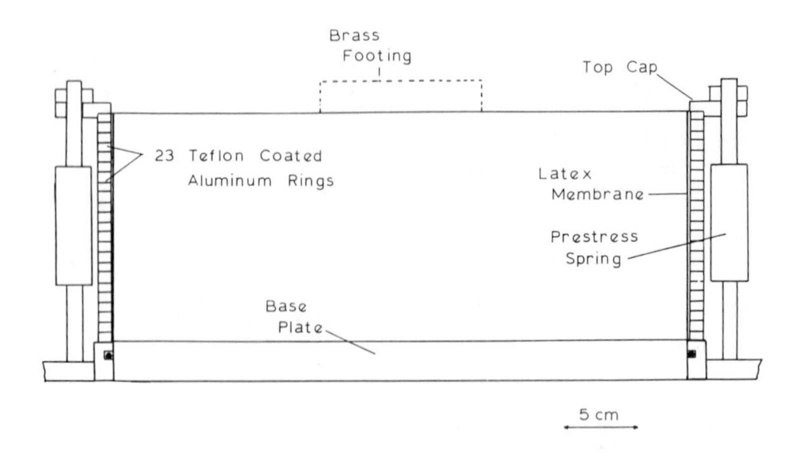

FIGURE 1 CROSS-SECTION OF STACKED-RING APPARATUS.

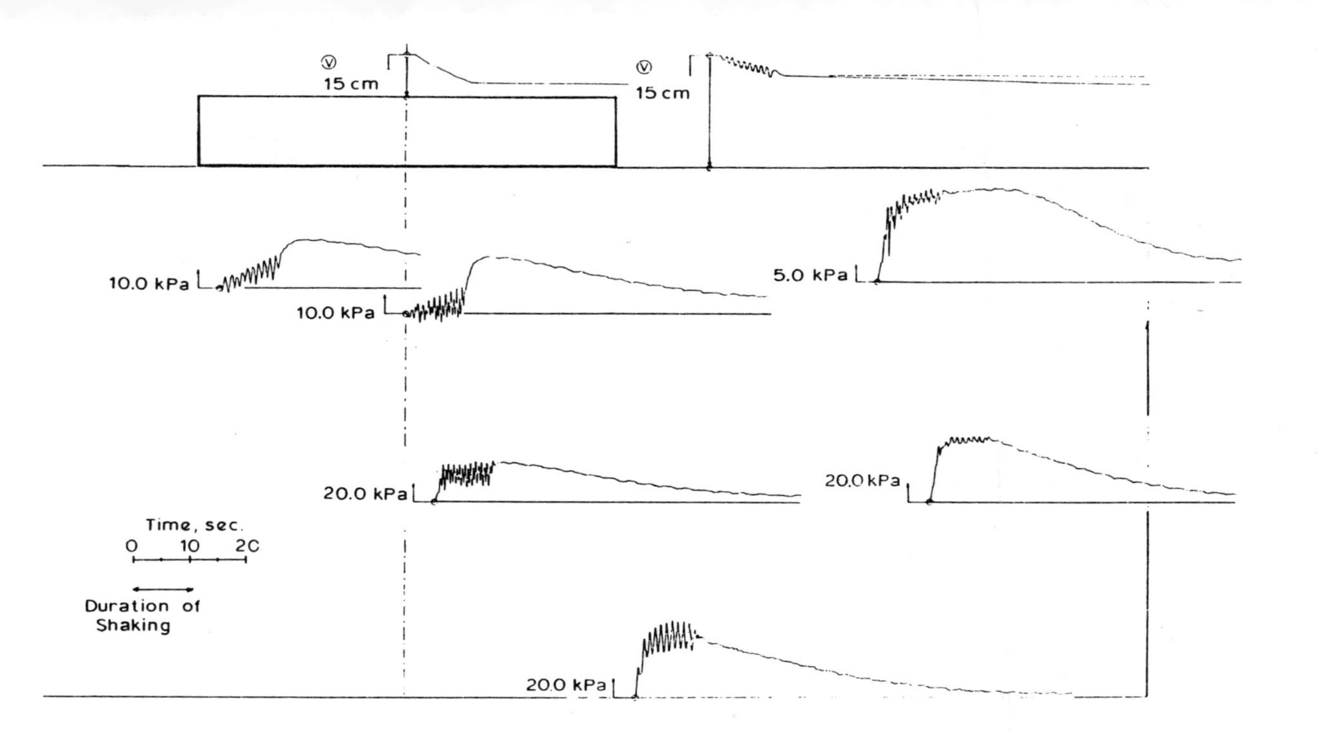

FIGURE 2 TYPICAL RESULTS FROM TEST WITH PEAK ACCELERATION OF 0.17g.

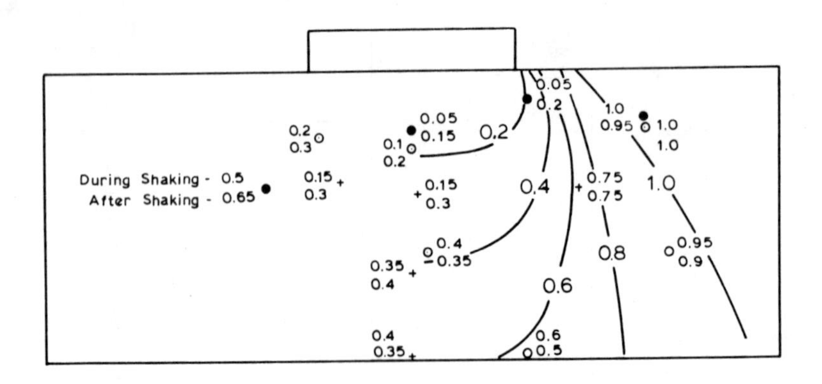

FIGURE 3 COMPOSITE OF PORE PRESSURE RATIOS.

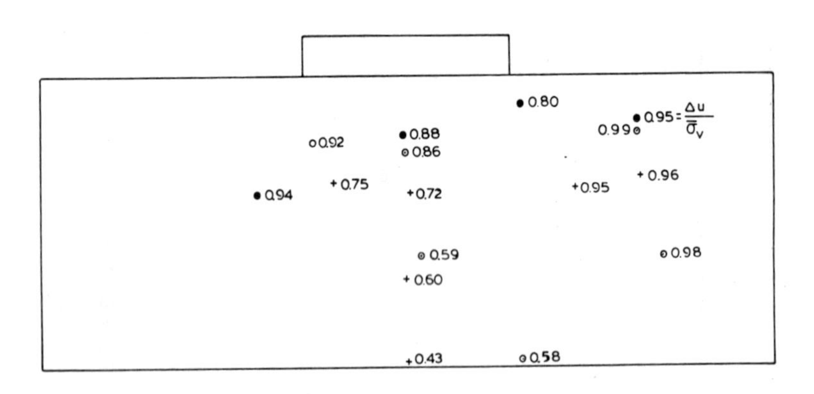

FIGURE 4 COMPOSITE OF PORE PRESSURE RATIOS WITH $\bar{\sigma}_V$ FROM GEOSTATIC STRESS ONLY.

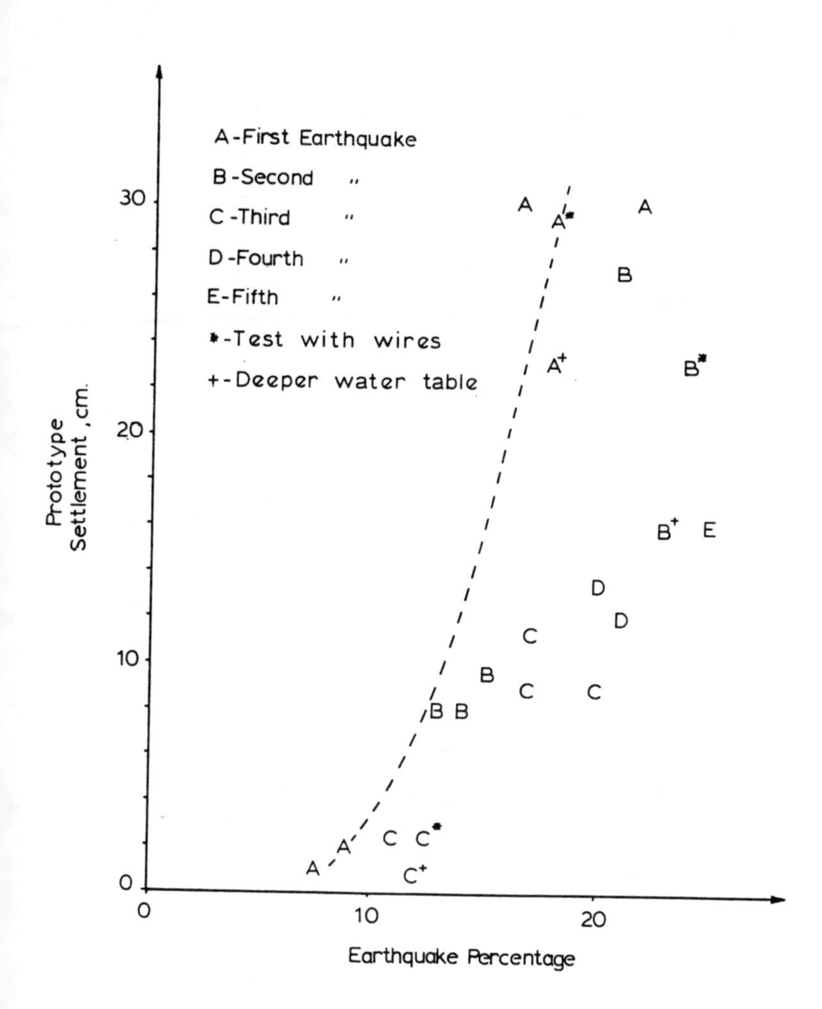

FIGURE 5 SETTLEMENT OF STRUCTURES VS. PEAK BASE ACCELERATION.

Soil Dynamics & Earthquake Engineering Conference / Southampton / 1982.07.13-15

Evaluation of Kinematic Interaction of Soil-Foundation Systems by a Stochastic Model

MASARU HOSHIYA
Muhashi Institute of Technology, Tokyo, Japan

KIYOSHI ISHII
Shimizu Construction Co.Ltd., Tokyo, Japan

INTRODUCTION

At the present practice, the earthquake response analysis of structures is generally based on an assumption that the ground just beneath a foundation vibrates in the same phase and with the same amplitude everywhere. However, it is very doubtful that a structure vibrates as if it were set on a shaking table. As the foundation slab, which is relatively stiff compared with the soil, works to constrain such ground motions, short period components of the ground motions whose wave length is short to the dimension of the slab are naturally weakened. Consequently the slab has the effect of a kind of low pass filter on the ground motions. This effect is called "Kinematic interaction" (Kausel, 1977). In the past, many studies have been made on the kinematic interaction both theoretically and experimentally (Yamahara, 1970, Iguchi, 1973, Newmark, 1977, Bernreuter, 1977, Ray, 1978, Wolf, 1979, Harada, 1979 and Shioya, 1980), and characteristics of the kinematic interaction have become clearer. However, since the effect of kinematic interaction is so complicated to depend totally on a geometrical pattern of a foundation slab, ground characteristics and the excitation, further researches are needed for incorporation of the effect into dynamic design practice.

This paper formulates the kinematic interaction of embedded rectangular foundations by the random vibration theory, and discusses the effect by examining field data obtained in earthquakes. Since the formulation is based on the fact that statistical correlation of ground motions at different points decreases as the distance between the different points increase, and much so when components of high frequency are contained in ground motions, characteristics of mutual correlations of different points' motions are first investigated. Then the formulation of the kinematic interaction is investigated. In order to justify a low pass filter by this stochastic model, earthquake records observed at a large scale inground tank and a foundation, whose material is cement-mixed soil-improved-ground, are analyzed as examples of deep and shallow embedded foundations respectively. Also the filtering effect on microtremor records observed at a reinforced concrete 4-story school building are investigated as another example of more general structures.

CHARACTERISTICS OF GROUND MOTIONS AT DIFFERENT POINTS

To clarify the description that ground motions at different points have the spatial variation, a few earthquake ground motion records are herein examined. Since many studies have been already made on the characteristics of ground motions at vertically different points, this section concentrates on the characteristics of ground motions at horizontally different points. Figure 1 shows records of earthquake ground motions at three horizontally different points located with the interval of 15m in a line and at the depth of 5.3m in the very soft ground in which the velocity of the shear wave is approximately 21.5m/s and the thickness of the layer is 10m. It is clear that long period components of ground motions have nearly same phases, but the short period components do not have always same phases. The maximum acceleration values at these three points differ remarkably.

Another example in Fig.2 shows earthquake ground motions of a soft rock at three different points; A, B and C in a straight line horizontally, of which the distances between points A and B and between points B and C are 40m and 38.5m respectively. The ground motions at three points seem to be almost identical both in the phase and amplitude except for small differences in the phase of the short period components. However coherence analyses indicated considerable differences between these ground motions. The results in Fig.3 show the averages of coherence functions and phase angles based on the eighteen horizontal components of earthquake records of nine different earthquakes. Figure 3 shows a typical example of statistical characteristics of ground motions at the different points, and the following important characteristics are observed: The coherence function and the phase angles can be divided into three parts of frequency domains of 0 - 2 Hz, 2 - 5 Hz and 5 - 10 Hz respectively based on well-remarked characteristics. In the frequency domain of 0 - 2 Hz, the phase angles are neary zero, and the coherences vary proportionally with the decrease of the frequency. This may be due to the relative sensitivity of seismometers, which do not have a fine sensitivity on the frequencies bellow 0.5 Hz because of the pendulum having the natural frequency of 5 Hz, and also due to the less intensity of this frequency domain components for total intensity of the ground motions. Therefore, the real statistical correlations in the frequency domain of 0 - 2 Hz are expected to resemble the statistical correlations of the frequency domain of 2 - 5 Hz as described bellow. In the frequency domain of 2 - 5 Hz, the phase angles are nearly zero, whereas the coherences vary inversely with the increase of the frequency. The average of the coherences at the frequency of 2Hz is about unity, and it becomes 0.8 at 5Hz. On the other hand the standard deviation of coherences at the frequencies of 2Hz to 5Hz increases from 0.0 to 0.2. Such tendency suits to our common images that the statistical correlations of the ground motions at the different points vary inversely with the increase of the frequency and the variations vary proportionally with the increase of the frequency. In the frequency domain of 5 - 10 Hz, the averages of the coherence function vary between 0.4 and 0.8, and the standard deviations of the coherence function are about 0.3. The averages of the phase angles are nearly zero, but the standard deviations of the phase angles amount to the angle of 45°. The main characteristics of the statistical correlations in the frequency domain of 5 - 10 Hz are the independency between the frequency components, and the less correlations compared with the statistical correlations in the frequency domain of 2 - 5 Hz.

The characteristics of the statistical correlations of the ground motions at the two points A and C, which locate at 78.5m apart, are similar to the characteristics of the statistical correlations of the ground motions at the two points A and B, whose distance is 40m, except for the slight decrease of the correlations (Ishii, 1981).

Although the characteristics of the statistical correlations of ground motions at different points are very complex, it is possible to make a simple stochastic model that represents a part of the characteristics described above.
A stochastic model of ground motions at the different points formulated in the next section retains the characteristics of the frequency domain of 2 - 5 Hz shown hereabove and basic characteristics of the ground motions at the different points. Thus it can be considered as a very realistic idealization.

FORMULATION OF KINEMATIC INTERACTION BY A STOCHASTIC MODEL

Consider to formulate the expression of the kinematic interaction of an embedded rectangular shape foundation model shown in Fig.4, where x- and y-coordinates in the horizontal plane and z-coordinate in the vertical direction are introduced for the analysis. Since a rigorous analytical solution for the embedded foundation does not exist at present due to the complicated boundary conditions that must be satisfied in theoretical formlation, and semi-analytical methods such as the FEM approach as the numerical techniques are not suitable to include the random vibration theory, a very simple technique, in which the complicated wave scatterings around the foundation are ignored, is used here to evaluate the motions of the foundation (Yamahara, 1970 and Iguchi, 1973).

As the rigid foundation constrains the surounding soil motions along the boundary, the motion of the foundation $u(t)$ is estimated from the average of the ground motions $u_o(x,y,z,t)$ on the foundation-soil interface as

$$u(t) = \frac{1}{S}\int_S u_o(x,y,z,t)ds \tag{1}$$

in which, S denotes the total area of the foundation-soil interface. When the foundation-soil interface is divided into five surfaces, Si, i=1.···, 5 as shown in Fig.4, Eq.(1) becomes

$$u(t) = \frac{1}{S}\sum_{i=1}^{5} \int_{s_i} u_o(x,y,z,t)ds_i \tag{2}$$

The autocorrelation function of $u(t)$ is given by

$$R_u(\tau) = E\left[u(t)\,u(t+\tau)\right] \tag{3}$$

$$= \frac{1}{S^2}\sum_{i=1}^{5}\sum_{j=1}^{5} \int_{s_i}\int_{s_j} E\left[u_o(x_i,y_i,z_i,t)\,u_o(x_j,y_j,z_j,t+\tau)\right]ds_i ds_j \tag{4}$$

Then, the power spectral density function of $u(t)$ is given by a Fourier transform of Eq.(4) as

$$G_u(\omega) = \frac{1}{2\pi}\int_{-\infty}^{\infty} R_u(\tau)\, e^{-i\omega\tau}d\tau \tag{5}$$

$$= \frac{1}{S^2}\sum_{i=1}^{5}\sum_{j=1}^{5} \int_{s_i}\int_{s_j} \frac{1}{2\pi}\left\{\int_{-\infty}^{\infty} E\left[u_o(x_i,y_i,z_i,t)\,u_o(x_j,y_j,z_j,t+\tau)\right]e^{-i\omega\tau}\,d\tau\right\}ds_i ds_j \tag{6}$$

$$= \frac{1}{S^2}\sum_{i=1}^{5}\sum_{j=1}^{5} \int_{s_i}\int_{s_j} G(x_i,y_i,z_i,x_j,y_j,z_j,\omega)\, ds_i ds_j \tag{7}$$

in which $G(x_i,y_i,z_i,x_j,y_j,z_j,\omega)$ denotes the cross spectral density function of the ground motions between two points (x_i,y_i,z_i) and (x_j,y_j,z_j). The cross spectral density function $G(x_i,y_i,z_i,x_j,y_j,z_j,\omega)$ is assumed to be a form of

$$G(x_i,y_i,z_i,x_j,y_j,z_j,\omega) = e^{-\alpha\left\{|x_i-x_j|+|y_i-y_j|+|z_i-z_j|\right\}} \cos\beta\,(L_3-z_i)\cos\beta(L_3-z_j)\; G_{u_o}(\omega) \tag{8}$$

where

$$\alpha = (a+b\omega)/V_s \qquad \beta = \omega/V_s \tag{9}$$

, and ω, Vs and $G_{u_o}(\omega)$ are a circular frequency, a shear wave velocity and the power spectral density function of the ground motion $\dot{u}_o(t)$ respectively, and a and b ($a,b \geqq 0$) are parameters determined experimentally. The cross spectral density function given by Eq.(8) decreases in an exponential manner as the distance between two points increases. The variable α varies linearly with $\omega/Vs(=\beta)$ which is the wave number of the shear wave ground motion, in particular a=0.0 and b>0.0 as used later. And, the cross spectral density function, modelled on the vertically propagating shear waves, has the phase differences that vary proportionally to the difference of depths between two points and the wave number β.

The power spectral density function of the foundation motion $G_u(\omega)$ is obtained after tedious integrations, and substituting Eqs.(8) and (9) into Eq.(7) as

$$G_u(\omega) = \frac{1}{S^2}(A_1A_2C_3 + 2A_2A_3 + 2A_1A_3 + 4B_1A_2B_3 + 4A_1B_2B_3 + 8B_1B_2A_3 + 2C_1A_2A_3 + 2A_1C_2A_3)\, G_{u_o}(\omega) \qquad (10)$$

where,

$$\left.\begin{aligned}
&S = L_1L_2 + 2L_2L_3 + 2L_1L_3 \\
&A_1 = 2(L_1\alpha + e^{-L_1\alpha} - 1)/\alpha^2 \qquad A_2 = 2(L_2\alpha + e^{-L_2\alpha} - 1)/\alpha^2 \\
&B_1 = (1 - e^{-L_1\alpha})/\alpha \qquad B_2 = (1 - e^{-L_2\alpha})/\alpha \\
&C_1 = e^{-L_1\alpha} \qquad C_2 = e^{-L_2\alpha} \\
&A_3 = \frac{\alpha}{\beta(\alpha^2 + \beta^2)}(\sin\beta L_3 \cos\beta L_3 + \beta L_3) \\
&\qquad - \frac{1}{(\alpha^2 + \beta^2)^2}(\alpha^2 + \alpha^2\cos^2\beta L_3 - \beta^2\sin^2\beta L_3) \\
&\qquad + \frac{2\alpha}{(\alpha^2 + \beta^2)^2} e^{-\alpha L_3}(\alpha\cos\beta L_3 - \beta\sin\beta L_3) \\
&B_3 = \frac{1}{(\alpha^2 + \beta^2)}\cos\beta L_3(\alpha\cos\beta L_3 + \beta\sin\beta L_3 - \alpha e^{-\alpha L_3}) \\
&C_3 = \cos^2\beta L_3
\end{aligned}\right\} \qquad (11)$$

The transfer function for the translation of the ground motions to the motion of the foundation is given by the following relationship,

$$G_u(\omega) = |H(\omega)|^2\, G_{u_o}(\omega) \qquad (12)$$

in which $H(\omega)$ means the filtering effect of the foundation slab for the travelling ground motions.

In cases of the particular foundation type, Eq.(10) becomes a much more simple form as follows;

(a) the square bottom embedded foundation (L_1=L_2)

$$G_u(\omega) = \frac{1}{S^2}(A_1^2C_3 + 4A_1A_3 + 8A_1B_1B_3 + 8B_1^2A_3 + 4A_1C_1A_3)\, G_{u_o}(\omega) \qquad (13)$$

(b) the not-embedded foundation (L_3=0)

$$G_u(\omega) = \frac{1}{S^2}(A_1 A_2)\, G_{u_o}(\omega) \tag{14}$$

(c) the two dimensional plane foundation (x-and z-plane)

$$G_u(\omega) = \frac{1}{L^2}(A_2 C_3 + 2A_3 + 4B_1 B_3 + 2C_2 A_3)\, G_{u_o}(\omega) \tag{15}$$

where

$$L = 2L_3 + L_1 \tag{16}$$

The one dimensional solution for Eq.(14) is the same as the formulation by Matsushima (Matsushima, 1977).

The parameter b in Eq.(9) lies in the interval (0.2, 0.4) from the results by the method of least squares on the coherence functions of observed records assuming a=0.0 (Ishii, 1981). The solutions, when a=0.0 and b=0.15 as described later in detail, agree generally well with the experimental solutions.

APPLICATIONS OF A STOCHASTIC MODEL ON FIELD DATA

In this section applications of a stochastic model are investigated by comparing calculated results with experimental results on three different types of structure.

Large-Scale Inground Tank

Earthquake accelerograms have been recorded for a large-scale inground cylindrical tank of reinforced concrete (Ishii, 1982). A diameter and a height are 67.9m wide and 26.2m deep respectively, a thicknesses of side walls and a bottom slab are 1.8m and 5.5m respectively, and the embedded depth is 24.5m as shown in Fig.5. Soil properties are that the upper part above GL-14m consists of reclamation soil and alluvium, and the lower part below GL-14m of diluvium. The average shear wave velocities of each layers are 150m/s at GL±0m∿-14m, 380m/s at GL-14m∿-40m and a faster velocity than 450m/s under GL-40m. As pointed out previously, the kinematic interaction is regarded as an effect of input energy loss of massless rigid foundation. It would be possible to treat the inground tank as a massless rigid foundation which has only the effect of the kinematic interaction, since the inground tank has small unit weight of 0.7 to 1.0t/m^3.

Figure 6 shows the comparison of the calculated results with the experimental result of the transfer function, that is the translation of the ground motions to the motion of the foundation. In Fig.6, the heavy line shows the experimental result and the light or marked dotted lines show the calculated results for different values of b when a=0.0. Since the stochastic model is formulated for a rectangular slab, the equivalent area transformation method is used to obtain the equivalent square foundation model for the cylindrical inground tank. It is observed from Fig.6 that the calculated results agree approximately well with the experimental result. The nodes at frequencies 2Hz and 5.2Hz in the calculated results are due to the phase characteristics of the cross spectral density function given by Eq.(8), whereas the amplitude ratios of nodes in the calculated result at a=0.0 and b=0.0 are zero, because the ground motions that have no spatial variation perfectly cancel each other. The analytical result by the finite element method (FEM) is shown in Fig.7 (Ishii, 1982). The calculated results by a stochastic model agree relatively well with the analytical result by the FEM too.

Foundation of Cement-Mixed Soil-Improved-Ground

A foundation whose material is cement-mixed soil-improved-ground is investigated as an example of a shallow embedded foundation (Sawada, 1981). The foundation considered here is a cylindrical foundation of 30m diameter and 2m depth and embedded in the very soft ground. As the shear wave velocity about 220m/s of cement-mixed soil-improved-ground is much faster than the shear wave velocity 21.5m/s of the surrounding ground, the cement-mixed soil-improved-ground can be considered as a foundation.

Figure 9 shows the calculated results compared with the experimental result of the transfer function for the translation of the ground motions to the motion of the foundation. The experimental result was evaluated as the product of a transfer functions between the foundation motion and the ground motion at GL-5.5m, and a analytical transfer function between the ground motion at GL-5.5m and the ground surface motion. Here the experimental result must contain the effects of both the kinematic and the dynamic interactions of a soil-foundation system. Main characteristics of the effect of the dynamic interaction is generally that the transfer function for the translation of the ground motions to the motion of the foundation has a predominant peak, whose amplitude ratio exceeds the unity at the natural frequency of the foundation. However there is no predominant peak which exceeds the unity in the transfer function as shown in Fig.9, since the kinematic interaction has much more effect than the dynamic interaction. The calculated results agree relatively well with the experimental result as same as in the case of the inground tank. But at the frequency of 8.3Hz, Fig.9 indicated the large difference between the experimental and calculated results. This difference may be due to the simplification of the stochastic model of Eq.(8) particularly in the high frequency domain.

Reinforced Concrete 4-Story School Building

A reinforced concrete 4-story school building in Fig.10 is investigated as another example of more general structures, whose length, width and height are 56m, 11m and 16.5m respectively, and it has a footing foundation supported on wooden piles (Hoshiya, 1982). In the analysis longitudinal components of observed microtremor records measured at various points shown in Fig.10 together with observed accelerograms at the foundation and at the under ground (GL-16m) were used, since the dynamic behavior of the structure in the longitudinal direction is regarded simpler than the behavior in the transverse direction. It is noted that the one directional array of seismometers at points G1 to G5 in Fig.10 are installed in order to evaluate the effect of the kinematic interaction.

Obviously the observed microtremors at different points located with the same interval in a line have the spatial variation in the amplitude and in the high frequency range especially. However, if a fictional rigid massless foundation is supposed to set on the ground covering these points, then the motion of the foundation may be expressed as the average of these ground motions as follows.

$$u = \frac{1}{L}\int_0^L u_o(x)\,dx \tag{17}$$

, in which u, u_o and L are a motion of the foundation, a ground motion and a length of a one-dimensional foundation, respectively. The discrete form of Eq.(17) is

$$u = \frac{1}{L}\sum_{i=1}^{n} u_{oi}\,\Delta L \tag{18}$$

, in which ΔL equals L/n and n is a large positive integer. Here the transfer function between the motion u of a fictional foundation and the ground motion at any measurement point will represent the kinematic interaction of a one-dimensional foundation of length L.

The transfer function between the average of the microtremor records at points G1 $\sim$ G5 and the microtremor record at the point G3 was evaluated and shown with a solid curve in Fig.11. The calculated result of the one-dimensional foundation model, in which the length of a foundation L is 56m, the shear wave velocity V_s of a soil is 120m/s and parameters, a and b are 0.0 and 0.15 respectively, is also shown by a dotted curve in Fig.11. The calculated result envelopes the experimental result which shows the effect of the kinematic interaction such that the experimental result varies inversely with the increase of the frequency except for the concave at the frequency of 5Hz.

Finally Fig. 12 shows the calculated result compared with the experimental results of the transfer function for the translation of the ground motions to the motion of the foundation. In Fig.12, the calculated result of the effect of the kinematic interaction is given by Eq.(15), and the effect of the dynamic interaction is evaluated by the transfer function for the translation of the ground motion to the motion of the foundation, in which the structure is modelled by a spring-mass-dashpot model given in Table.1. The calculated result agreed sufficiently with the experimental results which have a peak at the frequency of 2.3Hz and shows a concave at the frequency of 3.8Hz. It is found that the effect of the kinematic interaction is weaker than the effect of the dynamic interaction in this particular case.

CONCLUSIONS

This study formulated the kinematic interaction of a embedded rectangular foundation by the random vibration theory. Since the formulation is based on the fact that statistical correlation of ground motions at different points decreases as the distance between the points increases, and much so when high frequencies are contained in ground motions, characteristics of mutual correlations of motions at different points are first investigated, then the formulation of the kinematic interaction is investigated.

In the formulation the statistical correlation of the ground motions at different points is represented by the cross spectral density function given by Eq.(8), which has the characteristics described above. The cross spectral density function, modelled on the vertically propagating shear waves, has the phase differences that vary proportionally to the difference of depths between two points and the wave number, as a most realistic idealization of ground motions. The parameters a and b contained in the cross spectral density function are determined experimentally.

In order to examine the effect of a low pass filter by this stochastic model, observed earthquake and microtremor records of three different type structures are analyzed. The results were obtained as follows:

(1) In the case of a large-scale inground tank, the calculated results agree sufficiently with the experimental result in the transfer function for the translation of the ground motions to the motion of the foundation.

(2) In the case of a foundation whose material is a cement-mixed soil-improved-ground, the calculated results agree approximately well with the experimental result except at the frequency of 8.3Hz in the transfer function for the translation of the ground motions to the motion of the foundation.

(3) In the case of a reinforced concrete 4-story school building, microtremor records at the different points in a horizontal line are investigated to evaluate the kinematic interaction of an one-dimentional foundation. The calculated result by the stochastic model envelopes the experimental result. In the transfer function for the translation of the ground motions to the motion of the foundation, the calculated result agreed sufficiently with the experimental results by taking into consideration the effect of the dynamic interaction by using a spring-mass-dashpot model.

REFERENCES

Bernreuter, D.L. (1977) Assesment of Seismic Wave Effects on Soil-Structure Interaction. Proc. of the 4th SMIRT., K2/14

Harada, T. (1979) Dynamic Soil-Structure Interaction Analysis by Continuum Formulation Method. Thises presented to the Univ. of Tokyo, in partial fulfilment of the requirements for the Degree of Doctor of Engineering.

Hoshiya, M. and Ishii, K. (1982) Potential Loss of Ground Motion Due to Kinematic Interaction In a 4-Story RC Building. Trans. of JSCE., (under examination)

Iguchi, M. (1973) Seismic Response with Consideration of Both Phase Difference of Ground Motion and Soil-Structure Interaction. Proc. of the Japan Earthquake Engineering Symposium, B-16.

Ishii, K. (1981) Discussions on Kinematic Interactions of Soil-Foundation System with Probability Model. Reports of the Research Laboratory of Shimizu Construction Co., Ltd. Vol. 34.

Ishii, K. and Yamahara, H. (1982) A Study on the Filtering Effect of Foundation Slab Based on Earthquake Records of a Inground Tank. Trans. of the Architectural Institute of Japan No.312

Kausel, E., Whitman, R.V., Elasabee, F. and Morray, J.P. (1977) Dynamic Analysis of Embedded Structures. Trans. of the 4th SMIRT., K2/6

Matsushima, Y. (1977) Stochastic Response of Structure Due to Spatially Variant Earthquake Excitations. Proc. of the 6th WCEE., Vol.11, 3-103~108

Newmark, N.M., Hall, W.J. and Morgan, J.R. (1977) Comparison of Building Response and Free Field Motion in Earthquakes. Proc. of the 6th WCEE., Vol.11, 3-01~06

Ray, D. and Jhaveri, D.P. (1978) Effective Seismic Input Through Rigid Foundation Filtering. Nuclear Engineering and Design, 45, 185-195

Sawada, Y., Yajima, H., Sasaki, S. and Esashi, Y. (1980) Aseismic Effect of Cement-Mixed Soil Improvement - Loss Mechanism of Earthquake Motion on Improved Grounds, Part.(1) -. The Report of Central Research Laboratory of Electric Power Industry, Report No. 380022

Shioya, K. and Yamahara, H. (1980) Study on Filtering Effect of Foundation Slab Based on Observational Records. Proc. of the 7th WCEE., 5, 181-188

Wolf, J.P. and Obernhuber, P. (1979) Travelling Wave Effects in Soil-Structure Interaction. Trans. of the 5th SMIRT., K5/1

Yamahara, H. (1970) Ground Motions During Earthquakes and the Input Loss of Earthquake Power to an Excitation of Building. Soil and Foundation, Vol.X, No.2, 145-161

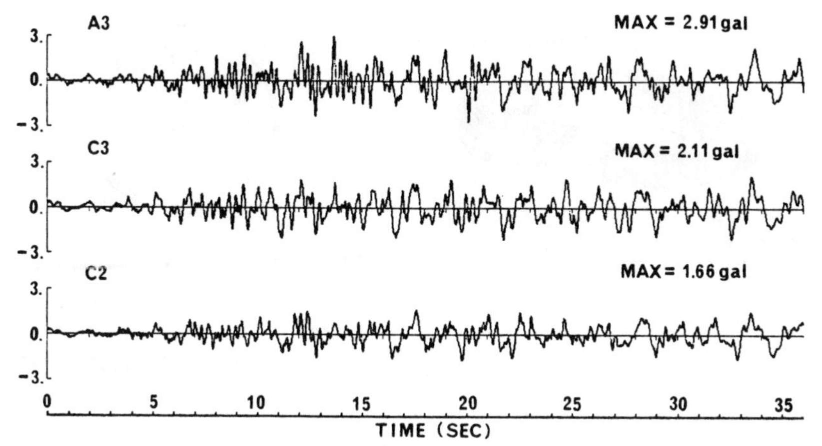

Fig.1 Observed Earthquake Records in a Very Soft Ground

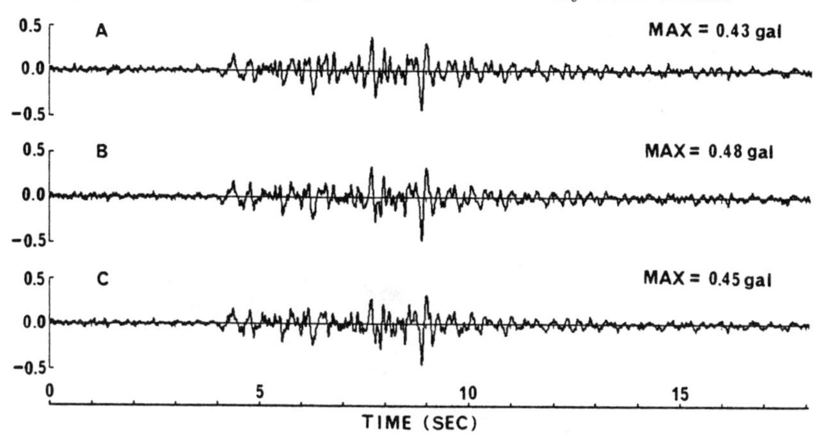

Fig.2 Observed Earthquake Records in a Soft Rock

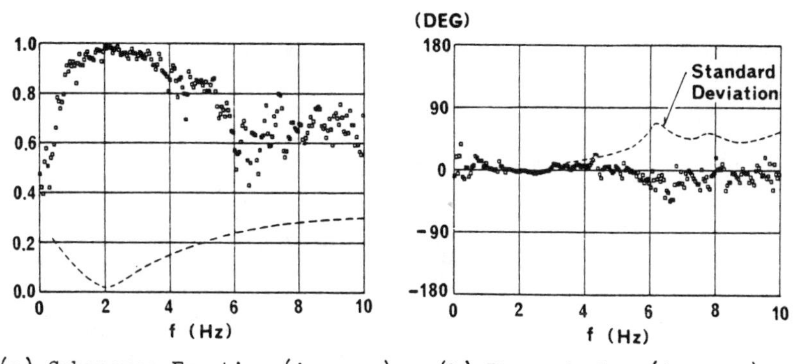

(a) Coherence Function (Average) (b) Phase Angles (Average)

Fig.3 Coherence Analysis of Earthquake Records in a Soft Rock

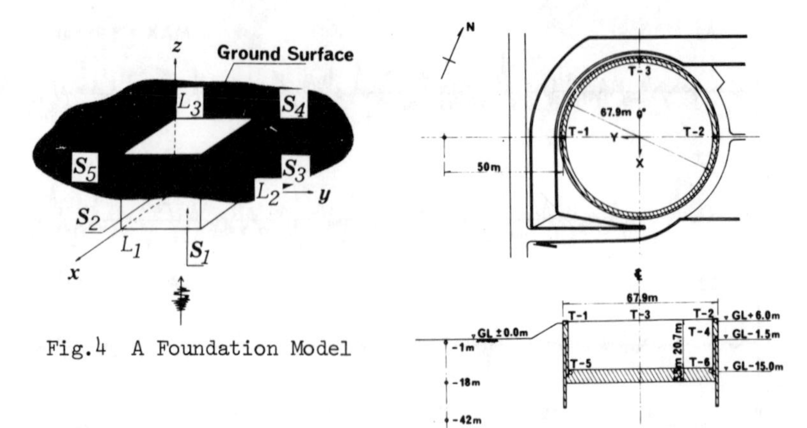

Fig.4 A Foundation Model

Fig.5 Large Scale Inground Tank

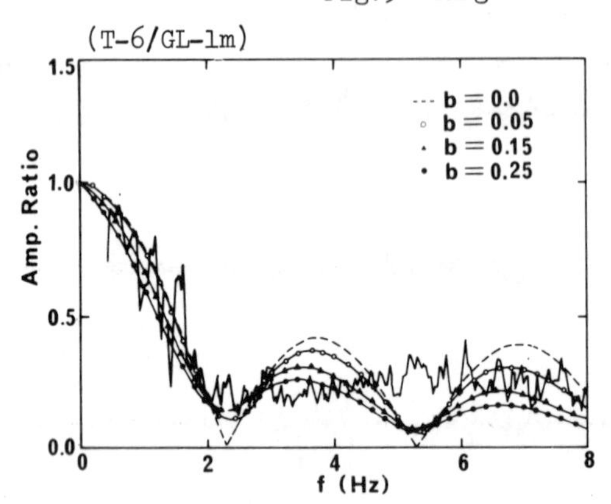

Fig.6 Transfer Function (Inground Tank)

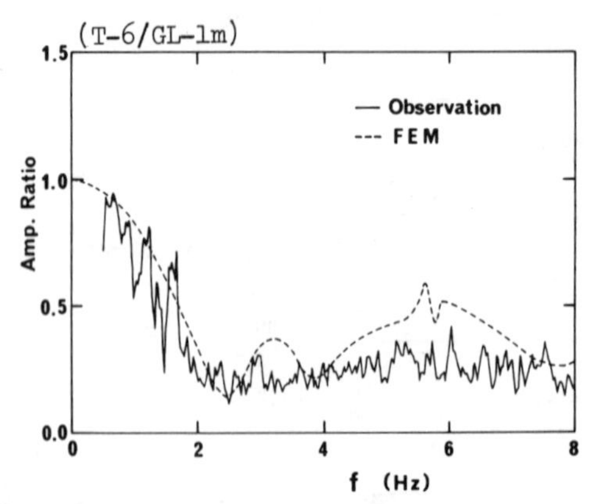

Fig.7 Transfer Function (Inground Tank)

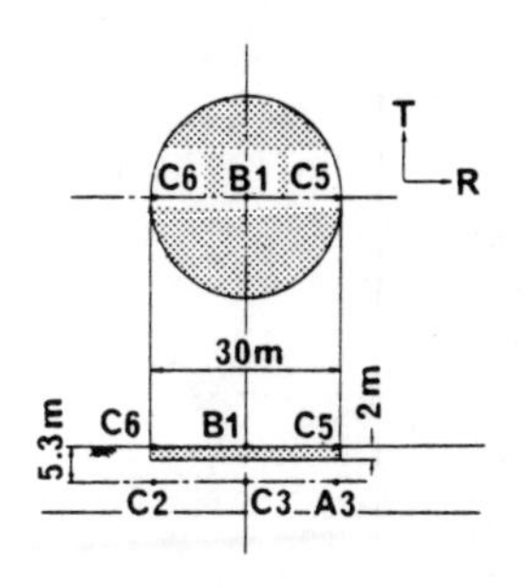

Fig.8 Foundation of Cement-Mixed Soil Improved Ground

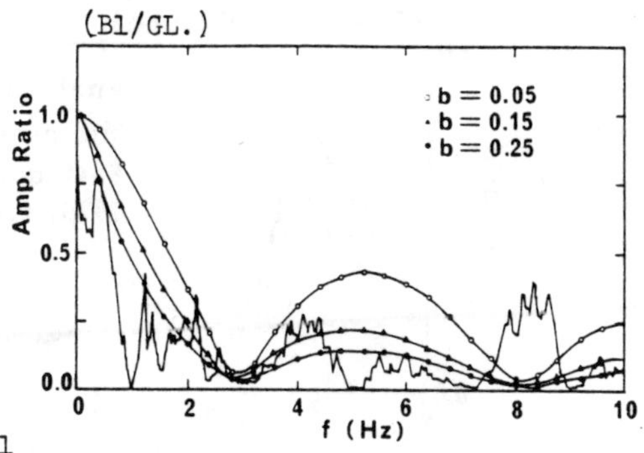

Fig.9 Transfer Function (Foundation of Cement-Mixed Soil Improved Ground)

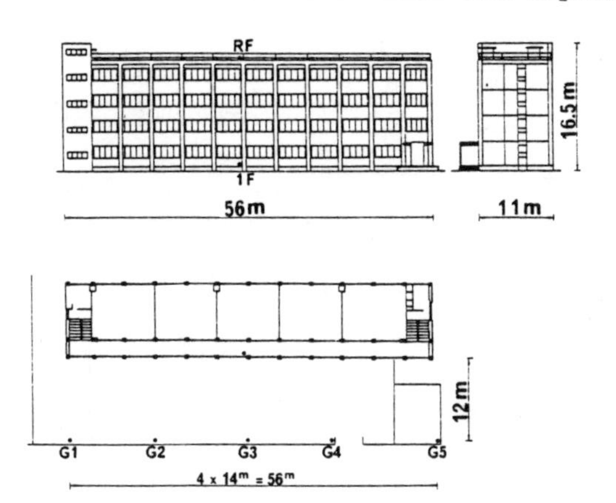

Fig.10 Reinforced Concrete 4-Story School Building

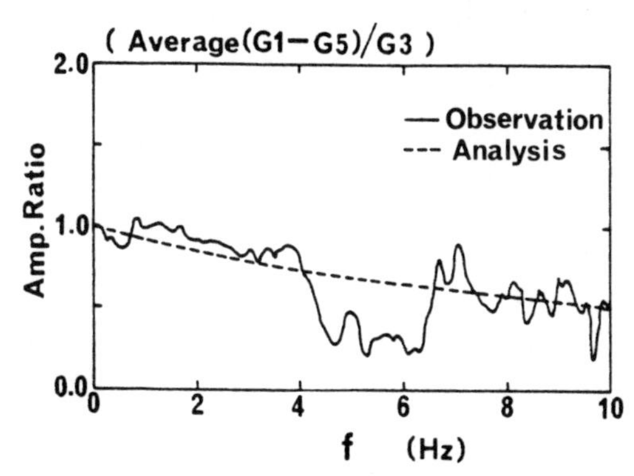

Fig.11 Transfer Function (4-Story School Building)

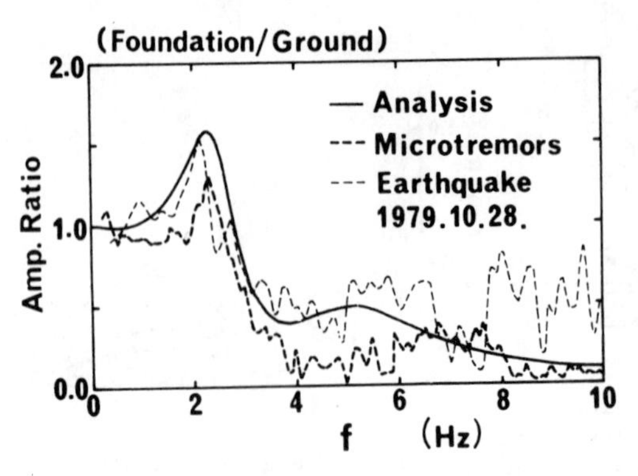

Fig.12 Transfer Function (4-Story School Building)

Table.1 A Spring-Mass-Dashpot Model
(4-Story School Building)

MODEL	WEIGHT	SPRING CONSTANT	DAMPING
	(t)	(10^3t/cm)	
W_1	W_1 = 526.9		
K_1		K_1 = 1.865	
W_2	W_2 = 582.1		MODAL DAMPING FACTOR
K_2		K_2 = 2.098	
W_3	W_3 = 593.5		h = 5.36%
K_3		K_3 = 2.331	
W_4	W_4 = 603.9		
K_4		K_4 = 2.564	
W_5	W_5 = 2555.		
K_x C_x		K_x = 1.779 $\times 10^3$t/cm	DAMPING COEFFICIENT C_x = 6.434 $\times 10^2$ts/cm

Soil Dynamics & Earthquake Engineering Conference / Southampton / 1982.07.13-15

Seismic Hazard Evaluation for Pore Pressure Increase in Soils

JEAN-LOU CHAMEAU
Purdue University, USA

INTRODUCTION

The cyclic loading which occurs during an earthquake is known to cause pore pressure increases in saturated sands which, in the extreme case, can lead to liquefaction. Ground failures and damage to structures due to liquefaction have been reported in many earthquakes throughout the world. With increasing awareness to this problem, liquefaction has become one of the significant hazards to be assessed in earthquake pronent regions. A practical representation of hazards associated with earthquakes is in the form of seismic hazard maps, which describe earthquake effects for a given geographical region. In this paper, procedures are set forth to accomplish this goal for the liquefaction hazard.

First, a probabilistic model is proposed to predict pore pressures under seismic loading. It employs a commonly used simplified approach to pore pressure prediction which is based on empirically observed relationships from laboratory tests. The model proposed herein incorporates uncertainties associated with the soil and loading parameters needed to describe them. The second major step involves incorporating the probabilistic model into a hazard analysis methodology which describes the hazard at a given site in terms of the level of pore pressure. Specifically, the result of the analysis is the probability of exceeding at least once a certain pore pressure ratio for a given seismic environment and a given time period. Finally, a methodology is outlined to develop contour maps of the probability of liquefaction for a given geographical area.

PROBABILISTIC DEVELOPMENT OF PORE PRESSURE

Experimental results (DeAlba, Seed and Chan, 1976) and analytical studies (Finn, Lee and Martin, 1977 and Chameau, 1980) indicate that the development of pore pressure in saturated sand during a stress cycle depends not only on the stress intensity, but also on the amount of pore pressure accumulated during the previous stress cycles. In this section a probabilistic model based on laboratory observed relationships is proposed to study this non-linear development of pore pressure under random loading. This model is part of a general probabilistic framework recently developed to assess the increase in pore pressure in sands due to seismic loading (Chameau, 1980). Another model using an analytical effective stress technique and a simulation procedure is also available to perform this assessment (Chameau and Clough, 1982).

The pore pressure development of sand from undrained laboratory experiments has been shown by Seed, et al (1976) to be defined by (1) a cyclic strength curve which expresses the number of cycles to zero effective stress, N_ℓ, as a function of the applied shear stress ratio, τ/σ_o', (Figure 1); and (2) a set of pore pressure generation curves which relate the normalized pore pressure ratio, u/σ_o', to the normalized number of cycles, N/N_ℓ (Figure 1). It is assumed that there is one curve available for each stress ratio τ/σ_o' of interest. Uncertainties in the soil characteristics may be taken into account by considering that for a given stress ratio, the number of cycles to zero effective stress is not a unique value, but is defined by a probability density function around its mean or median value. In order to simplify the mathematical expressions, the following notation is adopted in this section:

(a) The pore pressure ratio u/σ_o' is called R.

(b) The shear stress ratio τ/σ_o' is called S.

(c) The probability density function of the number of cycles to zero effective stress state N_ℓ is noted $f_{N_\ell}(n_\ell)$. In general, this distribution is also function of the shear stress ratio S.

(d) During each cycle, or interval between zero-crossings, the shear stress amplitude is described by the density function $f_\tau(\tau)$. In the following examples the shear stress is assumed to be either Rayleigh or exponentially distributed.

(e) The probability density function of the stress ratio is noted $f_S(s)$. This function can be derived from the knowledge of the probability density function of the shear stress.

(f) The functional relationship between the pore pressure ratio R, the stress ratio S, and the normalized number of cycles N/N_ℓ , as shown in Figure 1, is expressed as:

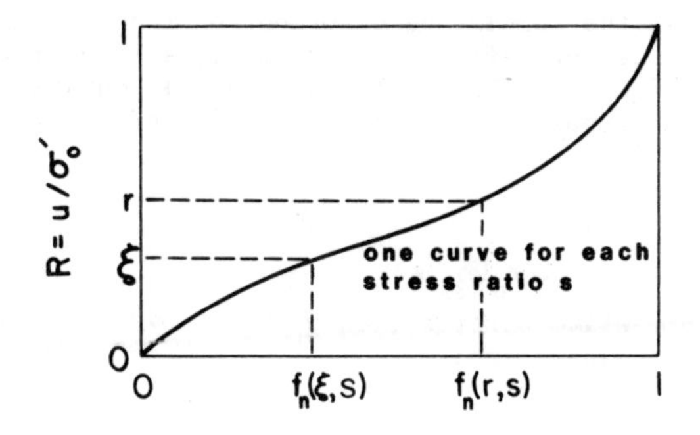

Pore Pressure Generation Curve

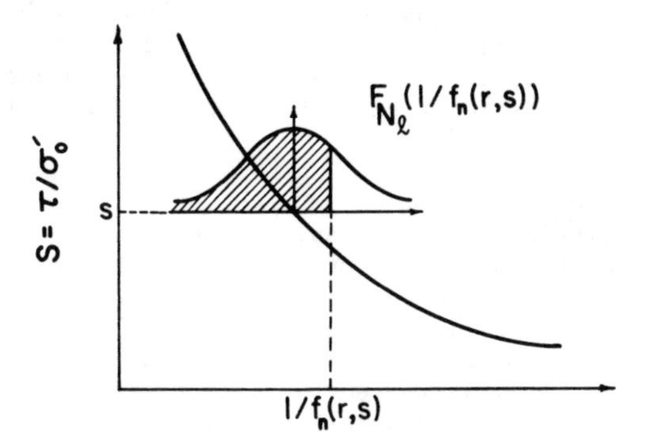

Cyclic Strength Curve

Figure 1. Pore Pressure Generation and Cyclic Strength Curves

$$N/N_{\ell} = f_n(R,\ S) \qquad (1)$$

The index n indicates that all the parameters of the relation are normalized.

<u>First Cycle of Loading</u>

To determine the cumulative distribution function of R, first consider the conditional distribution function of R given that the stress ratio S equals some particular value s. This function equals the probability that the normalized increment $1/N_{\ell}$ is less than or equal to $f_n(r, s)$:

$$F_{R/S}(r,\ s) = 1 - F_{N_{\ell}}\ (1/f_n(r,\ s)) \qquad (2)$$

where $F_{N_\ell}(n_\ell)$ is the cumulative distribution function of the number of cycles to zero effective stress N_ℓ, schematically represented in Figure 1. The right hand term in Equation 2 is obtained from the probability density function of N_ℓ, $f_{N_\ell}(n_\ell)$:

$$F_{R/S}(r, s) = \int_{1/f_n(r, s)}^{\infty} f_{N_\ell}(u)\, du \qquad (3)$$

The marginal cumulative distribution of R follows upon integration over all values of s:

$$F_R(r) = \int_0^{\infty} \int_{1/f_n(r,s)}^{\infty} f_{N_\ell}(u)\, f_S(s)\, du\, ds \qquad (4)$$

The double integration in Equation 4 completely describes the cumulative probability distribution function of the pore pressure ratio R for any value r of this ratio between 0 and 1. The probability density function of R is obtained by differentiation.

ith Cycle of Loading

The cumulative distribution $F_R(r)$ and the density function $f_R(r)$ of the pore pressure ratio at the end of the (i-1)th cycle will be used to derive the same functions at the end of cycle i. Similar to the first cycle analysis, the first step of the derivation is to consider the conditional distribution of R. The complexity is increased because this distribution depends not only on the shear stress ratio R but also on the accumulated pore pressure ratio at the end of cycle (i-1). If this accumulated ratio and the stress ratio have the particular values ξ and s, respectively, then the pore pressure ratio R will be less than r at the end of cycle i if, and only if, the increase in pore pressure is less than $\Delta r = r - \xi$. Δr can be related to the normalized increment $1/N_\ell$ as (Figure 1):

$$\frac{1}{N_\ell} = f_n(r, s) - f_n(\xi, s) \qquad (5)$$

For convenience we call this difference $\Delta f_n(r, \xi, s)$. Using this difference, the cumulative distribution of R follows by successive integrations (Chameau, 1980);

$$F_R^i(r) = \int_0^r \int_0^{\infty} \int_{1/\Delta f_n(r, \xi, s)}^{\infty} f_{N_\ell}(u)\, f_S(s)\, f_R^{i-1}(\xi)\, du\, ds\, d\xi \qquad (6)$$

where the superscripts i and i-1 stand for cycles i and i-1, respectively, and f_R^{i-1} is the probability density function of R at the end of the (i-1)th cycle. There is a dependence between cycles: the distribution function of the pore pressure at the end of cycle i is deduced from the knowledge of the probability density function of the pore pressure at the end of cycle (i-1). A computer program has been developed to numerically perform the foregoing integrations.

TYPICAL RESULTS

This section presents selected applications of the probabilistic model to an hypothetical level ground site. The soil is assumed to consist of sand having the characteristics of Monterey sand at a density of 54 percent, with the water table at a depth of two feet (0.6 meter). The pore pressure development is studied at a depth of 25 feet (7.6 meters). A cyclic strength curve obtained by DeAlba, et al (1976) is used in the analysis. The number of cycles to zero effective stress is considered to follow a lognormal distribution, and the level of uncertainty in the cyclic strength curve is taken as $\sigma_{LnN\ell} = 0.50$. This value is in the medium range of uncertainty values observed by previous investigators (Donovan, 1971 and Ferrito, Forest and Wu, 1979). The functional relationship between the pore pressure ratio R, the stress ratio S, and the normalized number of cycles N/N_ℓ is taken as:

$$N/N_\ell = 1 - \sin \frac{\pi}{2} \ (1 - R^{1/\beta})^{2\alpha} \qquad (7)$$

where α and β are functions of the soil properties and stress ratios and are obtained by regression analysis. This expression is similar to the one proposed by Seed, Martin and Lysmer (1976), but is more flexible. Typical α and β values range between 0.7 and 2.0, and 1.0 and 2.4, respectively (Chameau, 1980).

The last input necessary to the analysis is the earthquake induced shear stress distribution. This distribution describes the amplitude of the shear stress between two consecutive zero-crossings. Various probability functions, such as Beta, Gamma and exponential distributions, can be used to fit the distribution of the amplitudes of acceleration peaks in strong-motion records. In this example the peaks will be assumed exponentially distributed, since the exponential distribution is the simplest one and fits the data reasonably well (Mortgat, 1979). Consequently, the normalized shear stress $S(= \tau/\sigma_o')$ can also be considered exponentially distributed:

$$f_S(s) = \lambda_S \exp(- \lambda_S \ s) \qquad (8)$$

with

$$\lambda_S = \sqrt{2}\ g\ \sigma_o'/r_d\ \gamma_t\ H\ r_a \qquad (9)$$

where σ_o' is the initial effective stress, g the acceleration of gravity, γ_t the total unit weight, H the depth, r_d a flexibility factor (Seed and Idriss, 1971), and r_a the root mean square of the acceleration peaks. Values of the root mean square r_a have been computed for existing strong motion records by Mortgat (1979) and Zsutty and De Herrera (1979). In the present example the parameter r_a is equal to 15 in/sec^2 (38.2 cm/sec^2). This value of r_a is close to the values reported for each of the Taft, Hollister, First St., and Figueroa St. earthquake records. Interestingly, these four records have different number of cycles (zero-crossings), thus making it possible to assess the effect of duration on the potential for pore pressure development.

Using these data and Equations 4 and 6 the development of pore pressure is probabilistically defined. At the end of any zero-crossing of loading, the probability of exceeding a given level of excess pore pressure is determined using the computer code developed for this work. The results are expressed in Figure 2

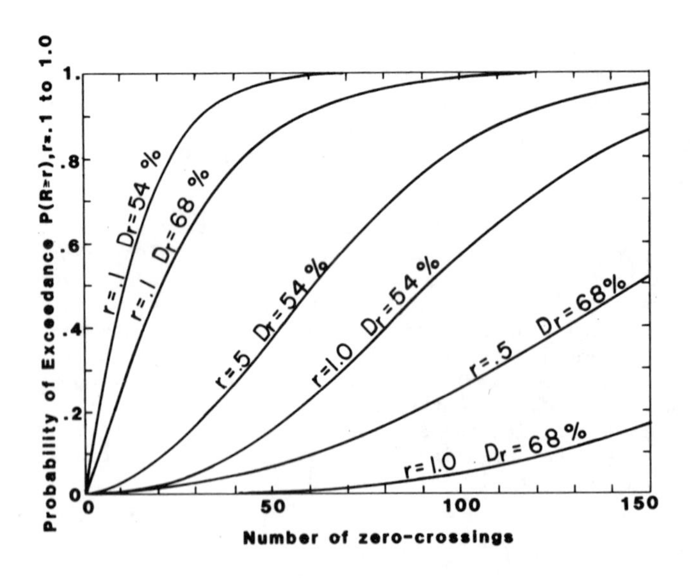

Figure 2. Relationship Between Probability of Exceedance of Pore Pressure Ratios and Number of Zero-Crossings of Shear Stress Loading for Monterey Sand.

where the probabilities of exceeding pore pressure ratios of 0.1, 0.5 and 1.0 are plotted versus the number of zero-crossings of loading. These data show:

- After 50 zero-crossings of loading the probabilities of exceeding pore pressure ratios of 0.1, 0.5 and 1.0 are 0.98, 0.37 and 0.16, respectively.

- After 150 zero-crossings of loading the probabilities of exceeding the same pore pressure ratios are 1.0, 0.97 and 0.87, respectively.

Concentrating on the case of pore pressure ratio of 1.0 (liquefaction) and considering that the number of zero-crossings for the Taft, Hollister, First St., and Figueroa St. records are 320, 94, 114 and 82, respectively, it is apparent that: The probability of reaching $R = 1.0$ during an event having characteristics similar to the Taft record is 1.0, while the probability is 0.52, 0.67 and 0.42 for events similar to the three other records, respectively. This clearly illustrates the effect of earthquake duration, expressed in number of zero-crossings, on the likelihood of pore pressure build-up.

Several sensitivity studies were also performed to assess the effects of the relative density, the level of uncertainty, and the root mean square of acceleration on the pore pressure probability curves. The details of these studies are presented in Chameau (1980). For example, probability curves are given in Figure 2 for a relative density of 68 percent, all other parameters being kept the same. As expected the probabilities of exceedance have been reduced relative to the previous case. The probability of exceeding $R = 1.0$ and $R = 0.50$ after 150 zero-crossings of loading are 0.18 and 0.52, respectively, as compared to 0.87 and 0.97 in the previous case. Similar results are presented in Figure 3 for sites located in San Francisco and for a 7.5 magnitude earthquake occurring on the San Andreas fault. Two different areas of the San Francisco waterfront fills are involved, one where the fill sand had a relative density as low as 40% and a second one with an average relative density of about 60%. As shown in Figure 3, the probability of exceeding a pore pressure ratio of 1.0 in a 7.5 magnitude event located 16 kilometers from the waterfront is only 17% for the medium density site, while it is 100% for the loose site. This emphasizes that the resistance to liquefaction for the medium dense site is much greater than that of the loose site. The probability of exceeding a given pore pressure ratio can also be shown to increase with increasing levels of uncertainty or root mean square of acceleration (Chameau, 1980). The probabilistic model has been used to evaluate the behavior of soil deposits where soil conditions and performance in earthquakes have been documented. These analyses indicate that the model correctly predicts the behavior of sites for which the field behavior during earthquakes is known (Chameau and Clough, 1982).

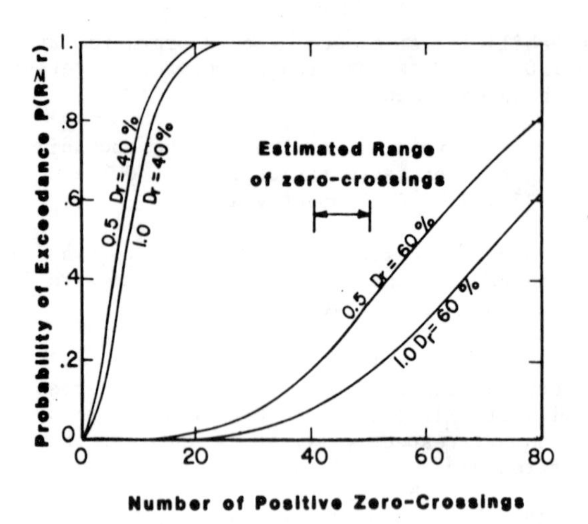

Figure 3. Relationship Between Probability of Exceedance of Pore Pressure Ratios of 0.5 and 1.0 and Number of Positive Zero-Crossings for the San Francisco Sites and a Magnitude 7.5 Earthquake.

SEISMIC HAZARD EVALUATION

Seismic hazard assessments are usually expressed in terms of site intensity, peak ground acceleration, root mean square of acceleration or response spectrum. Existing seismic hazard analysis procedures can be extended to assess the hazard in terms of a soil related parameter, the level of pore pressure. In this formulation the seismic hazard evaluation for root mean square of acceleration becomes the input to the seismic hazard evaluation for pore pressure. The annual probability of exceeding a level of pore pressure r at a site for a given seismic environment can be expressed as:

$$P(R \geq r) = \int_N \int_{r_a} P(R \geq r/r_a, N)\, f_{N/r_a}(r_a, N)\, f_{r_a}(r_a)\, dr_a\, dN \quad (10)$$

where: $P(R \geq r)$ is the annual probability of exceedance of the pore pressure ratio r.

$P(R \geq r/r_a, N)$ is the probability that the pore pressure ratio R will be greater than or equal to r given the root mean square r_a and the number of cycles (zero-crossings) N. This function is graphically represented by probability curves such as the curves in Figure 2.

f_{N/r_a} (N, r_a) is the condition probability density function of the number of zero-crossings N given the root mean square r_a.

f_{r_a} (r_a) is the probability density function of r_a which can be deduced from a root mean square versus return period graph (Mortgat, 1979).

The unknown term in Equation 10 is the conditional density function $f_{N/r_a}(r_a, N)$. Available studies and results to assess this relation are very sparse. Assumptions are necessary to define the function. The simplest assumption is to consider r_a and N to be statistically independent. This approach was used by Mortgat (1979) and is substantiated by previous work on PGA and duration (Yegian and Whitman, 1978). With this assumption Equation 10 reduces to:

$$\dot{P}(R \geq r) = \int_N \int_{r_a} P(R \geq r/r_a, N)\ f_N(N)\ f_{r_a}(r_a)\ dr_a\ dN \qquad (11)$$

where $f_N(N)$ is the probability density function of N, the number of zero-crossings. N can be assumed to be uniformly distributed between a minimum and maximum value. Based upon the limited data available the minimum and maximum values are about 50 and 250 to 350, respectively. Other distributions such as the Chi-square distribution can also be considered. Other potential assumptions are discussed by Chameau (1980). They include: (1) a weak dependence between N and r_a, where the range of N values slightly increases with r_a; (2) a global statistical relation between N, r_a and magnitude; and, (3) a weighted analysis of these assumptions where each assumption is given a subjective weight.

The results of a seismic hazard analysis for a site located close to San Francisco and having the characteristics of the hypothetical site (previous section) is given in Figure 4. Two pore pressure ratios were considered in the analysis, 0.50 and 1.0. The probability of at least one occurrence of the ratio R = 1.0 is 0.61 for a 50 year time period and 0.84 for a 100 year time period. For the ratio R = 0.50, the probabilities of at least one exceedance are 0.75 and 0.96 for the same time periods, respectively. Such high numbers are to be expected for a site close to the San Andreas fault. Parametric studies were also performed to describe how the assumptions, seismic environment, and site resistance affect the hazard assessment at a given site. As an example, the probability of at least one occurrence of R = 1.0 is plotted in Figure 4 for a relative density of 68 percent. The probability of at least one exceedance for 50 and 100 year time periods is significantly reduced.

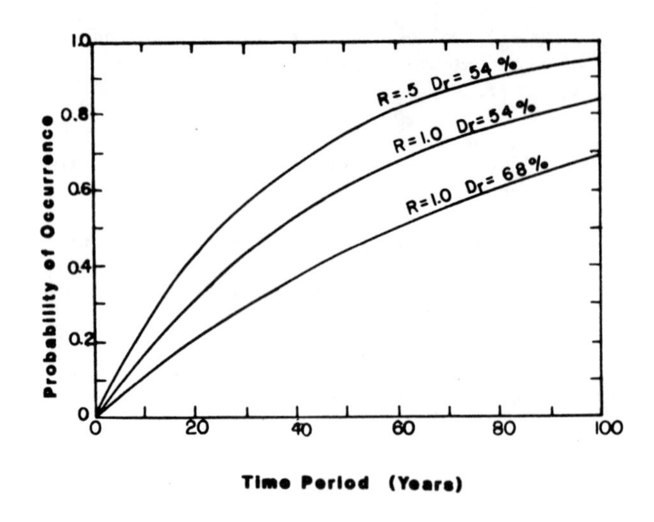

Figure 4. Seismic Hazard Evaluation for the San Francisco Sites

Seismic hazard maps or contour maps are usually developed for site intensity, peak ground acceleration or other ground motion parameters. The present methodology can extend the seismic hazard mapping techniques and define contour maps for a soil-related parameter, the level of pore pressure, or for a soil-related effect, liquefaction. By superimposing a grid on a given geographical area and performing the seismic hazard evaluation for each node of the grid, a seismic exposure map can be prepared for any desired probability of exceedance. An hypothetical example of such a map is given in Figure 5. Smooth contours connect the points of equal probability of liquefaction (pore pressure ratio of 1.0) for a given time period (100 years) and a given seismic environment (one area source and one live source). By repeating the analysis for other pore pressure ratios, iso-pore pressure maps could be developed. The hazard map indicates the likelihood that liquefiable materials,if present, may liquefy at a given site. This is an exposure map useful for planning purposes and identifying areas of substantial hazard. Additional geological and geotechnical information may be required to take into account specific local conditions and modify the hazard maps.

CONCLUSIONS

A probabilistic model is proposed to study the development of pore pressure in sands under random loading. The potential for

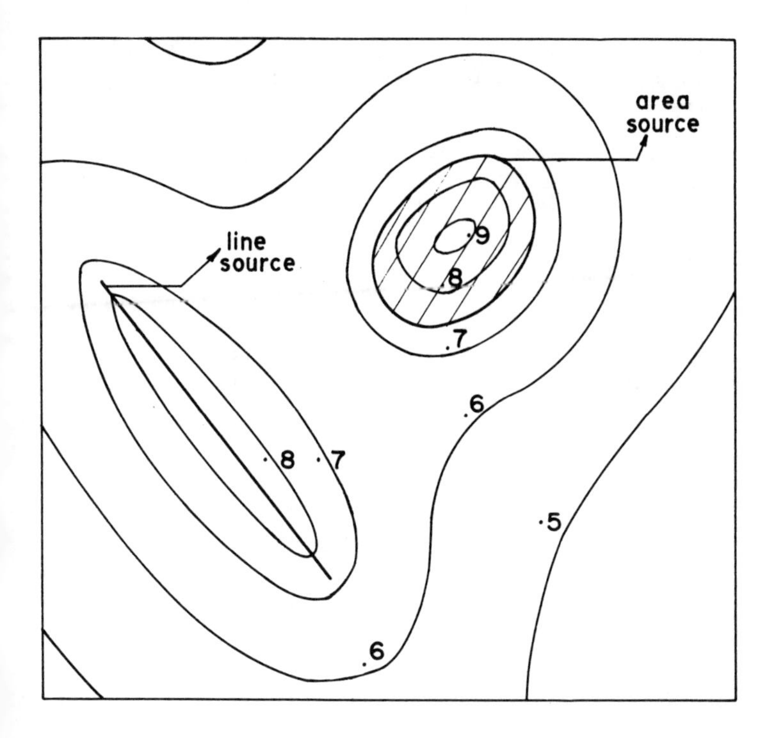

Figure 5. Liquefaction Hazard Map

pore pressure development is characterized empirically by a cycle strength curve and a set of pore pressure generation curves, both to be obtained in laboratory experiments. The probabilistic model is incorporated into a hazard analysis methodology. This results in the probability of exceeding at least once a certain pore pressure ratio for a given seismic environment and a given time period. The methodology can be used to generate seismic exposure maps for liquefaction. The primary results and conclusions drawn from this study are as follows:

1. The probabilistic pore pressure model provides a complete description of pore pressure development up to and including liquefaction.

2. The predicted results reflect the effect of earthquake duration showing that substantially different levels of pore pressure are likely to develop depending upon the duration.

3. The effect of uncertainty in specific soil parameters is most important for moderate intensity loading and short duration events.

4. The seismic hazard methodology can be very effectively used to compare the hazard potential of sites with different soil resistance and seismic environment.

ACKNOWLEDGEMENTS

Appreciation is expressed to Professor G. W. Clough of Stanford University for his advice and valuable suggestions. Financial support was provided by the National Science Foundation Grant ENV-17834 and the United States Geological Survey, Contract No. USGS 18380.

REFERENCES

Castro, G., "Liquefaction and Cyclic Mobility of Saturated Sands", Jour. of the Geot. Eng. Div., ASCE, Vol. 101, No. GT6, 1975.

Chameau, J. L., "Probabilistic and Hazard Analysis for Pore Pressure Increase in Soils Due to Seismic Loading", Ph.D. dissertation, Stanford University, Stanford, CA, October 1980.

Chameau, J. L., and Clough, G. W., "Probabilistic Pore Pressure Analysis for Seismic Loading", submitted for publication in the Jour. of the Geot. Eng. Div., ASCE, 1982.

Christian, J. T., and Swiger, W. F., "Statistics of Liquefaction and SPT Results", Jour. of the Geot. Eng. Div., ASCE, Vol. 101, No. GT11, Nov., 1975.

Clough, G. W. and Chameau, J. L., "A Study of the Behavior of the San Francisco Waterfront Fills Under Seismic Loading", Report No. 35, The John A. Blume Earthquake Engineering Center, Stanford, February, 1979.

Cornell, C. A., "Engineering Seismic Risk Analysis", BSSA 58(5), 1968.

De Alba, P., Seed, H. B., and Chan, C. K., "Sand Liquefaction in Large Scale Simple Shear Tests", Jour. of the Geot. Eng. Div., ASCE, Vol. 102, No. GT9, Sept. 1976.

Donovan, N. C., "A Stochastic Approach to the Seismic Liquefaction Problem", First National Conference on Applications of Statistics and Probability to Soil and Structural Engineering, Hong Kong. Sept. 1971.

Fardis, M. W., and Veneziano, D., "Probabilistic Liquefaction of Sands During Earthquakes", Report R79-14, Massachusetts Institute of Technology, March 1979.

Ferrito, J. M., Forrest, J. B., and Wu, G., "A Compilation of Cyclic Triaxial Liquefaction Test Data", Geotechnical Testing Journal, Vol. 2, No. 2, June 1979.

Finn, W. D. L., Lee, K. W., and Martin, G. R., "An Effective Stress Model for Liquefaction", Jour. of the Geot. Eng. Div., ASCE, Vol. 103, No. GT6, June 1977.

Haldar, A., and Tang, W. H., "Probabilistic Evaluation of Liquefaction Potential", Journal of the Geotechnical Engineering Division, ASCE, Vol. 105, No. GT2, Feb. 1979.

Mortgat, C. P., "A Probabilistic Definition of Effective Acceleration", Proceedings of the 2nd National Conference on Earthquake Engineering, Stanford, Aug. 1979.

Seed, H. B., and Idriss, I. M., "Simplified Procedure for Evaluating Soil Liquefaction Potential", Jour. of the Soil Mech. and Found. Div., ASCE, Vol. 97, No. SM9, Sept. 1971.

Seed, H. B., Martin, P. P., and Lysmer, J., "Pore Water Pressures during Soil Liquefaction", Jour. of the Geot. Eng. Div., ASCE, Vol. 102, No. GT4, April, 1976.

Seed, H. B., "Soil Liquefaction and Cyclic Mobility Evaluation for Level Ground during Earthquakes", Jour. of the Geot. Eng. Div., ASCE, Vol. 105, No. GT2, Feb. 1979.

Valera, J. E. and Donovan, N. C., "Soil Liquefaction Procedures - A Review", Jour. of the Geot. Eng. Div., ASCE, Vol. 101, No. GT6, June 1977.

Yegian, M. K., and Whitman, R. V., "Risk Analysis for Ground Failure by Liquefaction", Jour. of the Geot. Eng. Div., ASCE, Vol. 104, No. GT7, July, 1978.

Youd, T. L. and Hoose, S. N., "Liquefaction during 1906 San Francisco Earthquake", Journal of the Geotechnical Engineering Division, ASCE, Vol. 102, No. GT5, May 1976.

Zsutty, T., and De Herrera, M., "A Statistical Analysis of Accelerogram Based upon the Exponential Distribution Model", Proceedings, of the 2nd National Conference on Earthquake Engineering, Stanford, Aug. 1979.

Soil Dynamics & Earthquake Engineering Conference / Southampton / 1982.07.13-15

Seismic Hazard in Northeastern United States

H.K.ACHARYA, A.S.LUCKS & J.T.CHRISTIAN
Stone & Webster Engineering Corp., Boston, MA, USA

ABSTRACT

Seismic hazard has been evaluated at a number of locations in the northeastern United States using probabilistic methods. The analysis was carried out to examine the effects of significant variations in various seismic parameters, reflecting primarily the uncertainty in our knowledge of tectonic processes, inadequate historical data base, and absence of strong ground motion records. Seismic sources were considered in two ways:

. Using the tectonic province approach that is consistent with the procedure followed by the U. S. Nuclear Regulatory commission to evaluate nuclear power plants, and in which earthquakes not related to a particular geologic structure are assumed to be associated with a tectonic province.

. Using seismic zones that were identified solely from the analysis of patterns of historical seismicity.

The maximum earthquake for each province or zone was assumed to be either a) the historical maximum intensity or b) historical maximum intensity plus one unit. In both cases, several distant areas in which earthquakes of intensity ≥ VIII (MM) have occurred were considered additional seismic sources. Several published attenuation relationships were also considered. The effects of uncertainty in activity rates were examined by computing rates for several time intervals.

The uncertainity in attenuation relationships contributed significantly to the variation in seismic hazard estimates in the northeastern United States. For the sites studied, there was less than half order of magnitude increase in seismic hazard when the maximum magnitude earthquake was assumed to be historical maximum plus one unit rather than the historical maximum at lower site intensities.

Introduction

The low level of earthquake activity in eastern United States coupled with short instrumental history and absence of associated surface faulting leads to considerable uncertainty in the estimation of various seismicity parameters and therefore in the estimation of seismic hazard. McGuire (1977) examined the effects of these uncertainties and concluded that at several sites along the east coast of United States, events with intensities between VIII and IX (MM) had average annual probability of 10^{-4} of being equalled or exceeded. However these estimates were based on the assumption that the upper bound on earthquake intensity in eastern United States is randomly variable between the maximum observed in each seismic source and XII. Although such an assumption is useful for demonstrating the range of estimates of seismic hazard at a site, it is unrealistic to expect that earthquakes of intensity XII can occur everywhere in eastern United States. Acharya (1980a) has noted a spatial correlation between recent moderate earthquakes at depth greater than 10 km and damaging (intensity $\geqslant$ VIII (MM)) earthquakes and suggested that all seismically active areas in eastern United States may not experience damaging earthquakes, if present earthquakes are occurring at depths less than 10 km. It is therefore clearly desirable to reexamine the effects of uncertainties in seismicity parameters without the undue influence of including the probability of intensity XII earthquakes in all identified sources.

We have therefore estimated seismic hazard to four cities in the northeastern United States for a range of seismicity parameters. The four cities, Boston, Massachusetts; Washington, D.C., Buffalo, New York; and Charleston, West Virginia, shown in Figure 1 represent a variety of geographical and geologic settings. The seismic hazard model used in this study has been described in detail by Cornell, 1968; Cornell and Merz, 1975. A standard computer program (McGuire, 1976) was used for calculations. Because the historical earthquake record for the eastern U. S. describes ground motions primarily in terms of Modified Mercalli intensity, the seismic hazard was determined as the annual probability of exceeding a given intensity (MM) at each site.

The region considered for this study is bounded by latitudes N 34° to N 47° and longitudes W 66° to W 84° and includes most of the eastern United States and parts of southeastern Canada. A catalog of earthquakes with intensity $\geqslant$ V (MM) for the period 1871-1977 was compiled for this region by combining catalogs developed by Chiburis (1979), Nuttli (1979) and Bollinger (1975). Earthquakes in the eastern United States have been investigated by the Weather Bureau since 1873. Since the mid 19th Century, the eastern United States and southeastern Canada have been well populated and means of communication well established. Consequently, the earthquake catalog for this region is likely to be complete for earthquakes with intensity $\geqslant$ V (MM) after 1871. The earthquakes included in this catalog are plotted in Figure 1.

Seismic Sources

In a typical seismic hazard analysis, seismicity is defined by using geologic and tectonic data as well as observed earthquake locations. The region of study is divided up into seismic sources within which future seismic events are considered equally likely to occur at any location. In this study, two hypotheses are used for specifying seismic sources:

1) Tectonic Province Approach - Earthquakes not related to a particular geologic structure are assumed to be associated with the tectonic province in which they occur. The provinces are defined and boundaries delineated on the basis of the following criteria: a) style and degree of deformation, b) age of orogenic, igneous or tectonic activity and c) age of basement rocks. The boundaries of these provinces are drawn on the basis of studies by Rogers (1970), King (1969) and Hadley and Devine (1974). Portions of ten tectonic provinces are located in the study area and are shown in Figure 2. Specific structures that have a direct relationship to seismicity are also identified in Figure 2. They are the White Mountain Plutonic Zone, Monteregian Plutonic Zone and Clarendon - Linden Fault Zone. This approach is consistent with the procedures followed to evaluate earthquake hazard for nuclear power plants in accordance with Appendix A, 10 CFR 100.

2) Seismic Zone Approach - Many investigators have noted that patterns of seismicity determined from analysis of historical data cut across tectonic province boundaries. For example, Bollinger (1973) noted that even with consideration of epicentral errors, earthquakes in the southeastern United States occur in all four of the provinces in the area. Furthermore, Bollinger found that activity is oriented both parallel and transverse to the dominant northeast trending Appalachian structures. Hadley and Devine (1974) also noted that historical earthquake activity bears no consistent relation to tectonic provinces and subprovinces. Therefore, in this approach, seismic zones are identified from analysis of patterns of historical seismicity only (Bollinger, 1973; Chiburis, 1979; and Nuttli, 1979), and are assumed to represent deep seated causes. The White Mountain Plutonic Zone identified on the basis of seismicity alone is identical to the White Mountain tectonic province. These seismic zones are shown in Figure 3.

Five distant centers of seismicity have also been identified and are shown in Figures 2 and 3. These centers are Charleston, South Carolina; New Madrid, Missouri; the Wabash Valley in Indiana and Illinois; Anna, Ohio and LaMalbaie, Quebec. Although they are far from the four cities considered, large earthquakes in these sources can influence the annual probability for lower intensities. Therefore, the five areas were included in both cases as point sources.

Intensity - Activity Parameters

1) b value: Earthquake catalogs for different sections of eastern U. S. have been tested for completeness by Nuttli (1974), Bollinger (1973) and Chiburis (1979) and used to determine the b value or the slope of the log number of events vs intensity relationship. The b values for various seismic sources determined by these investigators are listed in Tables 1 and 2. It should be noted that β in thse tables refers to the relation between epicentral intensity and rate of occurrence expressed in natural logarithms.

2) Activity Rates

The earthquake epicenter map (Fig. 1) was overlain on the tectonic province map (Fig. 2) and seismic zones map (Fig. 3) to establish the rate of activity in each seismic source by counting the number of events which occurred in it during the following three different time periods:

Case	Time Period	Events
1	1871-1977	Intensity $\geqslant$ V (MM)
2	1949-1977	Intensity $\geqslant$ V (MM)
3	1929-1977	Intensity $\geqslant$ VI (MM)

Case 1 is intended to reflect the most recent 100 years of data even though the catalog may not be complete throughout the entire region. This time period coincides with the time period during which the U. S. Government agencies have been involved in the examination and reporting of earthquakes in the U. S. Bollinger (1973) has suggested that the earthquake catalog for the southeast may be complete for the last 100 years for intensity $\geqslant$ V (MM). As the northeast has been well populated over the last 100 years, it is likely that all events of intensity $\geqslant$ V have been reported for the northeast as well. Case 2 represents a time period for which the record has been shown to be complete for intensity $\geqslant$ V (Chiburis, 1979; Bollinger, 1973). Case 3 is an approximate 50 year period for which Chiburis and Bollinger have demonstrated completeness for events of intensity $\geqslant$ VI (MM). The activity rates for various seismic sources using both the tectonic province concept and seismic zone concept are listed for all three cases in Tables 1 and 2.

3) Maximum Intensity Earthquake:

The upper bound earthquake for each seismic source within the main study area was chosen, using each of the following two alternate methods: (i) The maximum intensity observed historically in that source, based on the entire catalog from 1534 to present. If the maximum intensity of the largest historical earthquake in a seismic zone/tectonic province was less than VII (MM), a maximum intensity of VII was used as an upper bound. (ii) For each source the upper bound was specified as maximum historical intensity plus one unit. In view of the low frequency of events in eastern U.S., it may be possible that events larger than historical time span may have occurred. That was the basis for McGuire's (1977) assumption that even events of intensity XII may occur anywhere. Based upon Acharya (1980 a, b) the maximum intensity in other seismic zones may not exceed VIII and accordingly the upper bound in these seismic zones is specified as maximum historical plus one unit.

The upper bound earthquake intensity for the five seismic sources located beyond the main study area was specified as follows:

Charleston	X (MM)	Historical Max.
New Madrid	XI (MM)	Historical Max.
Wabash Valley	IX (MM)	Nuttli (1979).
LaMalbaie	X (MM)	Historical Max.
Anna, Ohio	IX(MM)	Nuttli (1979).

4) Attenuation Relationships:

In the absence of strong ground motion data, attenuation of ground motion in eastern and central U. S. has been determined by several investigators using either actual intensity or isoseismal maps. The following three different attenuation relationships were used in this study:

1. Howell and Schultz (1975)

$I = Io + 4.578 - 1.36 \ln R$ $I = Io$ for $R \leq 29$ km

R is the epicentral distance in kms and $\sigma = 0.66$

2. Cornell and Merz (1975) developed two different relationships for soil and rock sites.

$I = Io + 3.211 - 1.31 \ln R$ $I = Io$ for $R \leq 16$ km $\sigma = 0.2$ rock sites

$I = Io + 3.72 - 1.3 \ln R$ $I = Io$ for $R \leq 11$ km, $\sigma = 0.5$ soil sites

3. McGuire (1976)

$I = Io + 3.08 - 1.34 \ln R$ $I = Io$ for $R \leq 10$ km $\sigma = 1.19$

The McGuire equation lies 1.0 to 1.5 units below the other relations because it is derived using individual intensity reports while other relationships were derived using distance to isoseismals. We believe that the use of individual reports is the appropriate method to predict intensity at a site. The Howell and Schultz (1975) relationship tends to be conservative since it is determined using isoseismal contours. Doing so results in the use of maximum radial distance to each isoseismal as the value for the distance parameter R. Incorporation of standard deviation of error into the calculations results therefore in higher computed intensities at a distant site then is suggested from actual historical data and therefore higher computed hazard at the site.

Method of Investigation: -

A total of 36 different combination of input data were prepared for analysis at each city. The different data sets were developed as follows:

a) Two different sets of seismic sources (tectonic provinces/ seismic zones) were specified and earthquakes assigned to these sources.

b) Three sets of activity rates were developed by considering different segments from the catalog.

c) Three different attenuation laws discussed earlier were used.

d) Maximum intensity earthquake in a seismic source was specified in two ways - one corresponding to historical maximum intensity in the source and one corresponding to historical maximum intensity in the source plus one unit.

Seismic hazard at all four cities, shown in Figure 1, was first computed by assuming tectonic province approach and historical maximum intensity in each source. The activity rates in various sources are much higher for the 1949-1977 period than the 1871-1977 period. We do not know whether this arises from better communications or reflects a real increase in activity. We therefore computed the annual probability of exceeding intensity V, VI and VII at a city for each attenuation relationship for all three rates and then computed the geometric mean of these risks to reflect average conditions. These results are shown in Table 3. A geometric mean of the results for all three attenuation relationships (average in Table 3) is then plotted in Figure 4.

Seismic hazard was then recomputed at all four cities using seismic zones approach. Average probabilities, computed as earlier, are shown in Table 4 and Figure 5. We will examine these results to understand the effect of uncertainties in (i) activity rate, (ii) attenuation relationships and (iii) maximum earthquake intensity in a source on estimated equthquake hazard at a site.

RESULT:

Figure 4 shows the average annual likelihood of equalling or exceeding a given intensity at all four cities using tectonic province approach. The likelihood of exceeding a given intensity is nearly same at all four cities. The four cities considered in this study are located in four different tectonic provinces (Figure 2) which exhibit low level of seismic activity (Table 1). It is therefore not surprising that the annual risk of exceeding a given intensity is essentially same at all four cites. Boston, shows higher annual hazard than the other cities and Charleston, shows somewhat lower annual hazard. The higher hazard at Boston, may be due to the proximity of White Mountain Plutonic zone with an associated maximum historical intensity of VIII. The characteristics of particular tectonic provinces appear to be inconsequential. Figure 4 also shows that at all four cities, the effect of increasing maximum intensity earthquake in a source is most pronounced at higher site intensities. For example, the difference in annual hazard, when maximum intensity is specified as historical maximum plus one unit as against when it is specified as historical maximum, is greater for site intensity VII (MM) than for site intensity V (MM). This difference is about an order of magnitude for site intensity VII at Buffalo, Washington and Charleston and slightly less at Boston.

Figure 5 shows that there is considerable variation in seismic hazard among four cities when seismic sources are defined in terms of seismic zones. Buffalo, shows higher seismic hazard than the other three cities. It is followed in decreasing order by Boston, Washington and Charleston. Buffalo is situated in the Attica - Buffalo - St. Catherines seismic zone (Figure 3) with historical maximum intensity VIII. Washington and Boston, are situated in the southern coastal zone with maximum historical intensity VII and Charleston, W. Va is located outside the seismic zones. The seismic hazards reflect this appropriately. Figure 5 shows that seismic hazard at Buffalo and Charleston increases only slightly when the maximum intensity earthquake in a source is specified as the historical maximum plus one unit as compared to specifying the historical maximum. On the other hand for Boston and Washington this difference is almost an order of magnitude for site intensity VII (MM). This suggests that the specification of maximum intensity - (historical maximum or historical maximum plus one unit) affects the seismic hazard for higher site intensities only when the seismic sources are large or when the dominant seismic source has a relatively low maximum historical earthquake.

A comparison of seismic hazard at these four cities in Figures 4 and 5 illustrates the effect of defining seismic sources in terms of tectonic provinces or seismic zones. Buffalo shows half to one order of magnitude higher seismic hazard when the seismic zone approach is used. Boston shows slightly greater seismic hazard with the seismic zone approach. On the other hand, Washington and Charleston do not show any significant difference between the seismic zone and tectonic province approach. The significantly higher hazard at Buffalo probably rises from the inclusion of Attica events in this zone in which Buffalo is located. This suggets that the different specifications of seismic sources will affect the seismic hazard at a site only when high intensity events occur in the same source in which the site is situated.

The effect of the attenuation relationship is clearly evident in Tables 3 and 4. The Howell and Schultz (1975) relationship leads to the smallest return period for all intensities at all sites. The Cornell and Merz (1975) relationship shows a wide variation based on site condition - rock or soil. The use of McGuire (1976) relationship leads to results close to the average and suggests that the use of individual intensity values in the determination of attenuation relationship leads to computed risk reflecting average conditions.

McGuire (1977) computed MM intensities at various sites on the U. S. east coast for 10^{-4} annual hazard using various hypothesis on the maximum intensity earthquake in a source. McGuire considered (i) historical maximum, (ii) random maximum IX intensity in all sources, (iii) random maximum XII intensity in all sources and (iv) the certainty that intensity XII events will eventually occur in all seismic sources. Two of the sites included in McGuire's study are Washington and Boston. Since we have used a different hypothesis for maximum intensity, a comparison with McGuire's results will provide an indication of the effect of this uncertainty. By extrapolating the curves in Figure 5 the computed intensities (MM) for both Boston and Washington for annual risk are about 7 and 8.3 for the historic maximum and historic maximum plus one unit, respectively. Based on the four hypothesis described above McGuire computed intensities 7.0, 8.0, 8.7 and 9.4 for Washington and 8.3, 8.4, 9.1 and 9.5 for Boston. Different assumptions about maximum intensity earthquake in a source can lead to a difference of more than two intensity units at a site for a given annual hazard. Based on the intensity acceleration relationship $\log a = 0.25I + 0.25$ developed by Murphy and O'Brien (1977) the acceleration at Washington and Boston can vary between 0.1g and 0.43g depending on the hypothesis used for assigning maximum intensity earthquake in a source. Clearly, along with the proper attenuation relationship this is a crucial parameter.

ACKNOWLEDGEMENT

We are thankful to J. Briedis, A. Foster, P. Trudeau, D. Creggar, R. Gillespie, F. Verok and R. Trudeau for help in organizing the data and to T. Y. Chang for checking our results. We also thank C. Stepp and E. C. Chiburis for reviewing the work and providing the earthquake catalog for northeastern U. S. Finally we thank Stone & Webster Engineering Corporation for support.

REFERENCES

Acharya, H. (1980a) Spatial correlation of large historical earthquakes and moderate deep shocks in eastern North America Geophys. Res. Lett., 7, 1061-1064.

Acharya, H. (1980b) Possible minimum depth of large historical earthquakes in eastern North America and its tectonic implications. Geophys. Res. Lett., 7, 619-620.

Bollinger, G. A. (1973) Seismicity of southeastern United States. Bull. Seis. Soc. Am., 63, 1785-1808.

Bollinger, G. A. (1975) Catalog of earthquake activity in southeastern United States. Virginia Polytechnic Institute.

Chiburis, E. C. (1979) Earthquake catalog for the northeastern U. S. and southeastern Canada (unpublished) personal communication.

Cornell, C. A. (1968) Engineering seismic risk analysis. Bull. Seis. Soc. Am., 58, 1583-1606.

Cornell, C. A. and Merz, H. A. (1975) A Seismic risk analysis of Boston. J. Structural Division Am. Soc. Civil Engineers, 10, 2027-2043.

Hadley, J. B. and Devine, J. F. (1974) Seismotectonic map of the United States, U. S. Geol. Swiv. Field studies, Map MF-620.

Howell, B. F. and Schultz, J. R. (1975) Attenuation of Modified Mercalli intensity with distance from the epicenter. Bull. Seis. Soc. Am., 65, 651-666.

King, P. B. (1969) The tectonics of North America. A discussion to accompany the Tectonic Map of North America. U. S. Geol. Surv. Prof. Paper 628.

McGuire, R. (1976) FORTRAN computer program for seismic risk analysis. U. S. Geol. Surv. Open File Rept. 76-67.

McGuire, R. (1977) Effects of uncertainty in seismicity on estimates of seismic hazard for the east coast of the United States. Bull. Seis. Soc. Am., 67, 827-848.

Murphy, J. R. and L. J. O'Brien. (1977) The correlation of peak ground acceleration amplitude with seismic intensity and other physical parameters, Bull. Seis. Soc. Am., 67, 877-917.

Nuttli, O. W. (1974) Magnitude - Recurrence relation for central Mississippi Valley earthquakes. Bull. Seis. Soc. Am., 64, 1189-1207.

Nuttli, O. W. (1979) Seismicity of the central United States in Geology in the siting of nuclear power plants -Reviews in Engineering Geology, Vol. IV, published by the Geological Soc. Am. 67-93.

Rodgers, J. (1970) The tectonics of the Apppalachians. Wiley Interscience, New York, 271 P.

TABLE 1
TECTONIC PROVINCE APPROACH
RECURRENCE AND ACTIVITY RATE INPUT PARAMETERS

Source	β	Annual Activity Rate 1929-1977	Annual Activity Rate 1949-1977	Annual Activity Rate 1871-1977	Maximum Historical Earthquakes
Appalachian Plateau	1.38	0.105	0.428	0.174	VII
Central Stable Region	1.38	0.032	0.303	0.194	VII
Atlantic Coastal Plain	1.38	0.042	0.268	0.194	VII
Grenville	0.97	0.407	0.589	0.585	VIII
Monteregion Plutons	0.97	0.042	0.232	0.109	VIII
Northern Valley & Ridge	1.38	0.083	0.143	0.094	VII
Piedmont - Blue Ridge	1.38	0.167	0.607	0.472	VII
S.E. New England - Maritime	0.97	0.011	0.196	0.142	VII
Southern Valley & Ridge	1.38	0.063	0.321	0.198	VIII
Superior	1.38	0.0	0.0	0.014	VII
White Mountain Plutons	0.97	0.052	0.215	0.170	VIII
Clarendon - Lindon	1.38	0.063	0.107	0.047	VIII
New England	0.97	0.094	0.66	0.364	VII
Anna, Ohio	1.38		0.124		IX
Wabash Valley	1.38		0.243		IX
New Madrid	1.38		0.624		XI
Charleston	1.38		0.120		X
La Malbaie	1.104		0.310		X

$\beta = b \ln 10$

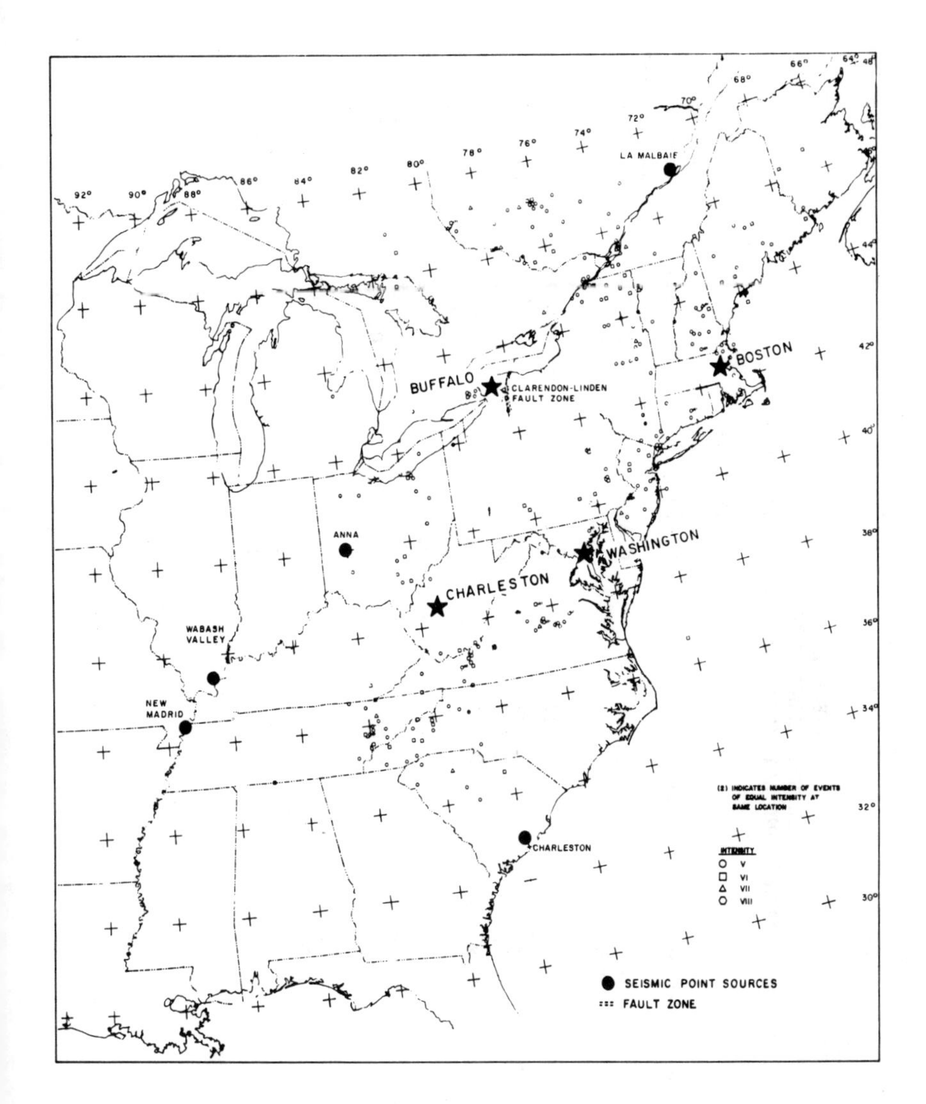
LA MALBAIE
BUFFALO
CLARENDON-LINDEN
FAULT ZONE
BOSTON
ANNA
WASHINGTON
CHARLESTON
WABASH
VALLEY
NEW
MADRID
CHARLESTON
(2) INDICATES NUMBER OF EVENTS
OF EQUAL INTENSITY AT
SAME LOCATION
INTENSITY
V
VI
VII
VIII
SEISMIC POINT SOURCES
FAULT ZONE
92°
90°
88°
86°
84°
82°
80°
78°
76°
74°
72°
70°
68°
66°
64°
48°
44°
42°
40°
38°
36°
34°
32°
30°

FIG. 1

TABLE 2
SEISMIC ZONE APPROACH
RECURRENCE AND ACTIVITY RATE INPUT PARAMETERS

Seismic Zone	β	Annual Activity Rate 1971-1977	Annual Activity Rate 1947-1977	Annual Activity Rate 1929-1977	Maximum Historical Earthquake
Western Quebec	0.97	0.698	0.893	0.396	VIII
Attica-Buffalo-St. Catherines	0.97	0.104	0.214	0.0625	VIII
Southern Coastal	0.97	0.491	0.607	0.0625	VII
White Mountain Plutons	0.97	0.198	0.286	0.0625	VIII
Northern Coastal	0.97	0.142	0.179	0.021	VII
New Brunswick	0.97	0.019	0.0	0.0	VII
Southern Appalachian	1.38	0.575	0.893	0.250	VIII
Central Virginia	1.38	0.151	0.179	0.042	VII
S. Carolina-Georgia	1.38	0.075	0.143	0.021	VII
Background	0.97	0.0024	0.0042	0.0014	VII
Anna, Ohio	1.38		0.124		IX
Wabash Valley	1.38		0.243		IX
New Madrid	1.38		0.624		XI
Charleston	1.38		0.120		X
La Malbaie	1.104		0.310		X

NOTE:

$\beta = b \ln 10$

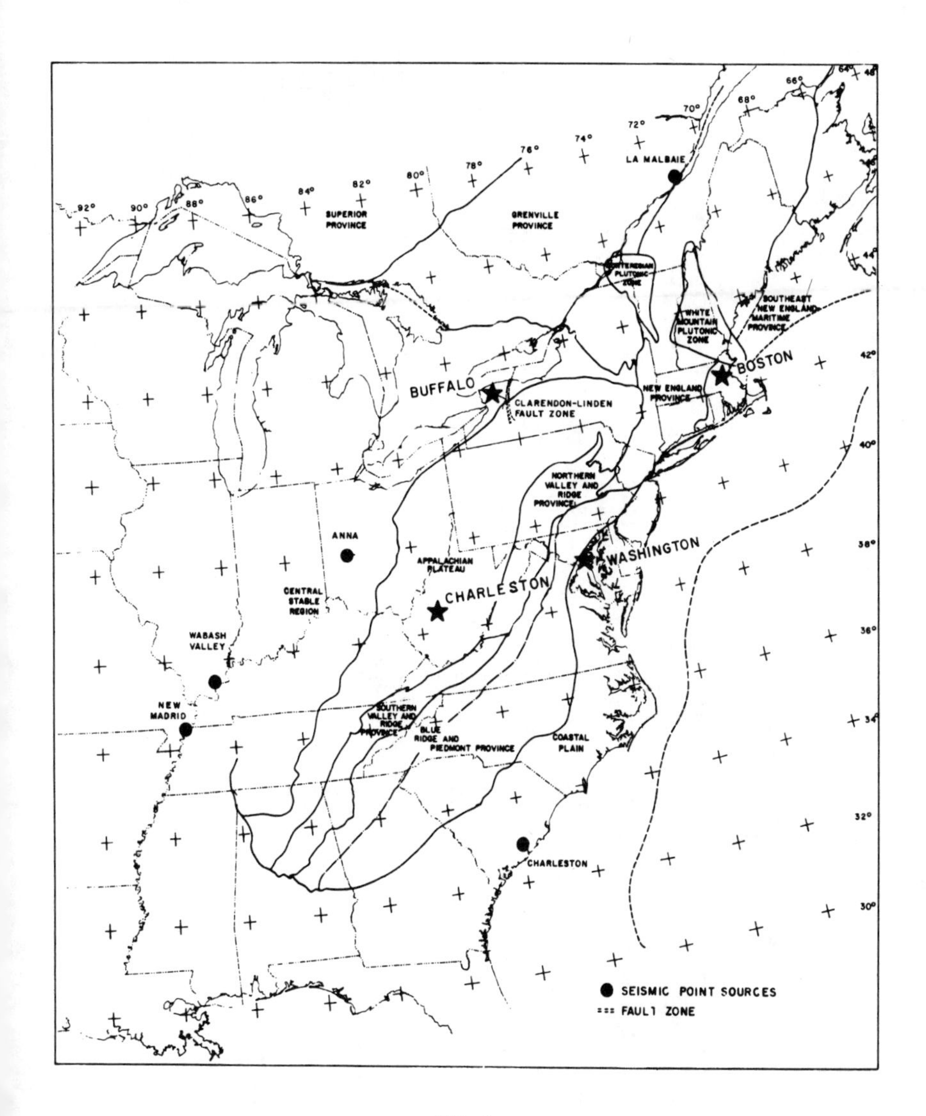
92°
90°
88°
86°
84°
82°
80°
78°
76°
74°
72°
70°
68°
66°
64°
46°
44°
42°
40°
38°
36°
34°
32°
30°
SUPERIOR PROVINCE
GRENVILLE PROVINCE
LA MALBAIE
PLUTONIC ZONE
WHITE MOUNTAIN PLUTONIC ZONE
SOUTHEAST NEW ENGLAND MARITIME PROVINCE
BOSTON
NEW ENGLAND PROVINCE
BUFFALO
CLARENDON-LINDEN FAULT ZONE
NORTHERN VALLEY AND RIDGE PROVINCE
ANNA
APPALACHIAN PLATEAU
WASHINGTON
CHARLESTON
CENTRAL STABLE REGION
WABASH VALLEY
NEW MADRID
SOUTHERN VALLEY AND RIDGE PROVINCE
BLUE RIDGE AND PIEDMONT PROVINCE
COASTAL PLAIN
CHARLESTON
SEISMIC POINT SOURCES
FAULT ZONE

FIG. 2

TABLE 3
GEOMETRIC MEAN RETURN PERIOD (YRS) FOR
DIFFERENT ATTENUATION RELATIONSHIPS
TECTONIC PROVINCE APPROACH

SITE	Attenuation Relationship	Historical Maximum			Historical Maximum Plus One Unit		
		V	VI	VII	V	VI	VII
Boston,	Howell & Schultz	22	120	1049	15	65	369
MA	Cornell & Merz	271	2840	11990407	121	806	12450
	McGuire	57	296	1876	40	175	909
	Average	69	465	6173	41	210	1610
Washington,	Howell & Schultz	22	143	1342	17	98	649
D.C.	Cornell & Merz	297	3472	1904762	192	1321	15384
	McGuire	43	237	1460	37	190	1060
	Average	66	490	15504	50	291	2198
Buffalo,	Howell & Schultz	21	137	1377	15	83	578
N.Y.	Cornell & Merz	333	9174	8695652	188	1399	39682
	McGuire	52	308	2232	39	205	1266
	Average	71	730	29940	48	287	3077
Charleston,	Howell & Schultz	26	189	2267	21	119	962
W. VA	Cornell & Merz	478	613	4784688	240	2480	27173
	McGuire	58	347	2427	49	261	1607
	Average	90	735	29762	63	425	3472

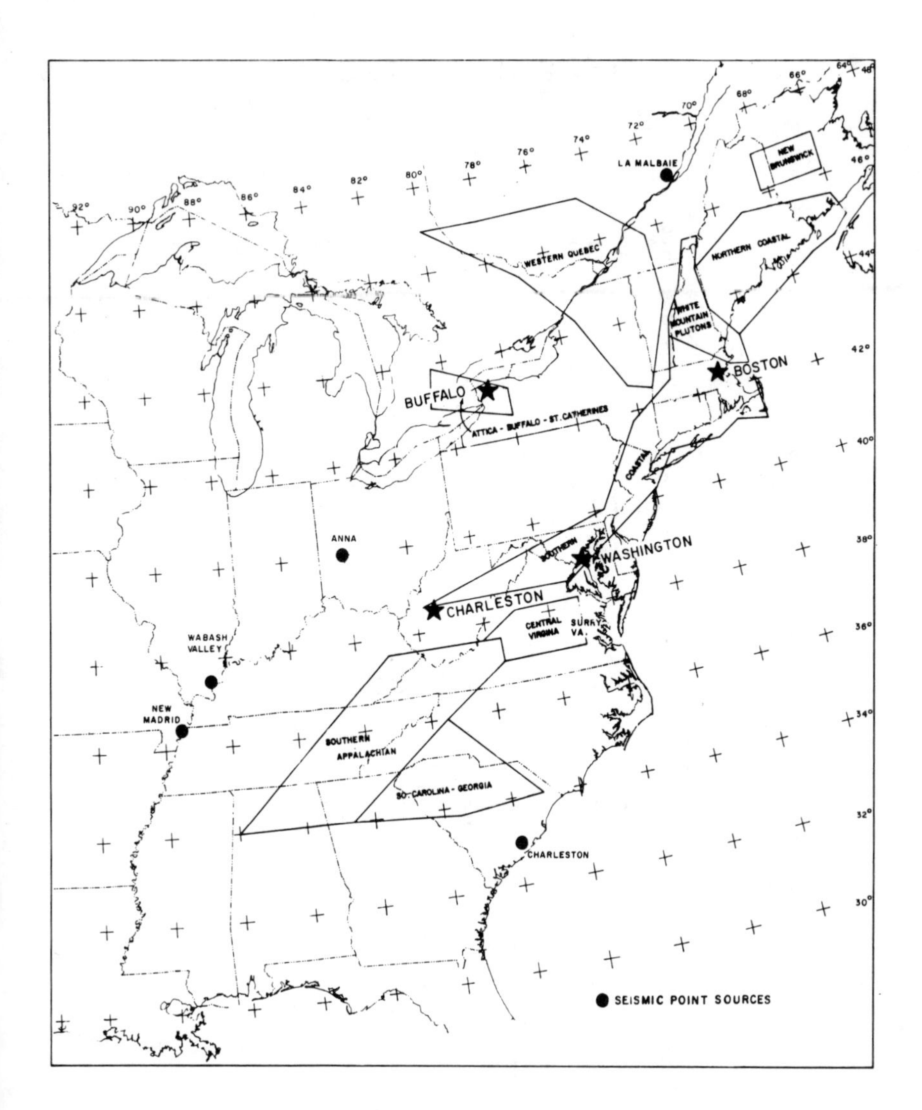
LA MALBAIE
NEW BRUNSWICK
WESTERN QUEBEC
NORTHERN COASTAL
WHITE MOUNTAIN PLUTONS
BOSTON
BUFFALO
ATTICA - BUFFALO - ST. CATHERINES
COASTAL
SOUTHERN
WASHINGTON
ANNA
CHARLESTON
CENTRAL VIRGINIA
SURRY VA.
WABASH VALLEY
NEW MADRID
SOUTHERN APPALACHIAN
SO. CAROLINA - GEORGIA
CHARLESTON
SEISMIC POINT SOURCES

FIG. 3

TABLE 4
GEOMETRIC MEAN RETURN PERIOD (YRS) FOR
DIFFERENT ATTENUATION RELATIONSHIPS
SEISMIC ZONES APPROACH

Site	Attenuation Relationship	Historical Maximum			Historical Maximum Plus One Unit		
		V	VI	VII	V	VI	VII
Boston,	Howell & Schultz	12	57	463	9	35	182
MA	Cornell & Merz	110	1258	909090	58	351	4348
	McGuire	29	133	741	23	91	420
	Average	33	212	6803	23	104	694
Washington,	Howell & Schultz	21	105	862	16	33	369
D.C.	Cornell & Merz	167	1348	925925	107	571	2217
	McGuire	45	198	1009	37	150	666
	Average	54	304	9346	40	141	820
Buffalo,	Howell & Schultz	9	28	109	8	25	78
N.Y.	Cornell & Merz	33	140	1130	29	93	394
	McGuire	17	54	199	15	47	152
	Average	17	59	291	15	47	168
Charleston,	Howell & Schultz	30	216	2857	23	133	1160
W. VA	Cornell & Merz	730	14065	11198208	299	5627	6024096
	McGuire	82	513	4000	63	348	2404
	Average	122	1160	52631	76	625	25641

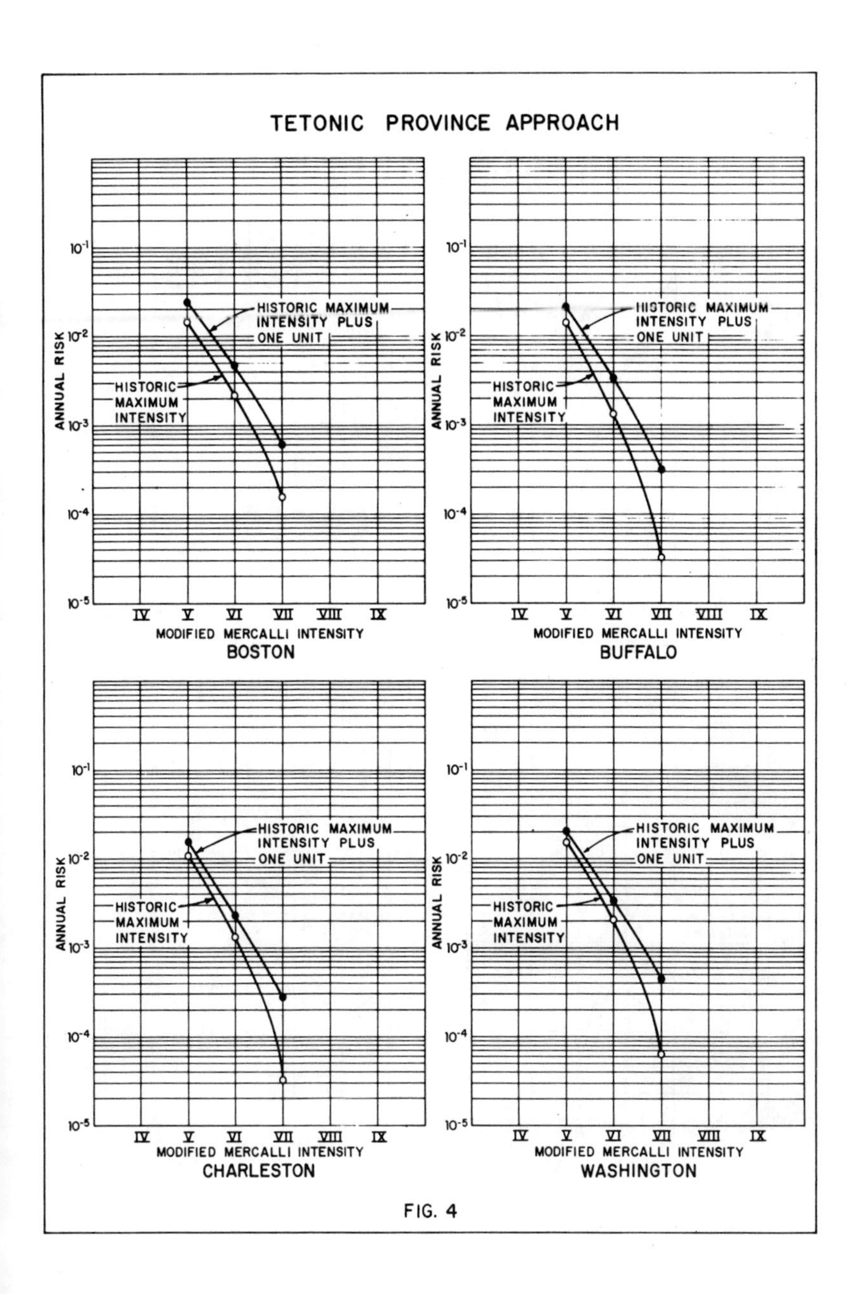
TETONIC PROVINCE APPROACH
ANNUAL RISK
10^{-1}
10^{-2}
10^{-3}
10^{-4}
10^{-5}
HISTORIC MAXIMUM INTENSITY PLUS ONE UNIT
HISTORIC MAXIMUM INTENSITY
IV V VI VII VIII IX
MODIFIED MERCALLI INTENSITY
BOSTON
BUFFALO
CHARLESTON
WASHINGTON

FIG. 4

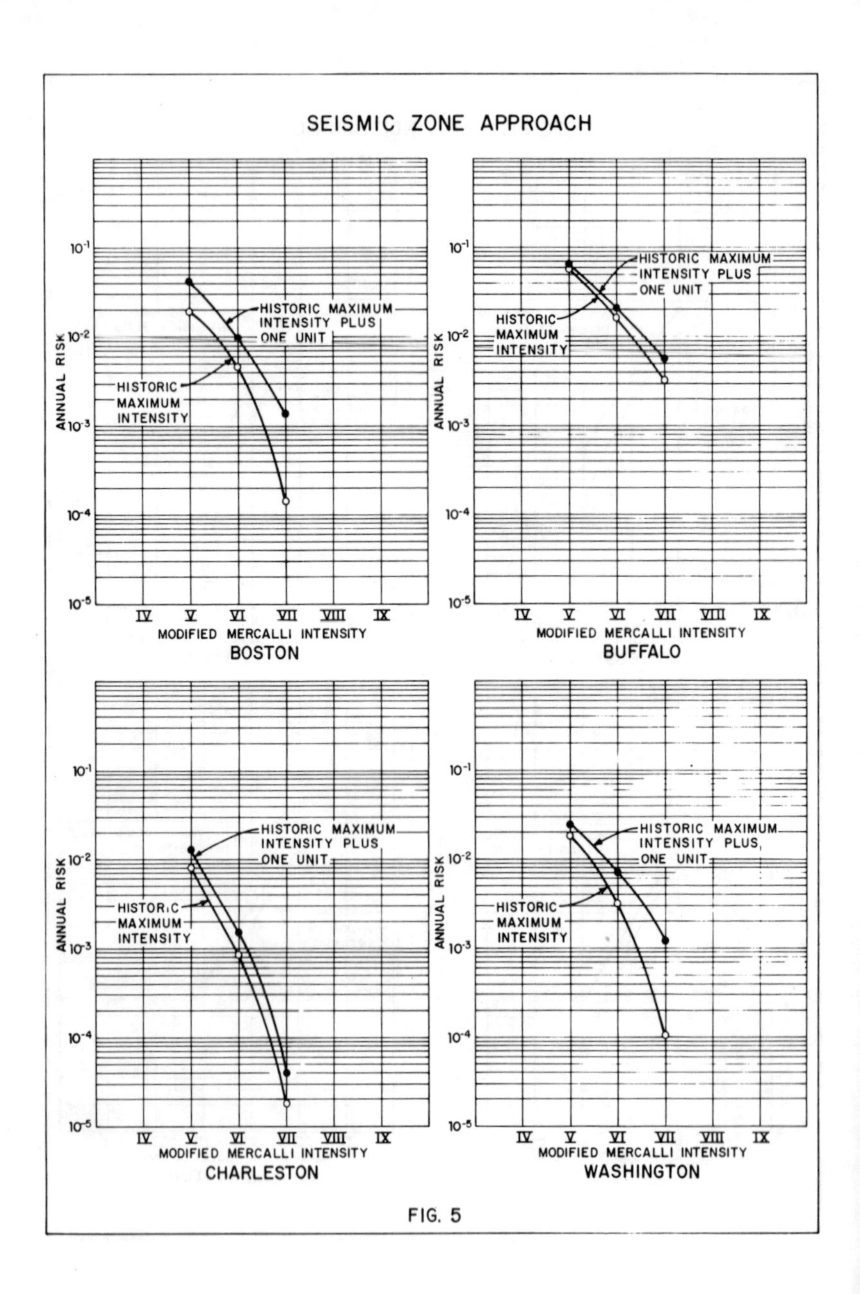
SEISMIC ZONE APPROACH
ANNUAL RISK
10^{-1}
10^{-2}
10^{-3}
10^{-4}
10^{-5}
HISTORIC MAXIMUM INTENSITY PLUS ONE UNIT
HISTORIC MAXIMUM INTENSITY
IV V VI VII VIII IX
MODIFIED MERCALLI INTENSITY
BOSTON
BUFFALO
CHARLESTON
WASHINGTON

FIG. 5